Airplane Flying Handbook

FAA-H-8083-3C

2021

U.S. Department of Transportation
FEDERAL AVIATION ADMINISTRATION
Flight Standards Service

Skyhorse Publishing

First published in 2021
First Skyhorse Publishing edition 2022

Skyhorse Publishing books may be purchased in bulk at special discounts for sales promotion, corporate gifts, fund-raising, or educational purposes. Special editions can also be created to specifications. For details, contact the Special Sales Department, Skyhorse Publishing, 307 West 36th Street, 11th Floor, New York, NY 10018 or info@skyhorsepublishing.com.

Skyhorse® and Skyhorse Publishing® are registered trademarks of Skyhorse Publishing, Inc.®, a Delaware corporation.

Visit our website at www.skyhorsepublishing.com.

10 9 8 7 6 5 4 3

Library of Congress Cataloging-in-Publication Data is available on file.

Cover design by Federal Aviation Administration

ISBN: 978-1-5107-7194-9
eBook ISBN: 978-1-5107-7195-6

Printed in China

Acknowledgments

The Airplane Flying Handbook was produced by the Federal Aviation Administration (FAA). The FAA wishes to acknowledge the following contributors:

Mr. Shane Torgerson for imagery of the Sedona Airport (Chapter 1)

Mr. Robert Frola for imagery of an Evektor-Aerotechnik EV-97 SportStar Max (Chapter 16)

Additional appreciation is extended to the General Aviation Joint Steering Committee (GA JSC) and the Aviation Rulemaking Advisory Committee's (ARAC) Airman Certification Standards (ACS) Working Group for their technical support and input.

Preface

The Airplane Flying Handbook provides basic knowledge that is essential for pilots. This handbook introduces basic pilot skills and knowledge that are essential for piloting airplanes. It provides information on transition to other airplanes and the operation of various airplane systems. It is developed by the Flight Standards Service, Airman Testing Standards Branch, in cooperation with various aviation educators and industry. This handbook is developed to assist student pilots learning to fly airplanes. It is also beneficial to pilots who wish to improve their flying proficiency and aeronautical knowledge, those pilots preparing for additional certificates or ratings, and flight instructors engaged in the instruction of both student and certificated pilots. It introduces the future pilot to the realm of flight and provides information and guidance in the performance of procedures and maneuvers required for pilot certification.

It is essential for persons using this handbook to become familiar with and apply the pertinent parts of 14 CFR and the Aeronautical Information Manual (AIM). The AIM is available online at www.faa.gov. The current Flight Standards Service airman training and testing material can be obtained from www.faa.gov.

This handbook supersedes FAA-H-8083-3B, Airplane Flying Handbook, dated 2016.

This handbook is available for download, in PDF format, from www.faa.gov.

This handbook is published by the United States Department of Transportation, Federal Aviation Administration, Airman Testing Standards Branch, P.O. Box 25082, Oklahoma City, OK 73125.

Comments regarding this publication should be emailed to AFS630comments@faa.gov.

Major Revisions

- Removed mandatory language or cited applicable regulations throughout handbook.

- Chapter 1 (Introduction to Flight Training) – Added information on the FAA Wings Program.

- Chapter 2 (Ground Operations) – Added a new graphic and information regarding detonation. Now uses the same marshalling graphic as the AMT General Handbook. Updated material on hand propping to match the material in the AMT General Handbook (it doesn't matter whether a pilot or mechanic is hand propping).

- Chapter 3 (Basic Flight Maneuvers) – Corrected G1000 and indications of slip and skid graphics.

- Chapter 4 (Energy Management) – All new chapter/material. Incremented the existing chapters 4-17 by 1 (now there are 18 chapters in total).

- Chapter 5 (Maintaining Aircraft Control) – Revised the order in which the material was presented.

- Chapter 7 (Ground Reference Maneuvers) – Corrected errors in text and graphics for eights on pylons.

- Chapter 9 (Approaches and Landings) – Added information concerning a forward slip to a landing and corrected Figure 9-6. Changed description associated with Crosswind Final Approach. Removed material on 360 degree power-off landing as this maneuver is not part of testing standard.

- Chapter 10 (Performance Maneuvers) – Added information on lazy eights.

- Chapter 11 (Night Operations) – Revised to align with material from CAMI.

- Chapter 13 (Transition to Multiengine Airplanes) – Incorporated the addendum. Corrected G1000 displays and force vectors on figures. Accelerated approach to stall minimum altitude revised to match the ACS. The 14 CFR part 23 certification standard used for many multiengine airplanes is now referred to a historical standard, since many of the previous requirements will not apply to newly certificated aircraft.

- Chapter 14 (Transition to Tailwheel Airplanes) – Made minor revision regarding handling characteristics.

- Chapter 15 (Transition to Turbopropeller-Powered Airplanes) – Addressed an NTSB recommendation regarding slow spool up time of split-shaft engines and corrected figure of fixed-shaft engine gauges.

- Chapter 16 (Transition to Jet-Powered Airplanes) – Removed extra information that appears unrelated to flying a turbojet and added information regarding energy management and distance versus altitude in a descent.

- Chapter 18 (Emergency Procedures) – Revised information regarding the safety of turning back after an engine failure after takeoff. Added a section on emergency response systems to include ballistic parachutes and autoland systems. Corrected figures of G1000 displays.

Airplane Flying Handbook (FAA-H-8083-3C)

Table of Contents

Chapter 3: Basic Flight Maneuvers

Chapter 4: Energy Management: Mastering Altitude and Airspeed Control

Chapter 5: Maintaining Aircraft Control: Upset Prevention and Recovery Training

Chapter 6: Takeoffs and Departure Climbs

Chapter 7: Ground Reference Maneuvers

Chapter 8: Airport Traffic Patterns

Chapter 9: Approaches and Landings

Chapter 11: Night Operations

Chapter 12: Transition to Complex Airplanes

Chapter 13: Transition to Multiengine Airplanes

Chapter 16: Transition to Jet-Powered Airplanes

Chapter 17: Transition to Light Sport Airplanes (LSA)

Chapter 18: Emergency Procedures

Glossary

Index

Airplane Flying Handbook (FAA-H-8083-3C)
Chapter 1: Introduction to Flight Training

Introduction

The overall purpose of primary and intermediate flight training, as outlined in this handbook, is the acquisition and honing of basic airmanship skills. *[Figure 1-1]* Airmanship is a broad term that includes a sound knowledge of and experience with the principles of flight; the knowledge, experience, and ability to operate an aircraft with competence and precision both on the ground and in the air; and the application of sound judgment that results in optimal operational safety and efficiency. *[Figure 1-2]* Learning to fly an aircraft has often been compared to learning to drive an automobile. This analogy is misleading. Since aircraft operate in a three-dimensional environment, they require a depth of knowledge and type of motor skill development that is more sensitive to this situation, such as:

- Coordination—the ability to use the hands and feet together subconsciously and in the proper relationship to produce desired results in the airplane.

- Timing—the application of muscular coordination at the proper instant to make flight, and all maneuvers, a constant, smooth process.

- Control touch—the ability to sense the action of the airplane and knowledge to determine its probable actions immediately regarding attitude and speed variations by sensing the varying pressures and resistance of the control surfaces transmitted through the flight controls.

- Speed sense—the ability to sense and react to reasonable variations of airspeed.

| Pre-Solo | Solo | Maneuvers | Cross-country | Checkride |

Figure 1-1. *Primary and intermediate flight training teaches basic airmanship skills and creates a good foundation for learners.*

An accomplished pilot demonstrates the knowledge and ability to:

- Assess a situation quickly and accurately and determine the correct procedure to be followed under the existing circumstance.

- Predict the probable results of a given set of circumstances or of a proposed procedure.

- Exercise care and due regard for safety.

- Accurately gauge the performance of the aircraft.

- Recognize personal limitations and limitations of the aircraft and avoid exceeding them.

- Identify, assess, and mitigate risk on an ongoing basis.

Figure 1-2. *Good airmanship skills include sound knowledge of the principles of flight and the ability to operate an airplane with competence and precision.*

The development of airmanship skills depends upon effort and dedication on the part of both the learner and the flight instructor, beginning with the very first training flight where proper habit formation begins with the learner being introduced to good operating practices.

Every airplane has its own particular flight characteristics. The purpose of primary and intermediate flight training, however, is not to learn how to fly a particular make and model airplane. The purpose of flight training is to develop the knowledge, experience, skills, and safe habits that establish a foundation and are transferable to any airplane. The pilot who has acquired necessary skills during training, and develops these skills by flying training-type airplanes with precision and safe flying habits, is able to easily transition to more complex and higher performance airplanes. Also note that the goal of flight training is a safe and competent pilot; passing required practical tests for pilot certification is only incidental to this goal.

Role of the FAA

The Federal Aviation Administration (FAA) is empowered by the U.S. Congress to promote aviation safety by prescribing safety standards for civil aviation. Standards are established for the certification of airmen and aircraft, as well as outlining operating rules. This is accomplished through the Code of Federal Regulations (CFR), formerly referred to as Federal Aviation Regulations (FAR). Title 14 of the CFR (14 CFR) is titled Aeronautics and Space with Chapter 1 dedicated to the FAA. Subchapters are broken down by category with numbered parts detailing specific information. *[Figure 1-3]* For ease of reference and since the parts are numerical, the abbreviated pattern 14 CFR part ___ is used (e.g., 14 CFR part 91).

This guidance is not legally binding in its own right and will not be relied upon by the FAA as a separate basis for affirmative enforcement action or other administrative penalty. Conformity with the guidance is voluntary only and nonconformity will not affect rights and obligations under existing statutes and regulations.

While the various subchapters and parts of 14 CFR provide general to specific guidance regarding aviation operations within the U.S., the topic of aircraft certification and airworthiness is spread through several interconnected parts of 14 CFR.

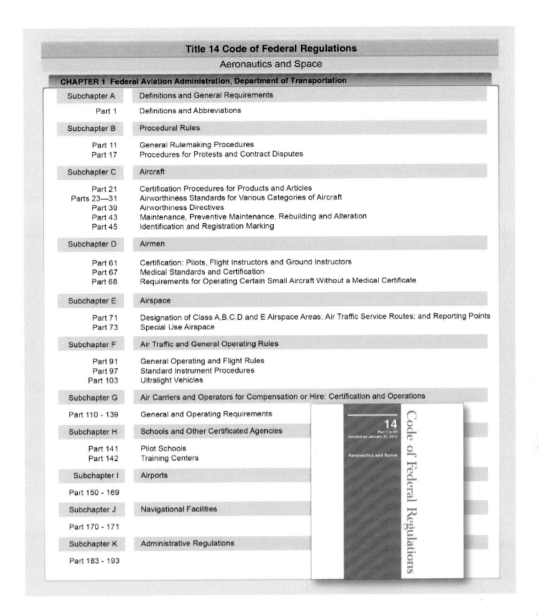

Figure 1-3. *Title 14 CFR, Chapter 1, Aeronautics and Space and subchapters.*

- 14 CFR part 21 prescribes procedural requirements for issuing airworthiness certificates and airworthiness approvals for aircraft and aircraft parts. A standard airworthiness certificate, FAA Form 8100-2 *[Figure 1-4]*, is required to be displayed in the aircraft in accordance with 14 CFR part 91, section 91.203(b). It is issued for aircraft type certificated in the normal, utility, acrobatic, commuter or transport category, and for manned free balloons. A standard airworthiness certificate remains valid as long as the aircraft meets its approved type design, is in a condition for safe operation and maintenance, and preventative maintenance and alterations are performed in accordance with 14 CFR parts 21, 43, and 91.

- 14 CFR part 39 is the authority for the FAA to issue Airworthiness Directives (ADs) when an unsafe condition exists in a product, aircraft, or part, and the condition is likely to exist or develop in other products of the same type design.

- 14 CFR part 43 prescribes rules governing the maintenance, preventive maintenance, rebuilding, and alteration of any aircraft having a U.S. airworthiness certificate. It also applies to the airframe, aircraft engines, propellers, appliances, and component parts of such aircraft.

- 14 CFR part 45 identifies the requirements for the identification of aircraft, engines, propellers, certain replacement and modification parts, and the nationality and registration marking required on U.S.-registered aircraft.

- 14 CFR part 91 outlines aircraft certifications and equipment requirements for the operation of aircraft in U.S. airspace. It also prescribes rules governing maintenance, preventive maintenance, and alterations. Also found in 14 CFR part 91 is the requirement to maintain records of maintenance, preventive maintenance, and alterations, as well as records of the 100-hour, annual, progressive, and other required or approved inspections.

UNITED STATES OF AMERICA
DEPARTMENT OF TRANSPORTATION-FEDERAL AVIATION ADMINISTRATION
STANDARD AIRWORTHINESS CERTIFICATE

1 NATIONALITY AND REGISTRATION MARKS	2 MANUFACTURER AND MODEL	3 AIRCRAFT SERIAL NUMBER	4 CATEGORY
N12345	Douglas DC-6A	43219	Transport

5 AUTHORITY AND BASIS FOR ISSUANCE

This airworthiness certificate is issued pursuant to 49 U.S.C. § 44704 and certifies that, as of the date of issuance, the aircraft to which issued has been inspected and found to conform to the type certificate therefore, to be in condition for safe operation, and has been shown to meet the requirements of the applicable comprehensive and detailed airworthiness code as provided by Annex 8 to the Convention on International Civil Aviation, except as noted herein.
Exceptions:

None

6 TERMS AND CONDITIONS

Unless sooner surrendered, suspended, revoked, or a termination date is otherwise established by the FAA, this airworthiness certificate is effective as long as the maintenance, preventative maintenance, and alterations are performed in accordance with Parts 21, 43, and 91 of the Federal Aviation Regulations, as appropriate, and the aircraft is registered in the United States.

DATE OF ISSUANCE	FAA REPRESENTATIVE		DESIGNATION NUMBER
01/20/2000	E.R. White	E.R. White	NE-XX

Any iteration, reproduction, or misuse of this certificate may be punishable by a fine not exceeding $1,000 or imprisonment not exceeding 3 years or both.
THIS CERTIFICATE MUST BE DISPLAYED IN THE AIRCRAFT IN ACCORDANCE WITH APPLICABLE FEDERAL AVIATION REGULATIONS.

FAA Form 8100-2 (04-11) Supersedes Previous Edition

Figure 1-4. *FAA Form 8100-2, Standard Airworthiness Certificate.*

While 14 CFR part 91, section 91.205 outlines the minimum equipment required for flight, the Airplane Flight Manual/Pilot's Operating Handbook (AFM/POH) lists the equipment required for the airplane to be airworthy. The equipment list found in the AFM/POH is developed during the airplane certification process. This list identifies those items that are required for airworthiness, optional equipment installed in addition to the required equipment, and any supplemental items or appliances.

Figure 1-5 shows an example of some of the required equipment, standard or supplemental (not required but commonly found in the aircraft) and optional equipment for an aircraft. The equipment list, originally issued by the manufacturer, is maintained by the Type Certificate Data Sheet (TCDS). An aircraft and its installed components and parts must conform to the original Type Certificate or approved altered conditions to meet the definition of airworthy in accordance with 14 CFR part 3.5.

Certification requirements for pilots, medical certificate requirements, and operating rules are found in the following parts:

- 14 CFR part 61 pertains to the certification of pilots, flight instructors, and ground instructors. It prescribes the eligibility, aeronautical knowledge, flight proficiency training, and testing requirements for each type of pilot certificate issued.

- 14 CFR part 67 prescribes the medical standards and certification procedures for issuing medical certificates for airmen and for remaining eligible for a medical certificate.

- 14 CFR part 68 contains requirements for operating certain small aircraft without a medical certificate.

- 14 CFR part 91 contains general operating and flight rules. The section is broad in scope and provides general guidance in the areas of general flight rules, visual flight rules (VFR), instrument flight rules (IFR), and as previously discussed aircraft maintenance, and preventive maintenance and alterations.

Sym:

Items in this listing are coded by a symbol indicating the status of the item. These codes are:

C Required item for FAA Certification.

S Standard equipment. Most standard equipment is applicable to all airplanes. Some equipment may be replaced by optional equipment.

O Optional equipment. Optional equipment may be installed in addition to or to replace standard equipment.

Qty: The quantity of the listed item in the airplane. A hyphen (-) in this column indicates that the equipment was not installed.

ATA Item	Description	SYM	QTY	Part Number	Unit Weight	Arm
34-08	GPS 1 Antenna	C	1	12744-001	0.4	136.2
34-09	GPS 2 Antenna	S	1	12744-001	0.4	110.3
34-10	Transponder Antenna	C	1	12739-001	0.1	105.0
34-11	VOR/LOC Antenna	C	1	12742-001	0.4	331.0
34-12	Turn coordinator, modified	C	1	11891-001	1.8	118.0
34-13	GMA 340 audio panel	S	1	12717-050	1.5	121.5
34-14	GNS 420 (GPS/COM/NAV)	O	1	12718-004	5.0	121.0
34-15	GNS 420 (GPS/COM/NAV)	C	1	12718-051	5.0	121.0
34-16	GNS 420 (GPS/COM/NAV)	O	1	12718-051	5.0	122.4
	EMax engine monitoring					
34-17	• Data acquisition unit	O	1	16692-001	2.0	118.0
34-18	• Monitor cabin harness	O	1	16695-005	2.0	108.0
	Sky watch option					
34-19	• Sky watch inverter	O	1	14484-001	0.5	118.0
34-20	• Sky watch antenna nsti	O	1	14480-001	2.3	150.5
34-21	• Sky watch track box	O	1	14477-050	10.0	140.0
	Stormscope option					
34-22	• Processor	O	1	12745-050	1.7	199.0
34-23	• Antenna	O	1	12745-070	0.9	191.0
	Transponder option					
34-24	• Mode A/C transponder	C	1	13587-001	1.6	124.9
34-25	• Mode S transponder	O	-	15966-050	2.6	121.0
	TAWS option					
34-26	• KGP 560 processor	O	1	15963-001	1.3	117.0
	XM satellite option					
34-27	• XM WX/radio receiver	O	1	16121-001	1.7	114.0
34-28	• XM radio remote control	O	1	16665-501	0.2	149.3
61	Propeller					
61-01	• Hartzell propeller installation	C	1	15319-00X	79.8	48.0
61-02	• McCauley propeller installation	O	1	15825-00X	78.0	50.0
61-03	• Propeller governor	C	1	15524-001	3.2	61.7
71	Power plant					
71-01	• Upper cowl	C	1	20181-003	10.5	78.4
71-02	• Lower cowl LH	C	1	20182-005	5.4	78.4
71-03	• Lower cowl RH	C	1	20439-005	5.4	78.4
71-03	• Engine baffling installation	C	1	15460-001	10.7	78.4

Figure 1-5. *Example of some of the required standard or supplemental and optional equipment for an aircraft.*

Flight Standards Service

The FAA's Flight Standards Service (FS) sets aviation standards for airmen and aircraft operations in the United States and for American airmen and aircraft around the world. Flight Standards is organized into four functional offices: Office of Safety Standards, Air Carrier Safety Assurance, General Aviation Safety Assurance, and Foundational Business.

The primary interface between FS and the general aviation community/general public is the local Flight Standards District Office (FSDO). The FSDOs are responsible for the certification and surveillance of certain air carriers, air operators, flight schools/training centers, airmen (pilots, flight instructors, mechanics and other certificate holders). FSDO inspectors also handle general aviation accident investigation at the request of, or in cooperation with, the National Transportation Safety Board.

Each FSDO is staffed by Aviation Safety Inspectors (ASIs) whose specialties include operations, maintenance, and avionics. General Aviation ASIs are highly qualified and experienced aviators. Once accepted for the position, an inspector will satisfactorily complete indoctrination training conducted at the FAA Academy. The indoctrination training coursework for a General Aviation Operations Inspector, which is oriented to the tasks to be performed by an ASI in the general aviation environment, includes classroom and flight training on pilot certification activities. Thereafter, the inspector will complete recurrent training on a regular basis. Among other duties, the ASI is responsible for administering FAA practical tests for pilot and flight instructor certificates and associated ratings. Questions concerning pilot certification and/or requests for other aviation information or services should be directed to the FSDO. For specific FSDO locations and telephone numbers, refer to www.faa.gov.

Role of the Pilot Examiner

Pilot and flight instructor certificates are issued by the FAA upon satisfactory completion of required knowledge and practical tests. The administration of practical tests is an FAA responsibility that may occur at the FSDO level. However, in order to satisfy the public need for pilot testing and certification services, the FAA delegates certain responsibilities, as the need arises, to private individuals who are not FAA employees. A Designated Pilot Examiner (DPE) is a private citizen who is designated as a representative of the FAA Administrator to perform specific (but limited) pilot certification tasks on behalf of the FAA and may charge a reasonable fee for doing so. Generally, a DPE's authority is limited to accepting applications and conducting practical tests leading to the issuance of specific pilot certificates and/or ratings. A DPE operates under the direct supervision of the FSDO that holds the examiner's designation file. A FSDO inspector is assigned to monitor the DPE's certification activities.

The FAA selects highly qualified individuals to be DPEs. These individuals have good industry reputations for professionalism, high integrity, a demonstrated willingness to serve the public, and adhere to FAA policies and procedures in certification matters. A DPE is expected to administer practical tests with the same degree of professionalism, using the same methods, procedures, and standards as an FAA ASI. Note that a DPE is not an FAA ASI. A DPE cannot initiate enforcement action, investigate accidents, or perform surveillance activities on behalf of the FAA. However, the majority of FAA practical tests at the recreational, private, and commercial pilot level are administered by DPEs.

Role of the Flight Instructor

The flight instructor is the cornerstone of aviation safety. The FAA has adopted an operational training concept that places the full responsibility for pilot training on the flight instructor. In this role, the instructor assumes the total responsibility for providing training in all the knowledge areas and skills necessary for pilots to operate safely and competently in the National Airspace System (NAS). This training includes airmanship skills, pilot judgment and decision-making, hazard identification, risk analysis, and good operating practices. (See Risk Management Handbook, FAA-H-8083-2). *[Figure 1-6]*

Figure 1-6. *The flight instructor is responsible for teaching and training.*

A flight instructor normally meets broad flying experience requirements, passes rigid knowledge and practical tests, and demonstrates the ability to apply recommended teaching techniques before being certificated.

A pilot training program is dependent on the quality of the ground and flight instruction given. A good flight instructor has a thorough understanding of the learning process, knowledge of the fundamentals of instruction, and the ability to communicate effectively with the learner.

A good flight instructor uses a syllabus and insists on correct techniques and procedures from the beginning of training so that the learner will develop proper habit patterns. The syllabus should embody the "building block" method of instruction in which the learner systematically progresses from the known to the unknown. The course of instruction should be laid out so that each new maneuver embodies the principles involved in the performance of those previously undertaken. Consequently, through each new subject introduced, the learner not only learns a new principle or technique, but also broadens their application of those previously learned and has their deficiencies in the previous maneuvers emphasized and made obvious. [Figure 1-7]

Lesson **Stalls**	Student _____	Date _____

Objective	• To familiarize the student with the stall warnings and handling characteristics of the airplane as it approaches a stall. To develop the student's skill in recognition and recovery from stalls.
Content	• Configuration of airplane for power-on and power-off stalls. • Observation of airplane attitude, stall warnings, and handling characteristics as it approaches a stall. • Control of airplane attitude, altitude, and heading. • Initiation of stall recovery procedures.
Schedule	• Preflight Discussion ....:10 • Instructor Demonstrations ....:25 • Student Practice ....:45 • Postflight Critique ....:10
Equipment	• Chalkboard or notebook for preflight discussion.
Instructor's actions	• Preflight—discuss lesson objective. • Inflight—demonstrate elements. Demonstrate power-on and power-off stalls and recovery procedures. Coach student practice. • Postflight—critique student performance and assign study material.
Student's actions	• Preflight—discuss lesson objective and resolve questions. • Inflight—review previous maneuvers including slow flight. Perform each new maneuver as directed. • Postflight—ask pertinent questions.
Completion standards	• Student should demonstrate competency in controlling the airplane at airspeeds approaching a stall. Student should recognize and take prompt corrective action to recover from power-on and power-off stalls.

This is a typical lesson plan for flight training which emphasizes stall recognition and recovery procedures.

Figure 1-7. *Sample lesson plan for stall training and recovery procedures.*

The flying habits of the flight instructor, both during flight instruction and as observed by learners when conducting other pilot operations, have a vital effect on safety. Learners consider their flight instructor to be a paragon of flying proficiency whose flying habits they, consciously or unconsciously, attempt to imitate. For this reason, a good flight instructor meticulously observes the safety practices taught to the learners. Additionally, a good flight instructor carefully observes all regulations and recognized safety practices during all flight operations.

A prospective pilot should know that there are other differences among flight instructors. Certain instructors who have performed at a high level have earned a Gold Seal Flight Instructor Certificate. This is not a requirement when looking for an instructor, but it is indication of an active and successful instructor. Top notch instructors also participate in the Pilot Proficiency Awards Wings Program (Wings program) to improve their proficiency and to serve as an example to learners who also benefit from program participation.

Generally, an individual who enrolls in a pilot training program is prepared to commit considerable time, effort, and expense in pursuit of a pilot certificate. A trainee may judge the effectiveness of the flight instructor and the overall success of the pilot training program solely in terms of being able to pass the requisite FAA practical test. A good flight instructor is able to communicate that evaluation through practical tests is a mere sampling of pilot ability that is compressed into a short period of time. The flight instructor's role is to train the "total" pilot.

Sources of Flight Training

The major sources of flight training in the United States include FAA-approved pilot schools and training centers, non-certificated (14 CFR part 61) flying schools, and independent flight instructors. FAA-approved schools are those flight schools certificated by the FAA as pilot schools under 14 CFR part 141. *[Figure 1-8]*

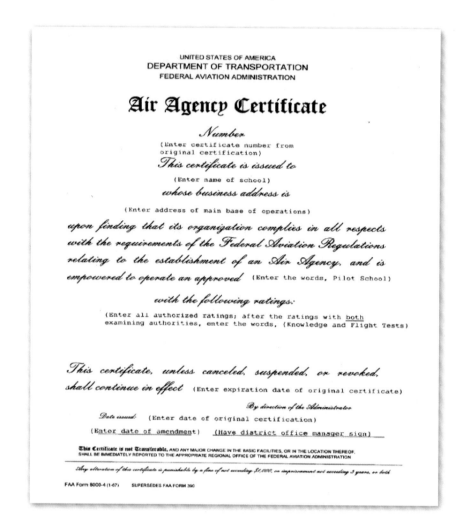

Figure 1-8. *FAA Form 8000-4, Air Agency Certificate.*

Application for part 141 certification is voluntary, and the school needs to meet specific requirements for personnel, equipment, maintenance, and facilities. The school operates each course offering in accordance with an established curriculum that includes a training course outline (TCO) approved by the FAA. Each TCO contains enrollment prerequisites, a detailed description of each lesson including standards and objectives, expected accomplishments and standards for each stage of training, and a description of the checks and tests used to measure each training course enrollee's accomplishments. An FAA-approved pilot school Air Agency certificate expires and needs to be renewed every 2 years.

Renewal is contingent upon proof of continued high quality instruction and a minimum level of instructional activity. Training at an FAA-certificated pilot school is structured and because of this structured environment, the graduates of these pilot schools are allowed to meet the certification experience requirements of 14 CFR part 61 with less flight time. Many FAA-certificated pilot schools have DPEs on staff to administer FAA practical tests. Some schools have been granted examining authority by the FAA. A school with examining authority for a particular course(s) has the authority to recommend its graduates for pilot certificates or ratings without further testing by the FAA. A list of FAA-certificated pilot schools and their training courses can be found at https://av-info.faa.gov/pilotschool.asp.

FAA-approved training centers are certificated under 14 CFR part 142. Training centers, like certificated pilot schools, operate in a structured environment with approved courses and curricula and stringent standards for personnel, equipment, facilities, operating procedures, and record keeping. Training centers certificated under 14 CFR part 142, however, specialize in the use of flight simulation (full flight simulators and flight training devices) in their training courses.

There are a number of flying schools in the United States that are not certified by the FAA. These schools operate under the provisions of 14 CFR part 61. Many of these non-certificated flying schools offer excellent training and meet or exceed the standards required of FAA-approved pilot schools. Flight instructors employed by non-certificated flying schools, as well as independent flight instructors, meet the same basic 14 CFR part 61 flight instructor requirements for certification and renewal as those flight instructors employed by FAA-certificated pilot schools. In the end, any training program is dependent upon the quality of the ground and flight instruction a learner receives.

Airman Certification Standards (ACS) and Practical Test Standards (PTS)

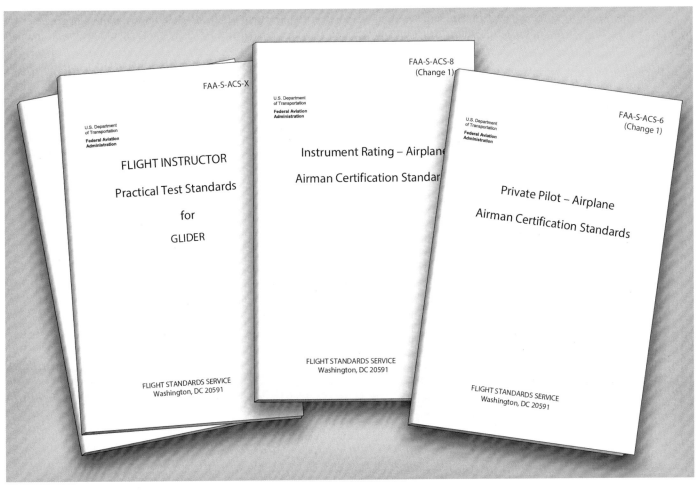

Figure 1-9. *Airman Certification Standards (ACS) developed by FAA*

Practical tests for FAA pilot certificates and associated ratings are administered by FAA inspectors and DPEs using FAA Airman Certification Standards (ACS) and Practical Test Standards (PTS), which contain structured areas of operation, tasks, and standards. *[Figure 1-9]* 14 CFR part 61, section 61.43 specifies that the practical test consists of the tasks specified in the areas of operation for the airman certificate or rating sought. To pass the test, the applicant demonstrates mastery of the aircraft performing each task successfully, proficiency and competency within the approved standards, and sound judgment.

It should be emphasized that the ACS and PTS are testing documents rather than teaching documents. Although the pilot applicant should be familiar with these books and refer to the standards they contain during training, the ACS and PTS are not intended to be used as a training syllabus. They contain the standards to which maneuvers/procedures on FAA practical tests should be performed and the FAA policies governing the administration of practical tests. An appropriately rated flight instructor is responsible for training a pilot applicant to acceptable standards in all subject matter areas, procedures, and maneuvers included in, and encompassed by, the tasks within each area of operation in the appropriate ACS and PTS. Flight instructors and pilot applicants should always remember that safe, competent piloting requires a commitment to learning, planning, and risk management that goes beyond rote performance of maneuvers. Descriptions of tasks and information on how to perform maneuvers and procedures are contained in reference and teaching documents, such as this handbook. A list of reference documents is contained in the appendices of each ACS and PTS. It is necessary that the latest version of the PTS and ACS, with all recent changes, be referenced for training. All recent versions and changes to the FAA ACS and PTS may be viewed or downloaded at www.faa.gov.

Safety Considerations

In the interest of safety and good habit pattern formation, there are certain basic flight safety practices and procedures that should be emphasized by the flight instructor, and adhered to by both instructor and learner, beginning with the very first dual instruction flight. These include, but are not limited to, collision avoidance procedures including proper scanning techniques and clearing procedures, runway incursion avoidance, stall awareness, positive transfer of controls, and flight deck workload management.

Collision Avoidance

All pilots should be alert to the potential for midair collision and impending loss of separation. The general operating and flight rules in 14 CFR part 91 set forth the concept of "see and avoid." This concept requires that vigilance shall be maintained at all times by each person operating an aircraft regardless of whether the operation is conducted under IFR or VFR. Pilots should also keep in mind their responsibility for continuously maintaining a vigilant lookout regardless of the type of aircraft being flown and the purpose of the flight. Most midair collision accidents and reported near midair collision incidents occur in good VFR weather conditions and during the hours of daylight. Most of these accident/incidents occur within 5 miles of an airport and/or near navigation aids. *[Figure 1-10]*

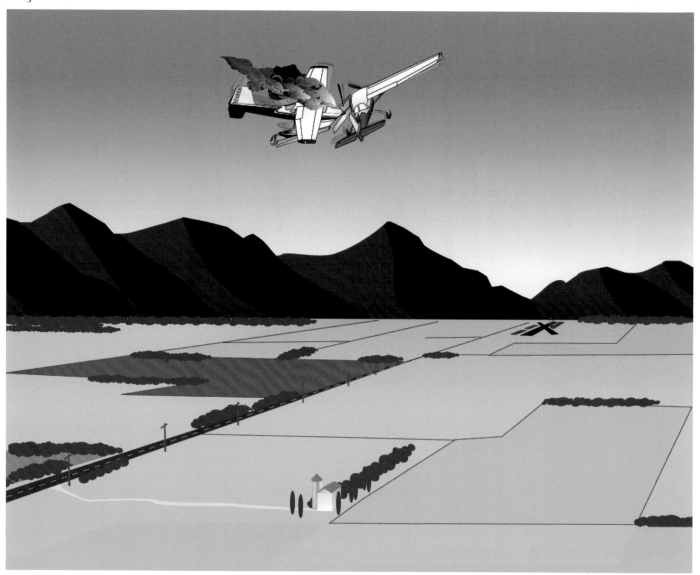

Figure 1-10. *Most midair collision accidents occur in good weather.*

The "see and avoid" concept relies on knowledge of the limitations of the human eye and the use of proper visual scanning techniques to help compensate for these limitations. Pilots should remain constantly alert to all traffic movement within their field of vision, as well as periodically scanning the entire visual field outside of their aircraft to ensure detection of conflicting traffic. Remember that the performance capabilities of many aircraft, in both speed and rates of climb/descent, result in high closure rates limiting the time available for detection, decision, and evasive action. *[Figure 1-11]*

Figure 1-11. *Proper scanning techniques can mitigate midair collisions. Pilots should be aware of potential blind spots and attempt to clear the entire area in which they are maneuvering.*

The probability of spotting a potential collision threat increases with the time spent looking outside, but certain techniques may be used to increase the effectiveness of the scan time. The human eyes tend to focus somewhere, even in a featureless sky. In order to be most effective, the pilot should shift glances and refocus at intervals. Most pilots do this in the process of scanning the instrument panel, but it is also important to focus outside to set up the visual system for effective target acquisition. Pilots should also realize that their eyes may require several seconds to refocus when switching views between items on the instrument panel and distant objects.

Proper scanning requires the constant sharing of attention with other piloting tasks, thus it is easily degraded by psychological and physiological conditions such as fatigue, boredom, illness, anxiety, or preoccupation.

Effective scanning is accomplished with a series of short, regularly-spaced eye movements that bring successive areas of the sky into the central visual field. Each movement should not exceed 10 degrees, and each area should be observed for at least 1 second to enable detection. Although horizontal back-and-forth eye movements seem preferred by most pilots, each pilot should develop a scanning pattern that is comfortable and adhere to it to assure optimum scanning.

Peripheral vision can be most useful in spotting collision threats from other aircraft. Each time a scan is stopped and the eyes are refocused, the peripheral vision takes on more importance because it is through this element that movement is detected. Apparent movement is usually the first perception of a collision threat and probably the most important because it is the discovery of a threat that triggers the events leading to proper evasive action. It is essential to remember that if another aircraft appears to have no relative motion, it is likely to be on a collision course. If the other aircraft shows no lateral or vertical motion, but is increasing in size, the observing pilot needs to take immediate evasive action to avoid a collision.

The importance of, and the proper techniques for, visual scanning should be taught at the very beginning of flight training. The competent flight instructor should be familiar with the visual scanning and collision avoidance information contained in AC 90-48, *Pilots' Role in Collision Avoidance*, and the Aeronautical Information Manual (AIM).

There are many different types of clearing procedures. Most are centered around the use of clearing turns. The essential idea of the clearing turn is to be certain that the next maneuver is not going to proceed into another aircraft's flightpath. Some pilot training programs have hard and fast rules, such as requiring two 90° turns in opposite directions before executing any training maneuver. Other types of clearing procedures may be developed by individual flight instructors. Whatever the preferred method, the flight instructor should teach the beginning learner an effective clearing procedure and insist on its use. The learner should execute the appropriate clearing procedure before all turns and before executing any training maneuver. Proper clearing procedures, combined with proper visual scanning techniques, are the most effective strategy for collision avoidance.

In case of pilot incapacitation, an installed Emergency Autoland (EAL) system may take control of an airplane, navigate to an airport, and land without additional human intervention. Currently, these systems take no evasive action in response to potential impact with another aircraft, although they transmit over the radio. Pilots should avoid the path of any aircraft under the control of an EAL or suspected as under the control of an EAL system. The *Emergency Procedures* chapter in this handbook contains additional information about these systems.

Runway Incursion Avoidance

A runway incursion is any occurrence at an airport involving an aircraft, vehicle, person, or object on the ground that creates a collision hazard or results in a loss of separation with an aircraft taking off, landing, or intending to land. The three major areas contributing to runway incursions are communications, airport knowledge, and flight deck procedures for maintaining orientation. *[Figure 1-12]*

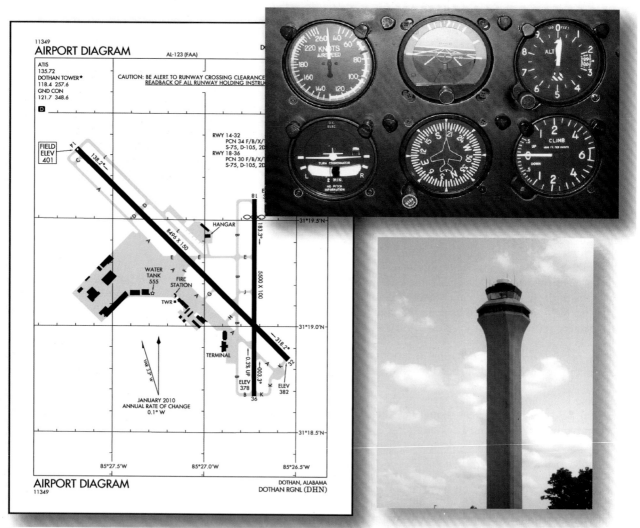

Figure 1-12. *Three major areas contributing to runway incursions are communications with air traffic control (ATC), airport knowledge, and flight deck procedures.*

Taxi operations require constant vigilance by the entire flight crew, not just the pilot taxiing the airplane. During flight training, the instructor should emphasize the importance of vigilance during taxi operations. Both the learner and the flight instructor need to be continually aware of the movement and location of other aircraft and ground vehicles on the airport movement area. Many flight training activities are conducted at non-tower controlled airports. The absence of an operating airport control tower creates a need for increased vigilance on the part of pilots operating at those airports. *[Figure 1-13]*

Figure 1-13. *Sedona Airport is one of the many airports that operate without a control tower.*

Planning, clear communications, and enhanced situational awareness during airport surface operations reduces the potential for surface incidents. Safe aircraft operations can be accomplished and incidents eliminated if the pilot is properly trained early on and throughout their flying career on standard taxi operating procedures and practices. This requires the development of the formalized teaching of safe operating practices during taxi operations. The flight instructor is the key to this teaching. The flight instructor should instill in the learner an awareness of the potential for runway incursion, and should emphasize the runway incursion avoidance procedures. For more information and a list of additional references, refer to Chapter 14 of the *Pilot's Handbook of Aeronautical Knowledge*.

Stall Awareness

14 CFR part 61, section 61.87 (d)(10) and (e)(10) require that a student pilot who is receiving training for a single-engine or multiengine airplane rating or privileges, respectively, log flight training in stalls and stall recoveries prior to solo flight. *[Figure 1-14]* During this training, the flight instructor should emphasize that the direct cause of every stall is an excessive angle of attack (AOA). The student pilot should fully understand that there are several flight maneuvers that may produce an increase in the wing's AOA, but the stall does not occur until the AOA becomes excessive. This critical AOA varies from 16°–20° depending on the airplane design. *[Figure 1-15]*

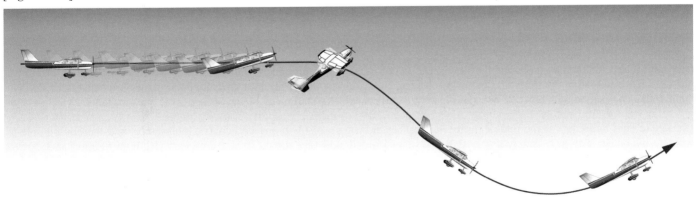

Figure 1-14. *All student pilots receive and log flight training in stalls and stall recoveries prior to their first solo flight.*

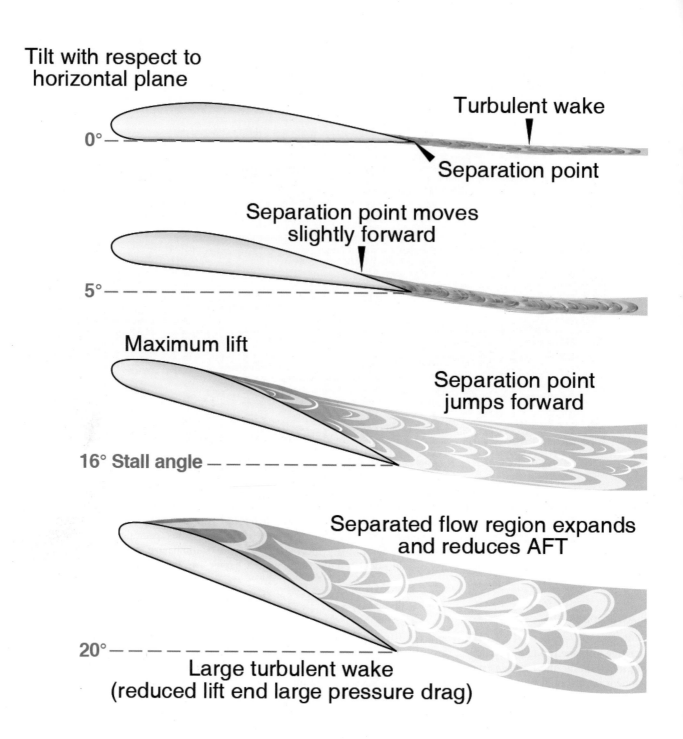

Figure 1-15. *Stalls occur when the airfoil's angle of attack reaches the critical point which can vary between 16° and 20°.*

The flight instructor should emphasize that low speed is not necessary to produce a stall. The wing can be brought to an excessive AOA at any speed. High pitch attitude is not an absolute indication of proximity to a stall. Some airplanes are capable of vertical flight with a corresponding low AOA. Most airplanes are quite capable of stalling at a level or near level pitch attitude.

The key to stall awareness is the pilot's ability to visualize the wing's AOA in any particular circumstance, and thereby be able to estimate his or her margin of safety above stall. This is a learned skill that should be acquired early in flight training and carried through the pilot's entire flying career.

The pilot should understand and appreciate factors such as airspeed, pitch attitude, load factor, relative wind, power setting, and aircraft configuration in order to develop a reasonably accurate mental picture of the wing's AOA at any particular time. It is essential to safety of flight that pilots take into consideration this visualization of the wing's AOA prior to entering any flight maneuver. Chapter 3, Basic Flight Maneuvers, discusses stalls in detail.

Use of Checklists

Checklists have been the foundation of pilot standardization and flight deck safety for years. *[Figure 1-16]* The checklist is a memory aid and helps to ensure that critical items necessary for the safe operation of aircraft are not overlooked or forgotten. Checklists need not be "do lists." In other words, the proper actions can be accomplished, and then the checklist used to quickly ensure all necessary tasks or actions have been completed with emphasis on the "check" in checklist. However, checklists are of no value if the pilot is not committed to using them. Without discipline and dedication to using the appropriate checklists at the appropriate times, the odds are on the side of error. Pilots who fail to take the use of checklists seriously become complacent and begin to rely solely on memory.

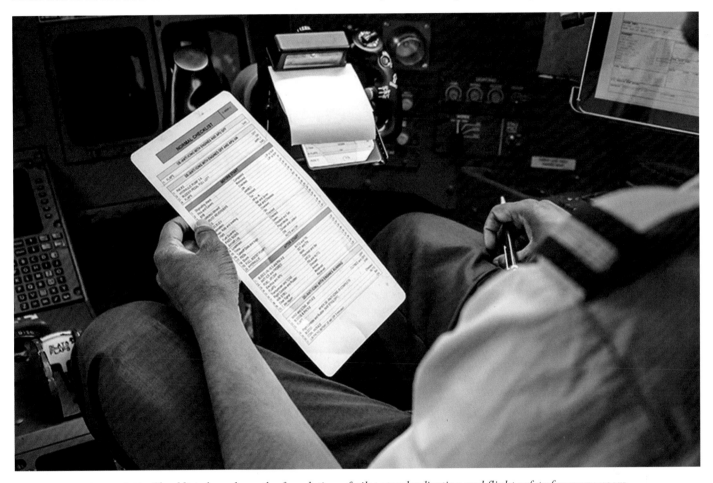

Figure 1-16. *Checklists have been the foundation of pilot standardization and flight safety for many years.*

The importance of consistent use of checklists cannot be overstated in pilot training. A major objective in primary flight training is to establish habit patterns that will serve pilots well throughout their entire flying career. The flight instructor should promote a positive attitude toward checklist usage, and the learner should realize its importance. At a minimum, prepared checklists should be used for the following phases of flight: *[Figure 1-17]*

- Preflight inspection
- Before engine start
- Engine starting
- Before taxiing
- Before takeoff
- After takeoff
- Cruise
- Descent
- Before landing
- After landing
- Engine shutdown and securing

Figure 1-17. *A sample checklist used by pilots.*

During flight training, there should be a clear understanding between the learner and flight instructor of who has control of the aircraft. Prior to any flight, a briefing should be conducted that includes the procedures for the exchange of flight controls. The following three-step process for the exchange of flight controls is highly recommended.

When a flight instructor wishes the learner to take control of the aircraft, he or she should say to the learner, "You have the flight controls." The learner should acknowledge immediately by saying, "I have the flight controls." The flight instructor should then confirm by again saying, "You have the flight controls." Part of the procedure should be a visual check to ensure that the other person actually has the flight controls. When returning the controls to the flight instructor, the learner should follow the same procedure the instructor used when giving control to the learner. The learner should stay on the controls until the instructor says, "I have the flight controls." There should never be any doubt as to who is flying the airplane at any time. Numerous accidents have occurred due to a lack of communication or misunderstanding as to who actually had control of the aircraft, particularly between learners and flight instructors. Establishing the above procedure during initial training ensures the formation of a very beneficial habit pattern.

Continuing Education

In many activities, the ability to receive feedback and continue learning contributes to safety and success. For example, professional athletes receive constant coaching. They practice various techniques to achieve their best. Medical professionals read journals, train, and master techniques to achieve better outcomes.

FAA WINGS Program

Compare continuous training and practice to 14 CFR part 61, section 61.56(c)(1) and (2), which allows for training and a sign-off within the previous 24 calendar months in order to act as a pilot in command. Many astute pilots realize that this regulation specifies a minimum requirement, and the path to enhanced proficiency, safety, and enjoyment of flying takes a higher degree of commitment such as using 14 CFR part 61, section 61.56(e). For this reason, many pilots keep their flight review up-to-date using the FAA WINGS program. The program provides continuing pilot education and contains interesting and relevant study materials that pilots can use all year round.

A pilot may create a WINGS account by logging on to www.faasafety.gov. This account gives the pilot access to the latest information concerning aviation technology and risk mitigation. It provides a means to document targeted skill development as a means to increase safety. As an added bonus, participants may receive a discount on certain flight insurance policies.

Chapter Summary

This chapter discussed some of the concepts and goals of primary and intermediate flight training. It identified and provided an explanation of regulatory requirements and the roles of the various entities involved. It also offered recommended techniques to be practiced and refined to develop the knowledge, proficiency, and safe habits of a competent pilot.

Airplane Flying Handbook (FAA-H-8083-3C)
Chapter 2: Ground Operations

Introduction

Experienced pilots place a strong emphasis on ground operations as this is where safe flight begins and ends. They know that hasty ground operations diminish their margin of safety. A smart pilot takes advantage of this phase of flight to assess various factors including the regulatory requirements, the pilot's readiness for pilot-in-command (PIC) responsibilities, the airplane's condition, the flight environment, and any external pressures that could lead to inadequate control of risk.

Flying an airplane presents many new responsibilities not required for other forms of transportation. Focus is often placed on the flying portion itself with less emphasis placed on ground operations. However pilots need to allow time for flight preparation. Situational awareness begins during preparation and only ends when the airplane is safely and securely returned to its tie-down or hangar, or if a decision is made not to go.

This chapter covers the essential elements for the regulatory basis of flight including:

1. An airplane's airworthiness requirements,
2. Important inspection items when conducting a preflight visual inspection,
3. Managing risk and resources, and
4. Proper and effective airplane surface movements using the AFM/POH and airplane checklists.

Preflight Assessment of the Aircraft

The visual preflight assessment mitigates airplane flight hazards. The preflight assessment ensures that any aircraft flown meets regulatory airworthiness standards and is in a safe mechanical condition prior to flight. Per 14 CFR part 3, section 3.5(a), the term "airworthy" means that the aircraft conforms to its type design and is in condition for safe operation. The owner/operator is primarily responsible for maintenance, but in accordance with 14 CFR part 91, section 91.7(a) and (b) no person may operate a civil aircraft unless it is in an airworthy condition and the pilot in command of a civil aircraft is responsible for determining whether the aircraft is in condition for safe flight. The pilot's inspection should involve the following:

1. Inspecting the airplane's airworthiness status.
2. Following the AFM/POH to determine the required items for visual inspection. *[Figures 2-1, 2-2, 2-3].*

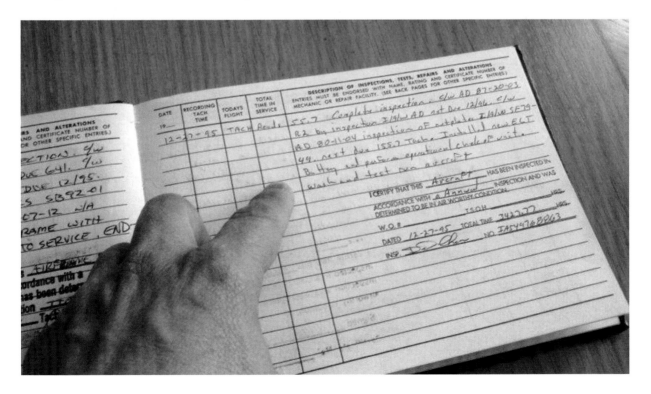

Figure 2-1. *Pilots should view the aircraft's maintenance logbook prior to flight to ensure the aircraft is safe to fly.*

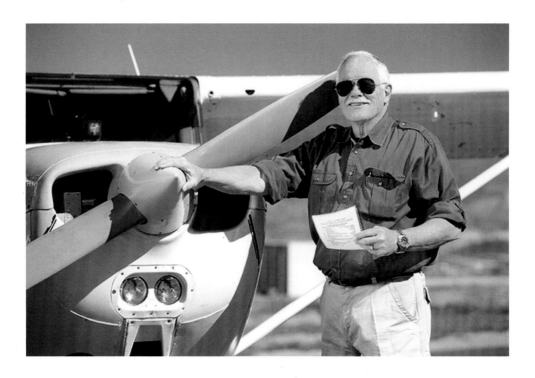

Figure 2-2. *A visual inspection of the aircraft before flight is an important step in mitigating airplane flight hazards.*

Figure 2-3. *Airplane Flight Manuals (AFM) and the Pilot Operating Handbook (POH) for each individual aircraft explain the required items for inspection.*

Each airplane has a set of logbooks that include airframe and engine, and in some cases, propeller and appliance logbooks, which are used to record maintenance, alteration, and inspections performed on a specific airframe, engine, propeller, or appliance. It is important that the logbooks be kept accurate, secure, and available for inspection. Airplane logbooks are not normally kept in the airplane. It should be a matter of procedure by the pilot to inspect the airplane logbooks or a summary of the airworthy status prior to flight to ensure that the airplane records of maintenance, alteration, and inspections are current and correct. *[Figure 2-4]* The following is required:

- Annual inspection within the preceding 12 calendar months (Title 14 of the Code of Federal Regulations (14 CFR) part 91, section 91.409(a))

- 100-hour inspection, if the aircraft is operated for hire (14 CFR part 91, section 91.409(b))

- Transponder certification within the preceding 24 calendar months (14 CFR part 91, section 91.413)

- Static system and encoder certification, within the preceding 24 calendar months, required for instrument flight rules (IFR) flight in controlled airspace (14 CFR part 91, section 91.411)

- 30-day VHF omnidirectional range (VOR) equipment check when using the VOR system of radio navigation for IFR flight (14 CFR part 91, section 91.171)

- Emergency locator transmitter (ELT) inspection within the last 12 months (14 CFR part 91, section 91.207(d))

- ELT battery due (14 CFR part 91, section 91.207(c))

- Current status of life limited parts per Type Certificate Data Sheets (TCDS) (14 CFR part 91, section 91.417)

- Status, compliance, logbook entries for airworthiness directives (ADs) (14 CFR part 91, section 91.417(a)(2)(v))

- Federal Aviation Administration (FAA) Form 337, Major Repair or Alteration (14 CFR part 91, section 91.417)

- Inoperative equipment (14 CFR part 91, section 91.213)

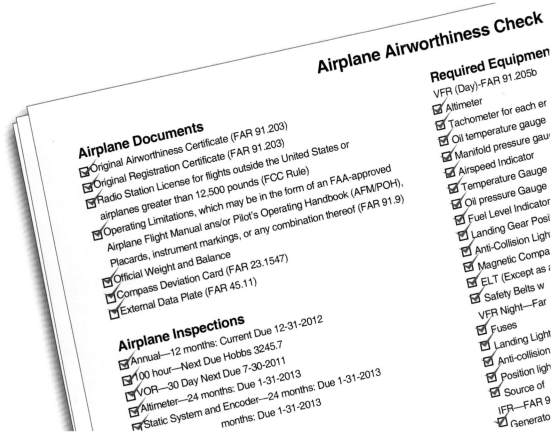

Figure 2-4. *A sample airworthiness checklist used by pilots to inspect an aircraft.*

A review determines if the required maintenance and inspections have been performed on the airplane. Any discrepancies need to be addressed prior to flight. Once the pilot has determined that the airplane's logbooks provide factual assurance that the airplane meets its airworthiness requirements, it is appropriate to inspect the airplane visually. The visual preflight inspection of the airplane should begin while approaching the airplane on the ramp. The pilot should make note of the general appearance of the airplane, looking for discrepancies such as misalignment of the landing gear and airplane structure. The pilot should also take note of any distortions of the wings, fuselage, and tail, as well as skin damage and any staining, dripping, or puddles of fuel or oils.

The pilot needs to determine that the following documents are, as appropriate, on board, attached, or affixed to the airplane:

- Current Airworthiness Certificate (14 CFR part 91, section 91.203)

- Current Registration Certificate (14 CFR part 91, section 91.203)

- Radio station license for flights outside the United States or airplanes greater than 12,500 pounds (Federal Communications Commission (FCC) rule)

- Operating limitations, which may be in the form of an FAA-approved AFM/POH, placards, instrument markings, or any combination thereof (14 CFR part 91, section 91.9)

- Current weight and balance data

- Compass correction card, if required under applicable airworthiness standards

- External data plate (14 CFR part 45, section 45.11)

Visual Preflight Assessment

The inspection should start with the cabin door. If the door is hard to open or close, does not fit snugly, or the door latches do not engage or disengage smoothly, the surrounding structure, such as the doorpost, should be inspected for misalignment, which could indicate structural damage. The visual preflight inspection should continue to the interior of the cabin or flight deck where carpeting should be inspected to ensure that it is serviceable, dry, and properly affixed; seat belts and shoulder harnesses should be inspected to ensure that they are free from fraying, latch properly, and are securely attached to their mounting fittings; seats should be inspected to ensure that the seats properly latch into the seat rails through the seat lock pins and that seat rail holes are not abnormally worn to an oval shape; *[Figure 2-5]* the windshield and windows should be inspected to ensure that they are clean and free from cracks, and crazing. A dirty, scratched, and/or a severely crazed window can result in near zero visibility due to light refraction at certain angles from the sun.

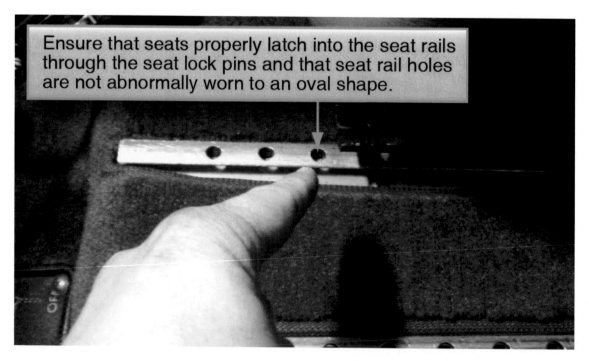

Figure 2-5. *Seats should be inspected to ensure that they are properly latched into the seat rails and checked for damage.*

The AFM/POH or a third party checklist based on the AFM/POH may be used to conduct the visual preflight inspection, and each manufacturer has a specified sequence for conducting the actions. In general, the following items are likely to be included in the AFM/POH preflight inspection:

- Landing gear control is DOWN, if applicable.

- Master, alternator, and magneto switches are OFF.

- Control column locks are REMOVED.

- Fuel selectors should be checked for proper operation in all positions, including the OFF position. Stiff fuel selectors or where the tank position is not legible or lacking detents are unacceptable.

- Trim wheels, which include elevator and may include rudder and aileron, are set for takeoff position.

- Mechanical air-driven gyro instruments should be inspected for signs of hazing on the instrument face, which may indicate leaks.

- Avionics master is OFF.

- Circuit breakers checked IN.

- Confirm that the landing gear handle is in the DOWN position, then turn the master switch ON. Note the fuel quantities on the fuel gauges and compare to the tank level by visual inspection. If so equipped, fuel pumps may be placed in the ON position to verify fuel pressure in the proper operating range.

- Other items may include checking that lights for both the interior and exterior airplane positions are operating and checking any annunciator panels.

- If the airplane has retractable gear, landing gear down and locked lights are checked green.

- Flight instruments should read as follows:

  - Airspeed should read zero.

  - The altimeter, when properly set to the current barometric setting, should indicate the field elevation within 75 feet for IFR flight.

  - If installed, the magnetic compass should indicate the airplane's direction accurately; and the compass correction card should be legible and complete. For conventional wet magnetic compasses, the instrument face should be clear and the instrument case full of fluid. A cloudy instrument face, bubbles in the fluid, or a partially filled case renders the compass unusable.

  - The vertical speed indictor (VSI) should read zero. If the VSI does not show a zero reading, a small screwdriver can be used to zero this instrument if not part of an electronic display. The mechanical VSI is the only flight instrument that a pilot has the prerogative to adjust. All others need to be adjusted by an FAA-certificated repairman or mechanic.

  - Avionics master switch ON to check avionics. Avionics master switch OFF, master switch OFF.

Aircraft equipped with Integrated Flight Deck (IFD) "glass-panel" avionics and supporting systems have specific requirements for checking prior to flight. Ground-based inspections may include verification that the flight deck reference guide is in the aircraft and accessible; checking of system driven removal of "Xs" over engine indicators; checking pitot/static and attitude displays; testing of low level alarms and annunciator panels; setting of fuel levels; and verification that the avionics cooling fans, if equipped, are functional. [Figure 2-6] The AFM/POH specifies how these preflight inspections are to take place. Since an advanced avionics aircraft preflight checklist may be extensive, pilots should allow time to ensure that all items are properly addressed.

Figure 2-6. *Ground-based inspections include verification that "Xs" on the instrument display are displayed until the sensor activates.*

Outer Wing Surfaces and Tail Section

Generally, the AFM/POH specifies a sequence for the pilot to inspect the aircraft that may sequence from the cabin entry access opening and then in a counterclockwise direction until the aircraft has been completely inspected. Besides the AFM/POH preflight assessment, the pilot should also develop awareness for potential areas of concern, such as signs of deterioration or distortion of the structure, whether metal or composite, as well as loose or missing rivets or screws.

In addition to items specified in the AFM/POH for inspection, the pilot should have an awareness for critical areas, such as spar lines, wing, horizontal, and vertical attach points including wing struts and landing gear attachment areas. The airplane skin should be inspected in these areas as load-related stresses are concentrated along spar lines and attach points. Spar lines are lateral rivet lines that extend across the wing, horizontal stabilizer, or vertical stabilizer. Pilots should pay close attention to spar lines looking for distortion, ripples, bubbles, dents, creases, or waves as any structural deformity may be an indication of internal damage or failure. Inspect around rivet heads looking for cracked paint or a black-oxide film that forms when a rivet works free in its hole. *[Figure 2-7]*

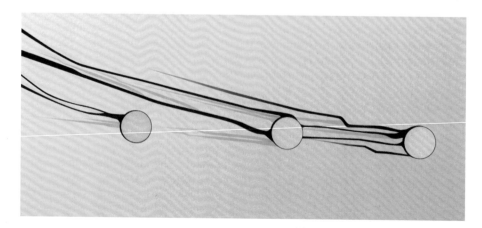

Figure 2-7. *Example of rivet heads where black oxide film has formed due to the rivet becoming loose in its hole.*

Additional areas that should be scrutinized are the leading edges of the wing, horizontal stabilizer, and vertical stabilizer. These areas may have been impact-damaged by rocks, ice, birds, and/or hangar rash incidents. Certain dents and dings may render the structure unairworthy. Some leading edge surfaces have aerodynamic devices, such as stall fences, slots, or vortex generators, and deicing equipment, such as weeping wings and boots. If these items exist on the airplane, the pilot should know their proper condition so that an adequate preflight inspection may occur.

On metal airplanes, wingtips, fairings, and non-structural covers may be fabricated out of thin fiberglass or plastic. These items are frequently affected by cracks radiating from screw holes or concentrated radii. Often, if any of these items are cracked, it is practice to "stop-drill" the crack to prevent crack progression. [Figure 2-8] Extra care should be exercised to ensure that these devices are in good condition without cracks that may render them unairworthy. Cracks that have continued beyond a stop-drilled location or any new adjacent cracks that have formed may lead to in-flight failure.

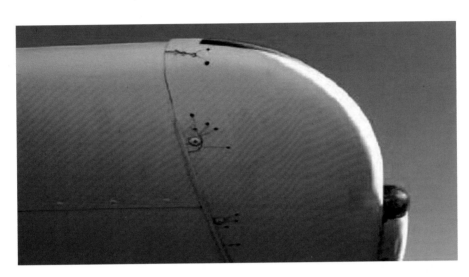

Figure 2-8. *Cracks radiating from screw holes that have been stop-drilled to prevent crack progression.*

Inspecting composite airplanes can be more challenging as the airplanes generally have no rivets or screws to aid the pilot in identifying spar lines and wing attach points. However, delamination of spar to skin or other structural problems may be identified by bubbles, fine hair-line cracks, or changes in sound when gently tapping on the structure with a fingertip. Anything out of place should be addressed by discussing the issue with a properly rated aircraft mechanic.

Fuel and Oil

While there are various formulations of aviation gasoline (AVGAS), only three grades are conventional: 80/87, 100LL, and 100/130. 100LL is the most widely available in the United States. AVGAS is dyed with a faint color for grade identification: 80/87 is dyed red; 100LL is dyed blue; and 100/130 is dyed green. All AVGAS grades have a familiar gasoline scent and texture. 100LL with its blue dye is sometimes difficult to identify unless a fuel sample is held up against a white background in reasonable white lighting.

Aircraft piston engines certified for grade 80/87 run satisfactorily on 100LL if approved as an alternate. The reverse is not true. Fuel of a lower grade should never be substituted for a required higher grade. Detonation will severely damage the engine in a very short period of time. Detonation, as the name suggests, is an explosion of the fuel-air mixture inside the cylinder. During detonation, the fuel/air charge (or pockets within the charge) explodes rather than burns smoothly. Because of this explosion, the charge exerts a much higher force on the piston and cylinder, leading to increased noise, vibration, and cylinder head temperatures. The violence of detonation also causes a reduction in power. Mild detonation may increase engine wear, though some engines can operate with mild detonation regularly. However, severe detonation can cause engine failure in minutes. [Figure 2-9] Because of the noise that it makes, detonation is "engine knock" or "pinging" in cars.

When approved for the specific airplane to be flown, automobile gasoline is sometimes used as a substitute fuel in certain airplanes. Its use is acceptable only when the particular airplane has been issued a Supplemental Type Certificate (STC) to both the airframe and engine.

Jet fuel is a kerosene-based fuel for turbine engines and a new generation of diesel-powered airplanes. Jet fuel has a stubborn, distinctive, non-gasoline odor and is oily to the touch. Jet fuel is clear or straw-colored, although it may appear dyed when mixed with AVGAS. Jet fuel has disastrous consequences when introduced into AVGAS-burning reciprocating airplane engines. A reciprocating engine operating on jet fuel may start, run, and power the airplane long enough for the airplane to become airborne, only to have the engine fail catastrophically after takeoff.

Figure 2-9. *An aircraft piston showing damage that occurred in just minutes as a result of detonation and overheating.*

Jet fuel refueling trucks and dispensing equipment are marked with JET-A placards in white characters on a black background. Because of the dire consequences associated with misfueling, fuel nozzles are specific to the type of fuel. AVGAS fuel filler nozzles are straight with a constant diameter. *[Figure 2-10]* However, jet fuel filler nozzles are flared at the end to prevent insertion into AVGAS fuel tanks. *[Figure 2-11]*

Figure 2-10. *An AVGAS fuel filler nozzle is straight with a constant diameter.*

Figure 2-11. *A jet fuel filler nozzle is flared at the end to prevent an inadvertent insertion into an AVGAS tank.*

Using the proper, approved grade of fuel is critical for safe, reliable engine operation. Without the proper fuel quantity, grade, and quality, the engine(s) will likely cease to operate. Therefore, it is imperative that the pilot visually verify that the airplane has the correct fuel quantity for the intended flight plus adequate and legal reserves, as well as inspect that the fuel is of the proper grade and that the quality of the fuel is acceptable. The pilot should always ensure that the fuel caps have been securely replaced following each fueling.

Many airplanes experience sensitivity to attitude when fueling for maximum capacity. Nosewheel or main landing gear strut extension, both high as well as low, and the slope of the ramp can significantly alter the attitude of the aircraft and therefore the fuel capacity. Always positively confirm the fuel quantity indicated on the fuel gauges by visually inspecting the level of fuel in each tank.

The pilot should be aware that fuel stains anywhere on the wing or any location where a fuel tank is mounted warrants further investigation—no matter how old the stains appear to be. Fuel stains are a sign of probable fuel leakage. On airplanes equipped with wet-wing fuel tanks, evidence of fuel leakage can be found along rivet lines. *[Figure 2-12]*

Figure 2-12. *Evidence of fuel leakage can be found along rivet lines.*

Checking for water and other sediment contamination is a key preflight item. Water tends to accumulate in fuel tanks from condensation, particularly in partially filled tanks. Because water is heavier than fuel, it tends to collect in the low points of the fuel system. Water can also be introduced into the fuel system from deteriorated gas cap seals exposed to rain or from the supplier's storage tanks and delivery vehicles. Sediment contamination can arise from dust and dirt entering the tanks during refueling or from deteriorating rubber fuel tanks or tank sealant. Deteriorating rubber from seals and sealant may show up in the fuel sample as small dark specks.

The best preventive measure is to minimize the opportunity for water to condense in the tanks. If possible, the fuel tanks should be completely filled with the proper grade of fuel after each flight, or at least filled after the last flight of the day. The more fuel that is in the tanks, the less room there is for condensation to occur. Keeping fuel tanks filled is also the best way to slow the aging of rubber fuel tanks and tank sealant.

Sufficient fuel should be drained from the fuel strainer quick drain and from each fuel tank sump to check for fuel grade/color, water, dirt, and odor. If water is present, it is usually in bubble or bead-like droplets, different in color (usually clear, sometimes muddy yellow to brown with specks of dirt), in the bottom of the sample jar. In extreme water contamination cases, consider the possibility that the entire fuel sample, particularly if a small sample was taken, is water. If water is found in the first fuel sample, continue sampling until no water and contamination appears. Significant and/or consistent water, sediment or contaminations are grounds for further investigation by qualified maintenance personnel. Each fuel tank sump should be drained during preflight and after refueling. The order of sumping the fuel system is often very important. Check the AFM/POH for specific procedures and order to be followed.

Checking the fuel tank vent is an important part of a preflight assessment. If outside air is unable to enter the tank as fuel is drawn into the engine, the eventual result is fuel starvation and engine failure. During the preflight assessment, the pilot should look for signs of vent damage and blockage. Some airplanes utilize vented fuel caps, fuel vent tubes, or recessed areas under the wings where vents are located. The pilot should use a flashlight to look at the fuel vent to ensure that it is free from damage and clear of obstructions. If there is a rush of air when the fuel tank cap is cracked, there could be a serious problem with the vent system.

Aviation oils are available in various single/multi-grades and mineral/synthetic-based formulations. It is important to use the approved and recommended oil for the engine at all times. The oil not only acts as a lubricant but also as a medium to transfer heat as a result of engine operation and to suspend dirt, combustion byproducts, and wear particles between oil changes. Therefore, the proper level of oil is required to ensure lubrication, effective heat transfer, and the suspension of various contaminants. The oil level should be checked during each preflight, rechecked with each refueling, and maintained to prevent the oil level from falling below the minimum required during engine operation.

During the preflight assessment, if the engine is cold, oil levels on the oil dipstick show higher levels than if the engine was warm and recently shutdown after a flight. When removing the oil dipstick, care should be taken to keep the dipstick from coming in contact with dirty or grimy areas. The dipstick should be inspected to verify the oil level. Typically, piston airplane engines have oil reservoirs with capacities between four and eight quarts, with six quarts being common. Aside from the level of oil, the oil's color also provides an insight as to its operating condition. Oils darken in color as the oil operating hours increase—this is common and expected as the oil traps contaminants. However, oils that rapidly darken in the first few hours of use after an oil change may indicate engine cylinder problems. Piston airplane engines consume a small amount of oil during normal operation. The amount of consumption varies on many factors; however, if consumption increases or suddenly changes, qualified maintenance personnel should investigate

It is suggested that the critical aspect of fuel and oil *not* be left to line service personnel without oversight of the pilot responsible for flight. While line personnel are aviation professionals, the pilot is responsible for the safe outcome of any flight. During refueling or when oil is added to an engine, the pilot should monitor and ensure that the correct quantity, quality, and grade of fuel and oil is added and that all fuel and oil caps have been securely replaced.

Landing Gear, Tires, and Brakes

The landing gear, tires, and brakes allow the airplane to maneuver from and return to the ramp, taxiway, and runway environment in a precise and controlled manner. The landing gear, tires, and brakes should be inspected to ensure that the airplane can be positively controlled on the ground. Landing gear on airplanes varies from simple fixed gear to complex retractable gear systems.

Fixed landing gear is a gear system in which the landing gear struts, tires, and brakes are exposed and lend themselves to relatively simple inspection. However, more complex airplanes may have retractable landing gear with multiple tires per landing gear strut, landing gear doors, over-center locks, springs, and electrical squat switches. Regardless of the system, the pilot should follow the AFM/POH during inspection to determine that the landing gear is ready for operation.

On many fixed-gear airplanes, inspection of the landing gear system can be hindered by wheel pants, which are covers used to reduce aerodynamic drag. It is still the pilot's responsibility to inspect the airplane properly. A flashlight helps the pilot in peering into covered areas. On low-wing airplanes, covered or retraceable landing gear presents additional effort required to crouch below the wing to inspect the landing gear properly.

The following provides guidelines for inspecting the landing gear system; however, the AFM/POH should be the pilot's reference for the appropriate procedures.

- The pilot, when approaching the airplane, should look at the landing gear struts and the adjacent ground for leaking hydraulic fluid that may be coming from struts, hydraulic lines from landing gear retraction pumps, or from the braking system. Landing gear should be relatively free from grease, oil, and fluid without any undue amounts. Any amount of leaking fluid is unacceptable. In addition, an overview of the landing gear provides an opportunity to verify landing gear alignment and height consistency.

- All landing gear shock struts should also be checked to ensure that they are properly inflated, clean, and free from hydraulic fluid and damage. All axles, links, collars, over-center locks, push rods, forks, and fasteners should be inspected to ensure that they are free from cracks, corrosion, and rust, and are in an airworthy condition.

- Tires should be inspected for proper inflation, an acceptable level of remaining tread, and normal wear pattern. Abnormal wear patterns, sidewall cracks, and damage, such as cuts, bulges, imbedded foreign objects, and visible cords, render the tire unairworthy. For airplanes that are flown by more than one pilot, what happened to the tires on previous flights becomes a significant unknown. Therefore, when possible, the airplane should be moved slightly to allow for evaluation of the complete tire circumference.

- Wheel hubs should be inspected to ensure that they are free from cracks, corrosion, and rust, that all fasteners are secure, and that the air valve stem is straight, capped, and in good condition.

- Brakes and brake systems should be checked to ensure that they are free from rust and corrosion and that all fasteners and safety wires are secure. Brake pads should have a proper amount of material remaining and should be secure. All brake lines should be secure, dry, and free of signs of hydraulic leaks, and devoid of abrasions and deep cracking.

- On tricycle gear airplanes, a shimmy damper is used to damp oscillations of the nose gear and should be inspected to ensure that it is securely attached, is free of hydraulic fluid leaks, and is in overall good condition. Some shimmy dampers do not use hydraulic fluid and instead use an elastomeric compound as the dampening medium. Nose gear links, collars, steering rods, and forks should be inspected to ensure the security of fasteners, minimal free play between torque links, crack-free components, and for proper servicing and general condition.

- On some conventional gear airplanes, those airplanes with a tailwheel or skid, the main landing gear may have bungee cords to help in absorbing landing loads and shocks. The bungee cords must be inspected for security and condition.

- Where the landing gear transitions into the airplane's structure, the pilot should inspect the attachment points and the airplane skin in the adjacent area—the pilot needs to inspect for wrinkled or other damaged skin, loose bolts, and rivets and verify that the area is free from corrosion.

Engine and Propeller

Properly managing the risks associated with flying requires that the pilot of the airplane identify and mitigate any potential hazards prior to flight to prevent, to the furthest extent possible, a hazard becoming a realized risk. The engine and propeller make up the propulsion system of the airplane—failure of this critical system requires a well-trained and competent pilot to respond with significant time constraints to what is likely to become a major emergency.

The pilot needs to ensure that the engine, propeller, and associated systems are functioning properly prior to operation. This starts with an overview of the cowling that surrounds the airplane engine. The pilot should look for loose, worn, missing, or damaged fasteners, rivets, and latches that secure the cowling around the engine and to the airframe. The pilot should be vigilant as fasteners and rivets can be numerous and surround the cowling requiring a visual inspection from above, the sides, and the bottom. Like other areas on the airframe, rivets should be closely inspected for looseness by looking for signs of a black oxide film around the rivet head. The pilot should pay attention to chipped or flaking paint around rivets and other fasteners as this may be a sign of a lack of security. Any cowling security issues need to be referred to a competent and rated airplane maintenance mechanic.

From the cowling, a general inspection of the propeller spinner, if so equipped, should be completed. Not all airplane/propeller combinations have a spinner, so adherence to the AFM/POH checklist is required. Spinners are subjected to great stresses and should be inspected to be free from dents, cracks, corrosion, and in proper alignment. Cracks may not only occur at locations where fasteners are used but also on the rear-facing spinner plate. In conditions where ice or snow may have entered the spinner around the propeller openings, the pilot should inspect the area to ensure that the spinner is internally free from ice. The engine/propeller/spinner is balanced around the crankshaft and a small amount of ice or snow can produce damaging vibrations. Cracks, missing fasteners, or dents result in a spinner that is unairworthy.

The propeller should be checked for blade erosion, nicks, cracks, pitting, corrosion, and security. On controllable pitch propellers, the propeller hub should be checked for oil leaks that tend to stream directionally from the propeller hub toward the tip. On airplanes so equipped, the alternator/generator drive belts should be checked for proper tension and signs of wear.

When inspecting inside the cowling, the pilot should check all surfaces for oil leaks or deterioration of oil and hydraulic lines, and make certain that the oil cap, filter, oil cooler, and drain plug are secure. The pilot should look for signs of fuel dye, which may indicate a fuel leak. Note that both fuel and oil stains may appear on a cowling inner surface. Observation may be difficult without the aid of a flashlight, so even during day operations, a flashlight is handy when peering into the cowling. The pilot should also check for loose or foreign objects inside the cowling, such as bird nests, shop rags, and/or tools. All visible wires and lines should be checked for security and condition. The exhaust system should be checked for white stains caused by exhaust leaks at the cylinder head or cracks in the exhaust stacks. The heat muffs, which provide cabin heating on some airplanes, should also be checked for general condition and signs of cracks or leaks. An isolated area of oxidized darkened paint on the engine may indicate an area experiencing excessive heat. If visible, the condition of the firewall may be checked for integrity.

The air filter should be checked to ensure that it is free from substantial dirt or restrictions, such as bugs, birds, nests, or other causes of airflow restriction. In addition, air filter elements are made from various materials. In all cases, the element should be free from decomposition and properly serviced.

Risk and Resource Management

Ground operations also include the pilot's assessment of the risk factors that contribute to safety of flight and the pilot's management of the resources, which may be leveraged to maximize the flight's successes. The Risk Management Handbook (FAA-H-8083-2) should be reviewed for a comprehensive discussion of this topic. A review of key points follows.

Approximately 85 percent of all aviation accidents have been determined by the National Transportation Safety Board (NTSB) to have been caused by "failure of the pilot to..." As such, a reduction of these failures is the fundamental cornerstone to risk and resource management. The risks involved with flying an airplane are very different from those experienced in daily activities, such as driving to work. Managing risks and resources requires a conscious effort that goes beyond the stick and rudder skills required to pilot the airplane.

Risk Management

Risk management is a formalized structured process for identifying and mitigating hazards and assessing the consequences and benefits of the accepted risk. A hazard is a condition, event, object, or circumstance that could lead to or contribute to an unplanned or undesired event, such as an incident or accident. It is a source of potential danger. Some examples of hazards are:

1. Marginal weather or environmental conditions
2. Lack of pilot qualification, currency, or proficiency for the intended flight.

Identifying the Hazard

Hazard identification is the critical first step of the risk management process. If pilots do not recognize and properly identify a hazard and choose to continue, the consequences of the risk involved is not managed or mitigated. In the previous examples, the hazard identification process results in the following assessment:

• Marginal weather or environmental conditions is an identified hazard because it may result in the pilot having a skill level that is not adequate for managing the weather conditions or requiring airplane performance that is unavailable.

• The lack of pilot training is an identified hazard because the pilot does not have experience to either meet the legal requirements or the minimum necessary skills to safely conduct the flight.

Risk

Risk is the future impact of a hazard that is not controlled or eliminated. It can be viewed as future uncertainty created by the hazard.

- If the weather or environmental conditions are not properly assessed, such as in a case where an airplane may encounter inadvertent instrument conditions, loss of airplane control may result.

- If the pilot's lack of training is not properly assessed, the pilot may be placed in flight regimes that exceed the pilot's stick-and-rudder capability.

Risk Assessment

Risk assessment determines the degree of risk and whether the degree of risk is worth the outcome of the planned activity. Once the planned activity is started, the pilot needs to consider whether to continue or not. A pilot should always have viable alternatives available in the event the original flight plan cannot be accomplished. Thus, hazard and risk are the two defining elements of risk management. A hazard can be a real or perceived condition, event, or circumstance that a pilot encounters. Risk assessment is a quantitative value weighted to a task, action, or event. When armed with the predicted risk assessment of an activity, pilots are able to manage and mitigate their risk.

In the example where marginal weather is the identified hazard, it is relatively simple to understand that the consequences of loss of control during any inadvertent encounter with instrument meteorological conditions (IMC) are likely to be severe for a pilot not prepared to fly on an instrument flight plan. A risk assessment for any such pilot in this example would determine that the risk is unacceptable and as a result, mitigation of the risk is required. Proper risk mitigation would require that flight be canceled or delayed until weather conditions were not conducive for inadvertent flight into instrument meteorological conditions.

Risk Identification

Identifying hazards and associated risk is key to preventing risk and accidents. If a pilot fails to search for risk, it is likely that he or she will neither see it nor appreciate it for what it represents. Unfortunately, in aviation, pilots seldom have the opportunity to learn from their small errors in judgment because even small mistakes in aviation are often fatal. In order to identify risk, the use of standard procedures is of great assistance. Several procedures are discussed in detail in the Risk Management Handbook (FAA-H-8083-2).

Risk Mitigation

Risk assessment is only part of the equation. After determining the level of risk, the pilot needs to mitigate the risk. For example, the VFR pilot flying from point A to point B (50 miles) in marginal flight conditions has several ways to reduce risk:

1. Wait for the weather to improve to good VFR conditions.
2. Take a pilot who is more experienced or who is certified as an instrument flight rules (IFR) pilot.
3. Delay the flight.
4. Cancel the flight.
5. Drive.

Resource Management

Familiarity with crew resource management (CRM) and single-pilot resource management (SRM) enables a crew or pilot to manage all available resources effectively and leads to a successful flight. In general aviation, SRM comes into play more often. The focus of SRM is on the single-pilot operation. SRM integrates the following:

- Situational Awareness
- Human Resource Management
- Task Management
- Aeronautical Decision-making (ADM)

Situational Awareness

Situational awareness is the accurate perception of operational and environmental factors that affect the flight. It is a logical analysis based upon the airplane, external support, environment, and the pilot. It is awareness on what is happening in and around the flight.

Human Resource Management

Human resource management requires an effective use of all available resources: human, equipment, and information.

Human resources include the essential personnel routinely working with the pilot to ensure safety of flight. These people include, but are not limited to: weather briefers, flight line personnel, maintenance personnel, crew members, pilots, and air traffic personnel. Pilots need to communicate effectively with these people. This is accomplished by using the key components of the communication process: inquiry, advocacy, and assertion. Pilots should recognize the need to seek enough information from these resources to make a valid decision. After the necessary information has been gathered, the pilot's decision should be passed on to those concerned, such as air traffic controllers, crewmembers, and passengers. The pilot may have to request assistance from others and be assertive to resolve some situations safely.

Equipment in many of today's aircraft includes automated flight and navigation systems. These automatic systems, while providing relief from many routine tasks, present a different set of problems for pilots. The automation intended to reduce pilot workload essentially removes the pilot from the process of managing the aircraft, thereby reducing situational awareness and leading to complacency. Information from these systems needs to be continually monitored to ensure proper situational awareness. Pilots should be aware of both equipment capabilities and equipment limitations in order to manage those systems effectively and safely.

Information workloads and automated systems, such as autopilots, need to be properly managed to ensure a safe flight. By planning ahead, a pilot can effectively reduce workload during critical phases of flight and prevent erosion of performance. The pilot who effectively manages his or her workload completes routine tasks as early as possible to preclude the possibility of becoming overloaded and stressed in the later, more critical stages of the flight.

Task Management

Pilots have a limited capacity for information. Once information flow exceeds the pilot's ability to process the information mentally, any additional information becomes unattended or displaces other tasks and information already being processed. In addition, distraction and fixation impede the ability to process information. For example, if a pilot becomes distracted and fixates on an instrument light failure, the unnecessary focus displaces capability and prevents appreciation of tasks of greater importance.

Aeronautical Decision-Making (ADM)

Flying safely requires the effective integration of three separate sets of skills: stick-and-rudder skills needed to control the airplane; skills related to proficient operation of aircraft systems; and ADM skills. The ADM process addresses all aspects of decision-making in the flight deck and identifies the steps involved in good decision-making. While the ADM process does not eliminate errors, it helps the pilot recognize errors and enables the pilot to manage the error to minimize its effects. These steps are:

1. Identifying personal attitudes hazardous to safe flight;
2. Learning behavior modification techniques;
3. Learning how to recognize and cope with stress;
4. Developing risk assessment skills;
5. Using all resources; and
6. Evaluating the effectiveness of one's own personal ADM skills.

Ground Operations

The airport ramp can be a complex environment with airport personnel, passengers, trucks, other vehicles, aircraft, and errant people and animals. The pilot is responsible for the operation of the airplane and should operate safely at all times. Ground operations subject the pilot to unique hazards, and mitigating those hazards requires proper planning and good situational awareness in the ground environment. A mitigation tactic involves reviewing the airport diagram prior to operating and having it readily available at all times. Whether departing to or from the ramp, the pilot needs to understand and capably manage the following:

1. Refueling operations
2. Passenger and baggage security and loading
3. Ramp and taxi operations
4. Standard ramp signals

During refueling operations, it is advisable that the pilot remove all passengers from the aircraft and witness the refueling to ensure that the correct fuel and quantity is dispensed into the airplane and that any caps and cowls are properly secured after refueling.

Passengers may have little experience with the open ramp of an airport. The pilot should ensure the safety of the passengers by cautioning them to move on the surface only as directed. If not under the pilot's direct supervision, passengers should have an escort to ensure their safety and ramp security. Baggage loading and security should also be supervised by the pilot. Unsecured baggage or improperly loaded baggage may adversely affect the center of gravity of the aircraft.

Ramp traffic may vary from a deserted open space to a complex environment with heavy corporate or military aircraft. Powerful aircraft may produce exhaust blast or rotor downwash, for example, which could easily cause a light airplane to become uncontrollable. Mitigating these hazards in a light airplane is important to starting off on a safe flight.

Some ramps may be staffed by personnel to assist the pilot in managing a safe departure from the ramp to the taxiway. *Figure 2-13* shows standard aircraft taxiing signals, such as those published in the Aeronautical Information Manual (AIM). There are other standard signals, such as those published in Advisory Circular 00-34, as revised, and by the Armed Forces. Furthermore, operation conditions in many areas may call for a modified set of taxi signals. The signals shown in *Figure 2-13* represent a minimum number of the most commonly used signals. Whether this set of signals or a modified set is used is not the most important consideration, as long as each flight operational center uses a suitable, agreed-upon set of signals.

Figure 2-13. *Standard hand signals used to assist pilots in managing a safe departure from the ramp to the taxiway or runway. Note that at night, the Emergency Stop signal is used for all stop indications .*

Engine Starting

Airplane engines vary substantially and specific procedures for engine starting should be accomplished in reference to the approved engine start checklist as detailed in the airplane's AFM/POH. However, some generally accepted hazard mitigation practices and procedures are outlined in this section.

Prior to engine start, the pilot needs to ensure that the ramp area surrounding the airplane is clear of persons, equipment, and other hazards that could come into contact with the airplane or the propeller. Also, the pilot should check what is behind the airplane prior to engine start as standard practice. A propeller or other engine thrust can accelerate objects to substantial velocities, causing damage to property, and injuring those on the ground. The pilot should mitigate the hazard of debris being blown into persons or property. At all times before engine start, the anti-collision lights should be turned on. For night operations, the position (navigation) lights should also be on. Finally, just prior to starter engagement, the pilot should always call "CLEAR" out of the side window and wait for a response from anyone who may be nearby *before* engaging the starter.

When activating the starter, the wheel brakes need to be depressed and one hand kept on the throttle to manage the initial starting engine speed. Ensuring that properly operating brakes are engaged prior to starter engagement prevents the airplane from rapidly lunging forward. After engine start, the pilot manipulates the throttle to set the engine revolutions per minute (rpm) to the AFM/POH-prescribed setting. In general, 1,000 rpm is recommended following engine start to allow oil pressure to rise and to minimize undue engine wear due to insufficient lubrication at high rpm. It is important to service an airplane engine with the proper grade of oil for the seasonal conditions and to apply engine preheat when temperatures approach and descend below freezing.

The oil pressure should be monitored after engine start to ensure that pressure is increasing toward the AFM/POH-specified value. The AFM/POH specifies an oil pressure range for the engine. If the limits are not reached and maintained, serious internal engine damage is likely. In most conditions, oil pressure should rise to at least the lower limit within 30 seconds. To prevent damage, the engine should be shut down immediately if the oil pressure does not rise to the AFM/POH values within the required time.

Engine starters are electric motors designed to produce rapid rotation of the engine crankshaft for starting. These electric motors are *not* designed for continuous duty. Their service life may be drastically shortened during a prolonged or difficult start as an excess buildup of heat can damage internal starter components. Avoid continuous starter operation for periods longer than 30 seconds without a cool down period of at least 30 seconds to 1 minute (some AFM/POH specify longer cool down routines). The smell of burning insulation from a starter may indicate that the recommended cranking time has been exceeded. After repeated unsuccessful start attempts, the pilot should seek advice from a qualified person to determine the cause for the difficulty.

Although quite rare, the starter motor may remain electrically and mechanically engaged after engine start. This can be detected by a continuous and very high current draw on the ammeter. Some airplanes also have a starter engaged warning light specifically for this purpose. The engine should be shut down immediately if this occurs.

The pilot should be attentive for sounds, vibrations, smells, or smoke that are not consistent with normal after-start operational experience. Any concerns should lead to a shutdown and further investigation.

Hand Propping

The procedures for hand propping should always be in accordance with the AFM/POH and performed only by persons who are competent with hand propping procedures. The consequences of the hazards associated with hand propping are serious to fatal.

Historically, when aircraft lacked electrical systems, it was necessary for pilots and ground personnel to "hand prop" an aircraft for starting. Today, most airplanes are equipped with electric starters, and the starter should be working if the airplane is airworthy. If not, a certificated Aviation Maintenance Technician should be called to make a repair. However, vintage airplanes may be encountered, and an airplane manufactured without an electric starter needs to be hand propped. Since a number of these airplanes have been produced, the procedures for hand propping are described in this section.

A few simple precautions help to avoid accidents when hand propping the engine. While touching a propeller, always assume that the ignition is on. The switches that control the magnetos operate on the principle of short-circuiting the current to turn the ignition off. If the switch is faulty, it can be in the "off" position and still permit current to flow in the magneto primary circuit. This condition could allow the engine to start when the switch is off.

Hand propping an aircraft is a hazardous procedure when done perfectly. Not mitigating the hazards associated with hand propping can lead to serious injury and a runaway airplane. A spinning propeller can be lethal should it strike someone. Persons not trained, not competent, or who do not understand how to mitigate the hazards associated with hand propping should *never* perform this procedure!

Hand propping requires a team of two properly trained people. Both individuals should be familiar with the airplane and hand propping techniques. The first person is responsible for directing the procedure including pulling the propeller blades through. The second person sits in the airplane to ensure that the brakes are set and to exercise controls as directed by the person pulling the propeller. When hand propping occurs, a person unfamiliar with the controls should ***never*** occupy the pilot's seat.

When hand propping is necessary, the ground surface near the propeller should be stable and free of debris. Loose gravel, wet grass, grease, mud, oil, ice, or snow might cause the person pulling the propeller through to slip into the rotating blades as the engine starts. Unless a firm footing is available, relocate the airplane to mitigate this hazardous consequence.

Both participants should discuss the procedure and agree on voice commands and expected actions. To begin the procedure, the fuel system and engine controls (tank selector, primer, pump, throttle, and mixture) are set for normal start. The ignition/magneto switch should be checked to be sure that it is OFF. Then, the descending propeller blade should be rotated so that it assumes a position slightly above the horizontal. The person doing the hand propping should face the descending blade squarely and stand slightly less than one arm's length from the blade. If a stance too far away were assumed, it would be necessary to lean forward in an unbalanced condition to reach the blade, which may cause the person to fall forward into the rotating blades when the engine starts. Allowing space for the person to be able to step away as the propeller is pulled down, and the engine starts, serves as safeguard in case the brakes fail.

The procedure and commands for hand propping are:

- Person out front says, "FUEL ON, SWITCH OFF, THROTTLE CLOSED, BRAKES SET."

- Pilot seat occupant, after making sure the fuel is ON, mixture is RICH, magneto switch is OFF, throttle is CLOSED, and brakes are SET, says, "FUEL ON, SWITCH OFF, THROTTLE CLOSED, BRAKES SET."

- Person out front, after pulling the propeller through to prime the engine says, "BRAKES AND CONTACT."

- Pilot seat occupant checks the brakes SET and turns the magnetos switch ON, then says, "BRAKES AND CONTACT."

The words CONTACT (magnetos ON) and SWITCH OFF (magnetos OFF) are used because they are significantly different from each other. Under noisy conditions or high winds, the words CONTACT and SWITCH OFF are less likely to be misunderstood than SWITCH ON and SWITCH OFF.

The propeller is swung by forcing the blade downward rapidly, pushing with the palms of both hands. If the blade is gripped tightly with the fingers, the person's body may be drawn into the propeller blades should the engine misfire, "kickback," or rotate momentarily in the opposite direction. As the blade is pushed down, the person should step backward, away from the propeller. If the engine does not start, the propeller should not be repositioned for another attempt until it is verified that the magneto switch is turned OFF. Excessive throttle opening after the engine has fired is the principal cause of backfiring during starting. Gradual opening of the throttle, while the engine is cold, reduces the potential for backfiring. Slow, smooth movement of the throttle assures correct engine operation.

Immediately after the engine starts, check the oil pressure indicator. If oil pressure does not show within 30 seconds, stop the engine and determine the trouble. If oil pressure is indicated, adjust the throttle to the aircraft manufacturer's specified rpm for engine warmup, which is usually between 1,000 to 1,300 rpm.

Most aircraft reciprocating engines are air-cooled and depend on the forward speed of the aircraft to maintain proper cooling. Therefore, particular care is necessary when operating these engines on the ground. During all ground running, operate the engine with the propeller in full low pitch and headed into the wind with the cowling installed to provide the best degree of engine cooling. Closely monitor the engine instruments at all times. Do not close the cowl flaps for engine warm-up, they need to be in the open position while operating on the ground. When warming up the engine, ensure that personnel, ground equipment that may be damaged, or other aircraft are not in the propeller wash.

When removing the wheel chocks or untying the tail after the engine starts, everyone involved should remember that the propeller is nearly invisible. Serious injuries and fatalities have occurred when people who have just started an engine walk or reach into the propeller arc to remove the chocks, reach the cabin, or when moving toward the tail of the airplane. Before the wheel chocks are removed, the throttle should be set to idle and the chocks approached only from the rear of the propeller. One should never approach the wheel chocks from the front or the side.

Taxiing

Taxiing is the controlled movement of the airplane under its own power while on the surface. Since an airplane is moved under its own power between a parking area and the runway, the pilot needs to understand and be proficient in taxi procedures.

A pilot should maintain situational awareness of the ramp, parking areas, taxiways, runway environment, and the persons, equipment and aircraft at all times. Without such awareness, safety may be compromised. Depending on the airport, the parking, ramp, and taxiways may or may not be controlled. As such, it is important that the pilot completely understands the operating environment. At small, rural airports these areas may be desolate with few aircraft and limited hazards; however, as the complexity of the airport increases so does the potential for hazards. Regardless of the complexity, some generally accepted procedures are appropriate.

- The pilot should be familiar with the parking, ramp, and taxi environment. This can be done by having an airport diagram, if available, out and in view at all times. *[Figure 2-14]*

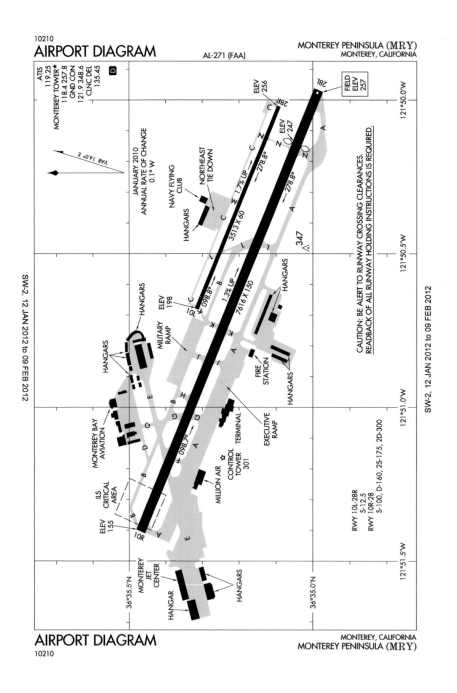

Figure 2-14. *Airport Diagram of Monterey Peninsula (MRY), Monterey, California.*

- Despite having familiarity with the airport, pilots should carefully review their complete taxi plan. For example, a pilot given the same taxi instructions by ATC, starts expecting those same instructions and might not realize that those instructions no longer apply. It only takes missing one instruction or turn to generate an accident. It is a human tendency to follow the same procedure over and over. This expectation bias has occurred to many pilots who did not stop and carefully consider and evaluate their taxi instructions.

- The pilot should be vigilant of the entire area around the airplane to ensure that the airplane clears all obstructions. If, at any time, there is doubt about a safe clearance from an object, the pilot should stop the airplane and check the clearance. It may be necessary to have the airplane towed or physically moved by a ground crew.

- When taxiing, the pilot's eyes should be looking outside the airplane scanning from side to side while looking both near and far to assess routing and potential conflicts.

- A safe taxiing speed should be maintained. The primary requirements for safe taxiing are positive control, the ability to recognize any potential hazards in time to avoid them, and the ability to stop or turn where and when desired, without undue reliance on the brakes. Pilots should proceed at a cautious speed on congested or busy ramps. Normally, the speed should be at the rate where movement of the airplane is dependent on the throttle. That is, slow enough so when the throttle is closed, the airplane can be stopped promptly.

- The pilot should place the aircraft on the taxiway center. Some taxiways have above-ground taxi lights and signage that could impact the airplane or propellers if the pilot does not exercise accurate control. When yellow taxiway centerline stripes are present, the pilot should visually place the centerline stripe so it is under the center of the airplane fuselage.

- When taxiing, the pilot should slow down before attempting a turn. Sharp high-speed turns place undesirable side loads on the landing gear and may result in tire damage or an uncontrollable swerve or a ground loop. Swerves are most likely to occur when turning from a downwind heading toward an upwind heading. In moderate to high-wind conditions, the airplane may weathervane increasing the swerving tendency.

Steering is accomplished with rudder pedals and brakes. To turn the airplane on the ground, the pilot should apply the rudder in the desired direction of turn and use the appropriate power or brake to control the taxi speed. The rudder pedal should be held in the direction of the turn until just short of the point where the turn is to be stopped. Rudder pressure is then released or opposite pressure is applied as needed.

More engine power may be required to start the airplane moving forward, or to start a turn, than is required to keep it moving in any given direction. When using additional power, the throttle should immediately be retarded once the airplane begins moving to prevent excessive acceleration.

The brakes should be tested for proper operation as soon as the airplane is put in motion. Applying power to start the airplane moving forward slowly, then retarding the throttle and simultaneously applying just enough pressure to one side, then the other to confirm proper function and reaction of both brakes. This is best if the airplane has individual left/right brakes to stop the airplane. If braking performance is unsatisfactory, the engine should be shut down immediately.

When taxiing at appropriate speeds in no-wind conditions, the aileron and elevator control surfaces have little or no effect on directional control of the airplane. These controls should not be considered steering devices and should be held in a neutral position.

When taxiing with a quartering headwind, the wing on the upwind side (the side that the wind is coming from) tends to be lifted by the wind unless the aileron control is held in that direction (upwind aileron UP). Moving the aileron into the UP position reduces the effect of the wind striking that wing, thus reducing the lifting action. This control movement also causes the downwind aileron to be placed in the DOWN position, thus a small amount of lift and drag on the downwind wing, further reducing the tendency of the upwind wing to rise. *[Figure 2-15]*

When taxiing with a quartering tailwind, the elevator should be held in the DOWN position, and the upwind aileron, DOWN. Since the wind is striking the airplane from behind, these control positions reduce the tendency of the wind to get under the tail and the wing and to nose the airplane over. The application of these crosswind taxi corrections helps to minimize the weathervaning tendency and ultimately results in easier steering. *[Figure 2-15]*

The presence of moderate to strong headwinds and/or a strong propeller slipstream creates lift on the horizontal tail surfaces and makes it necessary to control the pitch attitude while taxiing. The elevator control in nosewheel-type airplanes should be held in the neutral position, while in tailwheel-type airplanes, it should be held in the full aft position to hold the tail down unless the headwind gets very strong, which allows for an elevator position closer to neutral.

Downwind taxiing usually requires less engine power after the initial ground roll has begun, since the wind is pushing the airplane forward. To avoid overheating the brakes and controlling the airplane's speed when taxiing downwind, the pilot should keep engine power to a minimum. Rather than continuously riding the brakes to control speed, it is appropriate to apply brakes only occasionally. Other than sharp turns at low speed, the throttle should always be at idle before the brakes are applied. It is a common error to taxi with a power setting that requires controlling taxi speed with the brakes.

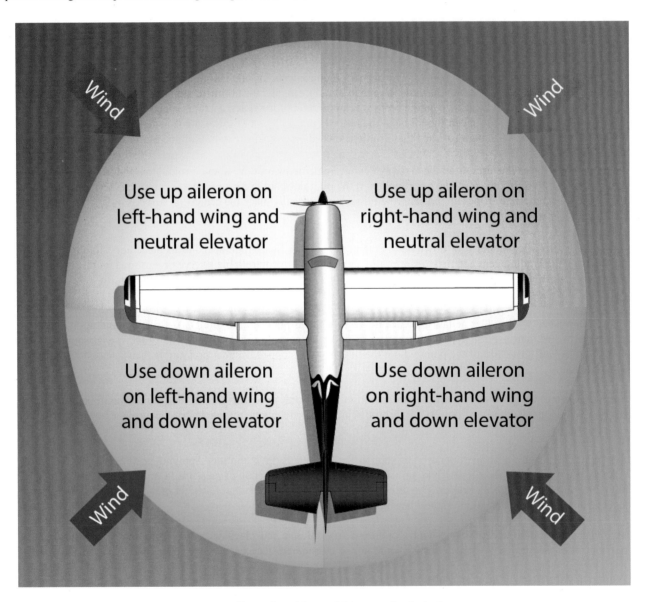

Figure 2-15. *Control positions of the nosewheel airplane.*

Normally, all turns should be started using the rudder pedal to steer the nosewheel. To tighten the turn after full pedal deflection is reached, the brake may be applied as needed. When stopping the airplane, it is always advisable to stop with the nosewheel straight ahead to relieve any side load on the nosewheel and to make it easier to start moving ahead. Note that certain makes and models have no nosewheel steering and the brakes need to be used to control any turns.

During crosswind taxiing, even the nosewheel-type airplane has some tendency to weathervane. However, the weathervaning tendency is less than in tailwheel-type airplanes because the main wheels are located behind the airplane's center of gravity, and the nosewheel's ground friction helps to resist the tendency. The nosewheel linkage from the rudder pedals provides adequate steering control for safe and efficient ground handling, and normally, only rudder pressure is necessary to correct for a crosswind.

Taxiing checklists are sometimes specified by the AFM/POH, and the pilot should accomplish any items that are required. If there are no specific checklist items, taxiing still provides an opportunity to verify the operation and cross-check of the flight instruments. In general, the flight instruments should indicate properly with the airspeed at or near zero (depending on taxi speed, wind speed and direction, and lower limit sensitivity); the attitude indicator should indicate pitch and roll level (depending on airplane attitude) with no flags; the altimeter should indicate the proper elevation within prescribed limits; the turn indicator should show the correct direction of turn with the ball movement toward the outside of the turn with no flags; the directional gyro should be set and crossed checked to the magnetic compass and verified accurate to the direction of taxi; and the vertical speed indicator (VSI) should read zero. These checks can be accomplished on conventional mechanical instrumented aircraft or those with glass displays.

Before-Takeoff Check

The before-takeoff check is the systematic AFM/POH procedure for checking the engine, controls, systems, instruments, and avionics prior to flight. Normally, the before-takeoff checklist is performed after taxiing to a run-up position near the takeoff end of the runway. Many engines require that the oil temperature reach a minimum value as stated in the AFM/POH before takeoff power is applied. Taxiing to the run-up position usually allows sufficient time for the engine to warm up to at least minimum operating temperature; however, the pilot should verify that the oil temperature is within the proper range prior to the application of high power.

A suitable location for run-up should be firm (a smooth, paved or turf surface if possible) and free of debris. Otherwise, the propeller may pick up pebbles, dirt, mud, sand, or other loose objects and hurl them backwards. This damages the propeller and may damage the tail of the airplane. Small chips in the leading edge of the propeller form stress risers or high stress concentrations. These are highly undesirable and may lead to cracks and possible propeller blade failure. The airplane should also be positioned clear of other aircraft and the taxiway. There should not be anything behind the airplane that might be damaged by the propeller airflow blasting rearward.

Before beginning the before-takeoff check, after the airplane is properly positioned for the run-up, it should be allowed to roll forward slightly to ensure that the nosewheel or tailwheel is in alignment with the longitudinal axis of the airplane.

While performing the before-takeoff check in accordance with the airplane's AFM/POH, the pilot divides attention between the inside and outside of the airplane. If the parking brake slips, or if application of the toe brakes is inadequate for the amount of power applied, the airplane could rapidly move forward and go unnoticed if pilot attention is fixed only inside the airplane. A good operational practice is to split attention from one item inside to a look outside.

Air-cooled engines generally are tightly cowled and equipped with baffles that direct the flow of air to the engine in sufficient volumes for cooling while in flight; however, on the ground, much less air is forced through the cowling and around the baffling. Prolonged ground operations may cause cylinder overheating long before there is an indication of rising oil temperature. To minimize overheating during engine run-up, it is recommended that the airplane be headed as nearly as possible into the wind and, if equipped, engine instruments that indicate cylinder head temperatures should be monitored. Cowl flaps, if available, should be set according to the AFM/POH.

Each airplane has different features and equipment and the before-takeoff checklist provided in airplane's AFM/POH should be used to perform the run-up. Many critical systems are checked and set during the before-takeoff check. Most airplanes have at least the following systems checked and set:

- Fuel System—set per the AFM/POH and verified ON and the proper and correct fuel tanks selected.

- Trim—set for takeoff position, which includes the elevator and may also include rudder and aileron trim.

- Flight Controls—checked throughout their entire operating range. This includes full aileron, elevator, and rudder deflection in all directions. Often, pilots do not exercise a full range of movement of the flight controls, which is not acceptable.

- Engine Operation—checked to ensure that temperatures and pressures are in their normal ranges; magneto or Full Authority Digital Engine Control (FADEC) operation on single or dual ignition are acceptable and within limits; and, if equipped, carburetor heat is functioning. If the airplane is equipped with a constant speed or feathering propeller, that its operation is acceptable, and the engine continues to run normally as the propeller is exercised.

- Electrical System—verified to ensure voltages are within operating range and that the system shows the battery system charging.

- Vacuum System—shows an acceptable level of vacuum, which is typically between 4.8 and 5.2 inches of mercury ("Hg) at 2,000 rpm. Refer to the AFM/POH for the manufacturer's values. It is important to ensure that mechanical gyroscopic instruments have adequate time to spool up to acceptable rpm in order for them to indicate properly. A hasty and quick taxi and run-up does not allow mechanical gyroscopic instruments to indicate properly and a departure into instrument meteorological conditions (IMC) is unadvisable.

- Flight Instruments–rechecked and set for the departure. Verify that the directional gyro and the magnetic compass are in agreement. If the directional gyro has a heading bug, it may be set to the runway heading that is in use or as assigned by air traffic control (ATC).

- Avionics–set with the appropriate frequencies, initial navigation sources and courses, autopilot preselects, transponder codes, and other settings and configurations based on the airplane's equipment and flight requirements.

- Takeoff Briefing–made out loud by the pilot even when no other person is there to listen. It should include a visual verification of the correct surface and direction to preclude a wrong surface departure. A sample takeoff briefing may be the following:

  "This will be normal takeoff (use normal, short, or soft as appropriate) from runway (use runway assigned), wind is from the (direction and speed), rotation speed is (use the specified or calculated manufacturer's takeoff or rotation speed (V_R)), an initial turn to (use planned heading) and climb to (use initial altitude in feet). The takeoff will be rejected for engine failure below V_R, applying appropriate braking, stopping ahead. Engine failure after V_R and with runway remaining, I will lower pitch, land, and apply appropriate braking, stopping straight ahead. Engine failure after V_R and with no runway remaining, I will lower pitch to best glide speed, no turns will be made prior to (insert appropriate altitude), land in the most suitable area, and apply appropriate braking, avoiding hazards on the ground as much as possible. I will only consider turning back to runway __ if I have reached at least __ feet AGL, which would be __ feet MSL. If time permits, fuel, ignition, and electrical systems will be switched off."

Takeoff Checks

The pilot should ensure that runway numbers on paved runways agree with magnetic compass and heading indicators before beginning takeoff roll. The last check as power is brought to full takeoff power includes:

1. Doors latched and windows closed as required?
2. Controls positioned to account for any crosswind?
3. Power correct?
4. Engine rpm normal?
5. Engine smooth?
6. Engine instruments normal and in green ranges?

After-Landing

During the after-landing roll, while maintaining airplane track over runway centerline with ailerons and heading down runway with rudder pedals, the airplane should be gradually slowed to normal taxi speed with normal brake pressure before turning off of the landing runway. Any significant degree of turn at faster speeds could result in subsequent damage to the landing gear, tires, brakes, or the airplane structure.

To give full attention to controlling the airplane during the landing roll, the after-landing checklist should be performed only after the airplane is brought to a complete stop beyond the runway holding position markings. There have been many cases where a pilot has mistakenly manipulated the wrong handle and retracted the landing gear, instead of the flaps, due to improper division of attention while the airplane was moving. However, this procedure may be modified if the manufacturer recommends that specific after-landing items be accomplished during landing rollout. For example, when performing a short-field landing, the manufacturer may recommend retracting the flaps on rollout to improve braking. In this situation, the pilot should make a positive identification of the flap control handle before retracting the flaps.

Clear of Runway and Stopped

Because of different configurations and equipment in various airplanes, the after-landing checklist within the AFM/POH should be used. Some of the items may include:

1. Power—set to the AFM/POH values such as throttle 1,000 rpm, propeller full forward, mixture leaned.
2. Fuel—may require switching tanks and fuel pumps switched off.
3. Flaps—set to the retracted position.
4. Cowl flaps—may be opened or closed depending on temperature conditions.
5. Trim—reset to neutral or takeoff position.
6. Lights—may be switched off if not needed, such as strobe lights.
7. Avionics—frequencies and transponder set for arrival airport taxi procedures.

Parking

Unless parking in a designated, supervised area, the pilot should select a location and heading that prevents propeller or jet blast of other airplanes from striking the airplane unnecessarily. Whenever possible, the airplane should be parked headed into the existing or forecast wind. Often airports have airplane tie downs located on ramp areas which may or may not be aligned with the wind or provide a significant choice in parking location. After stopping in the desired direction, the airplane should be allowed to roll straight ahead enough to straighten the nosewheel or tailwheel.

Engine Shutdown

The pilot should always use the procedures in the airplane's AFM/POH shutdown checklist for shutting down the engine and securing the airplane. Important items may include:

1. Parking Brake—set to ON.
2. Throttle—set to IDLE or 1,000 rpm.
3. If turbocharged, observe the manufacturer's spool down procedure.
4. Magneto Switch Test—momentarily check for proper grounding in the OFF position at idle rpm.
5. Propeller Control—set to HIGH rpm, if equipped.
6. Avionics—turn OFF.
7. Alternator—turn OFF.
8. Mixture—set to IDLE CUTOFF.
9. Magneto Switch—turn ignition switch to OFF when engine stops.
10. Install chocks (release parking brake in accordance with AFM/POH).
11. Master Switch—turn OFF.
12. Secure—install control locks and anti-theft security locks.

Post-Flight

A flight is not complete until the engine is shut down and the airplane is secured. A pilot should consider this an essential part of any flight.

Securing and Servicing

After engine shutdown and deplaning passengers, the pilot should accomplish a post-flight inspection. This includes a walk around to inspect the general condition of the aircraft. Inspect near and around the cowling for signs of oil or fuel streaks and around the oil breather for excessive oil discharge. Inspect under wings and other fuel tank locations for fuel stains. Inspect landing gear and tires for damage and brakes for any leaking hydraulic fluid. Inspect cowling inlets for obstructions.

Oil levels should be checked and quantities brought to AFM/POH levels. Fuel should be added based on the immediate use of the airplane. If the airplane is going to be inactive, it is a good operating practice to fill the fuel tanks to prevent water condensation from forming inside the tank. If another flight is planned, the fuel tanks should be filled based on the flight planning requirements for that flight.

The aircraft should be hangared or tied down, flight controls secured, and security locks in place. The type of tie downs may vary significantly from chains to well-worn ropes. Chains are not flexible and as such should not be made taut so as to allow the airplane some movement and prevent airframe structural damage. Tie down ropes are flexible and may be reasonably cinched to the airplane's tie down rings. Consider utilizing pitot tube covers, cowling inlet covers, rudder gust locks, window sunscreens, and propeller security locks to further enhance the safety and security of the airplane.

Hangaring is not without hazards to the airplane. The pilot should ensure that enough space is allocated to the airplane so it is free from any impact to the hangar, another aircraft, or vehicle. The airplane should be inspected after hangaring to ensure that no damage was imparted on the airplane.

Chapter Summary

This chapter places emphasis on determining the airworthiness of the airplane, preflight visual inspection, managing risk and pilot-available resources, safe surface-based operations, and the adherence to and proper use of the AFM/POH and checklists. The pilot should ensure that the airplane is in a safe condition for flight, and it meets all the regulatory requirements of 14 CFR part 91. A pilot also needs to recognize that flight safety includes proper flight preparation and having the experience to manage the risks associated with the expected conditions. An effective and continuous assessment and mitigation of the risks and appropriate utilization of resources goes a long way provided the pilot honestly evaluates their ability to act as PIC.

Airplane Flying Handbook (FAA-H-8083-3C)
Chapter 3: Basic Flight Maneuvers

Introduction

Airplanes operate in an environment that is unlike an automobile. Drivers tend to drive with a fairly narrow field of view and focus primarily on forward motion. Beginning pilots tend to practice the same. Flight instructors face the challenge of teaching beginning pilots about attitude awareness; which requires understanding the motions of flight. An airplane rotates in bank, pitch, and yaw while also moving horizontally, vertically, and laterally. The four fundamentals (straight-and-level flight, turns, climbs, and descents) are the principal maneuvers that control the airplane through the six motions of flight.

The Four Fundamentals

To master any subject, one should first master the fundamentals. For flying, this includes straight-and-level flight, turns, climbs, and descents. All flying tasks are based on these maneuvers, and an attempt to move on to advanced maneuvers prior to mastering the four fundamentals hinders the learning process.

Consider the following: a takeoff is a combination of a ground roll, which may transition to a brief period of straight-and-level flight, and a climb. After-departure includes the climb and turns toward the first navigation fix and is followed by straight-and-level flight. The preparation for landing at the destination may include combinations of descents, turns, and straight-and-level flight. In a typical general aviation (GA) airplane, the final approach ends with a transition from descent to straight-and-level while slowing for the touchdown and ground roll.

The flight instructor needs to impart competent knowledge of these basic flight maneuvers so that the beginning pilot is able to combine them at a performance level that at least meets the Federal Aviation Administration (FAA) Airman Certification Standards (ACS) or Practical Test Standards (PTS). As the beginning pilot progresses to more complex flight maneuvers, any deficiencies in the mastery of the four fundamentals are likely to become barriers to effective and efficient learning.

Effect and Use of Flight Controls

The airplane flies in an environment that allows it to travel up and down as well as left and right. Note that movement *up or down* depends on the flight conditions. If the airplane is right-side up relative to the horizon, forward control stick or wheel (elevator control) movement will result in a loss of altitude. If the same airplane is upside-down relative to the horizon that same forward control movement will result in a gain of altitude. *[Figure 3-1]* The following discussion considers the pilot's frame of reference with respect to the flight controls. *[Figure 3-2]*

Figure 3-1. *Basic flight controls and instrument panel.*

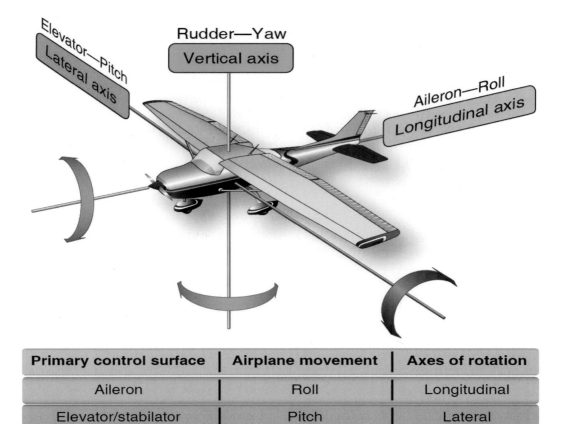

Rudder—Yaw
Vertical axis

Elevator—Pitch
Lateral axis

Aileron—Roll
Longitudinal axis

Primary control surface	Airplane movement	Axes of rotation
Aileron	Roll	Longitudinal
Elevator/stabilator	Pitch	Lateral
Rudder	Yaw	Vertical

Figure 3-2. *The pilot is always considered the referenced center of effect as the flight controls are used.*

With the pilot's hand:

- When pulling the elevator pitch control toward the pilot, which is an aft movement of the control wheel, yoke, control stick, or side stick controller (referred to as adding back pressure), the airplane's nose will rotate backwards relative to the pilot around the pitch (lateral) axis of the airplane. Think of this movement from the pilot's feet to the pilot's head.

- When pushing elevator pitch control toward the instrument panel, (referred to as increasing forward pressure), the airplane rotates the nose forward relative to the pilot around the pitch axis of the airplane. Think of this movement from the pilot's head to the pilot's feet.

- When right pressure is applied to the aileron control, which rotates the control wheel or yoke clockwise, or deflects the control stick or side stick to the right, the airplane's right wing banks (rolls) lower in relation to the pilot. Think of this movement from the pilot's head to the pilot's right hip.

- When left pressure is applied to the aileron control, which rotates the control wheel or yoke counterclockwise, or deflects the control stick or side stick to the left, the airplane's left wing banks (rolls) lower in relation to the pilot. Think of this movement from the pilot's head to the pilot's left hip.

With the pilot's feet:

- When forward pressure is applied to the right rudder pedal, the airplane's nose moves (yaws) to the right in relation to the pilot. Think of this movement from the pilot's left shoulder to the pilot's right shoulder.

- When forward pressure is applied to the left rudder pedal, the airplane's nose moves (yaws) to the left in relation to the pilot. Think of this movement from the pilot's right shoulder to the pilot's left shoulder.

While in flight, the control surfaces remain in a fixed position as long as all forces acting upon them remain balanced. Resistance to movement increases as airspeed increases and decreases as airspeed decreases. Resistance also increases as the controls move away from a streamlined position. While maneuvering the airplane, it is not the amount of control surface displacement the pilot needs to consider, but rather the application of flight control pressures that give the desired result.

The pilot should hold the pitch and roll flight controls (aileron and elevator controls, yoke, stick, or side-stick control) lightly with the fingers and not grab or squeeze them with the entire hand. When flight control pressure is applied to change a control surface position, the pilot should exert pressure on the aileron and elevator controls with the fingers only. This is an important concept and habit to learn. A common error with beginning pilots is that they grab the aileron and elevator controls with a closed palm with such force that sensitive feeling is lost. Pilots may wish to consider this error at the onset of training as it prevents the development of "feel," which is an important aspect of airplane control.

So that slight rudder pressure changes can be felt, both heels should support the weight of the pilot's feet on the floor with the ball of each foot touching the individual rudder pedals. The legs and feet should be relaxed. When using the rudder pedals, pressure should be applied smoothly and evenly by pressing with the ball of one foot. Since the rudder pedals are interconnected through springs or a direct mechanical linkage and act in opposite directions, when pressure is applied to one rudder pedal, foot pressure on the opposite rudder pedal should be relaxed proportionately.

In summary, during flight, the pressure the pilot exerts on the aileron and elevator controls and rudder pedals causes the airplane to move about the roll (longitudinal), pitch (lateral), and yaw (vertical) axes. When a control surface moves out of its streamlined position (even slightly), moving air exerts a force against that surface. It is this force that the pilot feels on the controls.

Feel of the Airplane

The ability to sense a flight condition, such as straight-and-level flight or a dive, without relying on instrumentation is often called "feeling the airplane." Examples of this "feel" may be sounds of the airflow across the airframe, vibrations felt through the controls, engine and propeller sounds and vibrations at various flight attitudes, and the sensations felt by the pilot through physical accelerations.

Humans sense "feel" through kinesthesis (the ability to sense movement through the body) and proprioception (unconscious perception of movement and spatial orientation). These stimuli are detected by nerves and by the semicircular canals of the inner ear. When properly developed, kinesthesis can provide the pilot with critical information about changes in the airplane's direction and speed; however, there are limits in kinesthetic sense when relied upon solely without visual information, as when flying in instrument meteorological conditions (IMC). Sole reliance on the kinesthetic sense ultimately leads to disorientation and loss of aircraft control.

Developing this "feel" takes time and exposure in a particular airplane. It only comes with dedicated practice at the various flight conditions so that a pilot's senses are trained by the sounds, vibrations, and forces produced by the airplane. The following are some important examples:

- Rushing air creates a distinctive noise pattern and as the level of sound increases, it likely indicates that the airplane's airspeed is increasing and that the pitch attitude is decreasing. As the noise decreases, the airplane's pitch attitude is likely increasing and its airspeed decreasing.

- The sound of the engine in cruise flight is different from that in a climb and different again when in a dive. In fixed-pitch propeller airplanes, when the airplane's pitch attitude increases, the engine sound decreases and as pitch attitude decreases, the engine sound increases.

- In a banked turn, the pilot is forced downward into the seat due to the resultant load factor. The increased G force of a turn feels the same as the pull up from a dive, and the decreased G force from leveling out feels the same as lowering the nose out of a climb.

Sources of actual "feel" are very important to the pilot. This actual feel is the result of acceleration, which is simply how fast velocity is changing. Acceleration describes the rate of change in both the magnitude and the direction of velocity. These accelerations impart forces on the airplane and its occupants during flight. The pilot can sense vertical forces through pressure changes into the seat or horizontal forces while being pushed from side to side in the seat if the airplane slips or skids. These forces need not be strong, only perceptible by the pilot, to be useful. An accomplished pilot who has excellent "feel" for the airplane is able to detect even the smallest accelerations.

The flight instructor should teach the difference between perceiving and reacting to sound, vibrations, and forces versus merely noticing them. It is this increased understanding that contributes to developing a "feel" for the airplane. A pilot who develops a "feel" for the airplane early in flight training is likely to have less difficulty during more advanced training.

Attitude Flying

An airplane's attitude is determined by the angular difference between a specific axis and the natural horizon. A false horizon can occur when the natural horizon is obscured or not readily apparent. This is an important concept because it requires the pilot to develop a pictorial sense of this natural horizon. Pitch attitude is the angle formed between the airplane's longitudinal axis, which extends from the nose to the tail of the airplane, and the natural horizon. Bank attitude is the angle formed by the airplane's lateral axis, which extends from wingtip to wingtip, and the natural horizon. *[Figures 3-3A and 3-3B]* Angular difference about the airplane's vertical axis (yaw) is an attitude relative to the airplane's direction of flight but not relative to the natural horizon.

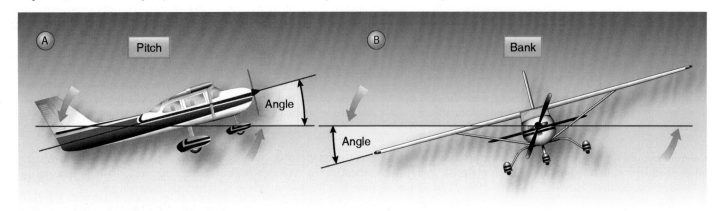

Figure 3-3. *(A) Pitch attitude is the angle formed between the airplane's longitudinal axis, which extends from the nose to tail of the airplane, and the natural horizon. (B) Bank attitude is the angle formed by the airplane's lateral axis, which extends from wingtip to wingtip, and the natural horizon.*

Controlling an airplane requires one of two methods to determine the airplane's attitude in reference to the horizon. When flying "visually" in visual meteorological conditions (VMC), a pilot uses their eyes and visually references the airplane's wings and cowling to establish the airplane's attitude to the natural horizon (a visible horizon). If no visible horizon can be seen due to clouds, whiteouts, haze over the ocean, night over a dark ocean, etc., it is IMC for practical and safety purposes. *[Figure 3-4]* When flying in IMC or when cross-checking the visual references, the airplane's attitude is controlled by the pilot referencing the airplane's mechanical or electronically-generated instruments to determine the airplane's attitude in relation to the natural horizon.

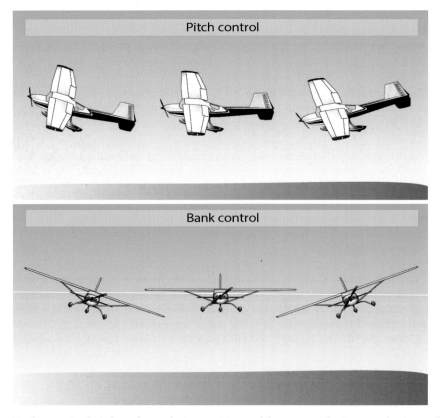

Figure 3-4. *Airplane attitude is based on relative positions of the nose and wings on the natural horizon.*

Airplane attitude control is composed of four components: pitch control, bank (roll) control, power control, and trim.

- Pitch control—controlling of the airplane's pitch attitude about the lateral axis by using the elevator to raise and lower the nose in relation to the natural horizon or to the airplane's flight instrumentation.

- Bank control—controlling of the airplane about the airplane's longitudinal axis by use of the ailerons to attain a desired bank angle in relation to the natural horizon or to the airplane's instrumentation.

- Power control—controlled by the throttle in most general aviation (GA) airplanes and is used when the flight situation requires a specific thrust setting or for a change in thrust to meet a specific objective.

- Trim control—used to relieve the control pressures held by the pilot on the flight controls after a desired attitude has been attained.

Note: Yaw control is used to cancel out the effects of yaw-induced changes, such as adverse yaw and effects of the propeller.

Integrated Flight Instruction

When introducing basic flight maneuvers to a beginning pilot, it is recommended that the "integrated" or "composite" method of flight instruction be used. This means the use of outside references and flight instruments to establish and maintain desired flight attitudes and airplane performance. When beginning pilots use this technique, they achieve a more precise and competent overall piloting ability. Although this method of airplane control may become second nature with experience, the beginning pilot needs to make a determined effort to master the technique. In all cases, a pilot's visual skills need to be sufficiently developed for long-term, safe, and effective aircraft control. *[Figure 3-5]*

90 percent of the time, the pilot's attention should be outside the flight deck. No more than 10 percent of the pilot's attention should be inside the flight deck.

Figure 3-5. *Integrated flight instruction teaches pilots to use both external and instrument attitude references.*

The basic elements of integrated flight instruction are as follows:

- The pilot visually controls the airplane's attitude in reference outside to the natural horizon. Approximately 90 percent of the pilot's attention should be devoted to outside visual references and scanning for airborne traffic. The process of visually evaluating pitch and bank attitude comes from a continuous stream of attitude information. When the pilot perceives that the attitude is other than desired, the pilot should make precise, smooth, and accurate flight control corrections to return the airplane to the desired attitude. Continuous visual checks of the outside references and immediate corrections made by the pilot minimize the chance for the airplane to deviate from the desired heading, altitude, and flightpath.

- The airplane's attitude is validated by referring to flight instruments and confirming performance. If the flight instruments display that the airplane's performance is in need of correction, the required correction needs to be determined and then precisely, smoothly, and accurately applied with reference to the natural horizon. The airplane's attitude and performance are then rechecked by referring to flight instruments. The pilot then maintains the corrected attitude by reference to the natural horizon.

- The pilot should monitor the airplane's performance by briefly checking the flight instruments. No more than about 10 percent of the pilot's attention should be inside the flight deck. The pilot should develop the skill to quickly analyze the appropriate flight instruments and then immediately return to the visual outside references to control the airplane's attitude.

The pilot should become familiar with the relationship between outside visual references to the natural horizon and the corresponding flight instrument indications. For example, a pitch attitude adjustment may require a movement of the pilot's reference point of several inches in relation to the natural horizon but correspond to a seemingly insignificant movement of the reference bar on the airplane's attitude indicator. Similarly, a deviation from a desired bank angle, which is obvious when referencing the airplane's wingtips or cowling relative to the natural horizon, may be imperceptible on the airplane's attitude indicator to the beginner pilot.

The most common error made by the beginner pilot is to make pitch or bank corrections while still looking inside. It is also common for beginner pilots to fixate on the flight instruments—a conscious effort is required by them to return to outside visual references. For the first several hours of instruction, flight instructors may choose to use flight instrument covers to develop a beginning pilot's skill or to correct a pilot's poor habit of fixating on instruments by forcing them to use outside visual references for aircraft control.

The beginning pilot, not being familiar with the intricacies of flight by references to instruments, including such things as instrument lag and gyroscopic precession, will invariably make excessive attitude corrections and end up "chasing the instruments." Airplane attitude by reference to the natural horizon, however, presents immediate and accurate indications many times larger than on any instrument. The beginning pilot should understand that anytime airplane attitude by reference to the natural horizon cannot be established or maintained, the situation has become a genuine emergency and that the use of integrated flight instruction does not prepare pilots for flight in IMC.

Straight-and-Level Flight

Straight-and-level flight is flight in which heading and altitude are maintained. The other fundamentals are derived as variations from straight-and-level flight, and the need to form proper and effective skills in flying straight and level should be understood. The ability to perform straight-and-level flight results from repetition and practice. A high level of skill results when the pilot perceives outside references, takes mental snap shots of the flight instruments, and makes effective, timely, and proportional corrections from unintentional slight turns, descents, and climbs.

Straight-and-level flight is a matter of consciously fixing the relationship of a reference point on the airplane in relation to the natural horizon. *[Figure 3-6]* The establishment of these reference points should be initiated on the ground as they depend on the pilot's seating position, height, and posture. The pilot should sit in a normal manner with the seat position adjusted, such that the pilot sees adequately over the instrument panel while being able to fully depress the rudder pedals without straining or reaching.

A flight instructor may use a dry erase marker or removable tape to make reference lines on the windshield or cowling to help the beginner pilot establish visual reference points. Vertical reference lines are best established on the ground, such as when the airplane is placed on a marked centerline, with the beginner pilot seated in proper position. Horizontal reference lines are best established with the airplane in flight, such as during slow flight and cruise configurations. The horizon reference point is always the same, no matter what altitude, since the point is always on the horizon, although the distance to the horizon will be further as altitude increases. There are multiple horizontal reference lines due to varying pitch attitude requirements; however, these teaching aids are generally needed for only a short period until the beginning pilot understands where and when to look while maneuvering the airplane.

Straight Flight

Maintaining a constant direction or heading is accomplished by visually checking the relationship of the airplane's wingtips to the natural horizon. Depending on whether the airplane is a high wing or low wing, both wingtips should be level and equally above or below the natural horizon. Any necessary bank corrections are made with the pilot's coordinated use of ailerons and rudder. *[Figure 3-7]* The pilot should understand that anytime the wings are banked, the airplane turns. The objective of straight flight is to detect and correct small deviations, necessitating minor flight control corrections. The bank attitude information can also be obtained from a quick scan of the attitude indicator (which shows the position of the airplane's wings relative to the horizon) and the heading indicator (which indicates if the airplane is off the desired heading).

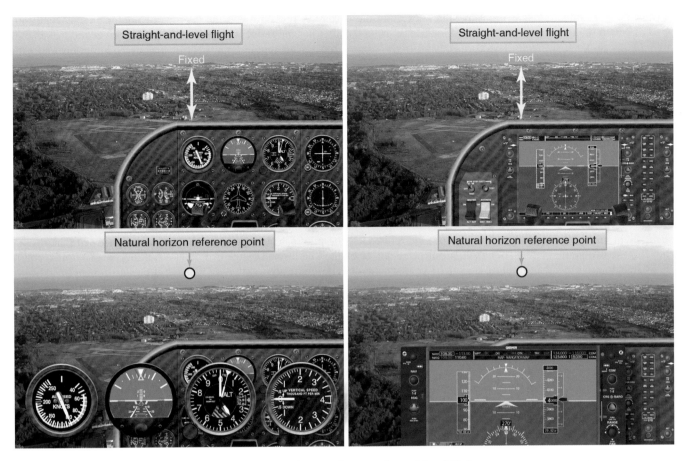

Figure 3-6. *Nose reference for straight-and-level flight.*

Figure 3-7. *Wingtip reference for straight-and-level flight.*

It is possible to maintain straight flight by simply exerting the necessary pressure with the ailerons or rudder independently in the desired direction of correction. However, the practice of using the ailerons and rudder independently is not correct and makes precise control of the airplane difficult. The correct bank flight control movement requires the coordinated use of ailerons and rudder. Straight-and-level flight requires almost no application of flight control pressures if the airplane is properly trimmed and the air is smooth. For that reason, the pilot should not form the habit of unnecessarily moving the flight controls. The pilot needs to learn to recognize when corrections are necessary and then to make a measured flight control response precisely, smoothly, and accurately.

Pilots may tend to look out to one side continually, generally to the left due to the pilot's left seat position and consequently focus attention in that direction. This not only gives a restricted angle from which the pilot is to observe but also causes the pilot to exert unconscious pressure on the flight controls in that direction. It is also important that the pilot not fixate in any one direction and continually scan outside the airplane, not only to ensure that the airplane's attitude is correct, but also to ensure that the pilot is considering other factors for safe flight. Continually observing both wingtips has advantages other than being the only positive check for leveling the wings. This includes looking for aircraft traffic, terrain and weather influences, and maintaining overall situational awareness.

Straight flight allows flying along a line. For outside references, the pilot selects a point on the horizon aligned with another point ahead. If those two points stay in alignment, the airplane will track the line formed by the two points. A pilot can also hold a course in VFR by tracking to a point in front of a compass or magnetic direction indicator, with only glances at the instrument or indicator to ensure being on course. The reliance on a surface point does not work when flying over water or flat snow covered surfaces. In these conditions, the pilot should rely on the magnetic heading indication.

Level Flight

In learning to control the airplane in level flight, it is important that the pilot be taught to maintain a light touch on the flight controls using fingers rather than the common problem of a tight-fisted palm wrapped around the flight controls. The pilot should exert only enough pressure on the flight controls to produce the desired result. The pilot should learn to associate the apparent movement of the references with the control pressures which produce attitude movement. As a result, the pilot can develop the ability to adjust the change desired in the airplane's attitude by the amount and direction of pressures applied to the flight controls without the pilot excessively referring to instrument or outside references for each minor correction.

The pitch attitude for level flight is first obtained by the pilot being properly seated, selecting a point toward the airplane's nose as a reference, and then keeping that reference point in a fixed position relative to the natural horizon. *[Figure 3-8]* The principles of attitude flying require that the reference point to the natural horizon position should be cross-checked against the flight instruments to determine if the pitch attitude is correct. If trending away from the desired altitude, the pitch attitude should be readjusted in relation to the natural horizon and then the flight instruments crosschecked to determine if altitude is now being corrected or maintained. In level flight maneuvers, the terms "increase the back pressure" or "increase pitch attitude" implies raising the airplane's nose in relation to the natural horizon and the terms "decreasing the pitch attitude" or "decrease pitch attitude" means lowering the nose in relation to the natural horizon. The pilot's primary reference is the natural horizon.

For all practical purposes, the airplane's airspeed remains constant in straight-and-level flight if the power setting is also constant. Intentional airspeed changes, by increasing or decreasing the engine power, provide proficiency in maintaining straight-and-level flight as the airplane's airspeed is changing. Pitching moments may also be generated by extension and retraction of flaps, landing gear, and other drag producing devices, such as spoilers. Exposure to the effect of the various configurations should be covered in any specific airplane checkout.

Common Errors

A common error of a beginner pilot is attempting to hold the wings level by only observing the airplane's nose. Using this method, the nose's short horizontal reference line can cause slight deviations to go unnoticed. However, deviations from level flight are easily recognizable when the pilot references the wingtips and, as a result, the wingtips should be the pilot's primary reference for maintaining level bank attitude. This technique also helps eliminate the potential for flying the airplane with one wing low and correcting heading errors with the pilot holding opposite rudder pressure. A pilot with a bad habit of dragging one wing low and compensating with opposite rudder pressure will have difficulty mastering other flight maneuvers.

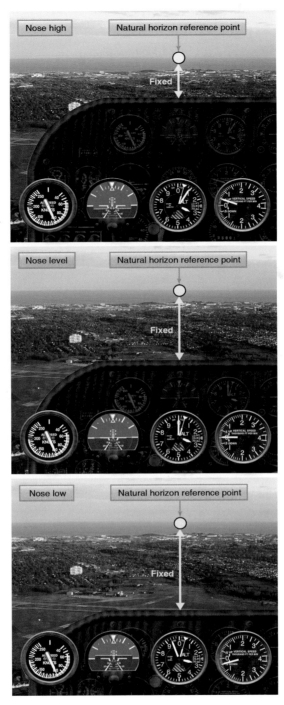

Figure 3-8. *Nose reference for level flight.*

Common errors include:

1. Attempting to use improper pitch and bank refeequent flights.
2. Forgetting the location of preselected reference points on subsequent flights.
3. Attempting to establish or correct airplane attitude using flight instruments rather than the natural horizon.
4. "Chasing" the flight instruments rather than adhering to the principles of attitude flying.
5. Mechanically pushing or pulling on the flight controls rather than exerting accurate and smooth pressure.
6. Not scanning outside the aircraft for other traffic and weather and terrain influences.
7. A tight palm grip on the flight controls resulting in a desensitized feeling of the hand and fingers.
8. Overcontrolling the airplane.
9. Habitually flying with one wing low or maintaining directional control using only the rudder control.
10. Failure to make timely and measured control inputs after a deviation from straight-and-level.
11. Inadequate attention to sensory inputs in developing feel for trence points on the airplane to establish attitude.

Trim Control

Trim control surfaces are required to offset any constant flight control pressure inputs provided by the pilot. For example, elevator trim is a typical trim in light GA airplanes and is used to null the pressure exerted by the pilot in order to maintain a particular pitch attitude. *[Figure 3-9]* This provides an opportunity for the pilot to divert attention to other tasks.

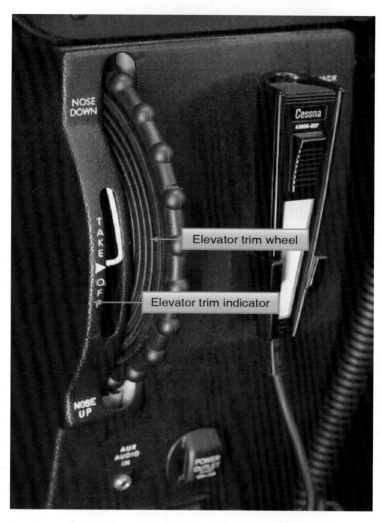

Figure 3-9. *Elevator trim is used in airplanes to null the pressure exerted by the pilot on the pitch flight control.*

Because of their relatively low power, speed, and cost constraints, not all light airplanes have a complete set (elevator, rudder, and aileron) of trim controls that are adjustable from inside the flight deck. Nearly all light airplanes are equipped with at least adjustable elevator trim. As airplanes increase in power, weight, and complexity, flight deck adjustable trim systems for the rudder and aileron may be available.

In airplanes where multiple trim axes are available, the rudder should be trimmed first. Rudder, elevator, and then aileron should be trimmed next in sequence. However, if the airspeed is varying, continuous attempts to trim the rudder and aileron produce unnecessary pilot workload and distraction. Attempts to trim the rudder at varying airspeeds are impractical in many propeller airplanes because of the built-in compensation for the effect of a propeller's left turning tendencies. The correct procedure is when the pilot has established a constant airspeed and pitch attitude, the pilot should then hold the wings level with aileron flight control pressure while rudder control pressure is trimmed out. Finally, aileron trim should be adjusted to relieve any aileron flight control pressure.

A properly trimmed airplane is an indication of good piloting skills. Any control forces that the pilot feels should be a result of deliberate flight control pressure inputs during a planned change in airplane attitude, not a result of forces being applied by the airplane. A common trim control error is the tendency for the pilot to overcontrol the airplane with trim adjustments. Attempting to fly the airplane with the trim is a common fault in basic flying technique even among experienced pilots. The airplane attitude should be established first and held with the appropriate flight control pressures, and then the flight control pressures trimmed out so that the airplane maintains the desired attitude without the pilot exerting flight control pressure.

Level Turns

A turn is initiated by banking the wings in the desired direction of the turn through the pilot's use of the aileron flight controls. Left aileron flight control pressure causes the left wing to lower in relation to the pilot. Right aileron flight control pressure causes the right wing to lower in relation to the pilot. In other words, to turn left, the pilot lowers the left wing with aileron by left stick. To turn right, the pilot lowers the right wing with right stick. Depending on bank angle and airplane engineering, at many bank angles, the airplane will continue to turn with ailerons neutralized. The sequence could be as follows:

1. Bank the airplane, adding either enough power or pitching up to compensate for the loss of vertical lift.
2. Neutralize controls as necessary to stop bank from increasing and hold desired bank angle.
3. Use the opposite stick (aileron) to return airplane to level.
4. Neutralize the ailerons (along with either power or pitch reduction) for level flight. *[Figure 3-10]*

Figure 3-10. *Level turn to the left.*

A turn is the result of the following:

- The ailerons bank the wings and determine the rate of turn for a given airspeed. Lift is divided into both vertical and horizontal lift components as a result of the bank. The horizontal component of lift moves the airplane toward the banked direction.

- The elevator pitches the nose of the airplane up or down in relation to the pilot and perpendicular to the wings. If the pilot does not add power, and there is sufficient airspeed margin, the pilot needs to slightly increase the pitch to increase wing lift enough to replace the wing lift being diverted into turning force so as to maintain the current altitude.

- The vertical fin on an airplane does not produce lift. Rather the vertical fin on an airplane is a stabilizing surface and produces no lift if the airplane is flying straight ahead. The vertical fin's purpose is to keep the aft end of the airplane behind the front end.

- The throttle provides thrust, which may be used for airspeed control and to vary the radius of the turn.

- The pilot uses the rudder to offset any adverse yaw developed by wing's differential lift and the engine/propeller. The rudder does not turn the airplane. The rudder is used to maintain coordinated flight.

For purposes of this discussion, turns are divided into three classes: shallow, medium, and steep.

- Shallow turns—bank angle is approximately 20° or less. This shallow bank is such that the inherent lateral stability of the airplane slowly levels the wings unless aileron pressure in the desired direction of bank is held by the pilot to maintain the bank angle.

- Medium turns—result from a degree of bank between approximately 20° and 45°. At medium bank angles, the airplane's inherent lateral stability does not return the wings to level flight. As a result, the airplane tends to remain at a constant bank angle without any flight control pressure held by the pilot. The pilot neutralizes the aileron flight control pressure to maintain the bank.

- Steep turns—result from a degree of bank of approximately 45° or more. The airplane continues in the direction of the bank even with neutral flight controls unless the pilot provides opposite flight control aileron pressure to prevent the airplane from overbanking. The actual amount of opposite flight control pressure used depends on various factors, such as bank angle and airspeed.

When an airplane is flying straight and level, the total lift is acting perpendicular to the wings and to the earth. As the airplane is banked into a turn, total lift is the resultant of two components: vertical and horizontal. *[Figure 3-11]* The vertical lift component continues to act perpendicular to the earth and opposes gravity. The horizontal lift component acts parallel to the earth's surface opposing centrifugal force. These two lift components act at right angles to each other, causing the resultant total lifting force to act perpendicular to the banked wing of the airplane. It is the horizontal lift component that begins to turn the airplane and not the rudder.

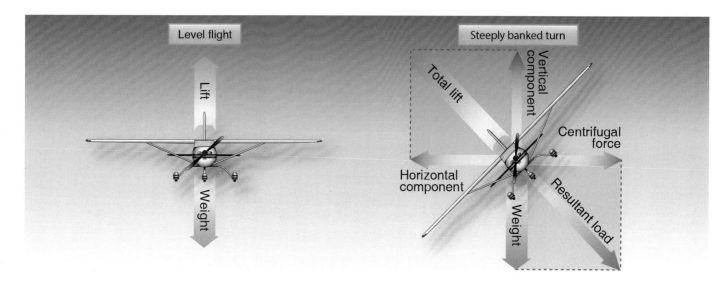

Figure 3-11. *When the airplane is banked into a turn, total lift is the resultant of two components: vertical and horizontal.*

In constant altitude, constant airspeed turns, it is necessary to increase the AOA of the wing when rolling into the turn by increasing back pressure on the elevator, as well to add power countering the loss of speed due to increased drag. This is required because total lift has divided into vertical and horizontal components of lift. In order to maintain altitude, the total lift (since total lift acts perpendicular to the wing) needs to be increased to meet the vertical component of lift requirements (to balance weight and load factor) for level flight.

The purpose of the rudder in a turn is to coordinate the turn. As lift increases, so does drag. When the pilot deflects the ailerons to bank the airplane, both lift and drag are increased on the rising wing and, simultaneously, lift and drag are decreased on the lowering wing. *[Figure 3-12]* This increased drag on the rising wing and decreased drag on the lowering wing results in the airplane yawing opposite to the direction of turn. To counteract this adverse yaw, rudder pressure is applied simultaneously with the aileron deflection in the desired direction of turn. This action is required to produce a coordinated turn. Coordinated flight is an important part of airplane control. Situations can develop when a pilot maintains certain uncoordinated flight control deflections, which create the potential for a spin. This is especially hazardous when operating at low altitudes, such as when operating in the airport traffic pattern.

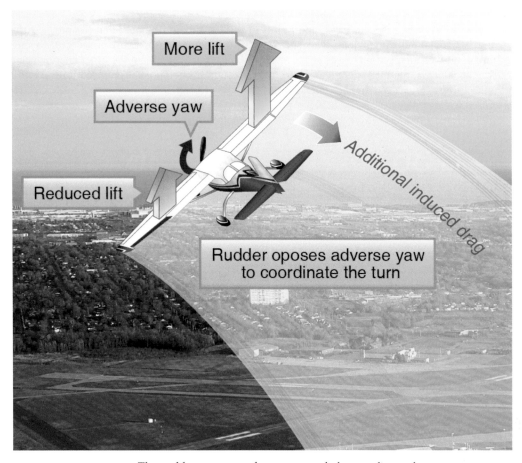

Figure 3-12. *The rudder opposes adverse yaw to help coordinate the turn.*

During uncoordinated flight, the pilot may feel that they are being pushed sideways toward the outside or inside of the turn. *[Figure 3-13]* The pilot feels pressed toward the outside of a turn during a skid and feels pressed toward the inside of a turn during a slip. The ability to sense a skid or slip is developed over time and as the "feel" of flying develops, a pilot should become highly sensitive to a slip or skid without undue reliance on the flight instruments.

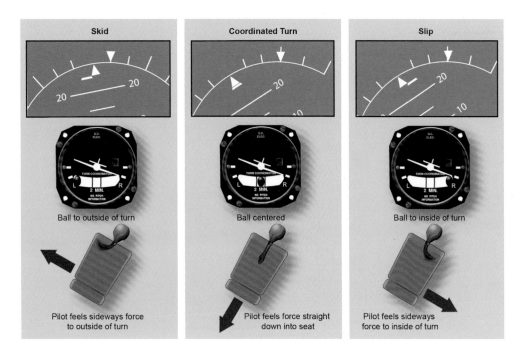

Figure 3-13. *Indications of a slip and skid.*

Turn Radius

To understand the relationship between airspeed, bank, and radius of turn, it should be noted that the rate of turn at any given true airspeed depends on the horizontal lift component. The horizontal lift component varies in proportion to the amount of bank. Therefore, the rate of turn at a given airspeed increases as the angle of bank is increased. On the other hand, when a turn is made at a higher airspeed at a given bank angle, the inertia is greater and the horizontal lift component required for the turn is greater, causing the turning rate to become slower. *[Figure 3-14]* Therefore, at a given angle of bank, a higher airspeed makes the radius of turn larger because the airplane turns at a slower rate.

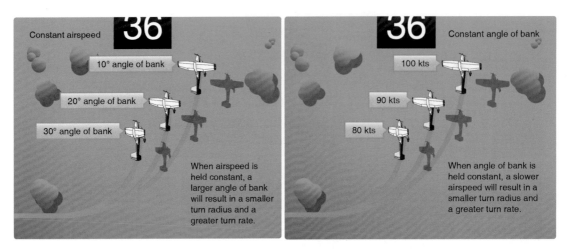

Figure 3-14. *Angle of bank and airspeed regulate rate and radius of turn.*

As the radius of the turn becomes smaller, a significant difference develops between the airspeed of the inside wing and the airspeed of the outside wing. The wing on the outside of the turn travels a longer path than the inside wing, yet both complete their respective paths in the same unit of time.

Therefore, the outside wing travels at a faster airspeed than the inside wing and, as a result, it develops more lift. This creates an overbanking tendency that needs to be controlled by the use of opposite aileron when the desired bank angle is reached. *[Figure 3-15]* Because the outboard wing is developing more lift, it also produces more drag. The drag causes a slight slip during steep turns that should be corrected by use of the rudder.

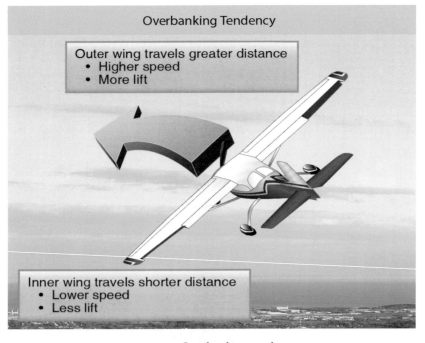

Figure 3-15. *Overbanking tendency.*

Establishing a Turn

On most light single-engine airplanes, the top surface of the engine cowling is fairly flat, and its horizontal surface to the natural horizon provides a reasonable indication for initially setting the degree of bank angle. *[Figure 3-16]* The pilot should then cross-check the flight instruments to verify that the correct bank angle has been achieved. Information obtained from the attitude indicator shows the angle of the wing in relation to the horizon.

Figure 3-16. *Visual reference for angle of bank.*

The pilot's seating position in the airplane is important as it affects the interpretation of outside visual references. A common problem is that a pilot may lean away from the turn in an attempt to remain in an upright position in relation to the horizon. This should be corrected immediately if the pilot is to properly learn to use visual references. *[Figure 3-17]*

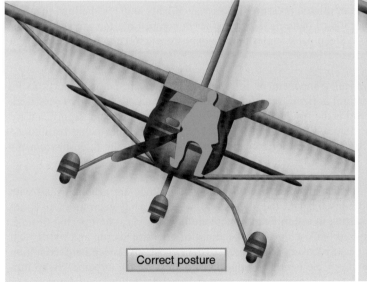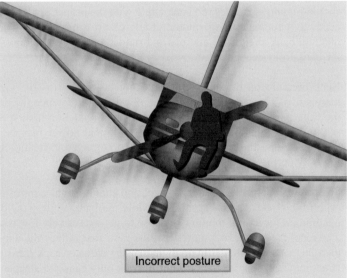

Figure 3-17. *Correct and incorrect posture while seated in the airplane.*

Because most airplanes have side-by-side seating, a pilot does not sit on the airplane's longitudinal axis, which is where the airplane rotates in roll. The pilot sits slightly off to one side, typically the left, of the longitudinal axis. Due to parallax error, this makes the nose of the airplane appear to rise when making a left turn (due to pilot lowering in relation to the longitudinal axis) and the nose of the airplane appear to descend when making right turns (due to pilot elevating in relation to the longitudinal axis). *[Figure 3-18]*

Beginning pilots should not use large aileron and rudder control inputs. This is because large control inputs produce rapid roll rates and allow little time for the pilot to evaluate and make corrections. Smaller flight control inputs result in slower roll rates and provide for more time to accurately complete the necessary pitch and bank corrections.

Figure 3-18. *Parallax view.*

Some additional considerations for initiating turns are the following:

- If the airplane's nose starts to move before the bank starts, the rudder is being applied too soon.

- If the bank starts before the nose starts turning or the nose moves in the opposite direction, the rudder is being applied too late.

- If the nose moves up or down when entering a bank, excessive or insufficient elevator back pressure is being applied.

After the bank has been established, all flight control pressures applied to the ailerons and rudder may be relaxed or adjusted, depending on the established bank angle, to compensate for the airplane's inherent stability or overbanking tendencies. The airplane should remain at the desired bank angle with the proper application of aileron pressure. If the desired bank angle is shallow, the pilot needs to maintain a small amount of aileron pressure into the direction of bank including rudder to compensate for yaw effects. For medium bank angles, the ailerons and rudder should be neutralized. Steep bank angles require opposite aileron and rudder to prevent the bank from steepening.

Back pressure on the elevator should not be relaxed as the vertical component of lift should be maintained if altitude is to be maintained. Throughout the turn, the pilot should reference the natural horizon, scan for aircraft traffic, and occasionally crosscheck the flight instruments to verify performance. A reduction in airspeed is the result of increased drag but is generally not significant for shallow bank angles. In steeper turns, additional power may be required to maintain airspeed. If altitude is not being maintained during the turn, the pitch attitude should be corrected in relation to the natural horizon and cross-checked with the flight instruments to verify performance.

Steep turns require accurate, smooth, and timely flight control inputs. Minor corrections for pitch attitude are accomplished with proportional elevator back pressure while the bank angle is held constant with the ailerons. However, during steep turns, it is not uncommon for a pilot to allow the nose to get excessively low resulting in a significant loss in altitude in a very short period of time. The recovery sequence requires that the pilot first reduce the angle of bank with coordinated use of opposite aileron and rudder and then increase the pitch attitude by increasing elevator back pressure. If recovery from an excessively nose-low, steep bank condition is attempted by use of the elevator only, it only causes a steepening of the bank and unnecessary stress on the airplane. Steep turn performance can be improved by an appropriate application of power to overcome the increase in drag. Depending on the purpose of a steep turn and the magnitude of control force needed, trimming additional elevator back pressure as the bank angle goes beyond 30° may assist the pilot during the turn.

Since the airplane continues turning as long as there is any bank, the rollout from the turn should be started before reaching the desired heading. The amount of lead required to rollout on the desired heading depends on the degree of bank used in the turn. A rule of thumb is to lead by one-half the angle of bank. For example, if the bank is 30°, lead the rollout by 15°. The rollout from a turn is similar to the roll-in except the flight controls are applied in the opposite direction. Aileron and rudder are applied in the direction of the rollout or toward the high wing. As the angle of bank decreases, the elevator pressure should be relaxed as necessary to maintain altitude. As the wings become level, the flight control pressures should be smoothly relaxed so that the controls are neutralized as the airplane returns to straight-and-level flight. If trim was used, such as during a steep turn, forward elevator pressure may be required until the trim can be adjusted. As the rollout is being completed, attention should be given to outside visual references, as well as the flight instruments to determine that the wings are being leveled and the turn stopped.

Because the elevator and ailerons are on one control, practice is required to ensure that only the intended pressure is applied to the intended flight control. For example, a beginner pilot is likely to unintentionally add pressure to the pitch control when the only bank was intended. This cross-coupling may be diminished or enhanced by the design of the flight controls; however, practice is the appropriate measure for smooth, precise, and accurate flight control inputs. For example, diving when turning right and climbing when turning left in airplanes is common with stick controls, because the arm tends to rotate from the elbow joint, which induces a secondary arc control motion if the pilot is not extremely careful. Likewise, lowering the nose is likely to induce a right turn, and raising the nose to climb tends to induce a left turn. These actions would apply for a pilot using the right hand to move the stick. Airplanes with a control wheel may be less prone to these inadvertent actions, depending on control positions and pilot seating. In any case, the pilot should retain the proper sight picture of the nose following the horizon, whether up, down, left, or right and isolate undesired motion.

Common errors in level turns are:

1. Failure to adequately clear in the direction of turn for aircraft traffic.
2. Gaining or losing altitude during the turn.
3. Not holding the desired bank angle constant.
4. Attempting to execute the turn solely by instrument reference.
5. Leaning away from the direction of the turn while seated.
6. Insufficient feel for the airplane as evidenced by the inability to detect slips or skids without flight instruments.
7. Attempting to maintain a constant bank angle by referencing only the airplane's nose.
8. Making skidding flat turns to avoid banking the airplane.
9. Holding excessive rudder in the direction of turn.
10. Gaining proficiency in turns in only one direction.
11. Failure to coordinate the controls.

Climbs and Climbing Turns

When an airplane enters a climb, excess lift needs to be developed to overcome the weight or gravity. This requirement to develop more lift results in more induced drag, which either results in decreased airspeed or an increased power setting to maintain a minimum airspeed in the climb. An airplane can only sustain a climb when there is sufficient thrust to offset increased drag; therefore, climb rate is limited by the excess thrust available.

The pilot should know the engine power settings, natural horizon pitch attitudes, and flight instrument indications that produce the following types of climb:

- Normal climb—performed at an airspeed recommended by the airplane manufacturer. Normal climb speed is generally higher than the airplane's best rate of climb. The additional airspeed provides for better engine cooling, greater control authority, and better visibility over the nose of the airplane. Normal climb is sometimes referred to as cruise climb.

- Best rate of climb (V_Y)—produces the most altitude gained over a given amount of time. This airspeed is typically used when initially departing a runway without obstructions until it is safe to transition to a normal or cruise climb configuration.

- Best angle of climb (V_X)—performed at an airspeed that produces the most altitude gain over a given horizontal distance. The best angle of climb results in a steeper climb, although the airplane takes more time to reach the same altitude than it would at best rate of climb airspeed. The best angle of climb is used to clear obstacles, such as a strand of trees, after takeoff. [Figure 3-19]

It should be noted that as altitude increases, the airspeed for best angle of climb increases and the airspeed for best rate of climb decreases. Performance charts contained in the Airplane Flight Manual or Pilot's Operating Handbook (AFM/POH) should be consulted to ensure that the correct airspeed is used for the desired climb profile at the given environmental conditions. There is a point at which the best angle of climb airspeed and the best rate of climb airspeed intersect. This occurs at the absolute ceiling at which the airplane is incapable of climbing any higher. [Figure 3-20]

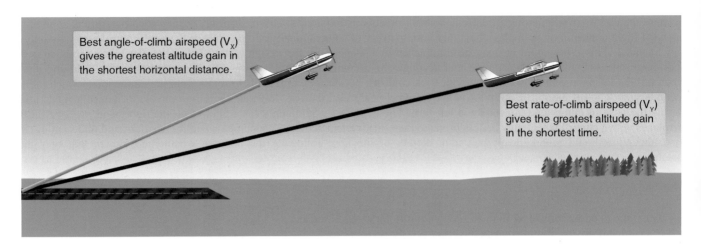

Best angle-of-climb airspeed (Vₓ) gives the greatest altitude gain in the shortest horizontal distance.

Best rate-of-climb airspeed (Vᵧ) gives the greatest altitude gain in the shortest time.

Figure 3-19. *Best angle of climb verses best rate of climb.*

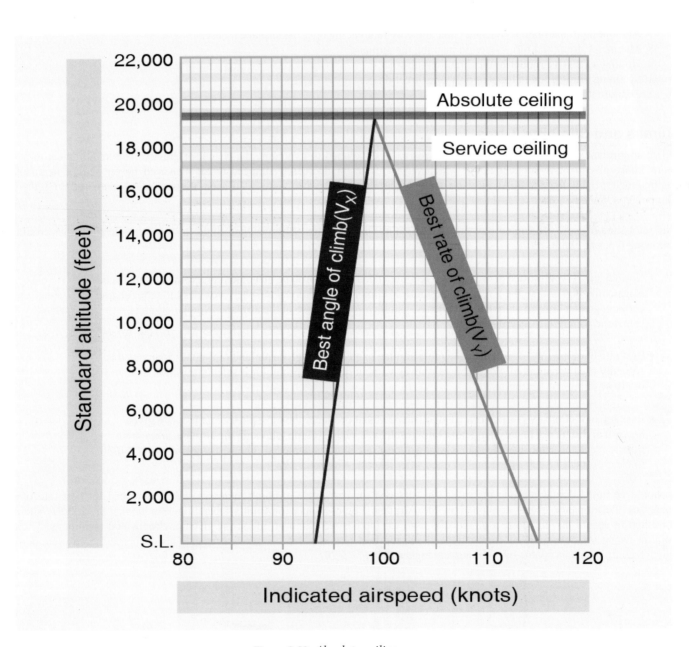

Figure 3-20. *Absolute ceiling.*

Establishing a Climb

A straight climb is entered by gently increasing back pressure on the elevator flight control to the pitch attitude referencing the airplane's nose to the natural horizon while simultaneously increasing engine power to the climb power setting. The wingtips should be referenced in maintaining the climb attitude while cross-checking the flight instruments to verify performance. In many airplanes, as power is increased, an increase in slipstream over the horizontal stabilizer causes the airplane's pitch attitude to increase more than desired. The pilot should be prepared for slipstream effects but also for the effect of changing airspeed and changes in lift. The pilot should be prepared to use the required flight control pressures to achieve the desired pitch attitude.

If a climb is started from cruise flight, the airspeed gradually decreases as the airplane enters a stabilized climb attitude. The thrust required to maintain straight-and-level flight at a given airspeed is not sufficient to maintain the same airspeed in a climb. Increase drag in a climb stems from increased lift demands made upon the wing to increase altitude. Climbing requires an excess of lift over that necessary to maintain level flight. Increased lift will generate more induced drag. That increase in induced drag is why more power is needed and why a sustained climb requires an excess of thrust.

For practical purposes gravity or weight is a constant. A vector diagram shows why more lift is necessary during a climb, as the vertical component of lift generated from the wings is no longer perpendicular to the wings and adds to drag. The total vertical force is increased by adding a vertical component of thrust from the powerplant, and the power should be advanced to the recommended climb power. On airplanes equipped with an independently controllable-pitch propeller, this requires advancing the propeller control prior to increasing engine power. Some airplanes may be equipped with cowl flaps to facilitate effective engine cooling. The position of the cowl flaps should be set to ensure cylinder head temperatures remain within the manufacturer's specifications.

Engines that are normally aspirated experience a reduction of power as altitude is gained. As altitude increases, air density decreases, which results in a reduction of power. The indications show a reduction in revolutions per minute (rpm) for airplanes with fixed pitch propellers; airplanes that are equipped with controllable propellers show a decrease in manifold pressure. The pilot should reference the engine instruments to ensure that climb power is being maintained and that pressures and temperatures are within the manufacturer's limits. As power decreases in the climb, the pilot continually advances the throttle or power lever to maintain specified climb settings.

The pilot should understand propeller effects during a climb and when using high power settings. The propeller in most airplanes rotates clockwise when seen from the pilot's position. As pitch attitude is increased, the center of thrust from the propeller moves to the right and becomes asymmetrical. This asymmetric condition is often called "P-factor." This is the result of the increased AOA of the descending propeller blade, which is the right side of the propeller disc when seen from the flight deck. As the center of propeller thrust moves to the right, a left turning yawing moment moves the nose of the airplane to the left. This is compensated by the pilot through right rudder pressure. In addition, torque that acts opposite to the direction of propeller rotation causes the airplane to roll to the left. Under these conditions, torque and P-factor cause the airplane to roll and yaw to the left. To counteract this, right rudder and aileron flight control pressures should be used. During the initial practice of climbs, this may initially seem awkward; however, after some experience the correction for propeller effects becomes instinctive.

As the airspeed decreases during the climb's establishment, the airplane's pitch attitude tends to lower unless the pilot increases the elevator flight control pressure. Nose-up elevator trim should be used so that the pitch attitude can be maintained without the pilot holding back elevator pressure. Throughout the climb, since the power should be fixed at the climb power setting, airspeed is controlled by the use of elevator pressure. The pitch attitude to the natural horizon determines if the pitch attitude is correct and should be cross-checked to the flight instruments to verify climb performance. *[Figure 3-21]*

Figure 3-21. *Climb indications.*

To return to straight-and-level flight from a climb, it is necessary to begin leveling-off prior to reaching the desired altitude. Level-off should begin at approximately 10 percent of the rate of climb. For example, if the airplane is climbing at 500 feet per minute (fpm), leveling off should begin 50 feet prior to reaching the desired altitude. The pitch attitude should be decreased smoothly and slowly to allow for the airspeed to increase. A loss of altitude may result if the pitch attitude is changed too rapidly without allowing the airspeed to increase proportionately.

After the airplane is established in level flight at a constant altitude, climb power should be retained temporarily so that the airplane accelerates to the cruise airspeed. When the airspeed reaches the desired cruise airspeed, the throttle setting and the propeller control, if equipped, should be set to the cruise power setting and the airplane re-trimmed.

Climbing Turns

In the performance of climbing turns, the following factors should be considered:

- With a constant power setting, the same pitch attitude and airspeed cannot be maintained in a bank as in a straight climb due to the increase in the total lift required. The airplane climbs at a slightly shallower climb angle because some of the lift is being used to turn the airplane.

- Steep bank angles significantly decreases the rate of climb. The pilot should establish and maintain an appropriate constant bank during the turn.

- The pilot should maintain a constant airspeed and constant rate of turn in both right and left turns. The coordination of all flight controls is a primary factor.

All the factors that affect the airplane during level constant-altitude turns affect the airplane during climbing turns. Compensation for the inherent stability of the airplane, overbanking tendencies, adverse yaw, propeller effects, reduction of the vertical component of lift, and increased drag needs to be managed by the pilot through the manipulation of the flight controls.

Climbing turns may be established by entering the climb first and then banking into the turn or climbing and turning simultaneously. During climbing turns, as in any turn, the loss of vertical lift should be compensated by an increase in pitch attitude. When a turn is coupled with a climb, the additional drag and reduction in the vertical component of lift need to be further compensated for by an additional increase in elevator back pressure. When turns are simultaneous with a climb, it is most effective to limit the turns to shallow bank angles. This provides for an efficient rate of climb. If a medium or steep banked turn is used, climb performance is degraded or possibly non-existent.

Common errors in the performance of climbs and climbing turns are:

1. Attempting to establish climb pitch attitude by primarily referencing the airspeed indicator and chasing the airspeed.
2. Applying elevator pressure too aggressively resulting in an excessive climb angle.
3. Inadequate or inappropriate rudder pressure during climbing turns.
4. Allowing the airplane to yaw during climbs usually due to inadequate right rudder pressure.
5. Fixation on the airplane's nose during straight climbs, resulting in climbing with one wing low.
6. Initiating a climbing turn without coordinated use of flight controls, resulting in no turn and a climb with one wing low.
7. Improper coordination resulting in a slip that counteracts the rate of climb, resulting in little or no altitude gain.
8. Inability to keep pitch and bank attitude constant during climbing turns.
9. Attempting to exceed the airplane's climb capability.
10. Using excessive forward elevator pressure during level-off resulting in a loss of altitude or excessive low G-force.

Descents and Descending Turns

When an airplane enters a descent, its attitude changes from level flight to flight with a descent profile. [Figure 3-22] In a descent, weight no longer acts solely perpendicular to the flightpath. Since induced drag is decreased as lift is reduced in order to descend, excess thrust will provide higher airspeeds. The weight/gravity force is about the same. This causes an increase in total thrust and a power reduction is required to balance the forces if airspeed is to be maintained.

Figure 3-22. *Descent indications.*

The pilot should know the engine power settings, natural horizon pitch attitudes, and flight instrument indications that produce the following types of descents:

- Partial power descent—the normal method of losing altitude is to descend with partial power. This is often termed cruise or en route descent. The airspeed and power setting recommended by the AFM/POH for prolonged descent should be used. The target descent rate should be 500 fpm. The desired airspeed, pitch attitude, and power combination should be preselected and kept constant.

- Descent at minimum safe airspeed—a nose-high, power-assisted descent condition principally used for clearing obstacles during a landing approach to a short runway. The airspeed used for this descent condition is recommended by the AFM/POH and is normally no greater than 1.3 VSO. Some characteristics of the minimum safe airspeed descent are a steeper-than-normal descent angle, and the excessive power that may be required to produce acceleration at low airspeed should "mushing" and/or an excessive rate of descent be allowed to develop.

- Emergency descent—some airplanes have a specific procedure for rapidly losing altitude. The AFM/POH specifies the procedure. In general, emergency descent procedures are high drag, high airspeed procedures requiring a specific airplane configuration (such as power to idle, propellers forward, landing gear extended, and flaps retracted), and a specific emergency descent airspeed. Emergency descent maneuvers often include turns.

Glides

A glide is a basic maneuver in which the airplane loses altitude in a controlled descent with little or no engine power. Forward motion is maintained by gravity pulling the airplane along an inclined path, and the descent rate is controlled by the pilot balancing the forces of gravity and lift. To level off from a partial power descent using a 1,000 feet per minute descent rate, the pilot should use 10 percent (100 feet in this example) as the distance above the desired level-off altitude to begin raising the nose and adding power to stop the descent and maintain airspeed.

Although glides are directly related to the practice of power-off accuracy landings, they have a specific operational purpose in normal landing approaches, and forced landings after engine failure. Therefore, it is necessary that they be performed more subconsciously than other maneuvers because most of the time during their execution, the pilot will be giving full attention to details other than the mechanics of performing the maneuver. Since glides are usually performed relatively close to the ground, accuracy of their execution and the formation of proper technique and habits are of special importance.

The glide ratio of an airplane is the distance the airplane travels in relation to the altitude it loses. For example, if an airplane travels 10,000 feet forward while descending 1,000 feet, its glide ratio is 10 to 1.

The best glide airspeed is used to maximize the distance flown. This airspeed is important when a pilot is attempting to fly during an engine failure. The best airspeed for gliding is one at which the airplane travels the greatest forward distance for a given loss of altitude in still air. This best glide airspeed occurs at the highest lift-to-drag ratio (L/D). *[Figure 3-23]* When gliding at airspeed above or below the best glide airspeed, drag increases. Any change in the gliding airspeed results in a proportional change in the distance flown. *[Figure 3-24]* As the glide airspeed is increased or decreased from the best glide airspeed, the glide ratio is lessened.

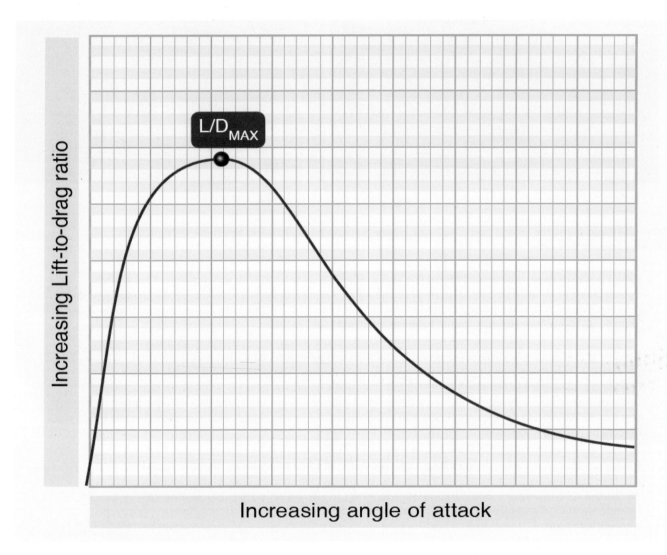

Figure 3-23. *L/D~MAX~.*

Figure 3-24. *Best glide speed provides the greatest forward distance*

Variations in weight do not affect the glide angle provided the pilot uses the proper airspeed. Since it is the L/D ratio that determines the distance the airplane can glide, weight does not affect the distance flown; however, a heavier airplane needs to fly at a higher airspeed to obtain the same glide ratio. For example, if two airplanes having the same L/D ratio but different weights start a glide from the same altitude, the heavier airplane gliding at a higher airspeed arrives at the same touchdown point in a shorter time. Both airplanes cover the same distance, only the lighter airplane takes a longer time.

Since the highest glide ratio occurs at maximum L/D, certain considerations should be given for drag-producing components of the airplane, such as flaps, landing gear, and cowl flaps. When drag increases, a corresponding decrease in pitch attitude is required to maintain airspeed. As the pitch is lowered, the glide path steepens and reduces the distance traveled. To maximize the distance traveled during a glide, all drag-producing components need to be eliminated if possible.

Wind affects the gliding distance. With a tailwind, the airplane glides farther because of the higher groundspeed. Conversely, with a headwind, the airplane does not glide as far because of the slower groundspeed. This is important for a pilot to understand and manage when dealing with engine-related emergencies and any subsequent forced landing.

During powered operations, the airplane design compensates for the effects of p-factor and propeller slipstream. While these effects disappear during a glide, the design compensation remains. During glides, it is likely that slight left rudder pressure will be required to maintain coordinated flight. In addition, the pilot needs to use greater deflection of the flight controls due to the relatively slow airflow over the control surfaces.

Minimum sink speed is used to maximize the time that the airplane remains in flight. It results in the airplane losing altitude at the lowest rate. Minimum sink speed occurs at a lower airspeed than the best glide speed. Flight at the minimum sink airspeed results in less distance traveled. Minimum sink speed is useful in flight situations where time in flight is more important than distance flown. An example is ditching an airplane at sea. Minimum sink speed is not an often published airspeed but generally is a few knots less than best glide speed.

In an emergency, such as an engine failure, attempting to apply elevator back pressure to stretch a glide back to the runway is likely to lead the airplane landing short and may even lead to loss of control if the airplane stalls. This leads to a cardinal rule of airplane flying: The pilot should not attempt to "stretch" a glide by applying back-elevator pressure and reducing the airspeed below the airplane's recommended best glide speed. The purpose of pitch control during the glide is to maintain the maximum L/D, which may require fore or aft flight control pressure to maintain best glide airspeed.

To enter a glide, the pilot should close the throttle and, if equipped, advance the propeller lever forward. With back pressure on the elevator flight control, the pilot should maintain altitude until the airspeed decreases to the recommended best glide speed. In most airplanes, as power is reduced, propeller slipstream decreases over the horizontal stabilizer, which decreases the tail-down force, and the airplane's nose tends to lower immediately. To keep pitch attitude constant after a power change, the pilot should counteract the pitch down with a simultaneous increase in elevator back pressure. This point is particularly important for fast airplanes as they do not readily lose their airspeed—any slight deviation of the airplane's nose downwards results in an immediate increase in airspeed. Once the airspeed has dissipated to best glide speed, the pitch attitude should be set to maintain that airspeed. This should be done with reference to the natural horizon and with a quick reference to the flight instruments. When the airspeed has stabilized, the airplane should be trimmed to eliminate any flight control pressures held by the pilot. Precision is required in maintaining the best glide airspeed if the benefits are to be realized.

A stabilized, power-off descent at the best glide speed is often referred to as normal glide. The beginning pilot should memorize the airplane's attitude and speed with reference to the natural horizon and note the sounds made by the air passing over the airplane's structure, forces on the flight controls, and the feel of the airplane. Initially, the learner may be unable to recognize slight variations in airspeed and angle of bank by vision or by the pressure required on the flight controls. The instructor should point out that an increase in sound levels denotes increasing speed, while a decrease in sound levels indicates decreasing speed. When a sound level change is perceived, the learner should cross-check the visual and pressure references. The learner should use all three airspeed references (sound, visual, and pressure) consciously until experience is gained, and then remain alert to any variation in attitude, feel, or sound.

After a solid comprehension of the normal glide is attained, the learner should be instructed in the differences between normal and abnormal glides. Abnormal glides are those glides conducted at speeds other than the best glide speed. Glide airspeeds that are too slow or too fast may result in the airplane not being able to make the intended landing spot, flat approaches, hard touchdowns, floating, overruns, and possibly stalls and an accident.

Gliding Turns

The absence of the propeller slipstream, , p-factor, loss of effectiveness of the various flight control surfaces at lower airspeeds, and designed-in aerodynamic corrections complicate the task of flight control coordination in comparison to powered flight for the learner. These principles should be thoroughly explained to the learner by the flight instructor.

Three elements in gliding turns that tend to force the nose down and increase glide speed are:

1. Decrease in lift due to the direction of the lifting force.
2. Excessive rudder inputs as a result of reduced flight control pressures.
3. The normal stability and inherent characteristics of the airplane to nose-down with the power off.

These three factors make it necessary to use more back pressure on the elevator than is required for a straight glide or a level turn, and they have an effect on control coordination. The rudder compensates for yawing tendencies when rolling in or out of a gliding turn; however, the required rudder pedal pressures are reduced as a result of the reduced forces acting on the control surfaces. A learner may apply excessive rudder pedal pressures based on experience with powered flight. This overcontrol of the aircraft may cause slips and skids and result in potentially hazardous flight control conditions.

Some examples of this hazard are:

- A low-level gliding steep turn during an engine failure emergency. If the rudder is excessively deflected in the direction of the bank while the pilot is increasing elevator back pressure in an attempt to retain altitude, the situation can rapidly turn into an unrecoverable spin.

- During a power-off landing approach. The pilot depresses the rudder pedal with excessive pressure that leads to increased lift on the outside wing, banking the airplane in the direction of the rudder deflection. The pilot may improperly apply the opposite aileron to prevent the bank from increasing while applying elevator back pressure. If allowed to progress, this situation may result in a fully developed cross-control condition. A stall in this situation almost certainly results in a rapid and unrecoverable spin.

Level-off from a glide is really two different maneuvers depending on the type of glide:

- First, in the event of a complete power failure, the best glide speed should be held until necessary to reconfigure for the landing. The pilot should plan for a steeper approach than usual. A 10 percent lead (100 feet if the descent rate is 1,000 feet per minute) factor should be sufficient to slow the descent before landing.

- Second, in the case of simulated power failure training, power should be applied as the 10 percent lead value appears on the altimeter. This allows a slow but positive power application to maintain or increase airspeed while the pilot raises the nose to stop the descent and re-trims the airplane as necessary.

The level-off from a practice glide should be started before reaching the desired altitude because of the airplane's downward inertia. The amount of lead depends on the rate of descent and the desired airspeed upon completion of the level off. For example, assume the aircraft is in a 500 fpm rate of descent, and the desired final airspeed is higher than the glide speed. The altitude lead should begin at approximately 100 feet above the target altitude. At the lead point, power should be increased to the appropriate level flight cruise power setting. The airplane's nose tends to rise as airspeed and power increase, and the pilot should smoothly control the pitch attitude such that the level-off is completed at the desired altitude and airspeed. When recovery is being made from a gliding turn to a normal glide, the back pressure on the elevator control, which was applied during the turn, needs to be decreased or the airplane may pitch up and experience a loss of airspeed. This error requires considerable attention and conscious control adjustment to re-establish a normal glide airspeed.

Common errors in the performance of descents and descending turns are:

1. Failure to adequately clear for aircraft traffic in the turn direction or descent.
2. Inadequate elevator back pressure during glide entry resulting in an overly steep glide.
3. Failure to slow the airplane to approximate glide speed prior to lowering pitch attitude.
4. Attempting to establish/maintain a normal glide solely by reference to flight instruments.
5. Inability to sense changes in airspeed through sound and feel.
6. Inability to stabilize the glide (chasing the airspeed indicator).
7. Attempting to "stretch" the glide by applying back-elevator pressure.
8. Skidding or slipping during gliding turns and not recognizing the difference in rudder forces with and without power.
9. Failure to lower pitch attitude during gliding turn entry resulting in a decrease in airspeed.
10. Excessive rudder pressure during recovery from gliding turns.
11. Inadequate pitch control during recovery from straight glide.
12. Cross-controlling during gliding turns near the ground.
13. Failure to maintain constant bank angle during gliding turns.

Chapter Summary

The four fundamental maneuvers of straight-and-level flight, turns, climbs, and descents are the foundation of basic airmanship. Effort and continued practice are required to master the fundamentals. It is important that a pilot consider the six motions of flight: bank, pitch, yaw and horizontal, vertical, and lateral displacement. In order for an airplane to fly from one location to another, it pitches, banks, and yaws while it moves over and above, in relationship to the ground, to reach its destination. The airplane should be treated as an aerodynamic vehicle that is subject to rigid aerodynamic laws. A pilot needs to understand and apply the principles of flight in order to control an airplane with the greatest margin of mastery and safety.

Airplane Flying Handbook (FAA-H-8083-3C)
Chapter 4: Energy Management: Mastering Altitude and Airspeed Control

Introduction

This chapter is all about managing the airplane's altitude and airspeed using an energy-centered approach. Energy management can be defined as the process of planning, monitoring, and controlling altitude and airspeed targets in relation to the airplane's energy state in order to:

1. Attain and maintain desired vertical flightpath-airspeed profiles.
2. Detect, correct, and prevent unintentional altitude-airspeed deviations from the desired energy state.
3. Prevent irreversible deceleration and/or sink rate that results in a crash.

Importance of Energy Management

Learning to manage the airplane's energy in the form of altitude and airspeed is critical for all new pilots. Energy management is essential for effectively achieving and maintaining desired vertical flight path and airspeed profiles, (e.g., constant airspeed climb) and for transitioning from one profile to another during flight, (e.g., leveling off from a descent).

Proper energy management is also critical to flight safety. Mistakes in managing the airplane's energy state can be deadly. Mismanagement of mechanical energy (altitude and/or airspeed) is a contributing factor to the three most common types of fatal accidents in aviation: loss of control in-flight (LOC-I), controlled flight into terrain (CFIT), and approach-and-landing accidents. Thus, pilots need to have:

1. An accurate mental model of the airplane as an energy system.
2. The competency to effectively coordinate control inputs to achieve and maintain altitude and airspeed targets.
3. The ability to identify, assess, and mitigate the risks associated with mismanagement of energy.

Viewing the Airplane as an Energy System

The total mechanical energy of an airplane in flight is the sum of its potential energy from altitude and kinetic energy from airspeed. The potential energy is expressed as *mgh*, and the kinetic energy as ½ *mV²*. Thus, the airplane's total mechanical energy can be stated as:

$$mgh + \tfrac{1}{2}\,mV^2$$

Where,

    m = mass
    g = gravitational constant
    h = height (altitude)
    V = velocity (airspeed)

A flying airplane is an "open" energy system, which means that the airplane can gain energy from some source (e.g., the fuel tanks) and lose energy to the environment (e.g., the surrounding air). It also means that energy can be added to or removed from the airplane's total mechanical energy stored as altitude and airspeed.

A Frame of Reference for Managing Energy State

At any given time, the energy state of the airplane is determined by the total amount and distribution of energy stored as altitude and airspeed. Note that the pilot's frame of reference for managing the airplane's energy state is airplane-centric—being a function of indicated altitude and indicated airspeed, and not height above the ground or groundspeed.

The indicated altitude displayed in the altimeter and its associated potential energy are based on the height of the airplane above a fixed reference point (mean sea level or MSL), not on the height above ground level (AGL), which changes with variations in terrain elevation. Likewise, the indicated airspeed displayed in the airspeed indicator and its associated kinetic energy are based on the speed of the airplane relative to the air, not on the speed relative to the ground below, which varies with changes in wind speed and direction.

Note that changes in indicated altitude and airspeed are attained through forces resulting from the pilot's direct manipulation of the controls. These direct control inputs determine the airplane's ability to climb/descend or accelerate/decelerate. In contrast, changes in AGL-altitude and groundspeed are affected by "external" factors, such as varying terrain elevation and wind, which the pilot cannot alter. Of course, the pilot should manipulate the airplane's energy in such way as to minimize any risks associated with terrain or wind. For example, the pilot may seek to manipulate energy state so as to maximize the airplane's energy gains and minimize energy loses when faced with rising terrain. A safer heading may also be an option.

Once airborne, the airplane gains energy from the force of engine thrust (T) and it loses energy from aerodynamic drag (D). The difference between energy in and out ($T - D$) is the net change, which determines whether total mechanical energy—stored as altitude and airspeed—increases, decreases, or remains the same.

When thrust exceeds drag ($T - D > 0$), the airplane's total mechanical energy increases. The pilot can store the surplus energy as increased altitude or airspeed. For example, if the pilot decides to put all the surplus energy into altitude, the airplane can climb at a constant airspeed. *[Figure 4-1A]* If the pilot opts to place all the surplus energy into airspeed, the airplane can accelerate while maintaining altitude. *[Figure 4-1B]*

When drag exceeds thrust, ($T - D < 0$), the airplane's total mechanical energy decreases. The pilot has two sources of stored energy to tap into. For example, the pilot may choose to let the airplane descend at a constant airspeed *[Figure 4-1C]* or slow down while maintaining altitude *[Figure 4-1D]* as stored energy is withdrawn to deal with the energy deficit. When energy gained equals that lost ($T - D = 0$), all thrust is spent on drag. In this case, the total amount of mechanical energy and its distribution over altitude and airspeed does not change. Both remain constant as the airplane maintains a constant altitude and airspeed. *[Figure 4-1E]*

Energy can also be exchanged between altitude and airspeed. For example, when a pilot trades airspeed for altitude, as altitude increases, airspeed decreases. In other words, when energy is exchanged, altitude and airspeed always change in opposite directions (absent any other energy or control inputs). As one goes up, the other one comes down. Also note that even though the distribution of energy over altitude and airspeed may change dramatically during energy exchange, the total amount of mechanical energy can remain the same at the end of the exchange maneuver *[Figure 4-1F]*, as long as thrust is adjusted to match drag as the latter varies with changes in airspeed.

Energy Transaction Examples	Net Energy Change (T – D)	Change in Stored Energy		Resulting Aircraft Condition
		Altitude	Airspeed	
A	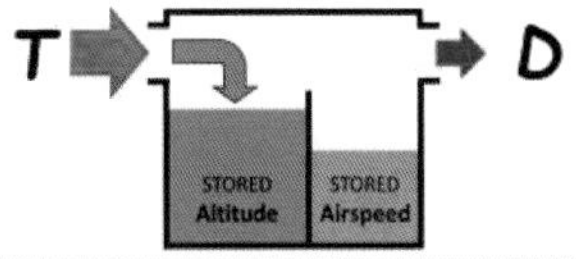 > 0	Increase	No change	Climb at constant airspeed
B	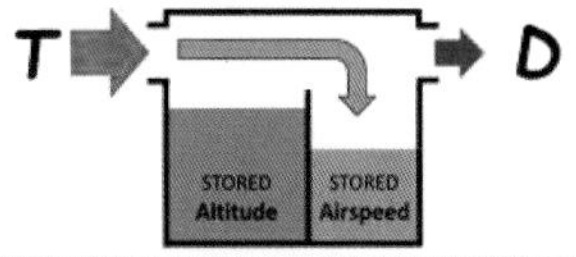 > 0	No change	Increase	Acceleration at constant altitude
C	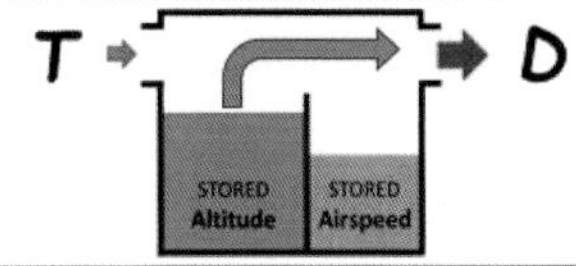 < 0	Decrease	No change	Descent at constant airspeed
D	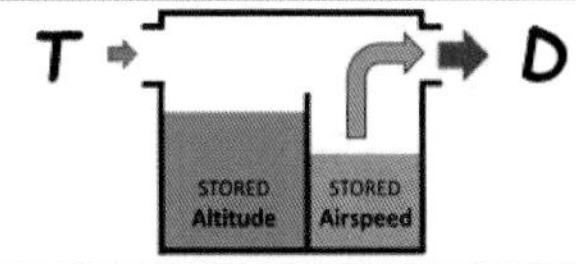 < 0	No change	Decrease	Deceleration at constant altitude
E	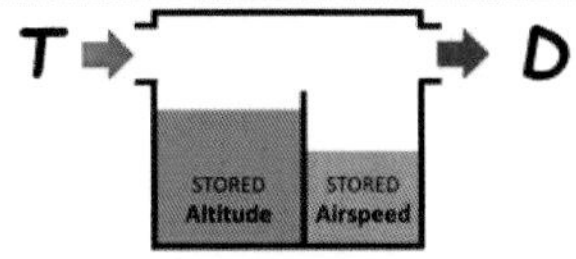 = 0	No change	No change	Constant altitude and airspeed
F	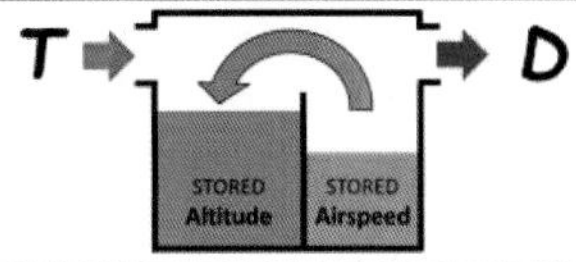 = 0	Increase	Decrease	Climb and deceleration

Figure 4-1 A-F. *Examples of typical energy transactions.*

Managing Energy is a Balancing Act

Since the airplane gains energy from engine thrust (T) and loses energy through aerodynamic drag (D), energy flows continuously into and out of the airplane while in flight. Usually measured as *Specific Excess Power* (P_S), or rate of energy change, the net energy flow is a direct function of the difference between thrust and drag.

$P_S = (T - D)V/W$

Where,

      T = Thrust
      D = Drag
      V = velocity (airspeed)
      W = aircraft weight

More importantly, there is a fundamental relationship between changes in the airplane's total energy resulting from this net energy flow on one hand, and changes in the energy stored as altitude and airspeed on the other. This fundamental relationship can be summarized through the airplane's energy balance equation. *[Figure 4-2]*

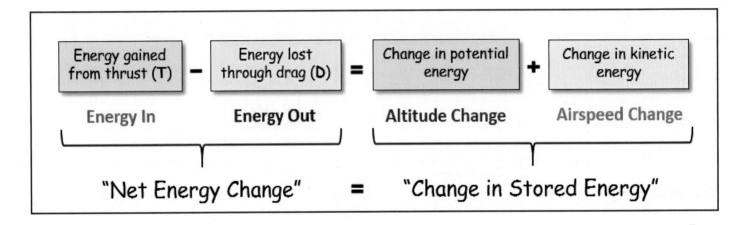

Figure 4-2. *The energy balance equation.*

The left side of the energy balance equation represents the airplane's net energy flow, while the right side reflects matching changes to the energy storage. Thus, changes to the airplane's total energy affect the left side of the equation, while the right side shows possible changes in energy distribution between altitude and airspeed.

Note that a change in total energy resulting from the difference between thrust and drag (left side) always matches the change in total energy redistributed over altitude and airspeed (right side). Although rate of energy change, expressed as specific excess power (P_S), varies during flight—becoming positive, negative, or zero—both sides of the equation are inexorably balanced regardless of whether the airplane is accelerating, decelerating, climbing, descending, or maintaining constant altitude and airspeed. *(Note: This simplified balance equation does not account for long-term changes in total mechanical energy caused by the reduction in aircraft weight as fuel is gradually burned in flight. Although the effect of weight loss on total energy becomes critical when solving long-term aircraft performance problems such as range and endurance, it is negligible when considering short-term flight control problems.)*

Of course, the pilot controls the change in total energy on the left side of the equation, as well as the distribution of any changes in energy over altitude and airspeed on the right side. How the pilot coordinates the throttle and elevator to achieve and maintain desired altitude and airspeed targets as well as avoid energy "crises" is at the core of energy management and is elaborated in the rest of the chapter.

Role of the Controls to Manage Energy State

An energy-centered approach clarifies the roles of the engine and flight controls beyond the simple "pitch for airspeed and power for altitude" by modeling how throttle and elevator inputs affect the airplane's total mechanical energy. From an energy perspective, the problem of controlling vertical flight path and airspeed becomes one of handling the airplane's energy state—the total amount of energy and its distribution over altitude and speed. Thus, rather than asking what controls altitude and what controls airspeed, a pilot can now ask what controls total energy and what controls its distribution over altitude and airspeed.

Primary Energy Role of the Throttle and Elevator

The throttle, by increasing or decreasing engine thrust against drag, regulates changes in total mechanical energy. As illustrated above, changing total energy is a function of both thrust and drag $(T - D)$. However, drag mainly varies long-term due to airspeed changes, or by using high lift/drag devices which can only increase drag. Therefore, changes in total energy are normally initiated by changing thrust, not drag. When the throttle setting makes thrust greater than drag, an increase of total mechanical energy is the result. When the throttle setting makes thrust less than drag, a decrease of total mechanical energy is the result. Once the desired path-speed profile is established, the throttle sets engine thrust to match the total energy demanded by vertical flight path and airspeed combined. The throttle then is the *total energy controller*.

On the other hand, the elevator is an energy exchanger and distribution device whose primary job is to allocate changes in total energy between vertical flight path and airspeed by adjusting pitch attitude. Here, once the chosen path-speed profile is achieved, the elevator sets the appropriate pitch attitude to maintain the demanded distribution of total energy over vertical flight path and airspeed. Thus, the elevator is the *energy distribution controller*.

The throttle and elevator then are really energy state controls—neither one controls altitude nor airspeed independently since these two variables are inherently coupled through the airplane's total mechanical energy. Instead, to control altitude and airspeed effectively, the pilot coordinates the use of both devices to manage the airplane's energy state.

The reservoir analogy *[Figure 4-3]* illustrates the energy-based role of the throttle and the elevator. In this analogy, the throttle controls the "valve" regulating the net total energy flow while the elevator controls the "valve" regulating the distribution of energy into and out of the altitude and airspeed "reservoirs." Referring back to the energy balance equation [*Figure 4-2*], it becomes clear then that the throttle controls the left side of the equation (total energy) and the elevator controls the right side (energy distribution).

As illustrated in *Figure 4-3*, when the throttle increases thrust above drag $(T - D > 0)$ the airplane gains total energy, and when the throttle reduces thrust below drag $(T - D < 0)$ the airplane loses total energy. The elevator then distributes this increase or decrease in total energy between altitude and airspeed. Finally, when the throttle adjusts thrust equal to drag $(T - D = 0)$, there is no change in total energy, but the energy stored as altitude and airspeed can be exchanged between the two reservoirs using the elevator, while total energy, at least short-term, remains constant.

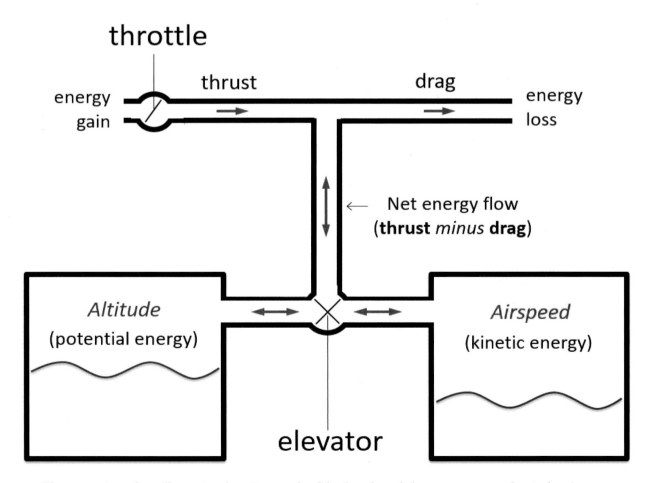

Figure 4-3. *The reservoir analogy illustrating the primary role of the throttle and elevator to manage the airplane's energy state.*

Additional Role for the Elevator

On the front side of the power required curve, where the airplane cruises at high speed (1 in *Figure 4-4*) and a low angle of attack (AOA) with little or no excess power or excess thrust (A in *Figure 4-4*), pulling back on the yoke or stick (elevator up) will result in a brief energy exchange climb, causing the airplane to slow down from 1 to 2 toward the center of the power curve *[Figure 4-4]*. This decrease in airspeed results in a reduction in total drag; hence available energy in the form of positive excess power ($P_S > 0$) where thrust exceeds drag ($T - D > 0$). With this excess power (B in *Figure 4-4*) the airplane can now climb at a constant airspeed or turn in level flight while maintaining a constant airspeed at an increased load factor.

On the backside of the power required curve, where the airplane flies at low speed (3 in *Figure 4-4*) and high AOA with little or no excess power or excess thrust (C in *Figure 4-4*), pushing forward on the yoke or stick (elevator down) will result in a brief energy exchange descent, causing the airplane to accelerate from 3 to 2 toward the center of the power curve *[Figure 4-4]*. This increase in airspeed results in a reduction in total drag; hence available energy in the form of positive excess power ($P_S > 0$) where thrust exceeds drag ($T - D > 0$). With this excess power (B in *Figure 4-4*) the airplane can now climb at a constant airspeed or turn in level flight while maintaining a constant airspeed at an increased load factor. This role of the elevator is critical to prevent unintentional, excessive deceleration or sink rate as illustrated later in the chapter (refer to *Preventing Irreversible Deceleration and/or Sink Rate* section).

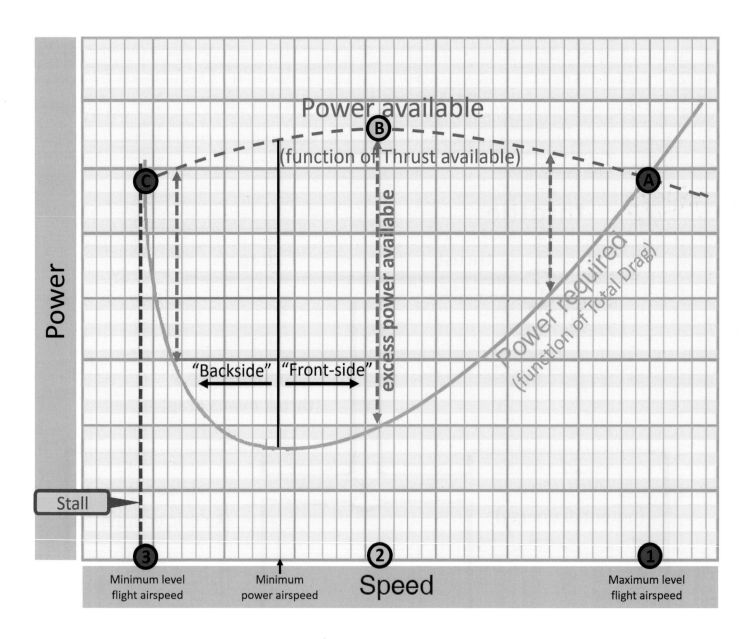

Figure 4-4. *The front side and backside of the power required curve, the power available curve, and the relative excess power available (power available - power required) at different speeds.*

While the elevator can assist the throttle in changing $T - D$ and P_S through changes in airspeed via energy exchange as described above, occasionally the elevator can directly increase the "D" in $T - D$ at any given speed during a level turn, thus helping the airplane rapidly bleed off total energy. As the airplane banks, load factor (lift/weight) increases because total lift has to increase to pull the airplane into the turn while simultaneously balancing its weight. This is accomplished by pulling back on the yoke (or stick) to increase AOA which results in increased induced drag and power required at any given speed. This action will quickly slow the airplane down and decrease total energy more rapidly than by just reducing the throttle setting to idle. This additional role of the elevator is shown on the power curve. *[Figure 4-5]*

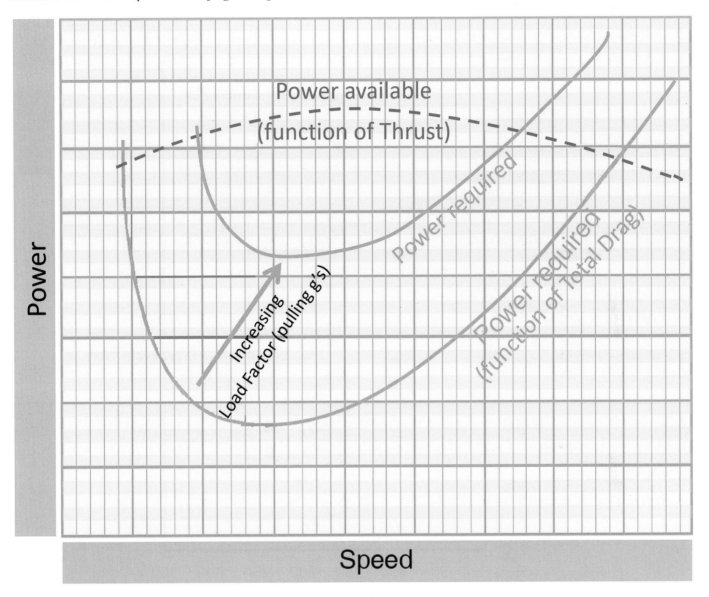

Figure 4-5. *The effect of increased load factor on total drag and power required at different airspeeds.*

Applying the respective role of the controls to manage the airplane's energy state leads to a set of simple "rules" for proper throttle-elevator coordination to effectively control vertical flight path and airspeed. What are these basic rules of energy control?

Rules of Energy Control

The central principle encapsulating the role of the throttle and elevator for managing the airplane's energy can be summed up as follows: coordinated throttle and elevator inputs control the airplane's energy state. Modifying a popular adage, the principle can be restated as "pitch plus power controls energy state." This central principle serves to guide a set of general energy control rules to achieve and maintain any desired vertical flight path and airspeed targets within the airplane's energy envelope.

Visualizing the Airplane's Ability to "Move" Between Energy States

To better understand the basic rules of energy control, a pilot needs to visualize an airplane's energy state and its ability to switch from one energy state to another. In other words, how does an airplane "move" from an initial altitude and airspeed to any other target altitude and airspeed within its flight envelope, and how does the pilot control the process? A map should help, and in this case, it charts the status of the aircraft in terms of energy.

In a navigation map, such as an aeronautical sectional chart, the geographic position of an airplane is determined by two variables—latitude and longitude. Likewise, in an "altitude-airspeed" or "energy" map the energy position of an airplane, its energy state, is defined by two variables—altitude and airspeed. *[Figure 4-6]*

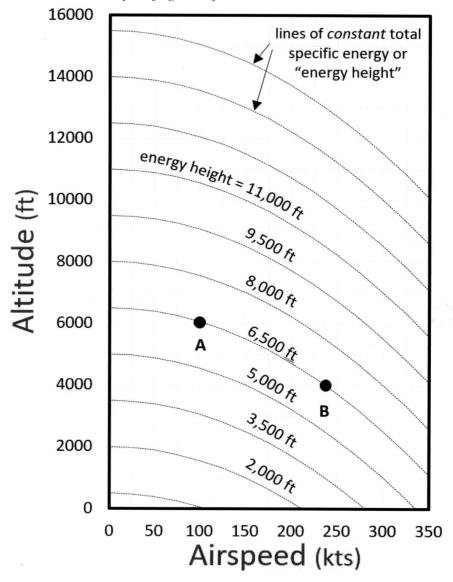

Figure 4-6. *The altitude-speed "map" showing lines of constant energy height.*

The position of an airplane in the altitude-airspeed map represents its total specific energy or E_S (which is simply the sum of its potential and kinetic energies divided by aircraft weight) as determined by its current altitude and airspeed.

$$E_S = h + \frac{V^2}{2g}$$

Where,

g = *gravitational constant*
h = *height (altitude)*
V = *velocity (airspeed)*

Since the total specific energy, E_S, has the units of height (e.g., feet), it is usually called energy height. It also gets this name from the fact that energy height is the *maximum* height that an airplane would reach from its current altitude, if it were to trade all its speed for altitude. *Figure 4-6* shows lines of constant total specific energy or energy height. Different positions of an airplane along a given energy height line have the same total energy regardless of their location on the line (e.g., A and B).

Thus, even though the airplane in point A is cruising at 100 knots and 6,000 feet, it has the same total specific energy expressed in height (6,500 feet) when cruising at 240 knots and 4,000 feet (B). This also means that the airplane in either position, A or B, would be able to "zoom" to the same maximum altitude of 6,500 feet by trading all its speed for altitude. The lines of constant energy height can be used as idealized trajectories to depict an airplane moving from one energy state to another solely through energy exchange (e.g., A to B). If the airplane rapidly exchanges altitude and airspeed, it would follow along the energy height line while, in the short term, maintaining constant total energy.

In addition to showing energy height lines, the energy map can also depict available specific excess power (P_S) contours, as well as energy trajectories of an airplane moving from one energy state to another. *[Figure 4-7]* The airplane can move along energy height lines by simply exchanging energy (e.g., A to B). However, to move across energy height lines, the airplane needs to increase or decrease total energy while distributing the energy change between altitude and airspeed. Thus, the ability of an airplane to go from one energy height to another (e.g., from A to positions C, D, or E) is a function of specific excess power (P_S), measured in rate of change in distance or height (e.g., feet per minute).

Examine the energy positions depicted in *Figure 4-7*. The airplane in position A is flying at 4,000 feet and 150 knots with a total energy equivalent to 5,000 feet. Since positions C, D and E are located at higher energy heights (11,000, 9,500, and 6,500 feet respectively), the only way for the airplane to reach them from position A is by increasing its total energy (i.e., increasing thrust above drag, or $P_S > 0$). The reverse is also true. If the airplane is in position C, D or E, the only way for it to get back to position A is by decreasing its total energy (i.e., decreasing thrust below drag, or $P_S < 0$). In other words, the rate at which the airplane can move from one energy height to another—e.g., how swiftly it can climb/descend at a steady speed, or accelerate/decelerate in level flight—is a function of specific excess power, which can be positive ($P_S > 0$) or negative ($P_S < 0$) depending on whether the airplane needs to move to an energy height that demands more or less total energy.

At the edge of the energy envelope, where available $P_S = 0$ at full throttle, the airplane can no longer climb while maintaining airspeed or accelerate without descending. Inside this envelope, inner contours increase in value, reaching a "peak" where available P_S is maximized. Notice that P_S at full throttle is maximized at a specific airspeed (V_Y) decreasing in value at slower or faster airspeeds. At V_Y then, the airplane can attain the maximum rate of climb while maintaining airspeed or the maximum acceleration without descending *[Figure 4-7]*.

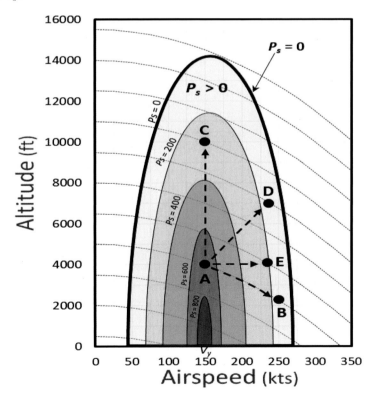

Figure 4-7. *Energy map depicting specific excess power (P_S) contours (shown in feet per minute) and energy trajectories for a hypothetical airplane.*

Three Basic Rules of Energy Control

An "energy-control" map can help visualize the basic energy control rules. *[Figure 4-8]* The energy-control map depicts not only the trajectories of an airplane transitioning from an arbitrary initial energy state (1) to other target states (2, 3, 4, 5, 6, and 7), but also the changes in energy caused by the throttle (blue/red arrows) and the elevator (green arrows). In other words, it allows pilots to visualize the basic control rules for moving an airplane from any state to another. The edge of the sustainable energy state envelope (where $P_S = 0$ at full throttle) is also illustrated.

Note that the line of constant total energy (dashed line) that divides the area in the map requiring more total energy (blue area) from that which requires less energy (red area) is depicted relative to the arbitrary initial energy state (1). The throttle adds (blue arrow) or subtracts (red arrow) the amount of total energy demanded by the new target energy state, while the elevator (green arrows) distributes the correct amount of total energy between potential and kinetic energies. By balancing the simultaneous actions of the controls, the airplane can follow the desired energy trajectory.

As illustrated in *Figure 4-8*, moving the airplane from position 1 to the energy states in 2 and 3 calls for a higher throttle setting to increase total energy by the same amount (in this example, positions 2 and 3 are located at the same higher-energy height). The difference between these two energy trajectories (1-to-2 and 1-to-3) lies in the way the total energy change is distributed by the elevator through changes in pitch attitude. As can be seen in *Figure 4-8*, changes in total energy by adjusting throttle setting (blue/red arrows) extend across lines of constant total energy (dashed line), while changes in energy distribution by adjusting the elevator deflection (green arrows) extend along the lines of constant total energy (equal energy height). Appropriate changes in total energy via the throttle and/or changes in energy distribution via the elevator, depicted by their respective energy "arrows," determine the direction of a given energy trajectory between two energy states. To visualize this effect, compare the trajectory from 1-to-2 with that from 1-to-3 and notice the way the corresponding elevator energy arrows (left green arrow = up-elevator; right green arrow = down-elevator) are positioned in relation to the throttle energy arrow (blue arrow = increased throttle).

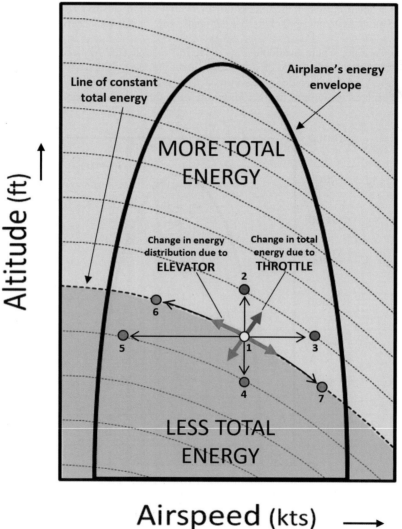

Figure 4-8. *The energy-control map helping to visualize the basic energy control rules.*

Thus, transitioning to a higher altitude at a constant speed (1-to-2) requires increased throttle and up-elevator to stay on speed, while transitioning to a faster airspeed at a constant altitude (1-to-3) demands increased throttle and (gradual) down-elevator to stay on path, re-trimming as needed to relieve elevator control pressures.

Transitioning to a lower altitude at a constant speed (1-to-4) requires decreased throttle and down-elevator to stay on speed, while transitioning to a slower airspeed at a constant altitude (1-to-5) demands decreased throttle and (gradual) up-elevator to stay on path, re-trimming as needed to relieve elevator control pressures.

Finally, transitioning to a higher altitude by trading speed for altitude (1-to-6) requires up-elevator without initially changing throttle setting, while transitioning to a faster airspeed by trading altitude for speed (1-to-7) requires down-elevator without initially changing throttle setting. In both cases, at the end of the energy exchange maneuver, the elevator will need to be re-trimmed and throttle setting adjusted to match drag at the new speed in order to maintain total energy constant while remaining at the new altitude-airspeed target.

As can be visualized in *Figure 4-8*, there are three general energy control rules for coordinating the throttle and elevator to move the airplane from one energy state to another:

Rule #1: *If* you want to move to a new energy state that demands more total energy, *then*:
    Throttle: increase throttle setting so that thrust is greater than drag, thus increasing total energy;

    Elevator: adjust pitch attitude as appropriate to distribute the total energy being gained over altitude and airspeed:

        a. To climb at constant speed, pitch up just enough to maintain the desired speed;
        b. To accelerate at constant altitude, gradually pitch down just enough to maintain path.

    Upon reaching new desired energy state, adjust pitch attitude and throttle setting as needed to maintain the new path-speed profile.

Rule #2: *If* you want to move to a new energy state that demands less total energy, *then*:

    Throttle: reduce throttle setting so that thrust is less than drag, thus decreasing total energy;

    Elevator: adjust pitch attitude as appropriate to distribute the total energy being lost over altitude and airspeed:

        a. To descend at constant speed, pitch down just enough to maintain the desired speed;
        b. To slow down at constant altitude, gradually pitch up just enough to maintain path.

    Upon reaching new desired energy state, adjust pitch attitude and throttle setting as needed to maintain the new path-speed profile.

Rule #3: *If* you want to move to a new energy state that demands no change in total energy, *then*:

    Throttle: do not change initially, but adjust to match drag at the end of maneuver as needed to maintain total energy constant;

    Elevator: adjust pitch attitude to exchange energy between altitude and airspeed:

        a. To trade speed for altitude, pitch up;
        b. To trade altitude for speed, pitch down.

    Upon reaching new desired energy state, adjust pitch attitude and throttle setting as needed to maintain the new path-speed profile.

Note that control rules 1 and 2 allow the elevator to distribute the change in total energy in different ways. For example, using rule 1.a the pilot may choose to adjust the pitch-up attitude to climb at a slower (or faster) airspeed. Other situations may require combining two control rules. One example is when, at maximum cruise airspeed in level flight, thrust has reached its maximum limit (i.e., $P_S = 0$) but the target energy state is at a higher altitude and total energy within the airplane's envelope. At maximum level airspeed, there is no excess thrust available to increase the airplane's total energy needed to climb. One solution is to initially trade kinetic for potential energy (rule 3.a), slowing down to an airspeed where drag is reduced below thrust, thus allowing the airplane to increase its total energy and climb at that slower airspeed (rule 1.a).

Mitigating Risks from Mismanagement of Energy

Besides learning the proper use of the controls for normal energy management tasks, pilots should be equipped with the ability to identify, assess, and mitigate two major risks associated with mismanagement of energy: 1) unwanted deviations from the desired energy state; and 2) unintentional, irreversible deceleration and/or sink rate causing depletion of mechanical energy. The first risk involves unintended altitude-airspeed deviations (refer to *Managing Energy Errors* section). The second risk entails unforeseen, continuous airspeed and/or altitude loss coupled with little or no available excess power in a given flight condition (refer to *Preventing Irreversible Deceleration and/or Sink Rate* section).

Two Energy Management Scenarios

Two flight scenarios illustrate the two major risks associated with failure to manage the airplane's energy state and how a pilot can identify, assess, and mitigate those risks.

Scenario 1

Unintentionally descending below the desired glideslope on final approach to landing and failing to make the proper correction. *[Figure 4-9]* To bring the airplane back to the desired glideslope, should the pilot pitch up, throttle up, or both?

Figure 4-9. *Descending below the desired glideslope.*

Scenario 2

Flying toward rising terrain and not being able to fly up and over it before impacting terrain. *[Figure 4-10]* Note the rising terrain all along the departure corridor. What can the pilot do to prevent an impending crash?

Figure 4-10. *Departing from Runway 33, Aspen/Pitkin County Airport (KASE), elevation 7,820 feet.*

For both scenarios, this section will demonstrate how proper energy management can provide the pilot with the skill to manage the associated risks and avoid tragic results.

Managing Energy Errors

In addition to learning effective techniques for maintaining stabilized path-speed profiles (e.g., tracking the glideslope) and transitioning from one profile to another during flight (e.g., leveling off from a descent), pilots should develop skills for managing unwanted deviations in vertical flight path and airspeed—returning the airplane to its target energy state. Since many inflight "energy crises" start as undetected, ignored or poorly managed path-speed deviations, pilots need the skills to recognize, correct and prevent these deviations.

Although the intention is to correct altitude and airspeed deviations, the pilot is always acting on the airplane's energy state. Thus, it is important to translate altitude-speed deviations into energy errors. *[Figure 4-11]* Because the airplane's total energy is distributed over altitude and airspeed, there are two types of energy errors: 1) total energy errors and 2) energy distribution errors.

		Airspeed		
		Slower	**Desired Airspeed**	**Faster**
Altitude	**Higher**	(1) Total Energy: OK Potential energy: high Kinetic Energy: low	(4) Total Energy: high Potential energy: high Kinetic Energy: OK	(7) Total Energy: very high Potential energy: high Kinetic Energy: high
	Desired Altitude	(2) Total Energy: low Potential energy: OK Kinetic Energy: low	**(5) Desired Energy State** Total Energy: OK Potential energy: OK Kinetic Energy: OK	(8) Total Energy: high Potential energy: OK Kinetic Energy: high
	Lower	(3) Total Energy: very low Potential energy: low Kinetic Energy: low	(6) Total Energy: low Potential energy: low Kinetic Energy: OK	(9) Total Energy: OK Potential energy: low Kinetic Energy: high

Figure 4-11. *An energy state matrix that translates the main altitude-speed deviations into energy errors relative to the desired energy state (5).*

Monitoring the altimeter (or other flight path reference) and airspeed indicator allows the pilot to distinguish these two types of energy errors. In total energy errors, the airplane has too much energy (blue boxes) or too little energy (red boxes). The pilot will notice that altitude and speed deviate in the same direction ("lower-and-slower" or "higher-and-faster"). On the other hand, in energy distribution errors the airplane may have the correct amount of total energy (green boxes) but its distribution over altitude and speed is incorrect. Here, altitude and speed deviate in opposite directions ("higher-and-slower" or "lower-and-faster"). In this case, the pilot deals with relative deviations—not absolute altitude and speed.

Following energy management principles, total energy errors are corrected by increasing or decreasing energy using the throttle, while energy distribution errors are corrected by exchanging energy between altitude and speed using the elevator. To correct a combination of total energy and distribution errors, both controls need to be used simultaneously. *Figure 4-12* summarizes the control skills needed to correct total energy and energy distribution errors.

Scenario 1 *[Figure 4-9]* is a good example to illustrate energy errors and the skills needed to correct and avoid them. *Figure 4-13* actually depicts three possible scenarios (B, C, and D) where an airplane on final approach to land has descended below its intended flight path. Should the pilot pitch up, throttle up, or both? It depends. The airplane is lower than desired, but the pilot should check the airspeed as well. Relative to the target airspeed, the actual speed may be slower (B), faster (D), or on target (C). In all three cases, the goal is to return the airplane to its correct energy state (A), following a deviation in altitude and/or airspeed.

Lower-and-slower (B) is fundamentally different from lower-and-faster (D). The former requires advancing the throttle forward to regain total energy (3 in *Figure 4-12*), while the latter requires pulling back on the yoke/stick to null the energy distribution error (9 in *Figure 4-12*).

Airspeed				
	Cautions when Very Slow	**Slower**	**Desired speed**	**Faster**
Higher	Relatively safe. Surplus altitude is available to gain speed by pushing the elevator forward.	(1) Exchange energy by pushing the elevator forward to accelerate and descent simultaneously. Maintain the throttle setting.	(4) Reduce throttle setting to reduce total energy. Use elevator to maintain correct airspeed and allow the airplane to descend.	(7) Reduce throttle setting significantly to decrease total energy. Pull back on elevator gradually to decelerate to correct airspeed and then descend.
Desired altitude	Risky. Consider gaining speed at the expense of some altitude initially to improve climb performance with full throttle.	(2) Increase throttle setting to gain total energy by accelerating. Use elevator to maintain the desired altitude.	(5) DESIRED ENERGY STATE Maintain both throttle and elevator (trim) settings.	(8) Reduce throttle setting to decelerate. Use elevator to maintain the desired altitude.
Lower	Dangerous! Apply full throttle to resolve this condition. Avoid pitching up, which would increase drag and reduce or impede climb performance when on the backside of the power required curve.*	(3) Increase throttle setting significantly to gain total energy. Push elevator forward gradually to accelerate to correct airspeed and then climb.	(6) Increase throttle setting to gain altitude and pull back on elevator to maintain correct airspeed.	(9) Exchange energy by pulling the elevator back to climb and decelerate simultaneously. Maintain the throttle setting.

(Altitude — left vertical label spanning the rows)

* Depending on aircraft type, as full throttle is applied at the start of correction maneuver, slight forward or aft elevator pressure may be needed to maintain a constant pitch attitude. As the airplane gains total energy, use the elevator to accelerate to correct airspeed and then climb.

Figure 4-12. *The control skills needed to correct total energy and energy distribution errors identified in Figure 4-11 with an additional column giving caution to the "very slow" condition where careful AOA management is needed in addition to energy management.*

Note that in the scenario depicted in B in *Figure 4-13*, advancing the throttle forward to increase energy would only succeed if excess thrust is available ($P_S > 0$). This may not be the case if the pilot has badly mismanaged energy and slowed down to a speed where induced drag is so high that even applying full throttle would result in no surplus energy (see column "Cautions When Very Slow" in *Figure 4-12*). Depending on the flight condition, available excess power at full throttle may be negative ($P_S < 0$). In this case, the only recourse is to first trade altitude for speed by pushing forward on the yoke/stick, reducing AOA and induced drag, and only then advancing the throttle forward to regain total energy. But if the airplane is too close to the ground, there may not be enough room to reverse the negative energy rate and prevent the airplane from striking the ground.

Now consider the scenario depicted in C in *Figure 4-13*, where the airplane has descended below the desired flight path but is flying at the correct speed. Here, even though there is no speed deviation, the pilot is faced with a combination of total energy and distribution errors. Regaining altitude without changing speed requires advancing the throttle forward while easing aft on the yoke/stick (6 in *Figure 4-12*). In other words, decoupling altitude and airspeed (i.e. changing one without changing the other) demands the use of both controls simultaneously.

In all cases, path and speed should be monitored carefully as they are corrected, adjusting pitch attitude and throttle setting as appropriate. Once short-term deviations are corrected, the airplane will need to be trimmed for long-term control to maintain the desired path-speed profile (5 in *Figure 4-12*).

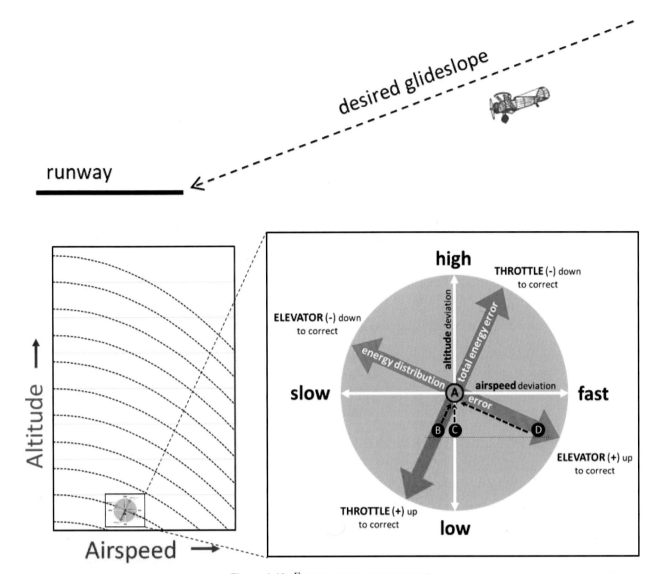

Figure 4-13. *Energy error management.*

The above approach-to-landing scenario is just one example illustrating the risk of mismanaging altitude-speed deviations. Pilots need to be able to identify, assess, and mitigate altitude and/or airspeed deviations during any phase of flight, including traffic pattern operations, take-offs and climbs, cruise flight, descending flight, and any procedure or maneuver involving turns.

Clearly, skills for promptly correcting path-speed deviations can enhance flight safety but the pilot should also be aware of the risk of unrecoverable depletion of the airplane's mechanical energy, especially as the airplane approaches the edges of its flight envelope where available excess power is zero.

Preventing Irreversible Deceleration and/or Sink Rate

During normal flight, the airplane experiences many instances of negative energy rates (negative specific excess power or $P_S < 0$) while decelerating at a constant altitude or descending at a constant airspeed; these are intended energy bleed rates. However, one of the greatest dangers from mismanaging the airplane's energy state is encountering unintended, excessive deceleration and/or sink rate coupled with little or no positive excess power available under a given flight condition. Failure to recover above a certain critical altitude results in depletion of mechanical energy. Regardless of what the pilot does past that point, the airplane will hit the ground.

To help pilots understand the risk of unintended energy depletion, let's take a closer look at Scenario 2 *[Figure 4-10]*. This flight scenario illustrates a situation that is all too common in general aviation: flying toward rising terrain and not being able to fly up and over it before impacting terrain.

As shown in *Figure 4-10*, there is rising terrain all along the departure corridor. The scenario is as follows:

1. A pilot of a normally-aspirated, twin-engine airplane departs out of Rocky Mountain Metropolitan airport (KBJC) in the morning on a nice summer day and flies into Aspen/Pitkin County airport (KASE).

2. The pilot enjoys the scenery around Aspen, eats lunch, and decides to return home in the early hot afternoon.

3. The pilot departs KASE off of runway 33. At full throttle/power the airplane takes longer to accelerate but rotates at the normal speed.

4. The pilot pitches to the normal pitch target, retracts the gear, and initiates a climb.

5. The pilot notices the airplane isn't performing as desired. The pilot checks to see if the gear is up and adjusts the mixtures to try and get a little more power.

6. The terrain is rising, the pilot gradually pitches up, and the airplane starts losing airspeed.

7. The airplane quits climbing.

8. The stall horn begins to sound.

The above scenario is hypothetical, but there have been very similar situations that have ended tragically.

The airplane in the above scenario has encountered an unintended deceleration and impending sink rate that could rapidly become irreversible. This can be shown in two ways, using the traditional power curve *[Figure 4-14]* and the energy map *[Figure 4-15]*:

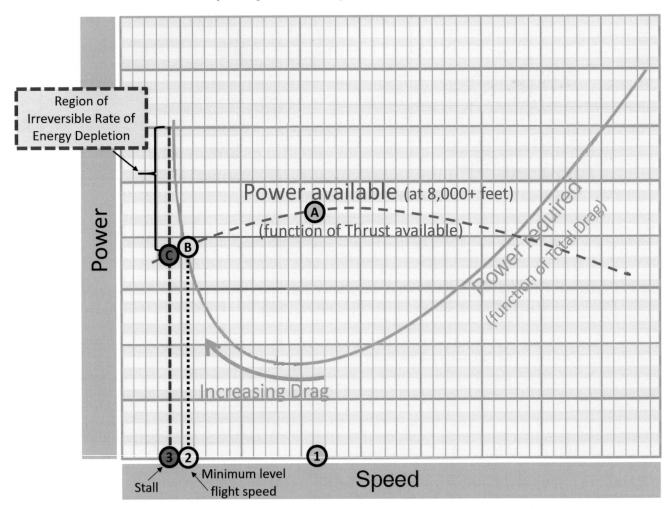

Figure 4-14. *The energy depletion scenario viewed in the power required and available curve. Compared with the power available curve depicted in Figure 4-4, note the lower power available curve at this high elevation (7,820 feet at the departure airport) and higher density altitude than standard during a hot afternoon.*

As illustrated in the airplane's power required and available curves *[Figure 4-14]*, the airplane slows down, going from speed 1 where it is climbing (A: power available greater than power required), to speed 2, where it stops climbing (B: power available equal to power required), and continuing to speed 3 where the stall horn sounds (C: power available less than power required). The energy map *[Figure 4-15]* tells the same story from a total mechanical energy standpoint: the airplane has positive P_S at point 1 and climbs to point 2 where it stops climbing since $P_S = 0$, then continues to point 3, where the $P_S < 0$ and the stall horn sounds.

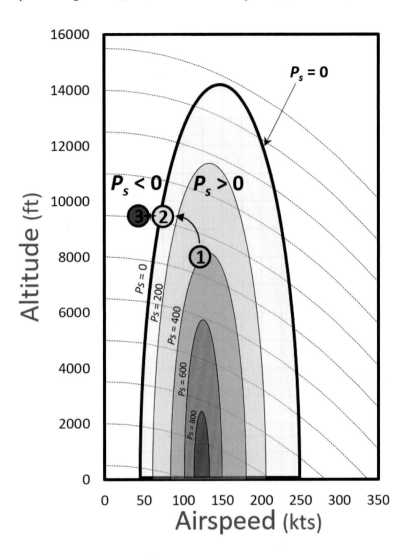

Figure 4-15. *The energy depletion scenario viewed in the energy map. Specific excess power (P_S) contours are labeled in units of feet per minute.*

The question then is: what does the pilot do to recover from this predicament? The answer is proper energy management. The airplane needs to move to a different place on the energy map that will allow the airplane to begin climbing. So, what does that mean?

- As can be seen in *Figure 4-12*, the pilot is in a scenario akin to that at the desired altitude, but with cautions when very slow.

- The pilot then has to do something that is not intuitive; consider gaining speed at the expense of some altitude initially to improve climbing performance with full throttle.

- Once the airplane accelerates to an airspeed in which the $P_S > 0$, it can begin to climb again.

The above recovery scenario is shown in the energy map *Figure 4-16*, which illustrates the important role of the elevator in assisting the pilot to recover from unintentional and dangerous deceleration and/or sink rate (refer to *Additional Role for the Elevator* section).

The airplane needs to gain speed at the expense of some altitude, moving from point 3 where the $P_S < 0$ to point 4 where the $P_S > 0$. The airplane can then initiate a constant airspeed climb to point 5, at the desired target altitude and airspeed *[Figure 4-16]*. Note that the desired target climb airspeed in the presence of rising terrain may be V_X, the speed for best angle of climb. V_X is slightly slower than V_Y, the speed for best rate of climb, and will result in a lower climb rate but steepest climb angle. Once the airplane has recovered from the unintentional airspeed loss and begins climbing at V_X, the pilot should assess the situation and make an important decision to mitigate further risk—either continue climbing or do something else. Should the airplane not have the needed performance to safely clear the rising terrain on its intended course, the pilot has at least another available option: make a 180 degree turn and return to land at the departure airport until temperature and density altitude conditions improve.

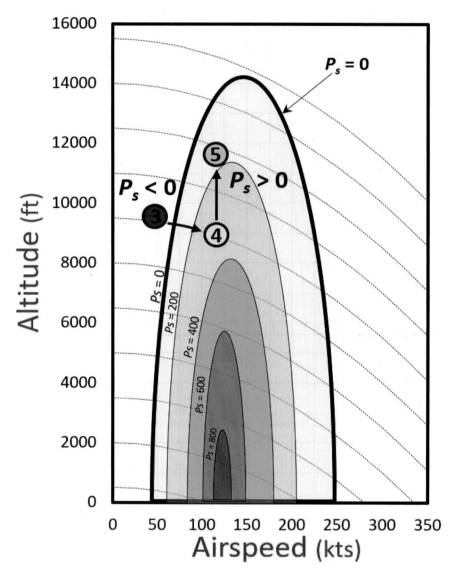

Figure 4-16. *The energy loss scenario recovery viewed in the energy map. Specific excess power (P_S) contours are labeled in units of feet per minute.*

The above rising terrain scenario is just one example illustrating the risk of irreversible deceleration and/or sink rate. Pilots need to be aware that unintentional depletion of mechanical energy can happen in various instances, especially as the airplane approaches the slow edge of its energy envelope at low altitude, where available specific excess power (P_S) is zero. Examples include unstable/slow approaches to landing; high-drag go-arounds where the pilot neglects to raise the gear and/or flaps; and steeper-than-normal turns in the traffic pattern. Note that irreversible sink rates do not necessarily involve exceeding the critical AOA resulting in a stall and spin. The airplane can be unstalled and still experience unrecoverable sink rates near the high-speed edge of its energy envelope, where available specific excess power (P_S) is also zero. Two examples are high-speed steep spirals following botched steep level turns, and high-speed dives too close to the ground.

The bottom line? Should the airplane ever experience unintended excessive negative energy rates with little or no excess power available under a given flight condition, the pilot needs to use proper energy management allowing a prompt recovery and a suitable follow-up action.

Review of Terms and Definitions

The terms and definitions specific to this chapter appear below.

Aircraft Energy Management

The process of planning, monitoring and controlling altitude and airspeed targets in relation to the airplane's energy state. Note that this definition is concerned with managing mechanical energy (altitude and airspeed) and addresses the safety (flight control) side of energy management. It does not address the efficiency (aircraft performance) side of energy management, which is concerned with how efficiently the engine generates mechanical energy from fuel and how efficiently the airframe spends that energy in flight.

Energy System

A flying airplane is an *open* energy system. That means that the airplane can gain energy from some source (e.g., fuel) and lose energy to the environment (e.g., surrounding air). In addition, energy can be added to or removed from the airplane's total mechanical energy stored as altitude and airspeed.

Total Mechanical Energy

Sum of the energy in altitude (potential energy) and the energy in airspeed (kinetic energy).

Kinetic Energy

Amount of energy due to the airspeed, expressed as $\frac{1}{2}mV^2$, where m = airplane's mass, and V = airspeed.

Potential Energy

Amount of energy due to the altitude, expressed as mgh, where m = airplane's mass, g = gravitational constant, and h = altitude.

Energy State

The airplane's total mechanical energy and its distribution between altitude and airspeed.

Energy Exchange

Trading one form of energy (e.g., altitude) for another form (e.g., airspeed).

Energy Balance Equation

According to this equation, the net transfer of mechanical energy into and out of the airplane (a function of thrust minus drag) is always equal to the change in its total mechanical energy (a function of altitude and airspeed). Note that this simplified definition does not account for long-term changes in total mechanical energy caused by the reduction in aircraft weight as fuel is gradually burned in flight.

Power Available

The airplane's rate of energy gain due to maximum available engine thrust at a given airspeed. Expressed as TV, where T = engine thrust and V = airspeed. Usually measured in horsepower, foot-pound per minute, or foot-pound per second.

Power Required

The airplane's rate of energy loss due to total drag at a given airspeed. Expressed as DV, where D = total drag and V = airspeed. Usually measured in horsepower, foot-pound per minute, or foot-pound per second.

Specific Excess Power (P_S)

Measured in feet per minute or feet per second, it represents rate of energy change—the ability of an airplane to climb or accelerate from a given flight condition. Available specific excess power is found by dividing the difference between power available and power required by the airplane's weight.

Energy Height or Total Specific Energy (E_S)

Measured in units of height (e.g., feet), it represents the airplane's total energy per unit weight. It is found by dividing the sum of potential energy and kinetic energy by the airplane's weight. It also represents the maximum height that an airplane would reach from its current altitude, if it were to trade all its speed for altitude.

Energy Error

An altitude and/or airspeed deviation from an intended target expressed in terms of energy. Depending on the airplane's total amount of energy and its distribution between altitude and airspeed, energy errors are classified as total energy errors, energy distribution errors, or a combination of both errors.

Total Energy Error

An energy error where the total amount of mechanical energy is not correct. The airplane has too much or too little total energy relative to the intended altitude-speed profile. When this error occurs, the pilot will observe that altitude and airspeed deviate in the *same* direction (e.g., higher and faster than desired; or lower and slower than desired). An example would be an airplane on final approach that is above the desired glide slope and at a faster airspeed than desired.

Energy Distribution Error

An energy error where the total mechanical energy is correct, but the distribution between potential (altitude) and kinetic energy (airspeed) is not correct relative to the intended altitude-speed profile. When this error occurs, the pilot will observe that altitude and airspeed deviate in *opposite* directions (e.g., higher and slower than desired; or lower and faster than desired). An example would be an airplane on final approach that is above the desired glide slope and at a slower airspeed than desired.

Irreversible Deceleration and/or Sink Rate

Unrecoverable depletion of mechanical energy as a result of continuous loss of airspeed and/or altitude coupled with insufficient excess power available under a given flight condition. Failure to recover above a certain critical AGL altitude results in the airplane hitting the ground regardless of what the pilot does.

Chapter Summary

Every pilot is an energy manager—managing energy in the form of altitude and airspeed from takeoff to landing. Proper energy management is essential for performing any maneuver as well as for attaining and maintaining desired vertical flightpath and airspeed profiles in everyday flying. It is also critical to flight safety since mistakes in managing energy state can contribute to loss of control inflight (LOC-I), controlled flight into terrain (CFIT), and approach and landing accidents. The objectives of this chapter are for pilots to: 1) gain an understanding of basic energy management concepts; 2) learn the energy role of the controls for managing the airplane's energy state; and 3) develop the ability to identify, assess, and mitigate risks associated with failure to manage the airplane's energy state.

Introduction

Safe pilots prevent loss of control in flight (LOC-I), which is the leading cause of fatal general aviation accidents in the U.S. and commercial aviation worldwide. LOC-I includes any significant deviation of an aircraft from the intended flightpath and it often results from an airplane upset. Maneuvering represents the most common phase of flight for general aviation LOC-I accidents; however, LOC-I accidents occur in all phases of flight.

To prevent LOC-I accidents, it is important for pilots to recognize and maintain a heightened awareness of situations that increase the risk of loss of control. Those situations include: uncoordinated flight, equipment malfunctions, pilot complacency, distraction, turbulence, and poor risk management. Attempting to fly in instrument meteorological conditions (IMC) when the pilot is not qualified or proficient is a common example of poor risk management. The *Emergency Procedures* chapter of this handbook contains specific information regarding unintended flight into IMC. Sadly, there are also LOC-I accidents resulting from intentional disregard for safety.

To maintain aircraft control when faced with these or other contributing factors, the pilot needs to be aware of situations where LOC-I can occur; recognize when an airplane is approaching a stall, has stalled, or is in an upset condition; and understand and execute the correct procedures to recover the aircraft.

Defining an Airplane Upset

The term "upset" was formally introduced by an industry work group in 2004 in the "Pilot Guide to Airplane Upset Recovery," which is a part of the "Airplane Upset Recovery Training Aid." The work group was primarily focused on large transport airplanes and sought to come up with one term to describe an "unusual attitude" or "loss of control," for example, and to generally describe specific parameters as part of its definition. Consistent with the Guide, the FAA considers an upset to be an event that unintentionally exceeds the parameters normally experienced in flight or training. These parameters are:

1. Pitch attitude greater than 25°, nose up
2. Pitch attitude greater than 10°, nose down
3. Bank angle greater than 45°
4. Within the above parameters, but flying at airspeeds inappropriate for the conditions

The reference to inappropriate airspeeds describes a number of undesired aircraft states, including stalls. However, stalls are directly related to angle of attack (AOA), not airspeed.

To develop the crucial skills to prevent LOC-I, a pilot may receive academic or on-aircraft upset prevention and recovery training (UPRT), which should include: slow flight, stalls, spins, and unusual attitudes.

Upset training places considerable emphasis on understanding and preventing an upset, so a pilot avoids such a situation. If an upset does occur, upset training also reinforces proper recovery techniques. A detailed discussion of UPRT follows, including core concepts, what the training should include, and what airplanes or kinds of simulation can be used for the training. A discussion of various maneuvers and how to execute them follows later in this chapter.

Upset Prevention and Recovery

An unusual attitude is commonly referenced as an unintended or unexpected attitude in instrument flight. These unusual attitudes are introduced to a pilot during student pilot training as part of basic attitude instrument flying and continue to be trained and tested as part of certification for an instrument rating, aircraft type rating, and an airline transport pilot certificate. A pilot is taught the conditions or situations that could cause an unusual attitude, with focus on how to recognize one, and how to recover from one.

Unusual Attitudes Versus Upsets

Given the upset definition, there are a few key distinctions between an unusual attitude and an upset. An upset:

- Includes stall events.

- Includes overspeeds or other inappropriate speeds for a given flight condition.

- Has defined parameters. For example, for training purposes an instructor could place the aircraft in a 30° bank with a nose-up pitch attitude of 15° and ask the student to recover and that would be considered an unusual attitude, but would not meet the upset parameters.

- Centers on unintentional situations that may lead to a startle effect. For example, during unusual attitude training, the pilot is often directed to close their eyes, and any element of surprise disappears.

The top four causal and contributing factors that have led to an upset and resulted in LOC-I accidents are:

1. Environmental factors
2. Mechanical factors
3. Human factors
4. Stall-related factors

Environmental Factors

Turbulence, or a large variation in wind velocity over a short distance, can cause upset and LOC-I. Maintain awareness of conditions that can lead to various types of turbulence, such as clear air turbulence, mountain waves, wind shear, and thunderstorms or microbursts. In addition to environmentally-induced turbulence, wake turbulence from other aircraft can lead to upset and LOC-I.

Icing can destroy the smooth flow of air over the airfoil and increase drag while decreasing the ability of the airfoil to create lift. Therefore, it can significantly degrade airplane performance, resulting in a stall if not handled correctly.

Mechanical Factors

Modern airplanes and equipment are very reliable, but anomalies do occur. Some of these mechanical failures can directly cause a departure from normal flight, such as asymmetrical flaps, malfunctioning or binding flight controls, and runaway trim.

Upsets can also occur if there is a malfunction or misuse of the autoflight system. Advanced automation may tend to mask the cause of the anomaly. Disengaging the autopilot and the autothrottles allows the pilot to directly control the airplane and possibly eliminate the cause of the problem. For these reasons the pilot should maintain proficiency to manually fly the airplane in all flight conditions without the use of the autopilot/autothrottles.

Although these and other in-flight anomalies may not be preventable, knowledge of systems and AFM/POH recommended procedures helps the pilot minimize their impact and prevent an upset. In the case of instrument failures, avoiding an upset and subsequent LOC-I may depend on the pilot's proficiency in the use of secondary instrumentation and partial panel operations.

Human Factors

VMC to IMC

Unfortunately, accident reports indicate that continued VFR flight from visual meteorological conditions (VMC) into marginal VMC and IMC is a factor contributing to LOC-I. A loss of the natural horizon substantially increases the chances of encountering vertigo or spatial disorientation, which can lead to upset.

IMC

When operating in IMC, maintain awareness of conditions.

Diversion of Attention

In addition to its direct impact, an in-flight anomaly or malfunction can also lead to an upset if it diverts the pilot's attention from basic airplane control responsibilities. Failing to monitor the automated systems, over-reliance on those systems, or incomplete knowledge and experience with those systems can lead to an upset. Diversion of attention can also occur simply from the pilot's efforts to set avionics or navigation equipment while flying the airplane.

Task Saturation

The margin of safety is the difference between task requirements and pilot capabilities. An upset and eventual LOC-I can occur whenever requirements exceed capabilities. For example, an airplane upset event that requires rolling an airplane from a near-inverted to an upright orientation may demand piloting skills beyond those learned during primary training. In another example, a fatigued pilot who inadvertently encounters IMC at night coupled with a vacuum pump failure, or a pilot fails to engage pitot heat while flying in IMC, could become disoriented and lose control of the airplane due to the demands of extended—and unpracticed—partial panel flight. Additionally, unnecessary low-altitude flying and impromptu demonstrations for friends or others on the ground could lead pilots to exceed their capabilities, with fatal results.

Sensory Overload/Deprivation

A pilot's ability to adequately correlate warnings, annunciations, instrument indications, and other cues from the airplane during an upset can be limited. Pilots faced with upset situations can be rapidly confronted with multiple or simultaneous visual, auditory, and tactile warnings. Conversely, sometimes expected warnings are not provided when they should be; this situation can distract a pilot as much as multiple warnings can.

The ability to separate time-critical information from distractions takes practice, experience, and knowledge of the airplane and its systems. Cross-checks are necessary not only to corroborate other information that has been presented, but also to determine if information might be missing or invalid. For example, a stall warning system may fail and therefore not warn a pilot of close proximity to a stall, so other cues need to be used to avert a stall and possible LOC-I. These cues include aerodynamic buffet, loss of roll authority, or inability to arrest a descent.

Spatial Disorientation

Spatial disorientation has been a significant factor in many airplane upset accidents. Accident data from 2008 to 2013 shows nearly 200 accidents associated with spatial disorientation with more than 70% of those being fatal. All pilots are susceptible to false sensory illusions while flying at night or in certain weather conditions. These illusions can lead to a conflict between actual attitude indications and what the pilot senses is the correct attitude. Disoriented pilots may not always be aware of their orientation error. Many airplane upsets occur while the pilot is engaged in some task that takes attention away from the flight instruments or outside references. Others perceive a conflict between bodily senses and the flight instruments, and allow the airplane to divert from the desired flightpath because they cannot resolve the conflict.

A pilot may experience spatial disorientation or perceive the situation in one of three ways:

1. Recognized spatial disorientation: the pilot recognizes the developing upset or the upset condition and is able to safely correct the situation.

2. Unrecognized spatial disorientation: the pilot is unaware that an upset event is developing, or has occurred, and fails to make essential decisions or take any corrective action to prevent LOC-I.

3. Incapacitating spatial disorientation: the pilot is unable to affect a recovery due to some combination of: (a) not understanding the events as they are unfolding, (b) lacking the skills required to alleviate or correct the situation, or (c) exceeding psychological or physiological ability to cope with what is happening.

For detailed information regarding causal factors of spatial disorientation, refer to Aerospace Medicine Spatial Disorientation and Aerospace Medicine Reference Collection, which provides spatial disorientation videos. The videos are available online at: www.faa.gov/about/office_org/headquarters_offices/avs/offices/aam/cami/library/online_libraries/aerospace_medicine/sd/videos/.

Surprise and Startle Response

Surprise is an unexpected event that violates a pilot's expectations and can affect the mental processes used to respond to the event. Startle is an uncontrollable, automatic muscle reflex, raised heart rate, blood pressure, etc., elicited by exposure to a sudden, intense event that violates a pilot's expectations.

This human response to unexpected events has traditionally been underestimated or even ignored during flight training. The reality is that untrained pilots often experience a state of surprise or a startle response to an airplane upset event. Startle may or may not lead to surprise. Pilots can protect themselves against a debilitating surprise reaction or startle response through scenario-based training, and in such training, instructors can incorporate realistic distractions to help provoke startle or surprise. To be effective the controlled training scenarios should have a perception of risk or threat of consequences sufficient to elevate the pilot's stress levels. Such scenarios can help prepare a pilot to mitigate psychological/physiological reactions to an actual upset.

Upset Prevention and Recovery Training (UPRT)

Upsets are not intentional flight maneuvers, except in maneuver-based training; therefore, they are often unexpected. The reaction of an inexperienced or inadequately trained pilot to an unexpected abnormal flight attitude is usually instinctive rather than intelligent and deliberate. Such a pilot often reacts with abrupt muscular effort, which is without purpose and even hazardous in turbulent conditions, at excessive speeds, or at low altitudes.

Without proper upset recovery training on interpretation and airplane control, the pilot can quickly aggravate an abnormal flight attitude into a potentially fatal LOC-I accident. Consequently, UPRT is intended to focus education and training on the prevention of upsets, and on recovering from these events if they occur. *[Figure 5-1]*

Figure 5-1. *Maneuvers that better prepare a pilot for understanding unusual attitudes and situations are representative of upset training.*

- Upset prevention refers to pilot actions to avoid a divergence from the desired airplane state. Awareness and prevention training serve to avoid incidents. Early recognition of an upset scenario coupled with appropriate preventive action often can mitigate a situation that could otherwise escalate into an LOC-I accident.

- Recovery refers to pilot actions that return an airplane that is diverging in altitude, airspeed, or attitude to a desired state from a developing or fully-developed upset. Recovery training serves to reduce accidents as a result of an unavoidable or inadvertently-encountered upset event. The pilot can learn to initiate a recovery to a normal flight mode immediately upon recognition of the developing upset condition. The pilot should ensure that control inputs and power adjustments applied to counter an upset are in direct proportion to the amount and rates of change of roll, yaw, and pitch, or airspeed so as to avoid overstressing the airplane unless ground contact is imminent.

UPRT Training Core Concepts

Airplane upsets are by nature time-critical events; they can also place pilots in unusual and unfamiliar attitudes that sometimes require counterintuitive control movements. Upsets have the potential to put a pilot into a life-threatening situation compounded by panic, diminished mental capacity, and potentially incapacitating spatial disorientation. Real-world upset situations often provide very little time to react, but exposure to such events during training can reduce surprise and mitigate confusion during an actual unexpected upset. The goal is to equip the pilot to promptly recognize an escalating threat pattern or sensory overload and quickly identify and correct an impending upset.

UPRT stresses that the first step is recognizing any time the airplane begins to diverge from the intended flightpath or airspeed. Pilots need to identify and determine what, if any, action should be taken. As a general rule, any time visual cues or instrument indications differ from basic flight maneuver expectations, the pilot should assume an upset and cross-check to confirm the attitude, instrument error or instrument malfunction.

To achieve maximum effect, it is crucial for UPRT concepts to be conveyed accurately and in a non-threatening manner. Reinforcing concepts through positive experiences significantly improves a pilot's depth of understanding, retention of skills, and desire for continued training. Also, training in a carefully structured environment allows for exposure to these events and can help the pilot react more quickly, decisively, and calmly when the unexpected occurs during flight. However, like many other skills, the skills needed for upset prevention and recovery are perishable and thus require continuous reinforcement through training.

UPRT in the airplane and flight simulation training device (FSTD) should be conducted in both visual and simulated instrument conditions to allow pilots to practice recognition and recovery under both situations. UPRT should allow them to experience and recognize some of the physiological factors related to each, such as the confusion and disorientation that can result from visual cues in an upset event. Training that includes recovery from bank angles exceeding 90 degrees could further add to a pilot's overall knowledge and skills for upset recognition and recovery. For such training, additional measures should be taken to ensure the suitability of the airplane or FSTD and that instructors are appropriately qualified.

Upset prevention and recovery training is different from aerobatic training. *[Figure 5-2]* In aerobatic training, the pilot knows and expects the maneuver, so effects of startle or surprise are missing. The main goal of aerobatic training is to teach pilots how to intentionally and precisely maneuver an aerobatic-capable airplane in three dimensions. The primary goal of UPRT is to help pilots overcome sudden onsets of stress to avoid, prevent, and recover from unplanned excursions that could lead to LOC-I.

Aerobatics vs. UPRT Flight Training Methods		
ASPECT OF TRAINING	**AEROBATICS**	**UPSET PREVENTION AND RECOVERY TRAINING**
Primary Objective	Precision maneuvering capability	Safe, effective recovery from aircraft upsets
Secondary Outcome	Improved manual aircraft handling skills	Improved manual aircraft handling skills
Aerobatic Maneuvering	Primary mode of training	Supporting mode of training
Academics	Supporting role	Fundamental component
Training Resources Utilized	Aircraft (few exceptions)	Aircraft or a full-flight simulator

Figure 5-2. *Some differences between aerobatic training and upset prevention and recovery training.*

Comprehensive UPRT builds on three mutually supportive components: academics, airplane-based training and, typically at the transport category type-rating training level, use of FSTDs. Each has unique benefits and limitations but, when implemented cohesively and comprehensively throughout a pilot's career, the components can offer maximum preparation for upset awareness, prevention, recognition, and recovery.

Academic Material (Knowledge and Risk Management)

Academics establish the foundation for development of situational awareness, insight, knowledge, and skills. As in practical skill development, academic preparation should move from the general to specific while emphasizing the significance of each basic concept. Although academic preparation is crucial and does offer a level of mitigation of the LOC-I threat, long-term retention of knowledge is best achieved when applied and correlated with practical hands-on experience.

The academic portion of UPRT should also address the prevention concepts surrounding aeronautical-decision making (ADM) and risk management (RM), and proportional counter response.

Prevention Through ADM and Risk Management

This element of prevention routinely occurs in a time scale of minutes or hours, revolving around the concept of effective ADM and risk management through analysis, awareness, resource management, and interrupting the error chain through basic airmanship skills and sound judgment. For instance, imagine a situation in which a pilot assesses conditions at an airport prior to descent and recognizes those conditions as being too severe to safely land the airplane. Using situational awareness to avert a potentially threatening flight condition is an example of prevention of an LOC-I situation through effective risk management. Pilots should evaluate the circumstances for each flight (including the equipment and environment), looking specifically for scenarios that may require a higher level of risk management. These include situations that could result in low-altitude maneuvering, steep turns in the pattern, uncoordinated flight, or increased load factors.

Another part of ADM is crew resource management (CRM) or single-pilot resource management (SRM). Both are relevant to the UPRT environment. When available, a coordinated crew response to potential and developing upsets can provide added benefits such as increased situational awareness, mutual support, and an improved margin of safety. Since an untrained crewmember can be the most unpredictable element in an upset scenario, initial UPRT for crew operations should be mastered individually before being integrated into a multi-crew, CRM environment. A crew should be able to accomplish the following:

1. Communicate and confirm the situation clearly and concisely;
2. Transfer control to the most situationally-aware crewmember;
3. Using standardized interactions, work as a team to enhance awareness, manage stress, and mitigate fear.

Prevention Through Proportional Counter-Response

In simple terms, proportional counter-response is the timely manipulation of flight controls and thrust, either as the sole pilot or crew as the situation dictates, to manage an airplane flight attitude or flight envelope excursion that was unintended or not commanded by the pilot.

The time-scale of this element of prevention typically occurs on the order of seconds or fractions of seconds, with the goal being the ability to recognize a developing upset and take proportionally-appropriate avoidance actions to preclude the airplane entering a fully-developed upset. Due to the sudden, surprising nature of this level of developing upset, there exists a high risk for panic and overreaction to ensue and aggravate the situation.

Recovery

Last but not least, the academics portion lays the foundation for development of UPRT skills by instilling the knowledge, procedures, and techniques required to accomplish a safe recovery. The airplane and FSTD-based training elements presented below serve to translate the academic material into structured practice. This can start with classroom visualization of recovery procedures and continue with repetitive skill practiced in an airplane, and then potentially further developed in the simulated environment.

In the event looking outside does not provide enough situational awareness of the airplane attitude, a pilot can use the flight instruments to recognize and recover from an upset. To recover from nose-high and nose-low attitudes, the pilot should follow the procedures recommended in the AFM/POH. In general, upset recovery procedures are summarized in *Figure 5-3*.

Upset Recovery Template

1. Disconnect the wing leveler or autopilot

2. Apply forward column or stick pressure to unload the airplane

3. Aggressively roll the wings to the nearest horizon

4. Adjust power as necessary by monitoring airspeed

5. Return to level flight

Figure 5-3. *Upset recovery template.*

Common Errors

Common errors associated with upset recoveries include the following:

1. Incorrect assessment of what kind of upset the airplane is in
2. Failure to disconnect the wing leveler or autopilot
3. Failure to unload the airplane, if necessary
4. Failure to roll in the correct direction
5. Inappropriate management of the airspeed during the recovery

Roles of FSTDs and Airplanes in UPRT

Training devices range from aviation training devices (e.g., basic and advanced) to FSTDs (e.g., flight training devices (FTD) and full flight simulators (FFS)) and have a broad range of capabilities. While all of these devices have limitations relative to actual flight, only the higher fidelity devices (i.e., Level C and D FFS) are a satisfactory substitution for developing UPRT skills in the actual aircraft. Except for these higher fidelity devices, initial skill development should be accomplished in a suitable airplane, and the accompanying training device should be used to build upon these skills. *[Figure 5-4]*

Figure 5-4. *A Level D full-flight simulator could be used for UPRT.*

Airplane-Based UPRT

Ultimately, the more realistic the training scenario, the more indelible the learning experience. Although creating a visual scene of a 110° banked attitude with the nose 30° below the horizon may not be technically difficult in a modern simulator, the learning achieved while viewing that scene from the security of the simulator is not as complete as when viewing the same scene in an airplane. Maximum learning is achieved when the pilot is placed in the controlled, yet adrenaline-enhanced, environment of upsets experienced while in flight. For these reasons, airplane-based UPRT improves a pilot's ability to overcome fear in an airplane upset event.

However, airplane-based UPRT does have limitations. The level of upset training possible may be limited by the maneuvers approved for the particular airplane, as well as by the flight instructor's own UPRT capabilities. For instance, UPRT conducted in the normal category by a typical flight instructor will necessarily be different from UPRT conducted in the aerobatic category by a flight instructor with expertise in aerobatics.

When considering upset training conducted in an aerobatic-capable airplane in particular, the importance of employing instructors with specialized UPRT experience in those airplanes cannot be overemphasized. Just as instrument or tailwheel instruction requires specific skill sets for those operations, UPRT demands that instructors possess the competence to oversee trainee progress, and the ability to intervene as necessary with consistency and professionalism. As in any area of training, the improper delivery of stall, spin, and upset recovery training often results in negative learning, which could have severe consequences not only during the training itself, but in the skills and mindset pilots take with them when they have passengers and place the lives of others at stake.

All-Attitude/All-Envelope Flight Training Methods

Sound UPRT encompasses operation in a wide range of possible flight attitudes and covers the airplane's limit flight envelope. This training is essential to prepare pilots for unexpected upsets. As stated at the outset, the primary focus of a comprehensive UPRT program is the avoidance of, and safe recovery from, upsets. Much like basic instrument skills, which can be applied to flying a vast array of airplanes, the majority of skills and techniques required for upset recovery are not airplane-specific. Just as basic instrument skills learned in lighter and lower performing airplanes are applied to more advanced airplanes, basic upset recovery techniques provide lessons that remain with pilots throughout their flying careers.

FSTD–based UPRT

UPRT can be effective in high fidelity devices (i.e., Level C and D FFS); however, instructors and pilots should be mindful of the technical and physiological boundaries when using a particular FSTD for upset training. This training is a current requirement for pilots seeking a multiengine airplane ATP certificate in accordance with 14 CFR part 61, section 61.156, and the training course must be FAA approved.

Coordinated Flight

Coordinated flight occurs whenever the pilot is proactively correcting for yaw effects associated with power (engine/propeller effects), aileron inputs, how an airplane reacts when turning, and airplane rigging. The airplane is in coordinated flight when the airplane's nose is yawed directly into the relative wind and the ball is centered in the slip/skid indicator (except for certain multiengine airplane operation with an engine failure). [Figure 5-5]

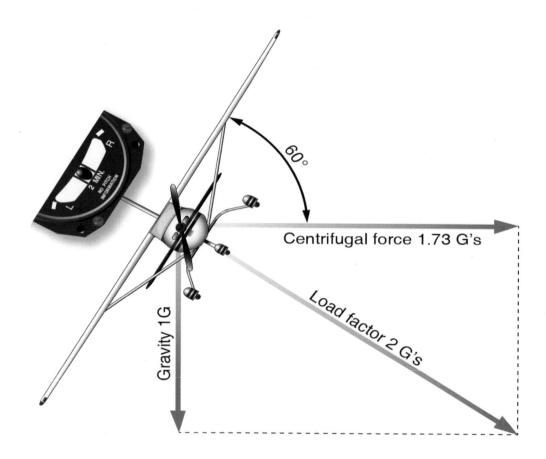

Figure 5-5. *Coordinated flight in a turn.*

Angle of Attack

The angle of attack (AOA) is the angle at which the chord of the wing meets the relative wind. The chord is a straight line from the leading edge to the trailing edge. At low angles of attack, the airflow over the top of the wing flows smoothly and produces lift with a relatively small amount of drag. As the AOA increases, lift as well as drag increases; however, above a wing's critical AOA, the flow of air separates from the upper surface and backfills, burbles, and eddies, which reduces lift and increases drag. This condition is a stall, which can lead to loss of control if the AOA is not reduced.

It is important for the pilot to understand that a stall is the result of exceeding the critical AOA, not of insufficient airspeed. The term "stalling speed" can be misleading, as this speed is often discussed when assuming 1G flight at a particular weight and configuration. Increased load factor directly affects stall speed (as well as do other factors such as gross weight, center of gravity, and flap setting). Therefore, it is possible to stall the wing at any airspeed, at any flight attitude, and at any power setting. For example, if a pilot maintains airspeed and rolls into a coordinated, level 60° banked turn, the load factor is 2G, and the airplane will stall at a speed that is 41 percent higher than the 1G stall speed. In that 2G level turn, the pilot has to increase AOA to increase the lift required to maintain altitude. At this condition, the pilot is closer to the critical AOA than during level flight and therefore closer to the higher stalling speed. Because "stalling speed" is not a constant number, pilots need to understand the underlying factors that affect it in order to maintain aircraft control in all circumstances.

Slow Flight

Flying at reduced airspeeds is normal in the takeoff/departure and approach/landing phases of flight. While pilots typically perform these operations at low airspeeds and close to the ground, pilots learn to maneuver an airplane in slow flight at a safe altitude. During slow flight, any further increase in angle of attack, increase in load factor, or reduction in power, will result in a stall warning (e.g., aircraft buffet, stall horn, etc.), and pilots should react to and correct for any stall indication. Note that stall training builds upon the knowledge and skill acquired from the slow flight maneuver and encompasses the period of time from the stall warning (e.g., aircraft buffet, stall horn, etc.) to the stall.

The objective of maneuvering in slow flight is to develop the pilot's ability to fly at low speeds and high AOAs. Through practice, the pilot becomes familiar with the feel, sound, and visual cues of flight in this regime, where there is a degraded response to control inputs and where it is more difficult to maintain a selected altitude. It is essential that pilots:

1. understand the aerodynamics associated with slow flight in various aircraft configurations and attitudes,
2. recognize airplane cues in these flight conditions,
3. smoothly manage coordinated flight control inputs while maneuvering without a stall warning, and
4. make prompt appropriate correction should a stall warning occur.

For pilot training and testing purposes, slow flight includes two main elements:

- Slowing to, maneuvering at, and recovering from an airspeed at which the airplane is still capable of maintaining controlled flight without activating the stall warning—5 to 10 knots above the 1G stall speed is a good target.

- Performing slow flight in configurations appropriate to takeoffs, climbs, descents, approaches to landing, and go-arounds.

Slow flight should be introduced with the target airspeed sufficiently above the stall to permit safe maneuvering, but close enough to the stall warning for the pilot to experience the characteristics of flight at a low airspeed. One way to determine the target airspeed is to slow the aircraft to the stall warning when in the desired slow flight configuration, pitch the nose down slightly to eliminate the stall warning, and add power to maintain altitude and note the airspeed.

When practicing slow flight, a pilot learns to divide attention between aircraft control and other demands. How the airplane feels at the slower airspeeds demonstrates that as airspeed decreases, control effectiveness decreases. For instance, reducing airspeed from 30 knots to 20 knots above the stalling speed will result in a certain loss of effectiveness of flight control inputs because of less airflow over the control surfaces. As airspeed is further reduced, the control effectiveness is further reduced and the reduced airflow over the control surfaces results in larger control movements being required to create the same response. Pilots sometimes refer to the feel of this reduced effectiveness as "sloppy" or "mushy" controls.

When flying above the minimum drag speed (L/D$_{MAX}$), more power is required to fly even faster. When flying at speeds below L/D$_{MAX}$, more power is required to fly even slower. Since slow flight will be performed well below L/D$_{MAX}$, the pilot should be aware that large power inputs or a reduction in AOA will be required to prevent the aircraft from decelerating. It is important to note that when flying below L/D$_{MAX}$ or *on the backside of the power curve*, as the AOA increases toward the critical AOA and the airplane's speed continues to decrease, small changes in the pitch control result in disproportionally large changes in induced drag and therefore changes in airspeed. As a result, pitch becomes a more effective control of airspeed when flying below L/D$_{MAX}$ and power is an effective control of the path.

It is also important to note that an airplane flying below L/D$_{MAX}$, exhibits a characteristic known as "speed instability" and the airspeed will continue to decay without appropriate pilot action. For example, if the airplane is disturbed by turbulence and the airspeed decreases, the airspeed may continue to decrease without the appropriate pilot action of reducing the AOA or adding power. [*Figure 5-6*]

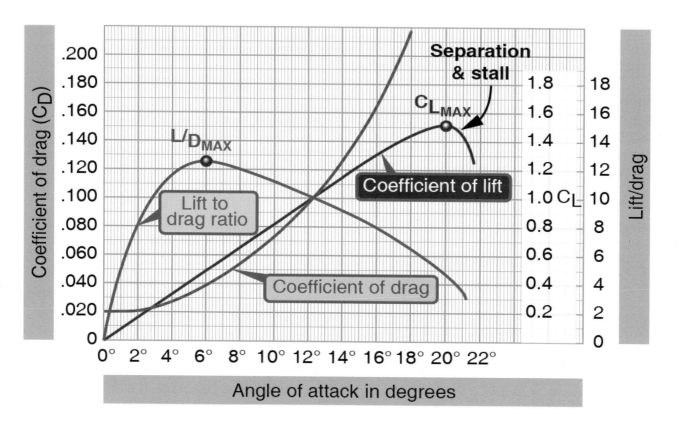

Figure 5-6. *Angle-of-attack in degrees.*

Performing the Slow Flight Maneuver

Slow flight training includes:

- Slowing the airplane smoothly and promptly from cruising to approach speeds without changes in altitude or heading, while increasing the angle of attack and setting the required power and trim.

- Configuration changes, such as extending the landing gear and adding flaps, while maintaining heading and altitude.

- Turning while maintaining altitude.

- Straight-ahead climbs and climbing medium-banked (approximately 20 degrees) turns, and straight-ahead power-off gliding descents and descending turns, which represent the takeoff and landing phases of flight.

Slow flight in a single-engine airplane should be conducted so the maneuver can be completed no lower than 1,500 feet AGL (3,000 for multiengine airplanes), or higher, if recommended by the manufacturer. In all cases, practicing slow flight should be conducted at an adequate height above the ground for recovery should the airplane inadvertently stall.

To begin the slow flight maneuver, the pilot should clear the area and gradually reduce thrust from cruise power and adjust the pitch to allow the airspeed to decrease while maintaining altitude. As the speed of the airplane decreases, there is a change in the sound of the airflow. As the speed approaches the target slow flight speed, which is an airspeed just above the stall warning in the desired configuration (i.e., approximately 5–10 knots above the stall speed for that flight condition), additional power will be needed to maintain altitude. During these changing flight conditions, the pilot should trim the airplane to compensate for changes in control pressures. If the airplane remains trimmed at the pre-maneuver cruising speed, strong aft (back) control pressure is needed on the elevator, which will make precise control difficult.

Slow flight is typically performed and evaluated in the landing configuration. Therefore, both the landing gear and the flaps should be extended to the landing position, as applicable. It is recommended the prescribed before-landing checks be completed to configure the airplane. The extension of gear and flaps typically occurs once cruise power has been reduced and at appropriate airspeeds to ensure limitations for extending those devices are not exceeded. Practicing this maneuver in other configurations, such as a clean or takeoff configuration, is also good training and may be evaluated on the practical test.

With an AOA just under the AOA which may cause an aerodynamic buffet or stall warning, the flight controls are less effective. [Figure 5-7] The elevator control is less responsive and larger control movements are necessary to retain control of the airplane. In propeller-driven airplanes, torque, slipstream effect, and P-factor may produce a strong left yaw, which requires right rudder input to maintain coordinated flight. The closer the airplane is to the 1G stall, the greater the amount of right rudder pressure required.

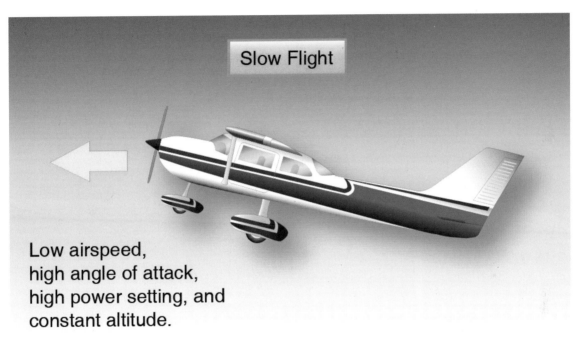

Slow Flight

Low airspeed, high angle of attack, high power setting, and constant altitude.

Figure 5-7. Slow flight—low airspeed, high angle of attack, high power, and constant altitude.

Maneuvering in Slow Flight

When the desired pitch attitude and airspeed have been established in straight-and-level slow flight, the pilot needs to maintain awareness of outside references and continually cross-check the airplane's instruments to maintain control. The pilot should note the feel of the flight controls, especially the airspeed changes caused by small pitch adjustments, and the altitude changes caused by power changes. The pilot should practice turns to determine the airplane's controllability characteristics at this low speed. During the turns, it will be necessary to increase power to maintain altitude. Abrupt or rough control movements during slow flight may result in a stall. For instance, abruptly raising the flaps while in slow flight can cause the plane to stall.

The pilot should also practice climbs and descents by adjusting the power when stabilized in straight-and-level slow flight. The pilot should note the increased yawing tendency at high power settings and counter it with rudder input as needed.

To exit the slow flight maneuver, add power. As airspeed and lift increase, apply forward control pressure to reduce the AOA and maintain altitude. Maintain coordinated flight, level the wings as necessary, and return to the desired flightpath. As airspeed increases, clean up the airplane by retracting flaps and landing gear, if they were extended, and adjust trim as needed. A pilot should anticipate the changes to the AOA as the landing gear and flaps are retracted to avoid a stall.

Common Errors

Common errors in the performance of slow flight are:

1. Failure to adequately clear the area
2. Inadequate back-elevator pressure as power is reduced, resulting in altitude loss
3. Excessive back-elevator pressure as power is reduced, resulting in a climb followed by rapid reduction in airspeed
4. Insufficient right rudder to compensate for left yaw
5. Fixation on the flight instruments
6. Failure to anticipate changes in AOA as flaps are extended or retracted
7. Inadequate power management
8. Inability to adequately divide attention between airplane control and orientation
9. Failure to properly trim the airplane
10. Failure to respond to a stall warning

Stalls

A stall is an aerodynamic condition which occurs when smooth airflow over the airplane's wings is disrupted, resulting in loss of lift. Specifically, a stall occurs when the AOA—the angle between the chord line of the wing and the relative wind—exceeds the wing's critical AOA. It is possible to exceed the critical AOA at any airspeed, at any attitude, and at any power setting. *[Figure 5-8]*

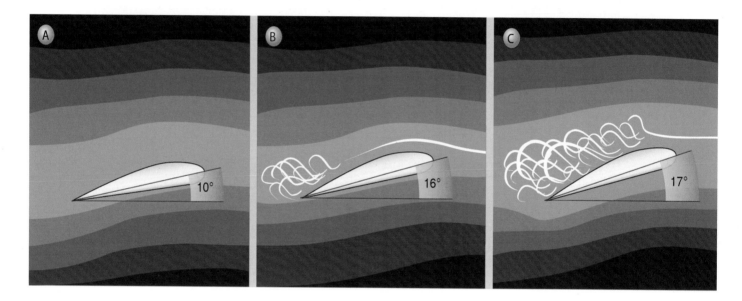

Figure 5-8. *Critical angle of attack and stall.*

For these reasons, it is important to understand factors and situations that can lead to a stall, and develop proficiency in stall recognition and recovery. Performing intentional stalls will familiarize the pilot with the conditions that result in a stall, assist in recognition of an impending stall, and develop the proper corrective response if a stall occurs. Stalls are practiced to two different levels:

• Impending Stall—an impending stall occurs when the AOA causes a stall warning, but has not yet reached the critical AOA. Indications of an impending stall can include buffeting, stick shaker, or aural warning.

• Full Stall—a full stall occurs when the critical AOA is exceeded. Indications of a full stall are typically that an uncommanded nose down pitch cannot be readily arrested, and may be accompanied by an uncommanded rolling motion. For airplanes equipped with stick pushers, their activation is also an indicator of a full stall.

Although it depends on the degree to which a stall has progressed, some loss of altitude is expected during recovery. The longer it takes for the pilot to recognize an impending stall, the more likely it is that a full stall will result. Intentional stalls should therefore be performed at an altitude that provides adequate height above the ground for recovery and return to normal level flight.

Stall Recognition

A pilot should recognize the flight conditions that are conducive to stalls and know how to apply the necessary corrective action. This level of proficiency involves learning to recognize an impending stall by sight, sound, and feel.

Stalls are usually accompanied by a continuous stall warning for airplanes equipped with stall warning devices. These devices may include an aural alert, lights, or a stick shaker all which alert the pilot when approaching the critical AOA. Most vintage airplanes, and many types of light-sport and experimental airplanes, do not have stall warning devices installed. However, certification standards permit manufacturers to provide the required stall warning either through the inherent aerodynamic qualities of the airplane (pre-stall buffeting) or through a stall warning device that gives a clear indication of the impending stall.

Other sensory cues for the pilot include:

- Feel—the pilot will feel control pressures change as speed is reduced. With progressively less resistance on the control surfaces, the pilot needs to use larger control movements to get the desired airplane response. The pilot will notice the airplane's reaction time to control movement increases.

- Vision—since the airplane can be stalled in any attitude, vision is not a foolproof indicator of an impending stall. However, maintaining pitch awareness is important.

- Hearing—as speed decreases, the pilot should notice a change in sound made by the air flowing along the airplane structure.

- Kinesthesia—the physical sensation (sometimes referred to as "seat of the pants" sensations) of changes in direction or speed is an important indicator to the trained and experienced pilot in visual flight. If this sensitivity is properly developed, it can warn the pilot of an impending stall.

Pilots should remember that a level-flight 1G published stalling speed is valid only:

1. In unaccelerated 1G flight
2. In coordinated flight (slip-skid indicator centered)
3. At one weight (typically maximum gross weight)
4. At a particular center of gravity (CG) (typically maximum forward CG)

Angle of Attack Indicators

An AOA indicator gives the pilot better situational awareness pertaining to the aerodynamic health of the airfoil. This can be referred to as stall margin awareness or knowing the existing margin between the current AOA and the critical AOA. While learning to recognize stalls without relying on stall warning devices is important, an AOA indicator provides an additional visual indication of the airplane's proximity to the critical AOA. The FAA along with the General Aviation Joint Steering Committee (GAJSC) is promoting the use of Angle of Attack (AOA) indicators to reduce the occurrence of loss of control in flight.

Without an AOA indicator, the AOA is "invisible" to pilots. These devices measure several parameters simultaneously and determine the current angle of attack providing a visual image to the pilot of the current AOA along with representation of the proximity to the critical AOA. These devices can give a visual representation of the energy management state of the airplane. The energy state of an airplane is the balance between airspeed, altitude, drag, and thrust and represents how efficiently the airfoil is operating. With this increased situational awareness pertaining to the energy condition of the airplane, the pilot has additional information to help prevent a loss of control scenario.

AOA indicators are increasingly affordable for GA airplanes. There are several different kinds of AOA indicators with varying methods for calculating AOA; therefore, proper installation and training on the use of these devices is important. AOA indicators measure several parameters simultaneously, determine the current AOA, and provide a visual image of the proximity to the critical AOA. *[Figure 5-9]* Some AOA indicators also provide aural indications, which can provide awareness to a change in AOA that is trending towards the critical AOA prior to installed stall warning systems. It's important to note that some indicators take flap position into consideration, but not all do.

While AOA indicators provide a simple visual representation of the current AOA and its proximity to the critical AOA, they are not without their limitations. These limitations should be understood by operators of GA airplanes equipped with these devices. Like advanced automation such as autopilots and moving maps, the misunderstanding or misuse of the equipment can have disastrous results. Some items that may limit the effectiveness of an AOA indicator are listed below:

1. Calibration techniques
2. Probes or vanes not being heated
3. The type of indicator itself
4. Flap setting
5. Wing contamination

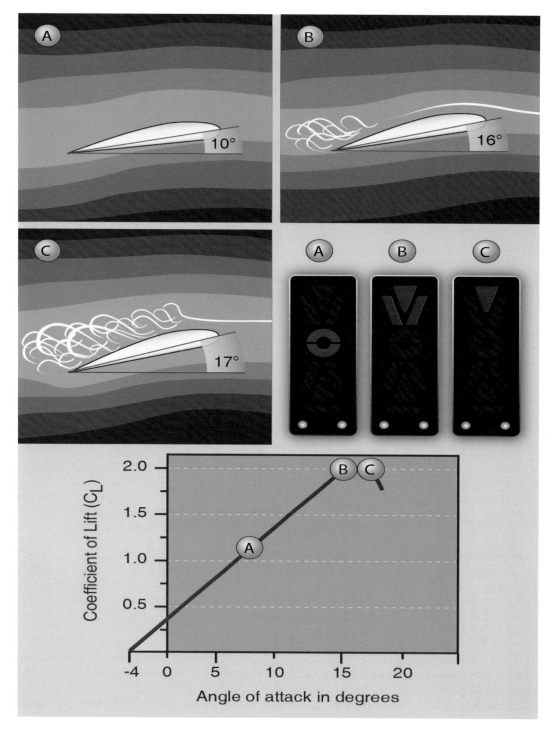

Figure 5-9. *A conceptual representation of an AOA indicator. It is important to become familiar with the equipment installed in a specific airplane.*

Installation of AOA indicators not required by type certification in GA airplanes has been streamlined by the FAA. The FAA established policy in February 2014 pertaining to non-required AOA systems and how they may be installed as a minor alteration, depending upon their installation requirements and operational utilization, and the procedures to follow for certification of these installations. For updated information on this, please reference the FAA website at www.faa.gov.

If airplane equipment includes an angle of attack indicator, the pilot should know how the particular device determines AOA, what the display indicates, and the appropriate response to any indication. Pilots are encouraged to conduct in-flight training to see the indications throughout various maneuvers, such as slow flight, stalls, takeoffs, and landings, and to practice the appropriate responses to those indications. It is also important to note that some items may limit the effectiveness of an AOA indicator (e.g., calibration techniques, wing contamination, unheated probes/vanes). Pilots flying an airplane equipped with an AOA indicator should refer to the pilot handbook information or contact the manufacturer for specific limitations applicable to that indicator type.

Ground and flight instructors should make every attempt to receive training from an instructor knowledgeable about AOA indicators prior to giving instruction pertaining to or in airplanes equipped with an AOA indicator. Pilot schools should incorporate training on AOA indicators in their syllabi whether their training aircraft are equipped with them or not.

Stall Characteristics

Different airplane designs can result in different stall characteristics. The pilot should know the stall characteristics of the airplane being flown and the manufacturer's recommended recovery procedures. Factors that can affect the stall characteristics of an airplane include its geometry, CG, wing design, and high-lift devices. Engineering design variations make it impossible to specifically describe the stall characteristics for all airplanes; however, there are enough similarities in small general aviation training-type airplanes to offer broad guidelines.

Most training airplanes are designed so that the wings stall progressively outward from the wing roots (where the wing attaches to the fuselage) to the wingtips. Some wings are manufactured with a certain amount of twist, known as washout, resulting in the outboard portion of the wings having a slightly lower AOA than the wing roots. This design feature causes the wingtips to have a smaller AOA during flight than the wing roots. Thus, the wing roots of an airplane exceed the critical AOA before the wingtips, meaning the wing roots stall first. Therefore, when the airplane is in a stalled condition, the ailerons should still have a degree of control effectiveness until/unless stalled airflow migrates outward along the wings. Although airflow may still be attached at the wingtips, a pilot should exercise caution using the ailerons prior to the reduction of the AOA because it can exacerbate the stalled condition. For example, if the airplane rolls left at the stall ("rolls-off"), and the pilot applies right aileron to try to level the wing, the downward-deflected aileron on the left wing produces a greater AOA (and more induced drag), and a more complete stall at the tip as the critical AOA is exceeded. This can cause the wing to roll even more to the left, which is why it is important to first reduce the AOA before attempting to roll the airplane.

The pilot should also understand how the factors that affect stalls are interrelated. In a power-off stall, for instance, the cues (buffeting, shaking) are less noticeable than in the power-on stall. In the power-off, 1G stall, the predominant cue may be the elevator control position (full up elevator against the stops) and a high descent rate.

Fundamentals of Stall Recovery

Depending on the complexity of the airplane, stall recovery could consist of as many as six steps. Even so, the pilot should remember the most important action to an impending stall or a full stall is to reduce the AOA. There have been numerous situations where pilots did not first reduce AOA, and instead prioritized power and maintaining altitude, which resulted in a loss of control. This section provides a generic stall recovery procedure for light general aviation aircraft adapted from a template developed by major airplane manufacturers and can be adjusted appropriately for the aircraft used. *[Figure 5-10]* However, a pilot should always follow the aircraft-specific manufacturer's recommended procedures if published and current.

Stall Recovery Template	
1. Wing leveler or autopilot	1. Disconnect
2. a) Pitch nose-down	2. a) Apply until impending stall indications are eliminated
b) Trim nose-down pitch	b) As needed
3. Bank	3. Wings Level
4. Thrust/Power	4. As needed
5. Speed brakes/spoilers	5. Retract
6. Return to the desired flight path	

Figure 5-10. *Stall recovery template.*

The recovery actions should be made in a procedural manner; they can be summarized in *Figure 5-10*. The following discussion explains each of the six steps:

1. Disconnect the wing leveler or autopilot (if equipped). Manual control is essential to recovery in all situations. Disconnecting this equipment should be done immediately and allow the pilot to move to the next crucial step quickly. Leaving the wing leveler or autopilot connected may result in inadvertent changes or adjustments to the flight controls or trim that may not be easily recognized or appropriate, especially during high workload situations.

2. a) Pitch nose-down control. Reducing the AOA is crucial for all stall recoveries. Push forward on the flight controls to reduce the AOA below the critical AOA until the impending stall indications are eliminated before proceeding to the next step.

   b) Trim nose-down pitch. If the elevator does not provide the needed response, pitch trim may be necessary. However, excessive use of pitch trim may aggravate the condition, or may result in loss of control or high structural loads.

3. Roll wings level. This orients the lift vector properly for an effective recovery. It is important not to be tempted to control the bank angle prior to reducing AOA. Both roll stability and roll control will improve considerably after getting the wings flying again. It is also imperative to proactively cancel yaw with proper use of the rudder to prevent a stall from progressing into a spin.

4. Add thrust/power. Power should be added as needed, as stalls can occur at high power or low power settings or at high airspeeds or low airspeeds. Advance the throttle promptly, but smoothly, as needed while using rudder and elevator controls to stop any yawing motion and prevent any undesirable pitching motion. Adding power typically reduces the loss of altitude during a stall recovery, but it does not eliminate a stall. The reduction in AOA is imperative. For propeller-driven airplanes, power application increases the airflow around the wing, assisting in stall recovery.

5. Retract speedbrakes/spoilers (if equipped). This will improve lift and the stall margin.

6. Return to the desired flightpath. Apply smooth and coordinated flight control movements to return the airplane to the desired flightpath being careful to avoid a secondary stall. However, be situationally aware of the proximity to terrain during the recovery and take the necessary flight control action to avoid contact with it.

The above procedure can be adapted for the type of aircraft flown. For example, a single-engine training airplane without an autopilot would likely only use four of the six steps. The first step is not applicable. The actual first step is the reduction of the AOA until the stall warning is eliminated. Use of pitch trim is less of a concern in a training airplane because most pilots can overpower the trim in these airplanes. Any improper trim can be corrected when returning to the desired flightpath. The next step is rolling the wings level followed by the addition of power as needed all while maintaining coordinated flight. If the airplane is not equipped with speedbrakes or spoilers, this step is also skipped. Returning to the desired flightpath concludes the recovery.

Similarly, a glider pilot does not have an autopilot; therefore, the first step is the reduction of AOA until the stall warning is eliminated. The pilot would then roll wings level while maintaining coordinated flight. Since there is no power to add, this step would not apply. Retracting speedbrakes or spoilers would be the next step for a glider pilot followed by returning to the desired flightpath.

Stall Training

Practice in both power-on and power-off stalls is important because it simulates stall conditions that could occur during normal flight maneuvers. It is important for pilots to understand the possible flight scenarios in which a stall could occur. Stall accidents usually result from an inadvertent stall at a low altitude, with the recovery not completed prior to ground contact. For example, power-on stalls are practiced to develop the pilot's awareness of what could happen if the airplane is pitched to an excessively nose-high attitude immediately after takeoff, during a climbing turn, or when trying to clear an obstacle. Power-off turning stalls develop the pilot's awareness of what could happen if the controls are improperly used during a turn from the base leg to the final approach. The power-off straight-ahead stall simulates the stall that could occur when trying to stretch a glide after the engine has failed, or if low on the approach to landing.

As in all maneuvers that involve significant changes in altitude or direction, the pilot should ensure that the area is clear of other air traffic at and below their altitude and that sufficient altitude is available for a recovery before executing the maneuver. It is recommended that stalls be practiced at an altitude that allows recovery no lower than 1,500 feet AGL for single-engine airplanes, or higher if recommended by the AFM/POH. Losing altitude during recovery from a stall is to be expected.

Approaches to Stalls (Impending Stalls), Power-On or Power-Off

An impending stall occurs when the airplane is approaching, but does not exceed the critical AOA. The purpose of practicing impending stalls is to learn to retain or regain full control of the airplane immediately upon recognizing that it is nearing a stall, or that a stall is likely to occur if the pilot does not take appropriate action. Pilot training should emphasize teaching the same recovery technique for impending stalls and full stalls.

The practice of impending stalls is of particular value in developing the pilot's sense of feel for executing maneuvers in which maximum airplane performance is required. These maneuvers require flight in which the airplane approaches a stall, but the pilot initiates recovery at the first indication, such as by a stall warning device activation.

Impending stalls may be entered and performed in the same attitudes and configurations as the full stalls or other maneuvers described in this chapter. However, instead of allowing the airplane to reach the critical AOA, the pilot should immediately reduce AOA once the stall warning device goes off, if installed, or recognizes other cues such as buffeting. The pilot should hold the nose-down control input as required to eliminate the stall warning. Then level the wings maintain coordinated flight, and then apply whatever additional power is necessary to return to the desired flightpath. The pilot will have recovered once the airplane has returned to the desired flightpath with sufficient airspeed and adequate flight control effectiveness and no stall warning. Performance of the impending stall maneuver is unsatisfactory if a full stall occurs, if an excessively low pitch attitude is attained, or if the pilot fails to take timely action to avoid excessive airspeed, excessive loss of altitude, or a spin.

Full Stalls, Power-Off

The practice of power-off stalls is usually performed with normal landing approach conditions to simulate an accidental stall occurring during approach to landing. However, power-off stalls should be practiced at all flap settings to ensure familiarity with handling arising from mechanical failures, icing, or other abnormal situations. Airspeed in excess of the normal approach speed should not be carried into a stall entry since it could result in an abnormally nose-high attitude.

To set up the entry for a straight-ahead power-off stall, airplanes equipped with flaps or retractable landing gear should be in the landing configuration. After extending the landing gear, applying carburetor heat (if applicable), and retarding the throttle sufficiently, the pilot holds the airplane at a constant altitude until the airspeed decelerates to normal approach speed. The airplane should then be smoothly pitched down to a normal approach attitude to maintain that airspeed. Wing flaps should be extended and pitch attitude adjusted to maintain the airspeed. Once in a normal approach, the pilot sets the power to idle.

When the approach attitude and airspeed have stabilized, the pilot should smoothly raise the airplane's nose to an attitude that induces a stall. Directional control should be maintained and wings held level by coordinated use of the ailerons and rudder. Once the airplane reaches an attitude that will lead to a stall, the pitch attitude is maintained with the elevator until the stall occurs. The stall is recognized by the full-stall cues previously described.

Recovery from the stall is accomplished by reducing the AOA, applying as much nose-down control input as required to eliminate the stall warning, leveling the wings, maintaining coordinated flight, and then applying power as needed. Right rudder pressure may be necessary to overcome the engine torque effects as power is advanced and the nose is being lowered. *[Figure 5-11]* If simulating an inadvertent stall on approach to landing, the pilot should initiate a go-around by establishing a positive rate of climb. Once in a climb, the flaps and landing gear should be retracted as necessary.

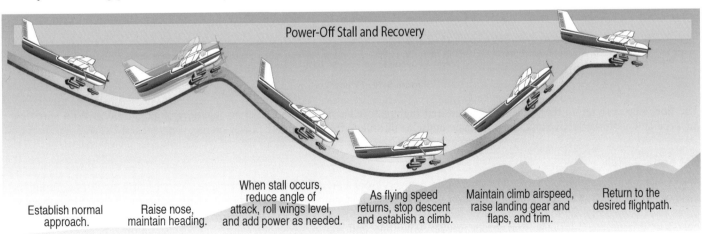

Power-Off Stall and Recovery

Establish normal approach.

Raise nose, maintain heading.

When stall occurs, reduce angle of attack, roll wings level, and add power as needed.

As flying speed returns, stop descent and establish a climb.

Maintain climb airspeed, raise landing gear and flaps, and trim.

Return to the desired flightpath.

Figure 5-11. *Power-off stall and recovery.*

Recovery from power-off stalls should also be practiced from shallow banked turns to simulate an inadvertent stall during a turn from base leg to final approach. During the practice of these stalls, the pilot should take care to ensure that the airplane remains coordinated and the turn continues at a constant bank angle until the full stall occurs. If the airplane is allowed to slip, the outer wing may stall first and move downward abruptly. In a skid, the bank angle may increase further to a potentially dangerous attitude. The recovery procedure is the same, regardless of whether one wing rolls off first. The pilot should apply as much nose-down control input as necessary to eliminate the stall warning, level the wings with ailerons, coordinate with rudder, and add power as needed. In the practice of turning stalls, no attempt should be made to stall or recover the airplane on a predetermined heading. However, to simulate a turn from base to final approach, the stall normally should be made to occur within a heading change of approximately 90°.

Full Stalls, Power-On

Power-on stall recoveries are practiced from straight climbs and climbing turns (15° to 20° bank) to help the pilot recognize the potential for an accidental stall during takeoff, go around, climb, or when trying to clear an obstacle. Airplanes equipped with flaps or retractable landing gear should normally be in the takeoff configuration; however, power-on stalls should also be practiced with the airplane in a clean configuration (flaps and gear retracted) to ensure practice with all possible takeoff and climb configurations. When practicing takeoff stall recovery, the airplane should be at maximum power, although for some airplanes it may be reduced to a setting that will prevent an excessively high pitch attitude.

To set up the entry for power-on stalls, the pilot establishes the airplane in the takeoff or climb configuration and slows the airplane to normal lift-off speed while continuing to clear the area of other traffic. Upon reaching the desired speed, the pilot sets takeoff power or the recommended climb power for the power-on stall (often referred to as a departure stall) while establishing a climb attitude. The purpose of reducing the airspeed to lift-off airspeed before the throttle is advanced to the recommended setting is to avoid an excessively steep nose-up attitude for a long period before the airplane stalls.

After establishing the climb attitude, the pilot should smoothly raise the nose to increase the AOA, and hold that attitude until the full stall occurs. As described in connection with the stall characteristics discussion, continual adjustments should be made to aileron pressure, elevator pressure, and rudder pressure to maintain coordinated flight while holding the attitude until the full stall occurs. In most airplanes, as the airspeed decreases the pilot should move the elevator control progressively further back while simultaneously adding right rudder and maintaining the climb attitude until reaching the full stall.

The pilot should recognize when the stall has occurred and take action without delay to prevent a prolonged stalled condition. The pilot should recover from the stall by immediately reducing the AOA and applying as much nose-down control input as required to eliminate the stall warning, level the wings with ailerons, coordinate with rudder, and smoothly advance the power as needed. Since the throttle is already at the climb power setting, this step may simply mean confirming the proper power setting. [Figure 5-12]

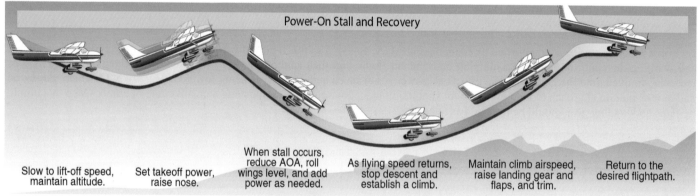

Power-On Stall and Recovery

Slow to lift-off speed, maintain altitude.

Set takeoff power, raise nose.

When stall occurs, reduce AOA, roll wings level, and add power as needed.

As flying speed returns, stop descent and establish a climb.

Maintain climb airspeed, raise landing gear and flaps, and trim.

Return to the desired flightpath.

Figure 5-12. *Power-on stall.*

The final step is to return the airplane to the desired flightpath (e.g., straight and level or departure/climb attitude). With sufficient airspeed and control effectiveness, the pilot may return the throttle to the appropriate power setting.

Secondary Stall

A secondary stall is so named because it occurs after recovery from a preceding stall. A normal recovery usually involves pointing the nose of the airplane toward the ground. However, if a stall should occur at low altitude, the pilot's natural impulse is to bring the nose up as soon as possible and to do so abruptly. This reaction is amplified as proximity to the ground increases. To demonstrate how this occurs at altitude, the pilot makes an abrupt recovery after one stall and exceeds the critical AOA a second time. Note that this stall may occur after any stall when the pilot does not sufficiently reduce the AOA by lowering the pitch attitude or attempts to break the stall by using power only. [Figure 5-13]

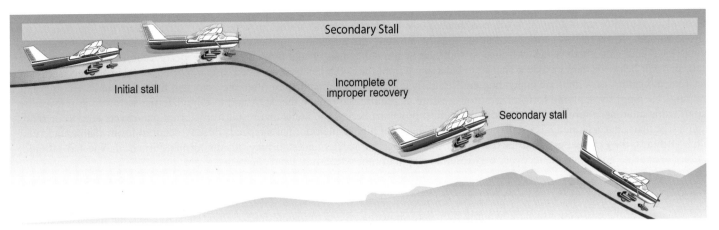

Figure 5-13. *Secondary stall.*

If a secondary stall occurs, the pilot should again perform the stall recovery procedures by applying nose-down elevator pressure as required to eliminate the stall warning, level the wings with ailerons, coordinate with rudder, and adjust power as needed. When the airplane is no longer in a stalled condition the pilot can return the airplane to the desired flightpath. For pilot certification, this is a demonstration-only maneuver. Only flight instructor applicants may be required to perform it on a practical test.

Accelerated Stalls

While pilots may understand the cause of an accelerated stall, it takes training to experience how these stalls develop and occur. The objectives of demonstrating an accelerated stall are to determine the stall characteristics of the airplane, experience stalls at speeds greater than the +1G stall speed, and develop the ability to instinctively recover at the onset of such stalls. This is a maneuver only commercial pilot and flight instructor applicants may be required to perform or demonstrate on a practical test. However, all pilots should be familiar with the situations that can cause an accelerated stall, how to recognize this type of stall, and how to execute the appropriate recovery should one occur.

At the same gross weight, airplane configuration, CG location, power setting, and environmental conditions, a given airplane consistently stalls at the same indicated airspeed provided the airplane is at +1G (i.e., steady-state unaccelerated flight). However, the airplane can also stall at a higher indicated airspeed when the airplane is subject to an acceleration greater than +1G, such as when turning, pulling up, or other abrupt changes in flightpath. Stalls encountered any time the G-load exceeds +1G are called "accelerated maneuver stalls." The accelerated stall would most frequently occur inadvertently during improperly executed turns, stall and spin recoveries, pullouts from steep dives, or when overshooting a base to final turn. An accelerated stall is typically demonstrated during steep turns.

A pilot should never practice accelerated stalls with wing flaps in the extended position due to the lower design G-load limitations in that configuration. Accelerated stalls should be performed with a bank of approximately 45°, and in no case at a speed greater than the airplane manufacturer's recommended airspeed, or the specified design maneuvering speed (V_A) or operating maneuvering speed (V_O).

It is important to be familiar with V_A or V_O, how it relates to accelerated stalls, and how it changes depending on the airplane's weight. V_A is the maximum speed at which the positive design load limit can be imposed either by gusts or full one-sided deflection with one control surface without causing structural damage. V_O is a historical operating limitation applicable to certain airplanes only. It represents the maximum speed where, at any given weight, the pilot may apply full control excursion without exceeding the design limit load factor. Performing accelerated stalls at speeds up to the applicable V_A or V_O, ensures the airplane will reach the critical AOA, which unloads the wing, before exceeding the design load limit. At speeds above V_A or V_O, the airplane can reach its design load limit at less than the critical AOA. This condition makes it possible to add additional load and overstress the airplane. Additional information on the effects of aircraft weight on stall speeds and structural limits while maneuvering is available in the "Aerodynamics of Flight" chapter of the *Pilot's Handbook of Aeronautical Knowledge* (FAA-H-8083-25).

There are two methods for performing an accelerated stall. The most common accelerated stall procedure starts from straight-and-level flight at an airspeed at or below V_A or V_O. The pilot rolls the airplane into a coordinated, level-flight 45° turn and then smoothly, firmly, and progressively increase the AOA through back elevator pressure until a stall occurs. Alternatively, the pilot rolls the airplane into a coordinated, level-flight 45° turn at an airspeed above V_A or V_O. After the airspeed slows to V_A or V_O, and at an airspeed 5 to 10 percent faster than the unaccelerated stall speed, the pilot progressively increases the AOA through back elevator pressure until a stall occurs. The increased back elevator pressure increases lift and the G load. The G load pushes the pilot's body down in the seat. The increased lift also increases drag, which may cause the airspeed to decrease. The pilot should know the published stall speed for 45° of bank, flaps up, before performing the maneuver. This speed is typically published in the AFM.

An airplane typically stalls during a level, coordinated turn similar to the way it does in wings-level flight, except that the stall buffet can be sharper. If the turn is coordinated at the time of the stall, the airplane's nose pitches away from the pilot just as it does in a wings-level stall since both wings will tend to stall nearly simultaneously. If the airplane is not properly coordinated at the time of stall, the stall behavior may include a change in bank angle until the AOA has been reduced. It is important to take recovery action at the first indication of a stall (if impending stall training/checking) or immediately after the stall has fully developed (if full stall training/checking) by applying forward elevator pressure as required to reduce the AOA and to eliminate the stall warning, level the wings using ailerons, coordinate with rudder, and adjust power as necessary. Stalls that result from abrupt maneuvers tend to be more aggressive than unaccelerated +1G stalls. Because they occur at higher-than-normal airspeeds or may occur at lower-than-anticipated pitch attitudes, they can surprise an inexperienced pilot. Since an accelerated stall may put the airplane in an unexpected attitude. Failure to execute an immediate recovery may result in a spin or other departure from controlled flight.

Cross-Control Stall

The objective of the cross-control stall demonstration is to show the effects of uncoordinated flight on stall behavior and to emphasize the importance of maintaining coordinated flight while making turns. This is a demonstration-only maneuver; only flight instructor applicants may be required to perform it on a practical test. However, all pilots should be familiar with the situations that can lead to a cross-control stall, how to recognize and avoid this stall, and how to recover should one occur.

The aerodynamic effects of the uncoordinated, cross-control stall can surprise the unwary pilot because this stall can occur with very little warning and can be deadly if it occurs close to the ground. The nose may pitch down, the bank angle may suddenly change, and the airplane may continue to roll to an inverted orientation, which is usually the beginning of a spin. It is therefore essential for the pilot to follow the stall recovery procedure by reducing the AOA until the stall warning has been eliminated, then roll wings level using ailerons, and coordinate with rudder inputs before the airplane enters a spiral or spin.

A cross-control stall occurs when the critical AOA is exceeded with aileron pressure applied in one direction and rudder pressure in the opposite direction, causing uncoordinated flight. A skidding cross-control stall is most likely to occur in the traffic pattern during a poorly planned and executed base-to-final approach turn. There may be an unrecognized tailwind component and higher groundspeed on the base leg, which causes the pilot to turn late or with inadequate bank. The airplane overshoots the runway centerline, and the pilot attempts to correct by increasing the bank angle, increasing back elevator pressure, and applying excess rudder in the direction of the turn (i.e., inside or bottom rudder pressure) to bring the nose around further to align it with the runway. The difference in lift between the inside and outside wing will increase, resulting in an unwanted increase in bank angle. At the same time, the nose of the airplane slices downward through the horizon. The natural reaction to this may be for the pilot to pull back on the elevator control, increasing the AOA toward critical. Should a stall be encountered with these inputs, the airplane may rapidly enter a spin. The safest action for an "overshoot" is to perform a go-around. At the relatively low altitude of a base-to-final approach turn, a pilot should be reluctant to use bank angles greater than 30 degrees and should not make a skidding turn if correcting for any overshoot.

Before performing this stall, the pilot should establish a safe altitude for entry and recovery in the event of a spin, and clear the area of other traffic while slowly retarding the throttle. The next step is to lower the landing gear (if equipped with retractable gear), close the throttle, and maintain altitude until the airspeed approaches the normal glide speed. To avoid the possibility of exceeding the airplane's limitations, the pilot should not extend the flaps. While the gliding attitude and airspeed are being established, the airplane should be retrimmed. Once the glide is stabilized, the airplane should be rolled into a medium-banked turn to simulate a final approach turn that overshoots the centerline of the runway.

During the turn, the pilot should smoothly apply excessive rudder pressure in the direction of the turn and hold the bank constant by applying opposite aileron pressure. At the same time, the pilot increases back elevator pressure to keep the nose from lowering. All of these control pressures should be increased until the airplane stalls. When the stall occurs, the pilot applies nose-down elevator pressure to reduce the AOA until the stall warning has been eliminated, removes the excessive rudder input and levels the wings, and adds power as needed to return to complete the recovery and return to the desired flightpath.

Elevator Trim Stall

The elevator trim stall demonstration shows what can happen when the pilot applies full power for a go-around without maintaining positive control of the airplane. [Figure 5-14] This is a demonstration-only maneuver; only flight instructor applicants may be required to perform it on a practical test. However, all pilots should be familiar with the situations that can cause an elevator trim stall, recognize its development, and take appropriate action to prevent it.

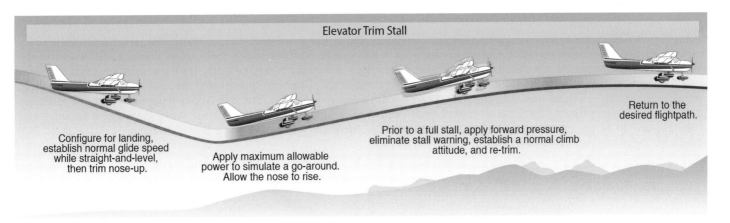

Figure 5-14. *Elevator trim stall.*

This situation may occur during a go-around procedure from a normal landing approach or a simulated, forced-landing approach, or immediately after a takeoff, with the trim set for a normal landing approach glide at idle power. The demonstration shows the importance of making smooth power applications, overcoming strong trim forces, maintaining positive control of the airplane to hold safe flight attitudes, and using proper and timely trim techniques. It also develops the pilot's ability to avoid actions that could result in this stall, to recognize when an elevator trim stall is approaching, and to take prompt and correct action to prevent a full stall condition. It is imperative to avoid the occurrence of an elevator trim stall during an actual go-around from an approach to landing.

At a safe altitude and after ensuring that the area is clear of other air traffic, the pilot should slowly retard the throttle and extend the landing gear (if the airplane is equipped with retractable gear). The next step is to extend the flaps to the one-half or full position, close the throttle, and maintain altitude until the airspeed approaches the normal glide speed.

When the normal glide is established, the pilot should trim the airplane nose-up for the normal landing approach glide. During this simulated final approach glide, the throttle is then advanced smoothly to maximum allowable power, just as it would be adjusted to perform a go-around.

The combined effects of increased propwash over the tail and elevator trim tend to make the nose rise sharply and turn to the left. With the throttle fully advanced, the pitch attitude increases above the normal climbing attitude. When it is apparent the airplane is approaching a stall, the pilot should apply sufficient forward elevator pressure to reduce the AOA and eliminate the stall warning before returning the airplane to the normal climbing attitude. The pilot will need to adjust trim to relieve the heavy control pressures and then complete the normal go around procedures and return to the desired flightpath. If taken to the full stall, recovery will require a significant nose-down attitude to reduce the AOA below its critical AOA, along with a corresponding significant loss of altitude.

Common Errors

Common errors in the performance of intentional stalls are:

1. Failure to adequately clear the area.
2. Over-reliance on the airspeed indicator and slip-skid indicator while excluding other cues after recovery.
3. Inadvertent accelerated stall by pulling too fast on the controls during a power-off or power-on stall entry.
4. Inability to recognize an impending stall condition.
5. Failure to take timely action to prevent a full stall during the conduct of impending stalls.
6. Failure to maintain a constant bank angle during turning stalls.
7. Failure to maintain proper coordination with the rudder throughout the stall and recovery.
8. Recovering before reaching the critical AOA when practicing the full stall maneuver.
9. Not disconnecting the wing leveler or autopilot, if equipped, prior to reducing AOA.
10. Recovery is attempted without recognizing the importance of pitch control and AOA.
11. Not maintaining a nose down control input until the stall warning is eliminated.
12. Pilot attempts to level the wings before reducing AOA.
13. Pilot attempts to recover with power before reducing AOA.
14. Failure to roll wings level after AOA reduction and stall warning is eliminated.
15. Inadvertent secondary stall during recovery.
16. Excessive forward-elevator pressure during recovery resulting in low or negative G load.
17. Excessive airspeed buildup during recovery.
18. Losing situational awareness and failing to return to desired flightpath or follow ATC instructions.

Spin Awareness

A spin is an aggravated stall condition that may result after a stall occurs. Mishandling of yaw control during a stall increases the likelihood of a spin entry. A spin results in the airplane following a downward corkscrew path. During a spin, the airplane rotates around its vertical axis affected by different lift and drag forces on each wing, and the airplane descends due to gravity, rolling, yawing, and pitching in a spiral path. *[Figure 5-15]* There are different types of spins. The spin type or types that occur in a particular airplane may be by airplane design, loading, control inputs, and density altitude. In all spins at least one of the wings is stalled. Refer to the airplane POH for spin recovery techniques appropriate to the make and model being flown. Techniques in the POH take precedence over information in this section.

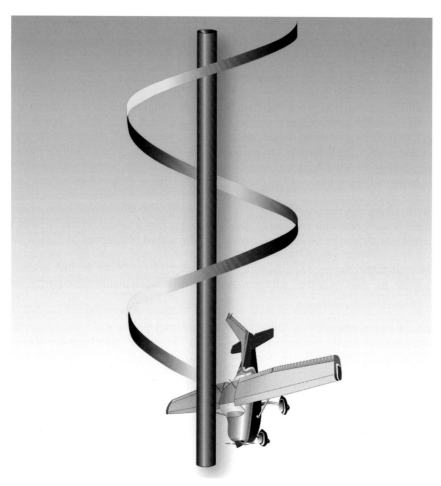

Figure 5-15. *Spin—an aggravated stall and autorotation.*

A spin occurs when at least one of the airplane's wings exceed the critical AOA (stall) with a sideslip or yaw acting on the airplane at, or beyond, the actual stall. An airplane will yaw not only because of incorrect rudder application but because of adverse yaw created by aileron deflection; engine/prop effects, including p-factor, torque, spiraling slipstream, and gyroscopic precession; and wind shear, including wake turbulence. If the yaw had been created by the pilot because of incorrect rudder use, the pilot may not be aware that a critical AOA has been exceeded until the airplane yaws out of control toward the lowering wing. A stall that occurs while the airplane is in a slipping or skidding turn can result in a spin entry and rotation in the direction of rudder application, regardless of which wingtip is raised. If the pilot does not immediately initiate stall recovery, the airplane may enter a spin.

Maintaining directional control and not allowing the nose to yaw before stall recovery is initiated is key to averting a spin. The pilot should apply the correct amount of rudder to keep the nose from yawing and the wings from banking.

Modern airplanes tend to be more reluctant to spin compared to older designs, however it is not impossible for them to spin. Mishandling the controls in turns, stalls, and uncoordinated slow flight can put even the most reluctant airplanes into an accidental spin. Proficiency in avoiding conditions that could lead to an accidental stall/spin situation, and in promptly taking the correct actions to recover to normal flight, is essential. An airplane needs to be stalled and yawed in order to enter a spin; therefore, continued practice in stall recognition and recovery helps the pilot develop a more instinctive and prompt reaction in recognizing an approaching spin. Upon recognition of a spin or approaching spin, the pilot should immediately execute spin recovery procedures.

Spin Procedures

The first rule for spin demonstration is to ensure that the airplane is approved for spins. Please note that this discussion addresses generic spin procedures; it does not cover special spin procedures or techniques required for a particular airplane. Safety dictates careful review of the AFM/POH and regulations before attempting spins in any airplane. The review should include the following items:

- The airplane's AFM/POH limitations section, placards, or type certification data to determine if the airplane is approved for spins

- Weight and balance limitations

- Recommended entry and recovery procedures

- The current 14 CFR part 91 parachute requirements

Also essential is a thorough airplane preflight inspection, with special emphasis on excess or loose items that may affect the weight, CG, and controllability of the airplane. It is also important to ensure that the airplane is within any CG limitations as determined by the manufacturer. Slack or loose control cables (particularly rudder and elevator) could prevent full anti-spin control deflections and delay or preclude recovery in some airplanes.

Prior to any intentional spin, clear the flight area above and below the airplane for other traffic. This task may occur while slowing the airplane for the spin entry. In addition, all spins should begin at an altitude high enough to complete recovery at or above 1,500 feet AGL. Note that the first turn in a spin results in an altitude loss of approximately 1,000 feet, while each subsequent turn loses about half that amount.

It may be appropriate to introduce spin training by first practicing both power-on and power-off stalls in a clean configuration. This practice helps familiarize the pilot with the airplane's specific stall and recovery characteristics. In all phases of training, the pilot should take care with handling of the power (throttle), and apply carburetor heat, if equipped, according to the manufacturer's recommendations

There are four phases of a spin: entry, incipient, developed, and recovery. *[Figure 5-16]*

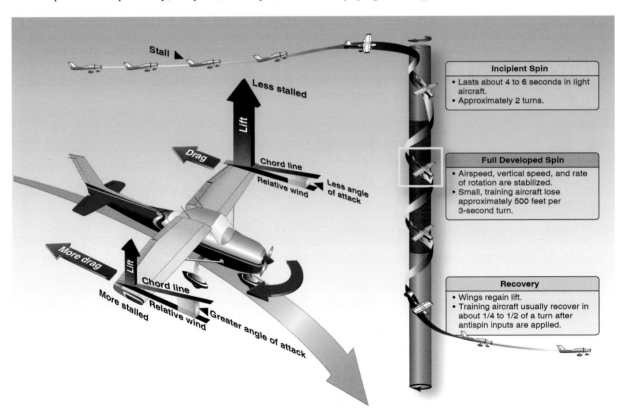

Figure 5-16. *Spin Entry and Recovery.*

Entry Phase

In the entry phase, the pilot intentionally or accidentally provides the necessary elements for the spin. The entry procedure for demonstrating a spin is similar to a power-off stall. During the entry, the pilot should slowly reduce power to idle, while simultaneously raising the nose to a pitch attitude that ensures a stall. As the airplane approaches a stall, the pilot smoothly applies full rudder in the direction of the desired spin rotation while applying full back (up) elevator to the limit of travel. Unless AFM/POH specifies otherwise, ailerons are maintained in the neutral position during the spin procedure.

Incipient Phase

The incipient phase occurs from the time the airplane stalls and starts rotating until the spin has fully developed. This phase may take two to four turns for most airplanes. In this phase, the aerodynamic and inertial forces have not achieved a balance. As the incipient phase develops, the indicated airspeed will generally stabilize at a low and constant airspeed and the symbolic airplane of the turn indicator should indicate the direction of the spin. The pilot should not use the slip/skid ball (inclinometer) to determine spin direction. The location of the instrument in the airplane determines how the ball will move rather than the direction of the spin. For example, the ball mounted on the left side of the airplane will always move to the left, even in spin with rotation to the right.

The pilot should initiate incipient spin recovery procedures prior to completing 360° of rotation. The pilot should apply full rudder opposite the direction of rotation. The turn indicator shows a deflection in the direction of rotation if disoriented.

Incipient spins that are not allowed to develop into a steady-state spin are the most commonly used maneuver in initial spin training and recovery techniques.

Developed Phase

The developed phase occurs when the airplane's angular rotation rate, airspeed, and vertical speed are stabilized in a flightpath that is nearly vertical. In the developed phase, aerodynamic forces and inertial forces are in balance, and the airplane's attitude, angles, and self-sustaining motions about the vertical axis are constant or repetitive, or nearly so. The spin is in equilibrium. It is important to note that some training airplanes will not enter into the developed phase but could transition unexpectedly from the incipient phase into a spiral dive. In a spiral dive the airplane will not be in equilibrium but instead will be accelerating and G load can rapidly increase as a result.

Recovery Phase

The recovery phase occurs when rotation ceases and the AOA of the wings is decreased below the critical AOA. This phase may last for as little as a quarter turn or up to several turns depending upon the airplane and the type of spin. To recover, the pilot applies control inputs to disrupt the spin equilibrium by stopping the rotation and unstalling the wing. To accomplish spin recovery, the pilot should always follow the manufacturer's recommended procedures. In the absence of the manufacturer's recommended spin recovery procedures and techniques, use the six-step spin recovery procedure in *Figure 5-17*. If the flaps and/or retractable landing gear are extended prior to the spin, they should be retracted as soon as practicable after spin entry.

Spin Recovery Template
1. Reduce the power (throttle) to idle
2. Position the ailerons to neutral
3. Apply full opposite rudder against the rotation
4. Apply positive, brisk, and straight forward elevator (forward of neutral)
5. Neutralize the rudder after spin rotation stops
6. Apply back elevator pressure to return to level flight and adjust power (throttle) as appropriate

Figure 5-17. *Spin recovery template.*

The following discussion explains each of the six steps a pilot should follow for spin recovery:

1. Reduce the power (throttle) to idle. Power aggravates spin characteristics. It can result in a flatter spin attitude and usually increases the rate of rotation.

2. Position the ailerons to neutral. Ailerons may have an adverse effect on spin recovery. Aileron control in the direction of the spin may accelerate the rate of rotation, steepen the spin attitude and delay the recovery. Aileron control opposite the direction of the spin may cause flattening of the spin attitude and delayed recovery; or may even be responsible for causing an unrecoverable spin. The best procedure is to ensure that the ailerons are neutral.

3. Apply and hold full opposite rudder against the rotation until the rotation stops. Rudder tends to be the most important control for recovery in typical single-engine airplanes, and its application should be brisk and full opposite to the direction of rotation. Avoid slow and overly cautious opposite rudder movement during spin recovery, which can allow the airplane to spin indefinitely, even with anti-spin inputs. A brisk and positive technique results in a more positive spin recovery.

4. Apply positive, brisk, and straight-forward elevator (forward of neutral). This step should be taken immediately after full rudder application. Do not wait for the rotation to stop before performing this step. The forceful movement of the elevator decreases the AOA and drives the airplane toward unstalled flight. In some cases, full forward elevator may be required for recovery. Hold the controls firmly in these positions until the spinning stops. (Note: If the airspeed is increasing, the airplane is no longer in a spin. In a spin, the airplane is stalled, and the indicated airspeed should therefore be relatively low and constant and should not be accelerating.)

5. Neutralize the rudder after spin rotation stops. Failure to neutralize the rudder at this time, when airspeed is increasing, causes a yawing or sideslipping effect.

6. Apply back elevator pressure to return to level flight and adjust power as appropriate. Be careful not to apply excessive back elevator pressure after the rotation stops and the rudder has been neutralized. Excessive back elevator pressure can cause a secondary stall and may result in another spin. Avoid exceeding the G-load limits and airspeed limitations during the pull out.

Again, it is important to remember that the spin recovery procedures and techniques described above are recommended for use only in the absence of the manufacturer's procedures. The pilot must always be familiar with the manufacturer's procedures for spin recovery.

Intentional Spins

If the manufacturer does not specifically approve an aircraft for spins, intentional spins are not authorized by the CFRs or suggested by this handbook. The official sources for determining whether the spin maneuver is approved are:

- Type Certificate Data Sheets or the aircraft specifications

- The limitation section of the FAA-approved AFM/ POH regarding and limiting gross weight, CG range, or amount of fuel

- On a placard located in clear view of the pilot in the airplane (e.g., "NO ACROBATIC MANEUVERS INCLUDING SPINS APPROVED")

In airplanes placarded against spins, there is no assurance that recovery from a fully-developed spin is possible. Unfortunately, accident records show occurrences in which pilots intentionally ignored spin restrictions. Despite the installation of placards prohibiting intentional spins in these airplanes, some pilots and even some flight instructors attempt to justify the maneuver, rationalizing that the spin restriction results from a "technicality" in the airworthiness standards. They believe that if the airplane was spin tested during its certification process, no problem should result from demonstrating or practicing spins.

Such pilots overlook the fact that certification of normal category single-engine airplanes that occurred in accordance with 14 CFR part 23, section 23.221(a) (which still applies to aircraft certified under that regulation) only required the airplane to recover from a one-turn spin or a three-second spin, whichever takes longer, in not more than one additional turn after initiation of the first control action for recover, or demonstrate compliance with the optional spin resistant requirements of that section. In other words, many of these airplanes were never required to recover from a fully developed spin. 14 CFR part 23, section 23.2150 states the current certification requirements pertaining to spin characteristics for airplanes certified under that regulation going forward. In all airplanes placarded against spins, there is absolutely no assurance that recovery from a fully developed spin is possible under any circumstances. The pilot of an airplane placarded against intentional spins should assume that the airplane could become uncontrollable in a spin.

Weight and Balance Requirement Related to Spins

In airplanes that are approved for spins, compliance with weight and balance requirements is important for safe performance and recovery from the spin maneuver. Pilots should know that even minor weight or balance changes can affect the airplane's spin recovery characteristics. Such changes can either degrade or enhance the spin maneuver and/or recovery characteristics. For example, the addition of weight in the aft baggage compartment, or additional fuel, may still permit the airplane to be operated within CG, but could seriously affect the spin and recovery characteristics. An airplane that may be difficult to spin intentionally in the utility category (restricted aft CG and reduced weight) could have less resistance to spin entry in the normal category (less restricted aft CG and increased weight). This situation arises from the airplane's ability to generate a higher AOA. An airplane that is approved for spins in the utility category but loaded in accordance with the normal category may not recover from a spin that is allowed to progress beyond one turn.

Common Errors

Common errors in the performance of intentional spins are:

1. Failure to apply full rudder pressure (to the stops) in the desired spin direction during spin entry
2. Failure to apply and maintain full up-elevator pressure during spin entry, resulting in a spiral
3. Failure to achieve a fully-stalled condition prior to spin entry
4. Failure to apply full rudder (to the stops) briskly against the spin during recovery
5. Failure to apply sufficient forward-elevator during recovery
6. Waiting for rotation to stop before applying forward-elevator
7. Failure to neutralize the rudder after rotation stops, possibly resulting in a secondary spin
8. Slow and overly cautious control movements during recovery
9. Excessive back-elevator pressure after rotation stops, possibly resulting in secondary stall
10. Insufficient back-elevator pressure during recovery resulting in excessive airspeed

Spiral Dive

A spiral dive, a nose-low upset, is a descending turn during which airspeed and G-load can increase rapidly and often results from a botched turn. In a spiral dive, the airplane is flying very tight circles, in a nearly vertical attitude and will be accelerating because it is no longer stalled. Pilots typically get into a spiral dive during an inadvertent IMC encounter, most often when the pilot relies on kinesthetic sensations rather than on the flight instruments. A pilot distracted by other sensations can easily enter a slightly nose-low, wing-low, descending turn and, at least initially, fail to recognize this error. Especially in IMC, it may be only the sound of increasing speed that makes the pilot aware of the rapidly developing situation. Upon recognizing the steep nose-down attitude and steep bank, the startled pilot may react by pulling back rapidly on the yoke while simultaneously rolling to wings-level. This response can create aerodynamic loads capable of causing airframe structural damage and/or failure.

The following discussion explains each of the five steps a pilot should use to recover from a spiral dive:

1. Reduce power (throttle) to idle. Immediately reduce power to idle to slow the rate of acceleration.

2. Apply some forward-elevator. Prior to rolling the wings level, it is important to unload the G-load on the airplane ("unload the wing"). This is accomplished by applying some forward-elevator pressure to return to about +1G. Apply just enough forward-elevator to ensure that you are not aggravating the spiral with aft-elevator. While generally a small input, this push has several benefits prior to rolling the wings level in the next step – the push reduces the AOA, reduces the G-load, and slows the turn rate while increasing the turn radius, and preventing a rolling pullout. The design limit of the airplane is exceeded more easily during a rolling pullout, so failure to reduce the G-load prior to rolling the wings level could result in structural damage or failure.

3. Roll to wings level using coordinated aileron and rudder inputs. Even though the airplane is in a nose-low attitude, continue the roll until the wings are completely level again before performing step four.

4. Gently raise the nose to level flight. It is possible that the airplane in a spiral dive might be at or even beyond V_{NE} (never exceed speed) speed. Therefore, control inputs are made slowly and gently at this point to prevent structural failure. Raise the nose to a climb attitude only after speed decreases to safe levels.

5. Increase power to climb power. Once the airspeed has stabilized to V_Y, apply climb power and climb back to a safe altitude.

In general, spiral dive recovery procedures are summarized in *Figure 5-18*.

Spiral Dive Recovery Template

1. Reduce power (throttle) to idle

2. Apply some forward elevator

3. Roll wings level

4. Gently raise the nose to level flight

5. Increase power to climb power

Figure 5-18. *Spiral dive recovery template.*

Common Errors

Common errors in the recovery from spiral dives are:

1. Failure to reduce power first
2. Mistakenly adding power
3. Attempting to pull out of dive without rolling wings level
4. Simultaneously pulling out of dive while rolling wings level
5. Not unloading the Gs prior to rolling level
6. Not adding power once climb is established

UPRT Summary

A significant point to note is that UPRT skills are both complex and perishable. Repetition is needed to establish the correct mental models, and recurrent practice/training is necessary as well. The context in which UPRT procedures are introduced and implemented is also an important consideration. The pilot should clearly understand, for example, whether a particular procedure has broad applicability, or is type-specific. To attain the highest levels of learning possible, the best approach starts with the broadest form of a given procedure, then narrows it down to type-specific requirements.

Chapter Summary

A pilot's most fundamental and important responsibility is to maintain aircraft control. Initial flight training thus provides skills to operate an airplane in a safe manner, generally within normal "expected" environments, with the addition of some instruction in upset and stall situations.

This chapter discussed the elements of basic airplane control, with emphasis on AOA. It offered a discussion of circumstances and scenarios that can lead to LOC-I, including stalls and airplane upsets. It discussed the importance of developing proficiency in slow flight, stalls, and stall recoveries, spin awareness and recovery, upset prevention and recovery, and spiral dive recovery.

Pilots need to understand that primary training cannot cover all possible contingencies that an airplane or pilot may encounter. They should seek recurrent/additional training for their normal areas of operation and seek appropriate training that develops their aeronautical skill set beyond the requirements for initial certification.

For additional considerations on performing some of these maneuvers in multiengine airplanes and turbojet-powered airplanes, refer to Chapters 12 and 15, respectively.

Additional advisory circular (AC) guidance is available at www.faa.gov:

1. AC 61-67 (as revised), Stall and Spin Awareness Training;
2. AC 120-109 (as revised), Stall Prevention and Recovery Training; and
3. AC 120-111 (as revised), Upset Prevention and Recovery Training.

Airplane Flying Handbook (FAA-H-8083-3C)
Chapter 6: Takeoffs and Departure Climbs

Introduction

About twenty percent of all yearly general aviation (GA) accidents occur during takeoff and departure climbs, and more than half of those accidents are the result of some sort of failure of the pilot. A significant number of takeoff accidents are the result of loss of control of the airplane. When compared to the entire profile of a normal flight, this phase of a flight is relatively short, but the pilot workload is intense. This chapter discusses takeoffs and departure climbs in airplanes under normal conditions and under conditions that require maximum performance.

Though it may seem relatively simple, the takeoff often presents the most hazards of any part of a flight. The importance of thorough knowledge of procedures and techniques coupled with proficiency in performance cannot be overemphasized.

The discussion in this chapter is centered on airplanes with tricycle landing gear (nose-wheel). Procedures for conventional gear airplanes (tail-wheel) are discussed in Chapter 14: Transition to Tailwheel Airplanes. The manufacturer's recommended procedures pertaining to airplane configuration, airspeeds, and other information relevant to takeoffs and departure climbs in a specific make and model airplane are contained in the Federal Aviation Administration (FAA)-approved Airplane Flight Manual and/or Pilot's Operating Handbook (AFM/POH) for that airplane. If any of the information in this chapter differs from the airplane manufacturer's recommendations as contained in the AFM/POH, the airplane manufacturer's recommendations take precedence.

Terms and Definitions

Although the takeoff and climb is one continuous maneuver, it will be divided into three separate steps for purposes of explanation: 1.) takeoff roll; 2.) lift-off; and 3.) initial climb after becoming airborne. Refer to *Figure 6-1* and the detail below.

- Takeoff roll (ground roll) is the portion of the takeoff procedure during which the airplane is accelerated from a standstill to an airspeed that provides sufficient lift for it to become airborne.

- Lift-off is when the wings are lifting the weight of the airplane off the surface. In most airplanes, this is the result of the pilot rotating the nose up to increase the angle of attack (AOA).

- The initial climb begins when the airplane leaves the surface and a climb pitch attitude has been established. Normally, it is considered complete when the airplane has reached a safe maneuvering altitude or an en route climb has been established.

Prior to Takeoff

Before going to the airplane, the pilot should check the POH/AFM performance charts to determine the predicted performance and decide if the airplane is capable of a safe takeoff and climb for the conditions and location. *[Figure 6-2]* High density altitudes reduce engine and propeller performance, increase takeoff rolls, and decrease climb performance. A more detailed discussion of density altitude and how it affects airplane performance can be found in the *Pilot's Handbook of Aeronautical Knowledge* (FAA-H-8083-25, as revised).

All run-up and pre-takeoff checklist items should be completed before taxiing onto the runway or takeoff area. As a minimum before every takeoff, all engine instruments should be checked for proper and usual indications, and all controls should be checked for full, free, and correct movement. The pilot should also consider available options if an engine failure occurs after takeoff. These options include the preferred direction for any emergency turns to landing sites based on the departure path, altitude, wind conditions, and terrain. In addition, the pilot should make certain that the approach and takeoff paths are clear of other aircraft. At nontowered airports, pilots should announce their intentions on the common traffic advisory frequency (CTAF) assigned to that airport. When operating from a towered airport, pilots need to contact the tower operator and receive a takeoff clearance before taxiing onto the active runway.

Taking off immediately behind another aircraft, particularly a large and heavy transport airplane, creates the risk of a wake turbulence encounter, and a possible loss of control. However, if an immediate takeoff behind a large heavy aircraft is necessary, the pilot should plan to minimize the chances of flying through an aircraft's wake turbulence by avoiding the other aircraft's flightpath or rotating prior to the point at which the preceding aircraft rotated. While taxiing onto the runway, the pilot should select ground reference points that are aligned with the runway direction to aid in maintaining directional control and alignment with the runway center line during the climb out. These may be runway centerline markings, runway lighting, distant trees, towers, buildings, or mountain peaks.

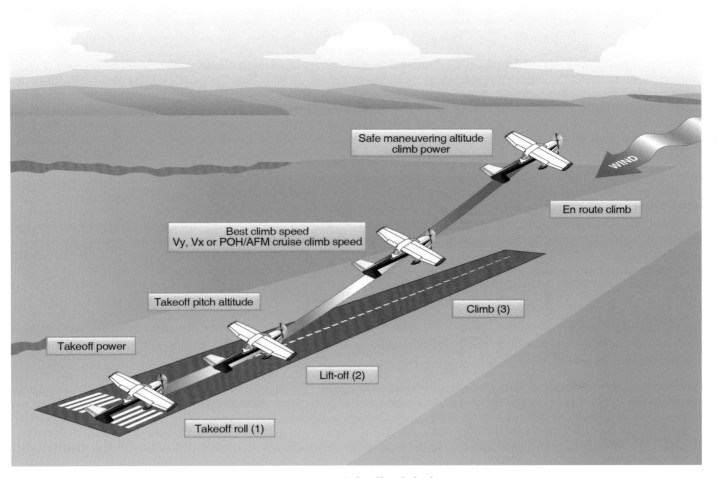

Figure 6-1. *Takeoff and climb.*

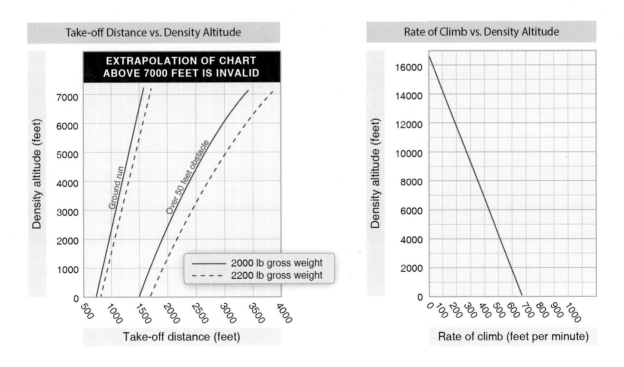

Figure 6-2. *Performance chart examples.*

Normal Takeoff

A normal takeoff is one in which the airplane is headed into the wind; there are times that a takeoff with a tail wind is necessary. However, the pilot should consult the POH/AFM to ensure the aircraft is approved for a takeoff with a tail wind and that there is sufficient performance and runway length for the takeoff. The pilot should also ensure that the takeoff surfaces are firm and of sufficient length to permit the airplane to gradually accelerate to normal lift-off and climb-out speed, and there are no obstructions along the takeoff path.

There are two reasons for making a takeoff as nearly into the wind as possible. First, since the airplane depends on airspeed, a headwind provides some of that airspeed even before the airplane begins to accelerate into the wind. Second, a headwind decreases the ground speed necessary to achieve flying speed. Slower ground speeds yield shorter ground roll distances and allow use of shorter runways while reducing wear and stress on the landing gear.

Takeoff Roll

For takeoff, the pilot uses the rudder pedals in most general aviation airplanes to steer the airplane's nose-wheel onto the runway centerline to align the airplane and nose-wheel with the runway. After releasing the brakes, the pilot should advance the throttle smoothly and continuously to takeoff power. An abrupt application of power may cause the airplane to yaw sharply to the left because of the torque effects of the engine and propeller. This is most apparent in high horsepower engines. As the airplane starts to roll forward, assure both feet are on the rudder pedals so that the toes or balls of the feet are on the rudder portions, not on the brake. Check the engine instruments for indications of a malfunction during the takeoff roll.

In nose-wheel type airplanes, pressures on the elevator control are not necessary beyond those needed to steady it. Applying unnecessary pressure only aggravates the takeoff and prevents the pilot from recognizing when elevator control pressure is actually needed to establish the takeoff attitude.

As the airplane gains speed, the elevator control tends to assume a neutral position if the airplane is correctly trimmed. At the same time, the rudder pedals are used to keep the nose of the airplane pointed down the runway and parallel to the centerline. The effects of engine torque and P-factor at the initial speeds tend to pull the nose to the left. The pilot should use whatever rudder pressure is needed to correct for these effects or winds. The pilot should use aileron controls into any crosswind to keep the airplane centered on the runway centerline. The pilot should avoid using the brakes for steering purposes as this will slow acceleration, lengthen the takeoff distance, and possibly result in severe swerving.

As the speed of the takeoff roll increases, more and more pressure will be felt on the flight controls, particularly the elevators and rudder. If the tail surfaces are affected by the propeller slipstream, they become effective first. As the speed continues to increase, all of the flight controls will gradually become effective enough to maneuver the airplane about its three axes. At this point, the airplane is being flown more than it is being taxied. As this occurs, progressively smaller rudder deflections are needed to maintain direction.

The feel of resistance to the movement of the controls and the airplane's reaction to such movements are the only real indicators of the degree of control attained. This feel of resistance is not a measure of the airplane's speed, but rather of its controllability. To determine the degree of controllability, the pilot should be conscious of the reaction of the airplane to the control pressures and immediately adjust the pressures as needed to control the airplane. The pilot should wait for the reaction of the airplane to the applied control pressures and attempt to sense the control resistance to pressure rather than attempt to control the airplane by movement of the controls.

A student pilot does not normally have a full appreciation of the variations of control pressures with the speed of the airplane. The student may tend to move the controls through wide ranges seeking the pressures that are familiar and expected and, as a consequence, over-control the airplane.

The situation may be aggravated by the sluggish reaction of the airplane to these movements. The flight instructor should help the student learn proper response to control actions and airplane reactions. The instructor should always stress using the proper outside reference to judge airplane motion. For takeoff, the student should always be looking far down the runway at two points aligned with the runway. The flight instructor should have the student pilot follow through lightly on the controls, feel for resistance, and point out the outside references that provide the clues for how much control movement is needed and how the pressure and response changes as airspeed increases. With practice, the student pilot should become familiar with the airplane's response to acceleration up to lift-off speed, corrective control movements needed, and the outside references necessary to accomplish the takeoff maneuver.

Lift-Off

Since a good takeoff depends on the proper takeoff attitude, it is important to know how this attitude appears and how it is attained. The ideal takeoff attitude requires only minimum pitch adjustments shortly after the airplane lifts off to attain the speed for the best rate of climb (V_Y). *[Figure 6-3]* The pitch attitude necessary for the airplane to accelerate to V_Y speed should be demonstrated by the instructor and memorized by the student. Flight instructors should be aware that initially, the student pilot may have a tendency to hold excessive back-elevator pressure just after lift-off, resulting in an abrupt pitch-up.

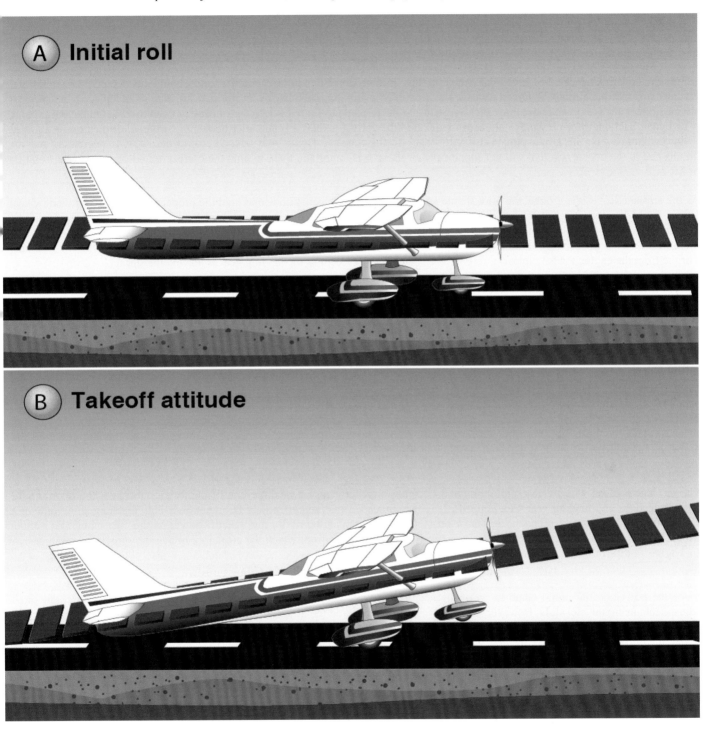

Figure 6-3. *Initial roll and takeoff attitude.*

Each type of airplane has a best pitch attitude for normal lift-off; however, varying conditions may make a difference in the required takeoff technique. A rough field, a smooth field, a hard surface runway, or a short or soft, muddy field all call for a slightly different technique, as will smooth air in contrast to a strong, gusty wind. The different techniques for those other-than-normal conditions are discussed later in this chapter.

When all the flight controls become effective during the takeoff roll in a nose-wheel type airplane, the pilot should gradually apply back-elevator pressure to raise the nose-wheel slightly off the runway, thus establishing the takeoff or lift-off attitude. This is the "rotation" for lift-off and climb. As the airplane lifts off the surface, the pitch attitude to hold the climb airspeed should be held with elevator control and trimmed to maintain that pitch attitude without excessive control pressures. The wings should be leveled after lift-off and the rudder used to ensure coordinated flight.

After rotation, the slightly nose-high pitch attitude should be held until the airplane lifts off. Rudder control should be used to maintain the track of the airplane along the runway centerline until any required crab angle in level flight is established. Forcing it into the air by applying excessive back-elevator pressure would only result in an excessively high-pitch attitude and may delay the takeoff. As discussed earlier, excessive and rapid changes in pitch attitude result in proportionate changes in the effects of torque, thus making the airplane more difficult to control.

Although the airplane can be forced into the air, this is considered an unsafe practice and should be avoided under normal circumstances. If the airplane is forced to leave the ground by using too much back-elevator pressure before adequate flying speed is attained, the wing's AOA may become excessive, causing the airplane to settle back to the runway or even to stall. On the other hand, if sufficient back-elevator pressure is not held to maintain the correct takeoff attitude after becoming airborne, or the nose is allowed to lower excessively, the airplane may also settle back to the runway. This would occur because the AOA is decreased and lift diminished to the degree where it will not support the airplane. It is important, then, to hold the correct attitude constant after rotation or lift-off.

As the airplane leaves the ground, the pilot should keep the wings in a level attitude and hold the proper pitch attitude. Outside visual scans should be intensified at this critical point to attain/maintain proper airplane pitch and bank attitude. Due to the minimum airspeed, the flight controls are not as responsive, requiring more control movement to achieve an expected response. A novice pilot often has a tendency to fixate on the airplane's pitch attitude and/or the airspeed indicator and neglect bank control of the airplane. Torque from the engine tends to impart a rolling force that is most evident as the landing gear is leaving the surface.

During takeoffs in a strong, gusty wind, it is advisable that an extra margin of speed be obtained before the airplane is allowed to leave the ground. A takeoff at the normal takeoff speed may result in a lack of positive control, or a stall, when the airplane encounters a sudden lull in strong, gusty wind, or other turbulent air currents. In this case, the pilot should allow the airplane to stay on the ground longer to attain more speed, then make a smooth, positive rotation to leave the ground.

Initial Climb

Upon liftoff, the airplane should be flying at approximately the pitch attitude that allows it to accelerate to V_Y. This is the speed at which the airplane gains the most altitude in the shortest period of time.

If the airplane has been properly trimmed for takeoff, some back-elevator pressure may be required to hold this attitude until the proper climb speed is established. Relaxation of any back-elevator pressure before this time may result in the airplane settling, even to the extent that it contacts the runway.

The airplane's speed will increase rapidly after it becomes airborne. Once a positive rate of climb is established, the pilot should retract the flaps and landing gear (if equipped). It is recommended that takeoff power be maintained until reaching an altitude of at least 500 feet above the surrounding terrain or obstacles. The combination of V_Y and takeoff power assures the maximum altitude gained in a minimum amount of time. This gives the pilot more altitude from which the airplane can be safely maneuvered in case of an engine failure or other emergency. The pilot should also consider flying at a lower pitch for cruise climb since flying at V_Y requires much quicker pilot response in the event of a powerplant failure to preclude a stall.

Since the power on the initial climb is set at the takeoff power setting, the airspeed should be controlled by making slight pitch adjustments using the elevators. However, the pilot should not fixate on the airspeed indicator when making these pitch changes, but should continue to scan outside to adjust the airplane's attitude in relation to the horizon. In accordance with the principles of attitude flying, the pilot should first make the necessary pitch change with reference to the natural horizon, hold the new attitude momentarily, and then glance at the airspeed indicator to verify if the new attitude is correct. Due to inertia, the airplane will not accelerate or decelerate immediately as the pitch is changed. It takes a little time for the airspeed to change. If the pitch attitude has been over or under corrected, the airspeed indicator will show a speed that is higher or lower than that desired. When this occurs, the cross-checking and appropriate pitch-changing process needs to be repeated until the desired climbing attitude is established. Pilots should remember the climb pitch will be lower when the airplane is heavily loaded, or power is limited by density altitude.

When the correct pitch attitude has been attained, the pilot should hold it constant while cross-checking it against the horizon and other outside visual references. The airspeed indicator should be used only as a check to determine if the attitude is correct.

After the recommended climb airspeed has been established and a safe maneuvering altitude has been reached, the pilot should adjust the power to the recommended climb setting and trim the airplane to relieve the control pressures. This makes it easier to hold a constant attitude and airspeed.

During initial climb, it is important that the takeoff path remain aligned with the runway to avoid drifting into obstructions or into the path of another aircraft that may be taking off from a parallel runway. A flight instructor should help the student identify two points inline ahead of the runway to use as a tracking reference. As long as those two points are inline, the airplane is remaining on the desired track. Proper scanning techniques are essential to a safe takeoff and climb, not only for maintaining attitude and direction, but also for avoiding collisions near the airport.

When the student pilot nears the solo stage of flight training, it should be explained that the airplane's takeoff performance will be much different when the instructor is not in the airplane. Due to decreased load, the airplane will become airborne earlier and climb more rapidly. The pitch attitude that the student has learned to associate with initial climb may also differ due to decreased weight, and the flight controls may seem more sensitive. If the situation is unexpected, it may result in increased anxiety that may remain until after the landing. Frequently, the existence of this anxiety and the uncertainty that develops due to the perception of an "abnormal" takeoff results in poor performance on the subsequent landing.

Common Errors

Common errors in the performance of normal takeoffs and departure climbs are:

- Failure to review AFM/POH and performance charts prior to takeoff.
- Failure to adequately clear the area prior to taxiing into position on the active runway.
- Abrupt use of the throttle.
- Failure to check engine instruments for signs of malfunction after applying takeoff power.
- Failure to anticipate the airplane's left turning tendency on initial acceleration.
- Overcorrecting for left turning tendency.
- Relying solely on the airspeed indicator rather than developing an understanding of visual references and tracking clues of airplane airspeed and controllability during acceleration and lift-off.
- Failure to attain proper lift-off attitude.
- Inadequate compensation for torque/P-factor during initial climb resulting in a sideslip.
- Over-control of elevators during initial climb-out and lack of elevator trimming.
- Limiting scan to areas directly ahead of the airplane (pitch attitude and direction), causing a wing (usually the left) to drop immediately after lift-off.
- Failure to attain/maintain best rate-of-climb airspeed (V_Y) or desired climb airspeed.
- Failure to employ the principles of attitude flying during climb-out, resulting in "chasing" the airspeed indicator.

Crosswind Takeoff

While it is usually preferable to take off directly into the wind whenever possible or practical, there are many instances when circumstances or judgment indicate otherwise. Therefore, the pilot must be familiar with the principles and techniques involved in crosswind takeoffs, as well as those for normal takeoffs. A crosswind affects the airplane during takeoff much as it does during taxiing. With this in mind, the pilot should be aware that the technique used for crosswind correction during takeoffs closely parallels the crosswind correction techniques used for taxiing.

Takeoff Roll

The technique used during the initial takeoff roll in a crosswind is generally the same as the technique used in a normal takeoff roll, except that the pilot needs to apply aileron pressure into the crosswind. This raises the aileron on the upwind wing, imposes a downward force on that wing to counteract the lifting force of the crosswind, and thus prevents the wing from rising. The pilot should remember that since the ailerons and rudder are deflected, drag will increase; therefore, less initial takeoff performance should be expected until the airplane is wings-level in coordinated flight in the climb.

While taxiing into takeoff position, it is essential that the pilot check the windsock and other wind direction indicators for the presence of a crosswind. If a crosswind is present, the pilot should apply full aileron pressure into the wind while beginning the takeoff roll. The pilot should maintain this control position, as the airplane accelerates, and until the ailerons become effective in maneuvering the airplane about its longitudinal axis. As the ailerons become effective, the pilot will feel an increase in pressure on the aileron control.

While holding aileron pressure into the wind, the pilot should use the rudder to maintain a straight takeoff path. *[Figure 6-4]* Since the airplane tends to weathervane into the wind while on the ground, the pilot will typically apply downwind rudder pressure. When the pilot increases power for takeoff, the resulting P-factor causes the airplane to yaw to the left. While this yaw may be sufficient to counteract the airplane's tendency to weathervane into the wind in a crosswind from the right, it may aggravate this tendency in a crosswind from the left. In any case, the pilot should apply rudder pressure in the appropriate direction to keep the airplane rolling straight down the runway.

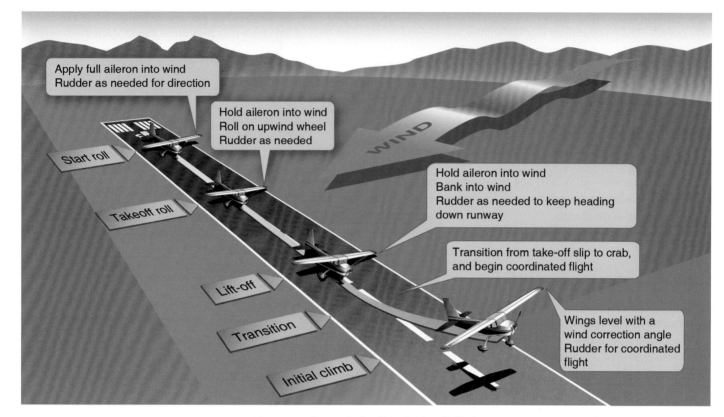

Figure 6-4. *Crosswind roll and takeoff climb.*

As the forward speed of the airplane increases, the pilot should apply sufficient aileron pressure into the crosswind to keep the wings level. The effect of the crosswind component will not completely vanish; therefore, the pilot needs to maintain some aileron pressure throughout the takeoff roll. If the upwind wing rises, the amount of wing surface exposed to the crosswind will increase, which may cause the airplane to lose lateral alignment with the runway centerline and to "skip." *[Figure 6-5]* The pilot uses rudder pressure to keep the airplane's longitudinal axis parallel to the runway centerline.

This "skipping" is usually indicated by a series of very small bounces caused by the airplane attempting to fly and then settling back onto the runway. During these bounces, the crosswind also tends to move the airplane sideways, and these bounces develop into side-skipping. This side-skipping imposes severe side stresses on the landing gear and may result in structural failure.

During a crosswind takeoff roll, it is important that the pilot hold sufficient aileron pressure into the wind not only to keep the upwind wing from rising but to hold that wing down so that the airplane sideslips into the wind enough to counteract drift immediately after lift-off.

Lift-Off

As the nose-wheel raises off of the runway, the pilot should hold aileron pressure into the wind. This may cause the downwind wing to rise and the downwind main wheel to lift off the runway first, with the remainder of the takeoff roll being made on that one main wheel. This is acceptable and is preferable to side-skipping.

If a significant crosswind exists, the pilot should hold the main wheels on the ground slightly longer than in a normal takeoff so that a smooth but very definite lift-off can be made. This allows the airplane to leave the ground under more positive control and helps it remain airborne while the pilot establishes the proper amount of wind correction. More importantly, this procedure avoids imposing excessive side-loads on the landing gear and prevents possible damage that would result from the airplane settling back to the runway while drifting.

As both main wheels leave the runway, the airplane begins to drift sideways with the wind, as ground friction is no longer a factor in preventing lateral movement. To minimize this lateral movement and to keep the upwind wing from rising, the pilot should establish and maintain the proper amount of crosswind correction prior to lift-off by applying aileron pressure into the wind. The pilot should also apply rudder pressure, as needed, to prevent weathervaning.

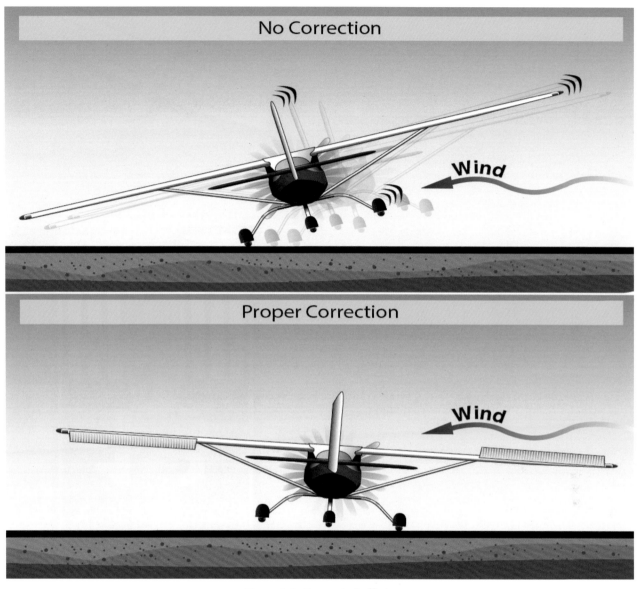

Figure 6-5. *Crosswind effect.*

Initial Climb

If a proper crosswind correction is applied, the aircraft will maintain alignment with the runway while accelerating to takeoff speed and then maintain that alignment once airborne. As takeoff acceleration occurs, the efficiency of the up-aileron will increase with aircraft speed causing the upwind wing to produce greater downward force and, as a result, counteract the effect of the crosswind. The yoke, having been initially turned into the wind, can be relaxed to the extent necessary to keep the aircraft aligned with the runway. As the aircraft becomes flyable and airborne, the wing that is upwind will have a tendency to be lower relative the other wing, requiring simultaneous rudder input to maintain runway alignment. This will initially cause the aircraft to sideslip. However, as the aircraft establishes its climb, the nose should be turned into the wind to offset the crosswind, wings brought to level, and rudder input adjusted to maintain runway alignment (crabbing). *[Figure 6-6]* Firm and positive use of the rudder may be required to keep the airplane pointed down the runway or parallel to the centerline. Unlike landing, the runway alignment (staying over the runway and its extended centerline) is paramount to keeping the aircraft parallel to the centerline. The pilot should then apply rudder pressure firmly and aggressively to keep the airplane headed straight down the runway. However, because the force of a crosswind may vary markedly within a few hundred feet of the ground, the pilot should check the ground track frequently and adjust the wind correction angle, as necessary. The remainder of the climb technique is the same used for normal takeoffs and climbs.

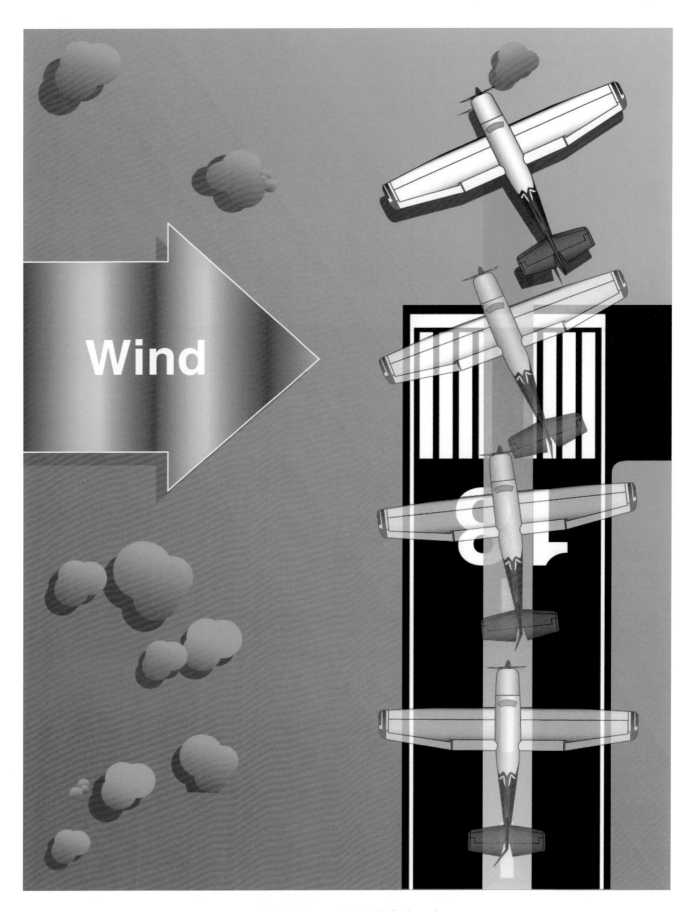

Figure 6-6. *Crosswind climb flightpath.*

The most common errors made while performing crosswind takeoffs include the following:

- Failure to review AFM/POH performance and charts prior to takeoff.
- Failure to adequately clear the area prior to taxiing onto the active runway.
- Using less than full aileron pressure into the wind initially on the takeoff roll.
- Mechanical use of aileron control rather than judging lateral position of airplane on runway from visual clues and applying sufficient aileron to keep airplane centered laterally on runway.
- Side-skipping due to improper aileron application.
- Inadequate rudder control to maintain airplane parallel to centerline and pointed straight ahead in alignment with visual references.
- Excessive aileron input in the latter stage of the takeoff roll resulting in a steep bank into the wind at lift-off.
- Inadequate drift correction after lift-off.

Ground Effect on Takeoff

Ground effect is a condition of improved performance encountered when the airplane is operating very close to the ground. Ground effect can be detected and normally occurs up to an altitude equal to one wingspan above the surface. *[Figure 6-7]* Ground effect is most significant when the airplane maintains a constant attitude at low airspeed at low altitude (for example, during takeoff when the airplane lifts off and accelerates to climb speed, and during the landing flare before touchdown).

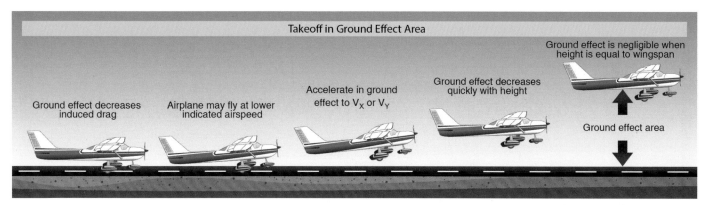

Figure 6-7. *Takeoff in ground effect area.*

When the wing is under the influence of ground effect, there is a reduction in upwash, downwash, and wingtip vortices. As a result of the reduced wingtip vortices, induced drag is reduced. When the wing is at a height equal to 1/4 the span, the reduction in induced drag is about 25 percent. When the wing is at a height equal to 1/10 the span, the reduction in induced drag is about 50 percent. At high speeds where parasite drag dominates, induced drag is a small part of the total drag. Consequently, ground effect is a greater concern during takeoff and landing.

At takeoff, the takeoff roll, lift-off, and the beginning of the initial climb are accomplished within the ground effect area. The ground effect causes local increases in static pressure, which cause the airspeed indicator and altimeter to indicate slightly lower values than they should and usually cause the vertical speed indicator to indicate a descent. As the airplane lifts off and climbs out of the ground effect area, the following occurs:

- The airplane requires an increase in AOA to maintain lift coefficient.
- The airplane experiences an increase in induced drag and thrust required.
- The airplane experiences a pitch-up tendency and requires less elevator travel because of an increase in downwash at the horizontal tail.
- The airplane experiences a reduction in static source pressure and a corresponding increase in indicated airspeed.

V_X is the speed at which the airplane achieves the greatest gain in altitude for a given distance over the ground. It is usually slightly less than V_Y, which is the greatest gain in altitude per unit of time. The specific speeds to be used for a given airplane are stated in the FAA-approved AFM/POH. The pilot should be aware that, in some airplanes, a deviation of 5 knots from the recommended speed may result in a significant reduction in climb performance; therefore, the pilot should maintain precise control of the airspeed to ensure the maneuver is executed safely and successfully.

Due to the reduced drag in ground effect, the airplane may seem to be able to take off below the recommended airspeed. However, as the airplane climbs out of ground effect below the recommended climb speed, initial climb performance will be much less than at V_Y or even V_X. Under conditions of high density altitude, high temperature, and/or maximum gross weight, the airplane may be able to lift off but will be unable to climb out of ground effect. Consequently, the airplane may not be able to clear obstructions. Lift-off before attaining recommended flight airspeed incurs more drag, which requires more power to overcome. Since the initial takeoff and climb is based on maximum power, reducing drag is the only option. To reduce drag, pitch should be reduced which means losing altitude. Pilots should remember that many airplanes cannot safely takeoff at maximum gross weight at certain altitudes and temperatures, due to lack of performance. Therefore, under marginal conditions, it is important that the airplane takes off at the speed recommended for adequate initial climb performance.

Ground effect is important to normal flight operations. If the runway is long enough or if no obstacles exist, ground effect can be used to the pilot's advantage by using the reduced drag to improve initial acceleration.

When taking off from an unsatisfactory surface, the pilot should apply as much weight to the wings as possible during the ground run and lift-off, using ground effect as an aid, prior to attaining true flying speed. The pilot should reduce AOA to attain normal airspeed before attempting to fly out of the ground effect areas.

Short-Field Takeoff and Maximum Performance Climb

When performing takeoffs and climbs from fields where the takeoff area is short or the available takeoff area is restricted by obstructions, the pilot should operate the airplane at the maximum limit of its takeoff performance capabilities. To depart from such an area safely, the pilot needs to exercise positive and precise control of airplane attitude and airspeed, so that takeoff and climb performance result in the shortest ground roll and the steepest angle of climb. *[Figure 6-8]* The pilot should consult and follow the performance section of the AFM/POH to obtain the power setting, flap setting, airspeed, and procedures prescribed by the airplane's manufacturer.

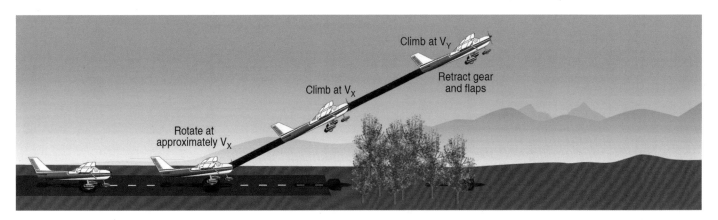

Figure 6-8. *Short-field takeoff.*

The pilot should have adequate knowledge in the use and effectiveness of the best angle-of-climb speed (V_X) and the best rate-of-climb speed (V_Y) for the specific make and model of airplane being flown in order to safely accomplish a takeoff at maximum performance.

Takeoff Roll

Taking off from a short field requires the takeoff to be started from the very beginning of the takeoff area. At this point, the airplane is aligned with the intended takeoff path. If the airplane manufacturer recommends the use of flaps, they are extended the proper amount before beginning the takeoff roll. This allows the pilot to devote full attention to the proper technique and the airplane's performance throughout the takeoff.

The pilot should apply takeoff power smoothly and continuously, without hesitation, to accelerate the airplane as rapidly as possible. Some pilots prefer to hold the brakes until the maximum obtainable engine revolutions per minute (rpm) are achieved before allowing the airplane to begin its takeoff run. However, it has not been established that this procedure results in a shorter takeoff run in all light, single-engine airplanes. The airplane is allowed to roll with its full weight on the main wheels and accelerate to the lift-off speed. As the takeoff roll progresses, the pilot should adjust the airplane's pitch attitude and AOA to attain minimum drag and maximum acceleration. In nose-wheel type airplanes, this involves little use of the elevator control since the airplane is already in a low-drag attitude.

Lift-Off

As V_X approaches, the pilot should apply back-elevator pressure until reaching the appropriate V_X attitude to ensure a smooth and firm lift-off, or rotation. Since the airplane accelerates more rapidly after lift-off, the pilot should apply additional back-elevator pressure to hold a constant airspeed. After becoming airborne, the pilot will maintain a wings-level climb at V_X until all obstacles have been cleared, or if no obstacles are present, until reaching an altitude of at least 50 feet above the takeoff surface. Thereafter, the pilot may lower the pitch attitude slightly and continue the climb at V_Y until reaching a safe maneuvering altitude. The pilot should always remember that an attempt to pull the airplane off the ground prematurely, or to climb too steeply, may cause the airplane to settle back to the runway or make contact with obstacles. Even if the airplane remains airborne, until the pilot reaches V_X, the initial climb will remain flat, which diminishes the pilot's ability to successfully perform the climb and/or clear obstacles. *[Figure 6-9]*

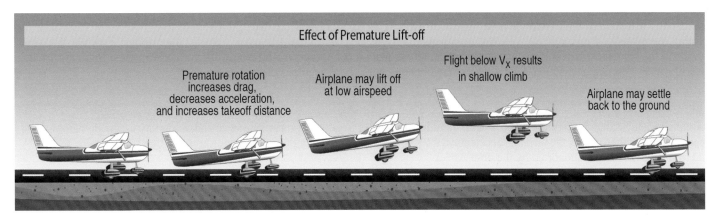

Figure 6-9. *Effect of premature lift-off.*

The objective is to rotate to the appropriate pitch attitude at (or near) V_X. The pilot should be aware that some airplanes have a natural tendency to lift off well before reaching V_X. In these airplanes, it may be necessary to allow the airplane to lift off in ground effect and then reduce pitch attitude to level until the airplane accelerates to V_X with the wheels just clear of the runway surface. This method is preferable to forcing the airplane to remain on the ground with forward elevator-control pressure until V_X is attained. Holding the airplane on the ground unnecessarily puts excessive pressure on the nose-wheel and may result in "wheel barrowing." It also hinders both acceleration and overall airplane performance.

Initial Climb

On short-field takeoffs, the landing gear and flaps should remain in takeoff position until the airplane is clear of obstacles (or as recommended by the manufacturer) and V_Y has been established. Until all obstacles have been cleared, the pilot should maintain focus outside the airplane instead of reaching for landing gear or flap controls or looking inside the airplane for any reason. When the airplane is stabilized at V_Y, the landing gear (if retractable) and flaps should be retracted. It is usually advisable to raise the flaps in increments to avoid sudden loss of lift and settling of the airplane. The pilot should next reduce the power to the normal climb setting or as recommended by the airplane manufacturer.

Common errors in the performance of short-field takeoffs and maximum performance climbs are:

- Failure to review AFM/POH and performance charts prior to takeoff.
- Failure to adequately clear the area.
- Failure to utilize all available runway/takeoff area.
- Failure to have the airplane properly trimmed prior to takeoff.
- Premature lift-off resulting in high drag.
- Holding the airplane on the ground unnecessarily with excessive forward-elevator pressure.
- Inadequate rotation resulting in excessive speed after lift-off.
- Inability to attain/maintain V_X.
- Fixation on the airspeed indicator during initial climb.
- Premature retraction of landing gear and/or wing flaps.

Soft/Rough-Field Takeoff and Climb

Takeoffs and climbs from soft fields require the use of operational techniques for getting the airplane airborne as quickly as possible to eliminate the drag caused by tall grass, soft sand, mud, and snow and may require climbing over an obstacle. The technique makes judicious use of ground effect to reduce landing gear drag and requires an understanding of the airplane's slow speed characteristics and responses. These same techniques are also useful on a rough field where the pilot should get the airplane off the ground as soon as possible to avoid damaging the landing gear.

Taking off from a soft surface or through soft surfaces or long, wet grass reduces the airplane's ability to accelerate during the takeoff roll and may prevent the airplane from reaching adequate takeoff speed if the pilot applies normal takeoff techniques. The pilot should be aware that the correct takeoff procedure for soft fields is quite different from the takeoff procedures used for short fields with firm, smooth surfaces. To minimize the hazards associated with takeoffs from soft or rough fields, the pilot should transfer the support of the airplane's weight as rapidly as possible from the wheels to the wings as the takeoff roll proceeds by establishing and maintaining a relatively high AOA or nose-high pitch attitude as early as possible. The pilot should lower the wing flaps prior to starting the takeoff (if recommended by the manufacturer) to provide additional lift and to transfer the airplane's weight from the wheels to the wings as early as possible. The pilot should maintain a continuous motion with sufficient power while lining up for the takeoff roll as stopping on a soft surface, such as mud or snow, might bog the airplane down.

Takeoff Roll

As the airplane is aligned with the takeoff path, the pilot should apply takeoff power smoothly and as rapidly as the powerplant can accept without faltering. As the airplane accelerates, the pilot should apply enough back-elevator pressure to establish a positive AOA and to reduce the weight supported by the nose-wheel.

When the airplane is held at a nose-high attitude throughout the takeoff run, the wings increasingly relieve the wheels of the airplane's weight as speed increases and lift develops, thereby minimizing the drag caused by surface irregularities or adhesion. If this attitude is accurately maintained, the airplane virtually flies itself off the ground, becoming airborne but at an airspeed slower than a safe climb speed because of ground effect. *[Figure 6-10]*

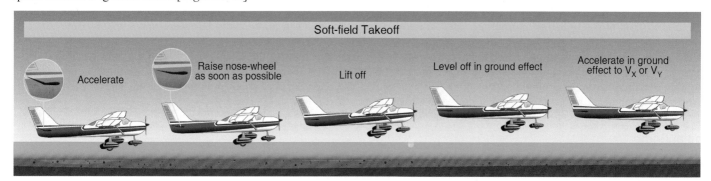

Figure 6-10. *Soft-field takeoff.*

Lift-Off

After the airplane becomes airborne, the pilot should gently lower the nose with the wheels clear of the surface to allow the airplane to accelerate to a minimum safe climb out speed, Immediately after the airplane becomes airborne and while it accelerates, the pilot should be aware that, while transitioning out of the ground effect area, the airplane will have a tendency to settle back onto the surface, even with full power applied. Therefore, it is essential that the airplane remain in ground effect until at least V_X is reached. This requires a good understanding of the control pressures, aircraft responses, visual clues, and acceleration characteristics of that particular airplane.

Initial Climb

After a positive rate of climb is established, and the airplane has accelerated to V_Y, the pilot should retract the landing gear and flaps, if equipped. If departing from an airstrip with wet snow or slush on the takeoff surface, the gear should not be retracted immediately so that any wet snow or slush can be air-dried. In the event an obstacle needs to be cleared after a soft-field takeoff, the pilot should perform the climb-out at V_X until the obstacle has been cleared. The pilot should then adjust the pitch attitude to V_Y and retract the gear and flaps. The power can then be reduced to the normal climb setting.

Common errors in the performance of soft/rough field takeoff and climbs are:

- Failure to review AFM/POH and performance charts prior to takeoff.
- Failure to adequately clear the area.
- Insufficient back-elevator pressure during initial takeoff roll resulting in inadequate AOA.
- Failure to cross-check engine instruments for indications of proper operation after applying power.
- Poor directional control.
- Climbing too high after lift-off and not leveling off low enough to maintain ground effect attitude.
- Abrupt and/or excessive elevator control while attempting to level off and accelerate after liftoff.
- Allowing the airplane to "mush" or settle resulting in an inadvertant touchdown after lift-off.
- Attempting to climb our of ground effect area before attaining sufficient climb speed.
- Failure to anticipate an increase in pitch attitude as the airplane climbs our of ground effect.

Rejected Takeoff/Engine Failure

Emergency or abnormal situations can occur during a takeoff that require a pilot to reject the takeoff while still on the runway. Circumstances such as a malfunctioning powerplant, inadequate acceleration, runway incursion, or air traffic conflict may be reasons for a rejected takeoff.

Prior to takeoff, the pilot should identify a point along the runway at which the airplane should be airborne. If that point is reached and the airplane is not airborne, immediate action should be taken to discontinue the takeoff. When properly planned and executed, the airplane can be stopped on the remaining runway without using extraordinary measures, such as excessive braking that may result in loss of directional control, airplane damage, and/or personal injury. The POH/AFM ground roll distances for take-off and landing added together provide a good estimate of the total runway needed to accelerate and then stop.

In the event a takeoff is rejected, the power is reduced to idle and maximum braking applied while maintaining directional control. If it is necessary to shut down the engine due to a fire, the mixture control should be brought to the idle cutoff position and the magnetos turned off. In all cases, the manufacturer's emergency procedure should be followed.

Urgency characterizes all power loss or engine failure occurrences after lift-off. In most instances, the pilot has only a few seconds after an engine failure to decide what course of action to take and to execute it.

In the event of an engine failure on initial climb-out, the pilot's first responsibility is to maintain aircraft control. At a climb pitch attitude without power, the airplane is at or near a stalling AOA. At the same time, the pilot may still be holding right rudder. The pilot should immediately lower the nose to prevent a stall while moving the rudder to ensure coordinated flight. The pilot should establish a controlled glide toward a plausible landing area, preferably straight ahead. Attempting to turn back to the takeoff runway should not be attempted unless the pilot previously trained for an emergency turn-back and sufficient altitude exists.

Noise Abatement

Aircraft noise problems are a major concern at many airports throughout the country. Many local communities have pressured airports into developing specific operational procedures that help limit aircraft noise while operating over nearby areas. As a result, noise abatement procedures have been developed for many of these airports that include standardized profiles and procedures to achieve these lower noise goals.

Airports that have noise abatement procedures provide information to pilots, operators, air carriers, air traffic facilities, and other special groups that are applicable to their airport. These procedures are available to the aviation community by various means. Most of this information comes from the Chart Supplements, local and regional publications, printed handouts, operator bulletin boards, safety briefings, and local air traffic facilities.

At airports that use noise abatement procedures, reminder signs may be installed at the taxiway hold positions for applicable runways to remind pilots to use and comply with noise abatement procedures on departure. Pilots who are unfamiliar with these procedures should ask the tower or air traffic facility for the recommended procedures. In any case, pilots should be considerate of the surrounding community while operating their airplane to and from such an airport. This includes operating as quietly, and safely as possible.

Chapter Summary

The takeoff and initial climb are relatively short phases required for every flight and are often taken for granted, yet 1 out of 5 accidents occur during this phase and half the mishaps are the result of pilot error. Becoming proficient in and applying the techniques and principles discussed in this chapter help pilots reduce their susceptibility to becoming a mishap statistic.

Airplane Flying Handbook (FAA-H-8083-3C)
Chapter 7: Ground Reference Maneuvers

Introduction

During initial training, pilots learn how various flight control pressure inputs affect the airplane. After achieving a sufficient level of competence, the pilot is ready to apply this skill and maintain the airplane, not only at the correct attitude and power configuration, but also along an appropriate course relative to objects on the ground. This skill is the basis for traffic patterns, survey, photographic, sight-seeing, aerial application (crop dusting), and various other flight profiles requiring specific flightpaths referenced to points on the surface.

Ground reference maneuvers are the principal flight maneuvers that combine the four fundamentals (straight-and-level, turns, climbs, and descents) into a set of integrated skills that the pilot uses in everyday flight activity. From every takeoff to every landing, a pilot exercises these skills to control the airplane. Therefore, a pilot needs to develop the proper coordination, timing, and attention in order to accurately and safely maneuver the airplane with regard to the required attitudes and ground references.

The pilot should be introduced by their instructor to ground reference maneuvers as soon as the pilot shows proficiency in the four fundamentals. Ground reference maneuvers call for manipulation of the flight controls using necessary control pressures to affect the airplane's attitude and position by using the outside natural horizon and ground-based references with brief periods of scanning the flight instruments.

Maneuvering by Reference to Ground Objects

Ground reference maneuvers train the pilot to accurately place the airplane in relationship to specific references and maintain a desired ground track. While vision is the most utilized sense, other senses are actively involved at different levels. For example, the amount of pressure needed to overcome flight control surface forces provides tactile feedback as to the airplane's airspeed and aerodynamic load.

It is a common error for beginning pilots to fixate on a specific reference, such as a single location on the ground or a spot on the natural horizon. A pilot fixating on any one reference loses the ability to determine rate, which significantly degrades a pilot's performance. By visually scanning across several references, the pilot learns how to determine the rate of closure to a specific point. In addition, the pilot should scan between several visual references to determine relative motion and to determine if the airplane is maintaining, or drifting to or from, the desired ground track. Consider a skilled automobile driver in a simple intersection turn; the driver does not merely turn the steering wheel some degree and hope that it will work out. The driver picks out several references, such as an island to their side, a painted lane line, or the opposing curb, and uses those references to make almost imperceptible adjustments to the amount of deflection on the steering wheel. At the same time, the driver adjusts the pressure on the accelerator pedal to smoothly join the new lane. In the same manner, multiple references are required to precisely control the airplane in reference to the ground.

Not all ground-based references are visually equal. Awareness of typical visual illusions helps a pilot select appropriate references. For example, larger objects or references may appear closer than they actually are when compared to smaller objects or references. Prevailing visibility has a significant effect on the pilot's perception of the distance to a reference. Excellent visibility with clear skies tends to make an object or reference appear closer than when compared to a hazy day with poor visibility. Rain can alter the visual image in a manner creating an illusion of being at a higher than actual altitude, and brighter objects or references may appear closer than dimmer objects. However, if using references of similar size and proportion, pilots find ground reference maneuvers easier to execute.

Ground-based references can be numerous. Examples include breakwaters, canals, fence lines, field boundaries, highways, railroad tracks, roads, pipe lines, power lines, water tanks, and many other objects; however, choices can be limited by geography, population density, infrastructure, or structures. The pilot should consider the type of maneuver being performed, altitude at which the maneuver will be performed, emergency landing requirements, density of structures, wind direction, visibility, and the type of airspace when selecting a ground-based reference.

Ground reference maneuvers develop a pilot's division of attention skill. A pilot needs to control the airplane's attitude while tracking a specific path over the ground. In addition, the pilot should be able to scan for hazards such as other aircraft, prepare for an emergency landing should the need arise, and scan the flight and engine instruments at regular intervals to ensure that a pending situation, such as decreasing oil pressure, does not turn into an unexpected incident.

Ground reference maneuvers place the airplane in a low altitude environment with associated hazards. Pilots should look for other aircraft, including helicopters, and look for obstructions such as radio towers and wires. In addition, pilots should consider engine failure and have one or more locations available for an emergency landing. Pilots should always clear the area with two 90° clearing turns looking to the left and the right, as well as above and below the airplane. The maneuver area should not cause disturbances and be well away from any open air assembly of persons, congested areas of a city, town, or settlement, or herd of livestock. Before performing any maneuver, the pilot should complete the required checklist items, make any radio announcements (such as on a practice area frequency), and safety clearing turns. As a general note, a ground reference maneuver should not exceed a bank angle of 45° or an airspeed greater than the maneuvering speed. As part of preflight planning, the pilot should determine the predicted (POH/AFM) stall speed at 50° or at the highest bank angle expected during the maneuver to assure there will be a safety margin above the stall speed during the maneuver.

Drift and Ground Track Control

Wind direction and velocity variations create the need for flightpath corrections during a ground reference maneuver. In a similar way that water currents affect the progress of a boat or ship, wind directly influences the path that the airplane travels in reference to the ground. Whenever the airplane is in flight, the movement of the air directly affects the actual ground track of the airplane.

For example, an airplane is traveling at 90 knots (90 nautical miles per hour) and the wind is blowing from right to left at 10 knots. The airplane continues forward at 90 knots but also travels left 10 nautical miles for every hour of flight time. If the airplane, in this example doubles its speed to 180 knots, it still drifts laterally to the left 10 nautical miles every hour. Unless in still air, traveling to a point on the surface requires compensation for the movement of the air mass.

Ground reference maneuvers are generally flown at altitudes between 600 and 1,000 feet above ground level (AGL). The pilot should consider the following when selecting the maneuvering altitude:

- The lower the maneuvering altitude, the faster the airplane appears to travel in relation to the ground.

- Drift should be easily recognizable from both sides of the airplane.

- The altitude should provide obstruction clearance of no less than 500 feet vertically above the obstruction and 2,000 feet horizontally.

- In the event of an engine failure, lower altitudes equate to less time to configure the airplane and reduced gliding distance before a forced landing.

- What specific altitude or altitude range does the testing standard call for?

Correcting Drift During Straight-and-Level Flight

When flying straight and level and following a selected straight-line direct ground track, the preferred method of correcting for wind drift is to angle the airplane sufficiently into the wind to cancel the effect of the sideways drift caused by the wind. The wind's speed, the angle between the wind direction and the airplane's longitudinal axis, and the airspeed of the airplane determine the required wind correction angle. For example, an airplane with an airspeed of 100 knots in an air mass moving at 20 knots directly from the side, should turn 12° into the wind to cancel the airplane's drift. If the wind in the above example is only 10 knots, the wind correction angle required to cancel the drift is six degrees. When the drift has been neutralized by heading the airplane into the wind, the airplane will fly the direct straight ground track.

To further illustrate this point, if a boat is crossing a river and the river's current is completely still, the boat could head directly to a point on the opposite shore on a straight course without any drift. However, rivers tend to have a downstream current that needs to be considered if the captain wants the boat to arrive at the opposite shore using a direct straight path. Any downstream current pushes the boat sideways and downstream at the speed of the current. To counteract this downstream movement, the boat needs to move upstream at the same speed as the river is moving the boat downstream. This is accomplished by angling the boat upstream to counteract the downstream flow. If done correctly, the boat follows a direct straight track across the river to the intended destination point. A slower forward speed of the boat or a faster river current requires a greater angle to counteract the drift. *[Figure 7-1]*

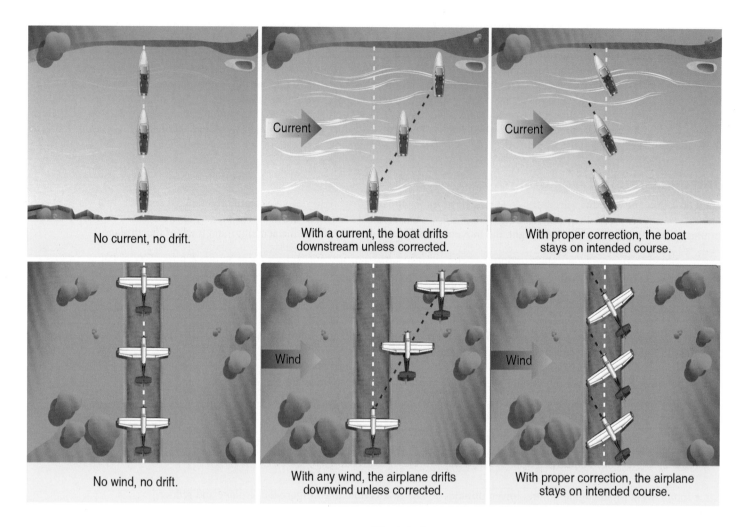

No current, no drift.

With a current, the boat drifts downstream unless corrected.

With proper correction, the boat stays on intended course.

No wind, no drift.

With any wind, the airplane drifts downwind unless corrected.

With proper correction, the airplane stays on intended course.

Figure 7-1. *Wind drift.*

As soon as the pilot lifts off the surface and levels the wings in a crosswind, the airplane begins tracking sideways. The force of the crosswind acts on the mass of the airplane, and the speed of drift increases up to the speed of the crosswind component. A wind that is directly to the right or the left (at a 90° angle) will cause the airplane to accelerate sideways at the same speed as the wind. When the wind is halfway between the side and the nose of the airplane (at a 45° angle), it causes a sideways drift up to just over 70 percent of the total speed of the wind. It should be understood that pilots do not calculate the required drift correction angles for ground reference maneuvers; they merely use the references and adjust the airplane's relationship to those references to cancel any drift. The groundspeed of the airplane is also affected by the wind. As the wind direction becomes parallel to the airplane's longitudinal axis, the magnitude of the wind's effect on the groundspeed is greater; as the wind becomes perpendicular to the longitudinal axis, the magnitude of the wind's effect on the groundspeed is less. In general, When the wind is blowing straight into the nose of the airplane, the groundspeed will be less than the airspeed. When the wind is blowing from directly behind the airplane, the groundspeed will be faster than the airspeed. In other words, when the airplane is headed upwind, the groundspeed is decreased; when headed downwind, the groundspeed is increased.

Constant Radius During Turning Flight

In a no-wind condition, a pilot may make a constant-radius turn over the ground using a fixed bank angle. If wind is present, however, a pilot will observe a change in the radius of a turn while maintaining that same constant bank angle. *[Figure 7-2]* As groundspeed increases, the observed radius of the turn increases. Conversely, as groundspeed decreases, the radius of the turn over the ground will decrease. For a ground-referenced constant-radius turn, the pilot compensates for changes in groundspeed by varying the bank angle throughout the turn. When groundspeed increases, the pilot banks more steeply to maintain a constant-radius turn over the ground. The converse is also true: when groundspeed decreases, the pilot uses a shallower bank.

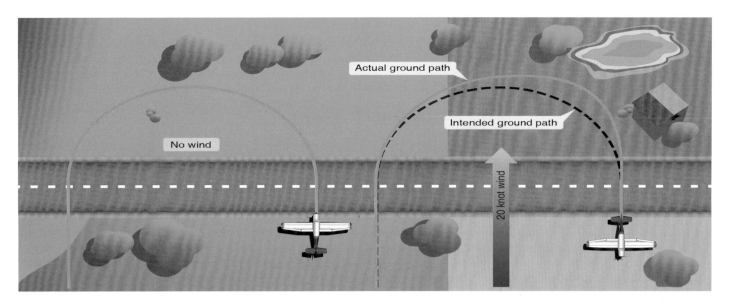

Figure 7-2. *Effect of wind during a turn.*

For a given true airspeed, the radius of turn in the air varies proportionally with the bank angle. To maintain a constant radius over the ground, the bank angle used is proportional to groundspeed. For example, an airplane is in the downwind position at 100 knots groundspeed. In this example, the wind is 10 knots, meaning that the airplane has an airspeed of 90 knots (for this discussion, assume true, calibrated, and indicated airspeed are all the same). If the pilot starts a turn using a 45° bank angle, the turn radius over the ground at that moment is approximately 890 feet. As the airplane turns, the groundspeed decreases and the bank angle needs to be reduced in order to maintain the same turn radius of 890 feet over the ground. At the upwind point of the turn, the bank angle should be approximately 33°. In another example, if the downwind is flown at an airspeed of 90 knots in a 10 knot tailwind with a desired turn radius of 2,000 feet, the bank angle would be approximately 24°. The bank angle flying upwind would be approximately 16°.

Put another way, at a higher groundspeed, there is less time to turn the airplane while trying to maintain a ground-referenced constant-radius turn. The pilot increases the bank angle in order to increase the rate of turn, and the increased rate of turn offsets the reduced time available to make the turn. Conversely, when flying at a lower groundspeed, the pilot reduces the angle of bank and rate of turn to compensate for the additional time taken while making the turn. With some experience, pilots may notice how wind direction affects the time needed for various segments of ground-referenced turns.

To demonstrate the effect that wind has on turns, the pilot should select a straight-line ground reference, such as a road or railroad track. *[Figure 7-3]* Choosing a straight-line ground reference that is parallel to the wind, the airplane would be flown into the wind and directly over the selected straight-line ground reference. Once a straight-line ground reference is established, the pilot makes a 360° constant medium-banked turn. As the airplane completes the 360° turn, it should return directly over the straight-line ground reference but downwind from the starting point. Choosing a straight-line ground reference that has a crosswind, and using the same 360° constant medium-banked turn, demonstrates how the airplane drifts away from the reference even as the pilot holds a constant bank angle. In both examples, the path over the ground is not circular, although in reference to the air, the airplane flew a perfect continuous radius.

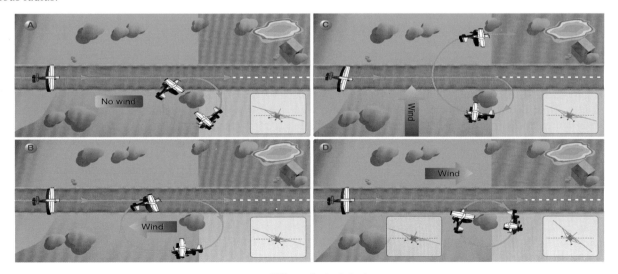

Figure 7-3. *Effect of wind during turn.*

In order to compensate for the effects of wind drift, the pilot adjusts the bank angle as the groundspeed changes throughout the turn. Where groundspeed is the fastest, such as when the airplane is headed downwind, the bank angle should be steepest. Where groundspeed is the slowest, such as when the airplane is headed upwind, the bank angle should be shallow. It is necessary to increase or decrease the angle of bank, which increases or decreases the rate of turn, to achieve the desired constant radius track over the ground.

Ground reference maneuvers should always be entered from a downwind position. This allows the pilot to establish the steepest bank angle required to maintain a constant radius ground track. If the bank is too steep, the pilot should immediately exit the maneuver and re-establish a lateral position that is further from the ground reference. The pilot should avoid bank angles in excess of 45°due to the increased stalling speed.

Tracking Over and Parallel to a Straight Line

The pilot should first be introduced to ground reference maneuvers by correcting for the effects of a crosswind over a straight-line ground reference, such as road or railroad tracks. If a straight road or railroad track is unavailable, the pilot should choose multiple references (three minimum) which line up along a straight path. The reference line should be suitably long so the pilot has sufficient time to understand the concepts of wind correction and practice the maneuver. Initially, the maneuver should be flown directly over the ground reference line with the pilot angling the airplane's longitudinal axis into the wind sufficiently such as to cancel the effect of drift. The pilot should scan between far ahead and close to the airplane to practice tracking multiple references.

When proficiency has been demonstrated by flying directly over the ground reference line, the pilot should then practice flying a straight parallel path that is offset from the ground reference. The offset parallel path should not be more than three-fourths of a mile from the reference line. The maneuver should be flown offset from the ground references with the pilot angling the airplane's longitudinal axis into the wind sufficiently to cancel the effect of drift while maintaining a parallel track.

Rectangular Course

A principal ground reference maneuver is the rectangular course. *[Figure 7-4]* The rectangular course is a training maneuver in which the airplane maintains an equal distance from all sides of the selected rectangular references. The maneuver is accomplished to replicate the airport traffic pattern that an airplane typically maneuvers while landing. While performing the rectangular course maneuver, the pilot should maintain a constant altitude, airspeed, and distance from the ground references. The maneuver assists the pilot in practicing the following:

- Maintaining a specific relationship between the airplane and the ground.

- Dividing attention between the flightpath, ground-based references, manipulating the flight controls, and scanning for outside hazards and instrument indications.

- Adjusting the bank angle during turns to correct for groundspeed changes in order to maintain constant-radius turns.

- Rolling out from a turn with the required wind correction angle to compensate for any drift caused by the wind.

- Establishing and correcting the wind correction angle in order to maintain the track over the ground.

- Preparing the pilot for the airport traffic pattern and subsequent landing pattern practice.

To fly the rectangular course, the pilot should first locate a square field, a rectangular field, or an area with suitable ground references on all four sides. Note that a square meets the definition of a rectangle. As previously mentioned, this area should be selected consistent with safe practices. The airplane should be flown parallel to and at an equal distance between one-half to three-fourths of a mile away from the field boundaries or selected ground references. The flightpath should be positioned outside the field boundaries or selected ground references so that the references may be easily observed from either pilot seat. It is not practical to fly directly above the field boundaries or selected ground references. The pilot should avoid flying close to the references, as this will require the pilot to turn using very steep bank angles, thereby increasing aerodynamic load factor and the airplane's stall speed, especially in the downwind to crosswind turn.

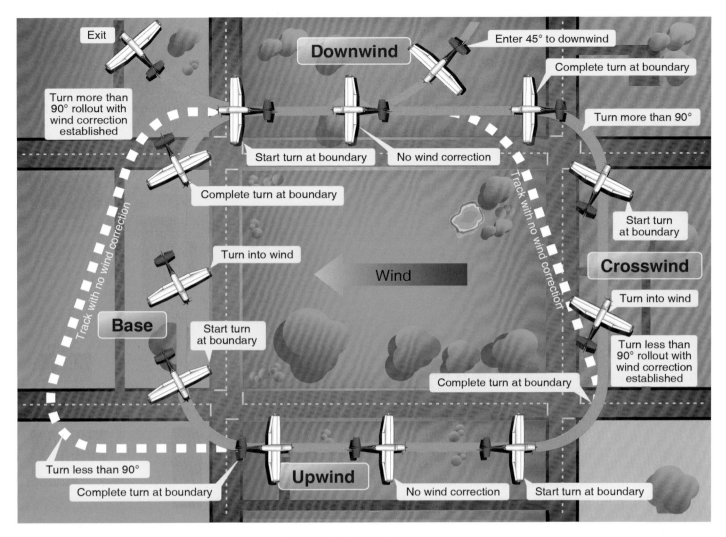

Figure 7-4. *Rectangular course.*

The entry into the maneuver should be accomplished downwind. This places the wind on the tail of the airplane and results in an increased groundspeed. There should be no wind correction angle if the wind is directly on the tail of the airplane; however, a real-world situation often results in some drift correction. The turn from the downwind leg onto the base leg is entered with a relatively steep bank angle. The pilot should roll the airplane into a steep bank with rapid, but not excessive, coordinated aileron and rudder pressures. As the airplane turns onto the following base leg, the tailwind lessens and becomes a crosswind; the bank angle is reduced gradually with coordinated aileron and rudder pressures. The pilot should be prepared for the lateral drift and compensate by turning more than 90° angling toward the inside of the rectangular course.

The next leg is where the airplane turns from a base leg position to the upwind leg. Ideally, on the upwind, the wind is directly on the nose of the airplane resulting in a direct headwind and decreased groundspeed; however, some drift correction may be necessary. The pilot should roll the airplane into a medium-banked turn with coordinated aileron and rudder pressures. As the airplane turns onto the upwind leg, the crosswind lessens and becomes a headwind, and the bank angle is gradually reduced with coordinated aileron and rudder pressures. Because the pilot was angled into the wind on the base leg, the turn to the upwind leg is less than 90°.

The next leg is where the airplane turns from an upwind leg position to the crosswind leg. The pilot should slowly roll the airplane into a shallow-banked turn, as the developing crosswind drifts the airplane into the inside of the rectangular course with coordinated aileron and rudder pressures. As the airplane turns onto the crosswind leg, the headwind lessens and becomes a crosswind. As the turn nears completion, the bank angle is reduced with coordinated aileron and rudder pressures. To compensate for the crosswind, the pilot maintains an angle into the wind, toward the outside of the rectangular course, which requires the turn to be less than 90°.

The final turn is back to the downwind leg, which requires a medium-banked angle and a turn greater than 90°. The groundspeed will be increasing as the turn progresses and the bank should be held and then rolled out in a rapid, but not excessive, manner using coordinated aileron and rudder pressures.

For the maneuver to be executed properly, the pilot should visually utilize the ground-based, nose, and wingtip references to properly position the airplane in attitude and in orientation to the rectangular course. In order to maintain a constant ground-based radius during the turns, each turn requires the bank angle to be adjusted to compensate for the changing groundspeed—the higher the groundspeed, the steeper the bank. If the groundspeed is initially higher and then decreases throughout the turn, the bank angle should progressively decrease throughout the turn. The converse is also true, if the groundspeed is initially slower and then increases throughout the turn, the bank angle should progressively increase throughout the turn until rollout is started. Also, the rate for rolling in and out of the turn should be adjusted to prevent drifting in or out of the course. When the wind is from a direction that could drift the airplane into the course, the banking roll rate should be slow. When the wind is from a direction that could drift the airplane to the outside of the course, the banking roll rate should be quick.

The following are the most common errors made while performing rectangular courses:

1. Failure to adequately clear the surrounding area for safety hazards, initially and throughout the maneuver.
2. Failure to establish a constant, level altitude prior to entering the maneuver.
3. Failure to maintain altitude during the maneuver.
4. Failure to properly assess wind direction.
5. Failure to establish the appropriate wind correction angle.
6. Failure to apply coordinated aileron and rudder pressure, resulting in slips and skids.
7. Failure to manipulate the flight controls in a smooth and continuous manner.
8. Failure to properly divide attention between airplane control and orientation with ground references.
9. Failure to execute turns with accurate timing.

Turns Around a Point

Turns around a point are a logical extension of both the rectangular course and S-turns across a road. The maneuver is a 360° constant radius turn around a single ground-based reference point. *[Figure 7-5]* The principles are the same in any turning ground reference maneuver—higher groundspeeds require steeper banks and slower ground speeds require shallower banks. The objectives of turns around a point are as follows:

• Maintaining a specific relationship between the airplane and the ground.

• Dividing attention between the flightpath, ground-based references, manipulating the flight controls, and scanning for outside hazards and instrument indications.

• Adjusting the bank angle during turns to correct for groundspeed changes in order to maintain a constant radius turn—steeper bank angles for higher ground speeds, shallow bank angles for slower groundspeeds.

• Improving competency in managing the quickly-changing bank angles.

• Establishing and adjusting the wind correction angle in order to maintain the track over the ground.

• Developing the ability to compensate for drift in quickly-changing orientations.

• Developing further awareness that the radius of a turn is correlated to the bank angle.

To perform a turn around a point, the pilot needs to complete at least one 360° turn; however, to properly assess wind direction, velocity, bank required, and other factors related to turns in wind, the pilot should complete two or more turns. As in other ground reference maneuvers, when wind is present, the pilot adjusts the airplane's bank and wind correction angle to maintain a constant radius turn around a point. In contrast to the ground reference maneuvers discussed previously, in which turns were approximately limited to either 90° or 180°, turns around a point are consecutive 360° turns, where pilot constantly adjusts the bank angle and the resulting rate of turn as the airplane sequences through the various wind directions. The pilot should make these adjustments by applying coordinated aileron and rudder pressure throughout the turn.

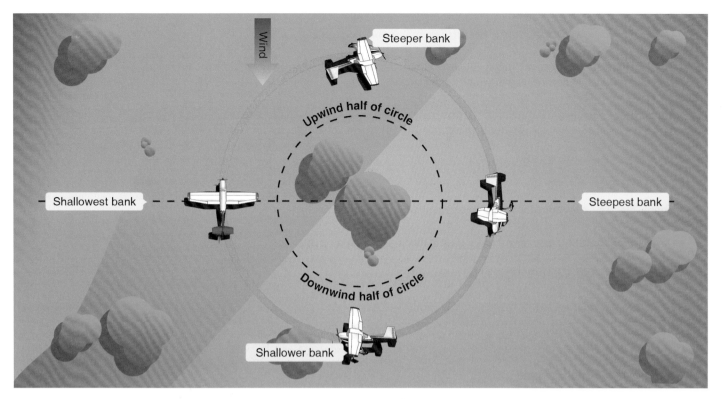

Figure 7-5. *Turns around a point.*

When performing a turn around a point, the pilot should select a prominent, ground-based reference that is easily distinguishable yet small enough to present a precise reference. The pilot should enter the maneuver downwind, where the groundspeed is at its fastest, at the appropriate radius of turn and distance from the selected ground-based reference point. In a high-wing airplane, the lowered wing may block the view of the ground reference point, especially in airplanes with side-by-side seating during a left turn (assuming that the pilot is flying from the left seat). To prevent this, the pilot may need to change the maneuvering altitude or the desired turn radius. The pilot should ensure that the reference point is visible at all times throughout the maneuver, even with the wing lowered in a bank.

Upon entering the maneuver, depending on the wind's speed, it may be necessary to roll into the initial bank at a rapid rate so that the steepest bank is set quickly to prevent the airplane from drifting outside of the desired turn radius. This is best accomplished by repeated practice and assessing the required roll in rate. Thereafter, the pilot should gradually decrease the angle of bank until the airplane is headed directly upwind. As the upwind becomes a crosswind and then a downwind, the pilot should gradually steepen the bank to the steepest angle upon reaching the initial point of entry.

During the downwind half of the turn, the pilot should progressively adjust the airplane's heading toward the inside of the turn. During the upwind half, the pilot should progressively adjust the airplane's heading toward the outside of the turn. Put another way, the airplane's heading should be ahead of its position over the ground during the downwind half of the turn and behind its position during the upwind half. Remember that the goal is to make a constant-radius turn over the ground and, because the airplane is flying through a moving air mass, the pilot should constantly adjust the bank angle to achieve this goal.

The following are the most common errors in the performance of turns around a point:

1. Failure to adequately clear the surrounding area for safety hazards, initially and throughout the maneuver.
2. Failure to establish a constant, level altitude prior to entering the maneuver.
3. Failure to maintain altitude during the maneuver.
4. Failure to properly assess wind direction.
5. Failure to properly execute constant-radius turns.
6. Failure to manipulate the flight controls in a smooth and continuous manner.
7. Failure to establish the appropriate wind correction angle.
8. Failure to apply coordinated aileron and rudder pressure, resulting in slips or skids.

S-Turns

An S-turn is a ground reference maneuver in which the airplane's ground track resembles two opposite but equal half-circles on each side of a selected ground-based straight-line reference. *[Figure 7-6]* This ground reference maneuver presents a practical application for the correction of wind during a turn. The objectives of S-turns across a road (or line) are as follows:

- Maintaining a specific relationship between the airplane and the ground.

- Dividing attention between the flightpath, ground-based references, manipulating the flight controls, and scanning for outside hazards and instrument indications.

- Adjusting the bank angle during turns to correct for groundspeed changes in order to maintain a constant-radius turn—steeper bank angles for higher groundspeeds, shallow bank angles for slower groundspeeds.

- Rolling out from a turn with the required wind correction angle to compensate for any drift cause by the wind.

- Establishing and correcting the wind correction angle in order to maintain the track over the ground.

- Developing the ability to compensate for drift in quickly-changing orientations.

- Arriving at specific points on required headings.

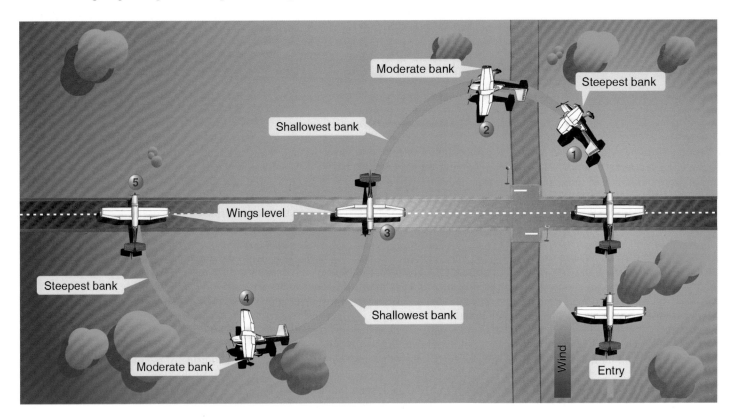

Figure 7-6. *S-turns.*

With the airplane in the downwind position, the maneuver consists of crossing a straight-line ground reference at a 90° angle and immediately beginning a 180° constant-radius turn. The pilot will then adjust the roll rate and bank angle for drift effects and changes in groundspeed, and re-cross the straight-line ground reference in the opposite direction just as the first 180° constant-radius turn is completed. The pilot will then immediately begin a second 180° constant-radius turn in the opposite direction, adjusting the roll rate and bank angle for drift effects and changes in groundspeed, again re-crossing the straight-line ground reference as the second 180° constant-radius turn is completed. If the straight-line ground reference is of sufficient length, the pilot may complete as many as can be safely accomplished.

In the same manner as the rectangular course, it is standard practice to enter ground-based maneuvers downwind where groundspeed is greatest. As such, the roll into the turn should be rapid, but not aggressive, and the angle of bank should be steepest when initiating the turn. As the turn progresses, the bank angle and the rate of rollout should be decreased as the groundspeed decreases to ensure that the turn's radius is constant. During the first turn, when the airplane is at the 90° point, it will be directly crosswind. In addition to the rate of rollout and bank angle, the pilot should control the wind correction angle throughout the turn.

Controlling the wind correction angle during a turn can be complex to understand. The concept may be understood by comprehending the difference between the number of degrees that the airplane has turned over the ground versus the number of degrees that the airplane has turned in the air. As an example, assume the airplane is exactly crosswind, meaning directly at a point that is 90° to the straight-lined ground reference. In this example, if the wind requires a 10° wind correction angle (for this example, this is a left turn with the crosswind from the left), the airplane would be at a heading that is 10° ahead when directly over the 90° ground reference point. In other words, the first 90° track over the ground would result in a heading change of 100° and the last 90° track over the ground would result in 80° of heading change.

As the turn progresses from a downwind position to an upwind position, the pilot should gradually decrease the bank angle with coordinated aileron and rudder pressure. The pilot should reference the airplane's nose, wingtips, and the ground references and adjust the rollout timing so that the wings become level just as the airplane crosses the straight-line ground reference at the proper heading, altitude, and airspeed. As the airplane re-crosses the straight-lined ground reference, the opposite turn begins—there should be no delay in rolling out from one turn and rolling into the next turn. Because the airplane is now upwind, the roll in should be smooth and gentle and the initial bank angle should be shallow. As the turn progresses, the wind changes from upwind, to crosswind, to downwind. In a similar manner described above, the pilot should adjust the bank angle to correct for changes in groundspeed. As the groundspeed increases, the pilot should increase the bank angle to maintain a constant-radius turn over the ground. At the 90° crosswind position, the airplane should also have the correct wind correction angle. As the airplane turns downwind and the groundspeed increases, the bank angle should be increased so that the rate of turn maintains a constant-radius turn.

The following are the most common errors made while performing S-turns across a road:

1. Failure to adequately clear surrounding area for safety hazards, initially and throughout the maneuver.
2. Failure to establish a constant, level altitude prior to entering the maneuver.
3. Failure to maintain altitude during the maneuver.
4. Failure to properly assess wind direction.
5. Failure to properly execute constant-radius turns.
6. Failure to manipulate the flight controls in a smooth and continuous manner when transitioning into turns.
7. Failure to establish the appropriate wind correction angle.
8. Failure to apply coordinated aileron and rudder pressure, resulting in slips or skids.

Elementary Eights

Elementary eights are a family of maneuvers in which each individual maneuver is one that the airplane tracks a path over the ground similar to the shape of a figure eight. There are various types of eights, progressing from the elementary to advanced types. Each eight is intended to develop a pilot's flight control coordination skills, strengthen their awareness relative to the selected ground references, and enhance division of attention so that flying becomes more instinctive than mechanical. Eights require a greater degree of focused attention to the selected ground references; however, the real significance of eights is that the pilot develops the ability to fly with precision.

Elementary eights include eights along a road, eights across a road, and eights around pylons. Each of these maneuvers is a variation of a turn around a point. Each eight uses two ground reference points about which the airplane turns first in one direction and then the opposite direction—like a figure eight.

Eights maneuvers are designed for the following purposes:

- Further development of the pilot's skill in maintaining a specific relationship between the airplane and the ground references.

- Improving the pilot's ability to divide attention between the flightpath and ground-based references, manipulation of the flight controls, and scanning for outside hazards and instrument indications during both turning and straight-line flight.

- Developing the pilot's skills to visualize each specific segment of the maneuver and the maneuver as a whole, prior to execution.

- Developing a pilot's ability to intuitively manipulate flight controls to adjust the bank angle during turns to correct for groundspeed changes in order to maintain constant-radius turns and proper ground track between ground references.

Eights Along a Road

An eight along a road is a ground reference maneuver in which the ground track consists of two opposite 360° adjacent turns. An imaginary line drawn through the center of each 360° turn is perpendicular to the straight-line ground reference (road, railroad tracks, fence line, pipeline right-of-way, etc.) as illustrated in Figure 7-7. Like the other ground reference maneuvers, the objective is to further develop division of attention while compensating for drift, maintaining orientation with ground references, and maintaining a constant altitude.

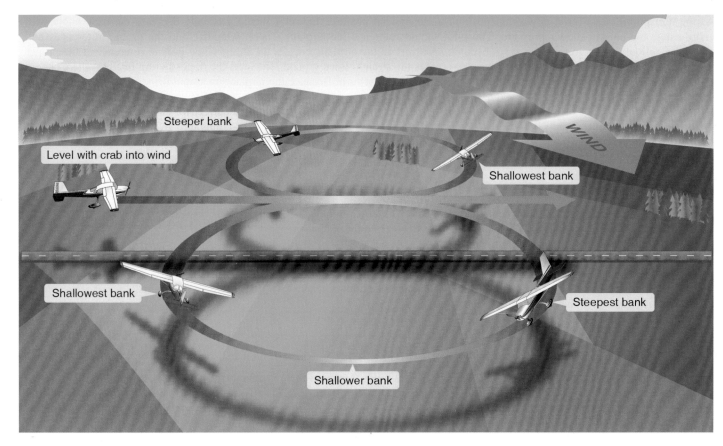

Figure 7-7. *Eights along a road.*

Although eights along a road may be performed with the wind blowing parallel or perpendicular to the straight-line ground reference, only the perpendicular wind situation is explained since the principles involved are common to each. The pilot should select a straight-line ground reference that is perpendicular to the wind and position the airplane parallel to and directly above the straight-line ground reference. Since this places the airplane in a crosswind position, the pilot should compensate for the wind drift with an appropriate wind correction angle.

The following description is illustrated in *Figure 7-7*. The airplane is initially in a crosswind position, perpendicular to the wind, and over the ground-based reference. The first turn should be to the right toward a downwind position starting with a steepening bank. When the entry is made into the turn, it requires that the turn begin with a medium bank and gradually steepen to its maximum bank angle when the airplane is directly downwind. As the airplane turns from downwind to crosswind, the bank angle needs to be gradually reduced since groundspeed is decreasing; however, 1/2 of the reduction in groundspeed occurs during the first 2/3 of the turn from downwind to crosswind.

The pilot needs to control the bank angle as well as the rate at which the bank angle is reduced so that the wind correction angle is correct. Assuming that the wind is coming from the right side of the airplane, the airplane heading should be slightly ahead of its position over the ground. When the airplane completes the first 180° of ground track, it is directly crosswind, and the airplane should be at the maximum wind correction angle.

As the turn is continued toward the upwind, the airplane's groundspeed is decreasing, which requires the pilot to reduce the bank angle to slow the rate of turn. If the pilot does not reduce the bank angle, the continued high rate of turn would cause the turn to be completed prematurely. Another way to explain this effect is—the wind is drifting the airplane downwind at the same time its groundspeed is slowing. If the airplane has a steeper-than-required bank angle, its rate of turn will be too fast and the airplane will complete the turn before it has had time to return to the ground reference.

When the airplane is directly upwind, which is at 270° into the first turn, the bank angle should be shallow with no wind correction. As the airplane turns crosswind again, the airplane's groundspeed begins increasing; therefore, the pilot should adjust the bank angle and corresponding rate of turn proportionately in order to reach the ground reference at the completion of the 360° ground track. The pilot may vary the bank angle to correct for any previous errors made in judging the returning rate and closure rate. The pilot should time the rollout so that the airplane is straight-and-level over the starting point with enough drift correction to hold it over the straight-line ground reference. Assuming that the wind is now from the left, the airplane should be banked at a left wind correction angle.

After momentarily flying straight-and-level with the established wind correction, along the ground reference, the pilot should roll the airplane into a medium bank-turn in the opposite direction to begin the 360° turn on the upwind side of the ground reference. The wind will decrease the airplane's groundspeed and drift the airplane back toward the ground reference; therefore, the pilot should decrease the bank slowly during the first 90° of the upwind turn in order to establish a constant radius. During the next 90° of turn, the pilot should increase the bank angle, since the groundspeed is increasing, to maintain a constant radius and establish the proper wind correction angle before reaching the 180° upwind position.

As the remaining 180° of turn continues, the wind becomes a tailwind and then a crosswind. Consistent with previous downwind and crosswind descriptions, the pilot should increase the bank angle as the airplane reaches the downwind position and decrease the bank angle as the airplane reaches the crosswind position. Further, the rate of roll-in and roll-out should be consistent with how fast the groundspeed changes during the turn. Remember, when turning from an upwind or downwind position to a crosswind position, 1/2 of the groundspeed change occurs during the first 2/3 of the 90° turn. The final 1/2 of the change in groundspeed occurs during the last 1/3 of the turn. In contrast, when turning from a crosswind position to an upwind or downwind position, the first 1/2 of the groundspeed change occurs during the first 1/3 of the 90° turn. The final 1/2 of the change in groundspeed occurs during the last 2/3 of the turn.

To successfully perform eights along a ground reference, the pilot should be able to smoothly and accurately coordinate changes in bank angle to maintain a constant-radius turn and counteract drift. The speed in which the pilot can anticipate these corrections directly affects the accuracy of the overall maneuver and the amount of attention that can be directed toward scanning for outside hazards and instrument indications.

Eights Across a Road

This maneuver is a variation of eights along a road and involves the same principles and techniques. The primary difference is that at the completion of each loop of the figure eight, the airplane should cross an intersection or a specific ground reference point. [Figure 7-8]

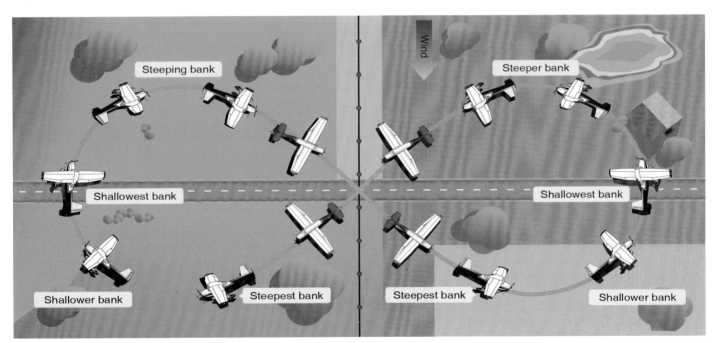

Figure 7-8. *Eights across a road.*

The loops should be across the road and the wind should be perpendicular to the loops. Each time the reference is crossed, the crossing angle should be the same, and the wings of the airplane should be level. The eights may also be performed by rolling from one bank immediately to the other, directly over the reference.

Eights Around Pylons

Eights around pylons is a ground reference maneuver with the same principles and techniques of correcting for wind drift as used in turns around a point and the same objectives as other ground track maneuvers. Eights around pylons utilizes two ground reference points called "pylons." Turns around each pylon are made in opposite directions to follow a ground track in the form of a figure 8. [Figure 7-9]

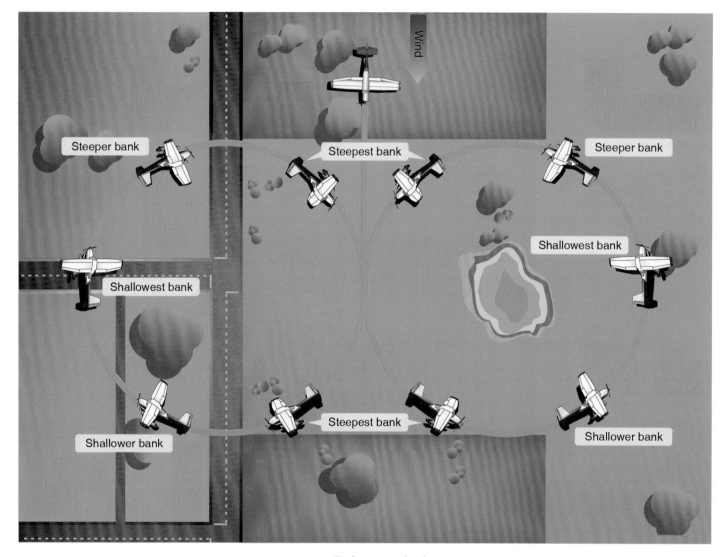

Figure 7-9. *Eights around pylons.*

The pattern involves flying downwind between the pylons and upwind outside of the pylons. It may include a short period of straight-and-level flight while proceeding diagonally from one pylon to the other. The pylons should be on a line perpendicular to the wind. The maneuver should be started with the airplane on a downwind heading while passing mid-way between the pylons. The distance between the pylons and the wind velocity determines the initial angle of bank required to maintain a constant turn radius from the pylons during each turn. The steepest banks are necessary just after each turn entry and just before the rollout from each turn where the airplane is headed downwind and the groundspeed is highest. The shallowest banks are when the airplane is headed directly upwind and the groundspeed is lowest.

As in other ground reference maneuvers, the rate at which the bank angle changes depends on the wind velocity. If the airplane proceeds diagonally from one turn to the other, the rollout from each turn needs to be completed on the proper heading with sufficient wind correction angle to ensure that after brief straight-and-level flight, the airplane arrives at the point where a turn of the same radius can be made around the other pylon. The straight-and-level flight segments should be tangent to both circular patterns.

Common Errors

Common errors in the performance of elementary eights are:

1. Failure to adequately clear the surrounding area for safety hazards, initially and throughout the maneuver.
2. Poor selection of ground references.
3. Failure to establish a constant, level altitude prior to entering the maneuver.
4. Failure to maintain adequate altitude control during the maneuver.
5. Failure to properly assess wind direction.
6. Failure to properly execute constant-radius turns.
7. Failure to manipulate the flight controls in a smooth and continuous manner.
8. Failure to establish the appropriate wind correction angles.
9. Failure to apply coordinated aileron and rudder pressure, resulting in slips or skids.
10. Failure to maintain orientation as the maneuver progresses.

Eights on Pylons

The eights on pylons is the most advanced and difficult of the ground-reference maneuvers. Because of the techniques involved, the eights on pylons are unmatched for developing intuitive control of the airplane. Similar to eights around pylons except altitude is varied to maintain a specific visual reference to the pivot points.

When performing eights on pylons, the pilot imagines there is a line parallel to the airplane's lateral axis that extends from the pilot's eyes to the pylon. Along this line, the airplane appears to pivot as it turns around the pylon. In other words, if a taut string extended from the pilot's eyes to the pylon, the string would remain parallel to lateral axis as the airplane makes a turn around the pylon. The goal of eights on pylons is to keep the line from the pilot's eyes to the pylon parallel to the lateral axis. The string should not be at an angle to the lateral axis while the airplane flies around the pylon. *[Figure 7-10]* When explaining eights on pylons, instructors sometimes use the term "wingtip" to represent the proper visual reference line to the pylon. This interpretation is not correct. High-wing, low-wing, swept-wing, and tapered-wing airplanes, as well as those with tandem or side-by-side seating, all present different angles from the pilot's eye to the wingtip. *[Figure 7-11]*

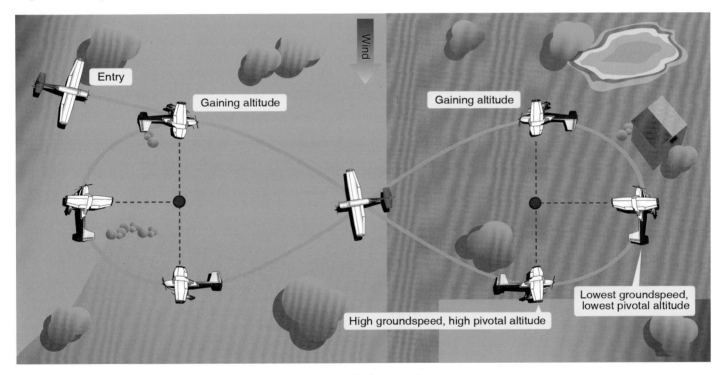

Figure 7-10. *Eights on pylons.*

The visual reference line, while not necessarily on the wingtip itself, may be positioned in relation to the wingtip (ahead, behind, above, or below), and differs for each pilot and from each seat in the airplane. This is especially true in tandem (fore and aft) seat airplanes. In side-by-side type airplanes, there is very little variation in the visual reference lines for different people, if those people are seated with their eyes at approximately the same level. Therefore, in the correct performance of eights on pylons, as in other maneuvers requiring a lateral reference, the pilot should use a visual reference line that, from eye level, parallels the lateral axis of the airplane.

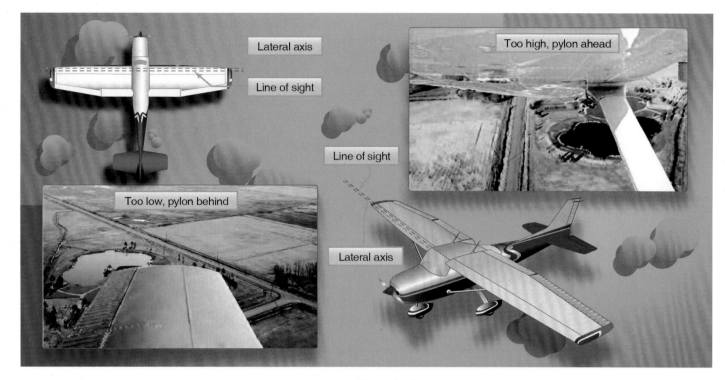

Figure 7-11. *Line of sight.*

The altitude that is appropriate for eights on pylons is called the "pivotal altitude" and is determined by the airplane's groundspeed. In previous ground-track maneuvers, the airplane flies a prescribed path over the ground and the pilot attempts to maintain the track by correcting for the wind. With eights on pylons, the pilot maintains lateral orientation to a specific spot on the ground. This develops the pilot's ability to maneuver the airplane accurately while dividing attention between the flightpath and the selected pylons on the ground.

An explanation of the pivotal altitude is also essential. First, a good rule of thumb for estimating the pivotal altitude is to square the groundspeed, then divide by 15 (if the groundspeed is in miles per hour) or divide by 11.3 (if the groundspeed is in knots), and then add the mean sea level (MSL) altitude of the ground reference. The pivotal altitude is the altitude at which, for a given groundspeed, the projection of the visual reference line to the pylon appears to pivot. Visually, a taut string, if extended from the pilot's eyes to the pylon, would remain parallel to lateral axis as the airplane makes a turn around the pylon. *[Figure 7-12]* The pivotal altitude does not vary with the angle of bank unless the bank is steep enough to affect the groundspeed.

Groundspeed		Approximate Pivotal Altitude
Knots	MPH	
87	100	670
91	105	735
96	110	810
100	115	885
104	120	960
109	125	1050
113	130	1130

Figure 7-12. *Speed versus pivotal altitude.*

Distance from the pylon affects the angle of bank. At any altitude above that pivotal altitude, the projected reference line appears to move rearward in a circular path in relation to the pylon. Conversely, when the airplane is below the pivotal altitude, the projected reference line appears to move forward in a circular path. *[Figure 7-13]* To demonstrate this, the pilot will fly at maneuvering speed and at an altitude below the pivotal altitude, and then place the airplane in a medium-banked turn. The projected visual reference line appears to move forward along the ground (pylon appears to move back) as the airplane turns. The pilot then executes a climb to an altitude well above the pivotal altitude. When the airplane is again at maneuvering speed, it is placed in a medium-banked turn. At the higher altitude, the projected visual reference line appears to move backward across the ground (pylon appears to move forward).

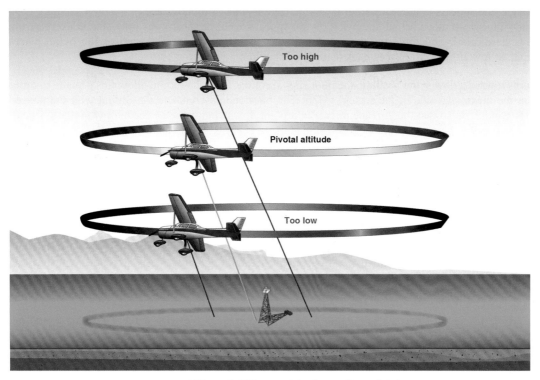

Figure 7-13. *Effect of different altitudes on line of sight.*

After demonstrating the maneuver at a high altitude, the pilot should reduce power and begin a descent at maneuvering speed in a continuing medium-bank turn around the pylon. The apparent backward movement of the projected visual reference line with respect to the pylon will slow down as altitude is lost and will eventually stop for an instant. If the pilot continues the descent below the pivotal altitude, the projected visual reference line with respect to the pylon will begin to move forward.

The altitude at which the visual reference line ceases to move across the ground is the pivotal altitude. If the airplane descends below the pivotal altitude, the pilot should increase power to maintain airspeed while regaining altitude to the point at which the projected reference line moves neither backward nor forward but actually pivots on the pylon. In this way, the pilot can determine the pivotal altitude of the airplane.

The pivotal altitude changes with variations in groundspeed. Since the headings throughout turns continuously vary from downwind to upwind, the groundspeed constantly changes. This results in the proper pivotal altitude varying slightly throughout the turn. The pilot should adjust for this by climbing or descending, as necessary, to hold the visual reference line on the pylons.

Selecting proper pylons is an important factor of successfully performing eights on pylons. They should be sufficiently prominent so the pilot can view them when completing the turn around one pylon and heading for the next. They should also be adequately spaced to provide time for planning the turns but not spaced so far apart that they cause unnecessary straight-and-level flight between the pylons. The distance between the pylons should allow for the straight-and-level flight segment to last from 3 to 5 seconds. The selected pylons should also be at the same elevation, since differences of over a few feet necessitate climbing or descending between each turn. The pilot should select two pylons along a line that lies perpendicular to the direction of the wind.

The pilot should estimate the pivotal altitude during preflight planning. Weather reports and consultation with other pilots flying in the area may provide both the wind direction and velocity. If the references are previously known (many flight instructors already have these ground-based references selected), the sectional chart will provide the MSL of the references, the Pilot's Operating Handbook (POH) provides the range of maneuvering airspeeds (based on weight), and the wind direction and velocity can be estimated to calculate the appropriate pivotal altitudes. The pilot should calculate the pivotal altitude for each position: upwind, downwind, and crosswind.

The pilot should begin the eight on pylons maneuver by flying diagonally crosswind between the pylons to a point downwind from the first pylon, so that the first turn can be made into the wind. As the airplane approaches a position where the pylon appears to be just ahead of the wingtip, the pilot should begin the turn by lowering the upwind wing to the point where the visual reference line aligns with the pylon. The reference line should appear to pivot on the pylon. As the airplane heads upwind, the groundspeed decreases, which lowers the pivotal altitude. As a result, the pilot should descend to hold the visual reference line on the pylon. As the turn progresses on the upwind side of the pylon, the wind becomes more of a crosswind. Since this maneuver does not require the turn to be completed at a constant radius, the pilot does not need to apply drift correction to complete the turn.

If the visual reference line appears to move ahead of the pylon (pylon appears to move back), the pilot should increase altitude. If the visual reference line appears to move behind the pylon (pylon appears to move ahead), the pilot should decrease altitude. Deflecting the rudder to yaw the airplane and force the wing and reference line forward or backward to the pylon places the airplane in uncoordinated flight, at low altitude, with steep bank angles and should not be attempted.

As the airplane turns toward a downwind heading, the pilot should rollout from the turn to allow the airplane to proceed diagonally to a point tangent on the downwind side of the second pylon. The pilot should complete the rollout with the proper wind correction angle to correct for wind drift, so that the airplane arrives at a point downwind from the second pylon that is equal in distance from the pylon as the corresponding point was from the first pylon at the beginning of the maneuver.

At this point, the pilot should begin a turn in the opposite direction by lowering the upwind wing to the point where the visual reference line aligns with the pylon. The pilot should then continue the turn the same way the corresponding turn was performed around the first pylon but in the opposite direction.

With prompt correction, and a very fine control pressures, it is possible to hold the visual reference line directly on the pylon even in strong winds. The pilot may make corrections for temporary variations, such as those caused by gusts or inattention, by reducing the bank angle slightly to fly relatively straight to bring forward a lagging visual reference line or by increasing the bank angle temporarily to turn back a visual reference line that has moved ahead. With practice, these corrections may become slight enough to be barely noticeable. It is important to understand that variations in pylon position are according to the apparent movement of the visual reference line. Attempting to correct pivotal altitude by the using the altimeter is ineffective.

Eights on pylons are performed at bank angles ranging from shallow to steep. [Figure 7-14] The pilot should understand that the bank chosen does not alter the pivotal altitude. As proficiency is gained, the instructor should increase the complexity of the maneuver by directing the learner to enter at a distance from the pylon that results in a specific bank angle at the steepest point in the pylon turn.

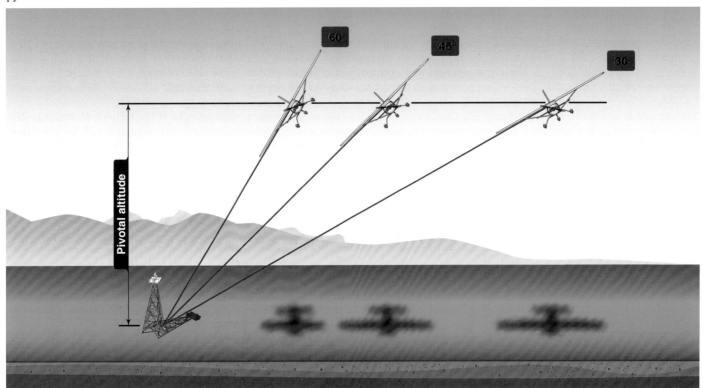

Figure 7-14. *Bank angle versus pivotal altitude.*

Common Errors

The most common error in attempting to hold a pylon is incorrect use of the rudder. When the projection of the visual reference line moves forward with respect to the pylon, many pilots tend to apply inside rudder pressure to yaw the wing backward. When the reference line moves behind the pylon, pilots tend to apply outside rudder pressure to yaw the wing forward. The pilot should use the rudder only for coordination.

Other common errors in the performance of eights on pylons are:

1. Failure to adequately clear the surrounding area for safety hazards, initially and throughout the maneuver.
2. Skidding or slipping in turns (whether trying to hold the pylon with rudder or not).
3. Excessive gain or loss of altitude.
4. Poor choice of pylons.
5. Not entering the pylon turns into the wind.
6. Failure to assume a heading when flying between pylons that will compensate sufficiently for drift.
7. Failure to time the bank so that the turn entry is completed with the pylon in position.
8. Abrupt control usage.
9. Inability to select pivotal altitude.

Chapter Summary

Ground reference maneuvers require planning and high levels of vigilance to ensure that the practice and performance of these maneuvers are executed where the safety to groups of people, livestock, communities, and the pilot is not compromised. While training to perform ground reference maneuvers, a pilot learns coordination, timing, and division of attention to maneuver the airplane accurately in reference to flight attitudes and specific ground references. After mastering ground reference maneuvers, the pilot should be able to command the airplane to specific pitch, roll, and yaw attitudes, correct for the effects of wind drift, and control the airplane's orientation in relation to ground-based references. While safety is paramount in all aspects of flying, ground reference maneuvers focus on mitigation of risk during low altitude flying. With these enhanced skills, the pilot also significantly improves their competency in everyday flight maneuvers, such as straight-and-level, turns, climbs, and descents.

Airplane Flying Handbook (FAA-H-8083-3C)
Chapter 8: Airport Traffic Patterns

Introduction

Airport traffic patterns ensure that air traffic moves into and out of an airport safely. The direction and placement of the pattern, the altitude at which it is to be flown, and the procedures for entering and exiting the pattern may depend on local conditions. Information regarding the procedures for a specific airport can be found in the Chart Supplements. General information on airport operations and traffic patterns can also be found in the Aeronautical Information Manual (AIM).

Airport Traffic Patterns and Operations

Just as roads and streets are essential for operating automobiles, airports or airstrips are essential for operating airplanes. Since flight begins and ends at an airport or other suitable landing field, pilots need to learn the traffic rules, traffic procedures, and traffic pattern layouts in use at various airports.

When an automobile is driven on congested city streets, it can be brought to a stop to give way to conflicting traffic. Airplane pilots do not have that option. Consequently, traffic patterns and traffic control procedures exist to minimize conflicts during takeoffs, departures, arrivals, and landings. The exact nature of each airport traffic pattern is dependent on the runway in use, wind conditions (which determine the runway in use), obstructions, and other factors.

Airports vary in complexity from small grass or sod strips to major terminals with paved runways and taxiways. Regardless of the type of airport, a pilot should know and abide by the applicable rules and operating procedures. In addition to checking the traffic pattern and operating procedures for airports of intended use in the Chart Supplements, pilots should know how to interpret any airport visual markings and signs that may be encountered. In total, the information provided to the pilot keeps air traffic moving with maximum safety and efficiency. However, the use of any traffic pattern, service, or procedure does not diminish the pilot's responsibility to see and avoid other aircraft from ramp-out to ramp-in.

When operating at an airport with an operating control tower, the pilot receives a clearance to approach or depart, as well as pertinent information about the traffic pattern by radio. The tower operator can instruct pilots to enter the traffic pattern at any point or to make a straight-in approach without flying the usual rectangular pattern. Many other deviations are possible if the tower operator and the pilot work together in an effort to keep traffic moving smoothly. Jets or heavy airplanes will frequently fly wider and/or higher patterns than lighter airplanes, and will sometimes make a straight-in approach for landing.

A pilot is not expected to have extensive knowledge of all traffic patterns at all airports, but if the pilot is familiar with the basic rectangular pattern, it is easy to make proper approaches and departures from most airports, regardless of whether or not they have control towers. However, if there is not a control tower, it is the pilot's responsibility to determine the direction of the traffic pattern, to comply with appropriate traffic rules, and to display common courtesy toward other pilots operating in the area. When operating at airports without a control tower, the pilot may not see all traffic. Therefore, the pilot should develop the habit of continued scanning even when air traffic appears light or nil. Adherence to the basic rectangular traffic pattern reduces the possibility of conflicts and reduces the probability of a midair collision.

Standard Airport Traffic Patterns

An airport traffic pattern includes the direction and altitude of the pattern and procedures for entering and leaving the pattern. Unless the airport displays approved visual markings indicating that turns should be made to the right, the pilot should make all turns in the pattern to the left.

Figure 8-1 shows a standard rectangular traffic pattern. The traffic pattern altitude is usually 1,000 feet above the elevation of the airport surface. The use of a common altitude at a given airport is the key factor in minimizing the risk of collisions at airports without operating control towers.

Aircraft speeds are restrained by 14 CFR part 91, section 91.117. When operating in the traffic pattern at most airports with an operating control tower, aircraft typically fly at airspeeds no greater than 200 knots (230 miles per hour (mph)). Sensible practice suggests flying at or below these speeds when operating in the traffic pattern of an airport without an operating control tower. In any case, the pilot should adjust the airspeed, when necessary, so that it is compatible with the airspeed of other aircraft in the traffic pattern.

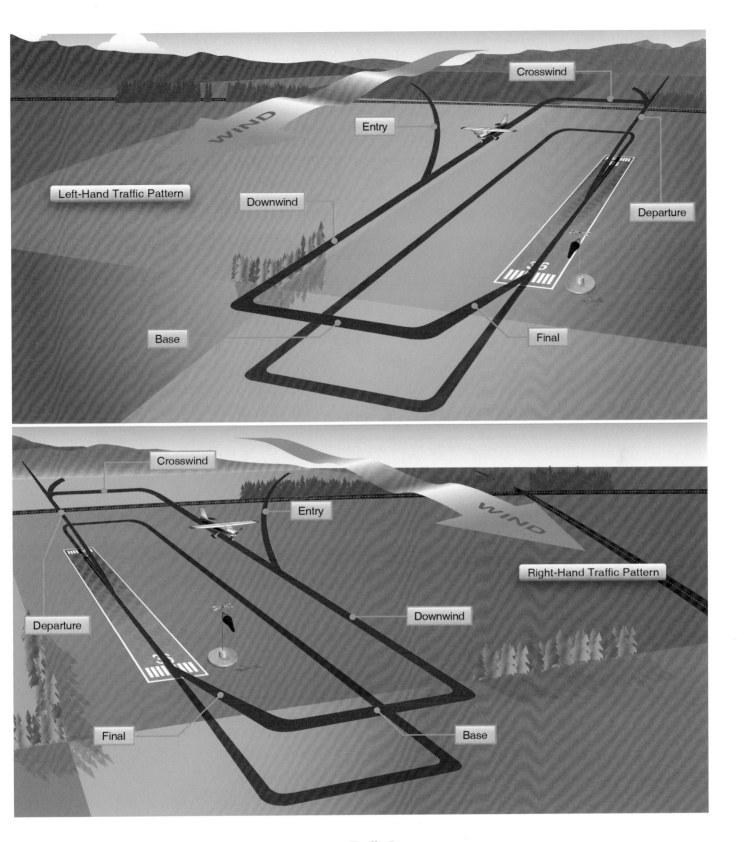

Figure 8-1. *Traffic Patterns*

When entering the traffic pattern at an airport without an operating control tower, inbound pilots are expected to observe other aircraft already in the pattern and to conform to the traffic pattern in use. The pilot should enter the traffic pattern at a point well clear of any other observed aircraft. If there are no other aircraft observed, the pilot should check traffic indicators and wind indicators on the ground to determine which runway and traffic pattern direction to use. *[Figure 8-2]* Many airports have L-shaped traffic pattern indicators displayed with a segmented circle adjacent to the runway. The short member of the L shows the direction in which traffic pattern turns are made when using the runway parallel to the long member. The pilot should check the indicators from a distance or altitude well above the traffic pattern in case any other aircraft are in the traffic pattern.

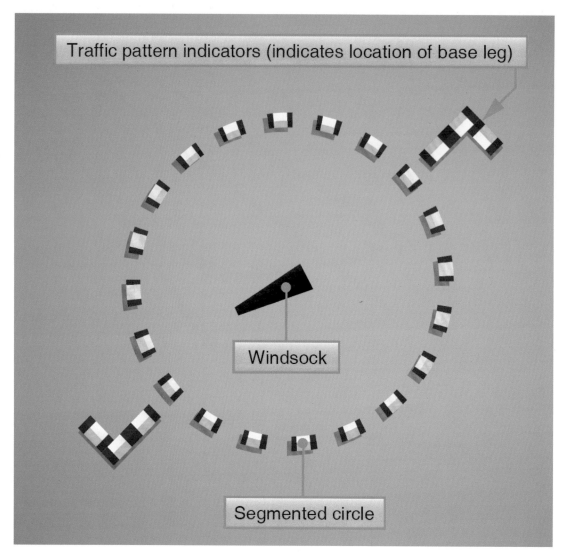

Figure 8-2. *Traffic pattern indicators.*

Consider the following points when arriving at an airport for landing:

- The pilot should be aware of the appropriate traffic pattern altitude before entering the pattern and remain clear of the traffic flow until established on the entry leg.

- The traffic pattern is normally entered at a 45° angle to the downwind leg, headed toward a point abeam the midpoint of the runway to be used for landing.

- The pilot should ensure that the entry leg is of sufficient length to provide a clear view of the entire traffic pattern and to allow adequate time for planning the intended path in the pattern and the landing approach.

- Entries into traffic patterns while descending create specific collision hazards and should be avoided.

The downwind leg is a course flown parallel to the landing runway, but in a direction opposite to the intended landing direction. This leg is flown approximately 1/2 to 1 mile out from the landing runway and at the specified traffic pattern altitude. When flying on the downwind leg, the pilot should complete all before-landing checks and extend the landing gear if the airplane is equipped with retractable landing gear. Pattern altitude is maintained until at least abeam the approach end of the landing runway. At this point, the pilot should reduce power and begin a descent. The pilot should continue the downwind leg past a point abeam the approach end of the runway to a point approximately 45° from the approach end of the runway, and make a medium-bank turn onto the base leg. Pilots should consider tailwinds and not descend too much on the downwind in order to have sufficient altitude to continue the descent on the base leg.

The base leg is the transitional part of the traffic pattern between the downwind leg and the final approach leg. Depending on the wind condition, the pilot should establish the base leg at a sufficient distance from the approach end of the landing runway to permit a gradual descent to the intended touchdown point. While on the base leg, the ground track of the airplane is perpendicular to the extended centerline of the landing runway, although the longitudinal axis of the airplane may not be aligned with the ground track if turned into the wind to counteract drift.

While on the base leg and before turning onto the final approach, the pilot should ensure that there is no close proximity to another aircraft already established on final approach. When two or more aircraft are approaching an airport for the purpose of landing, the aircraft at the lower altitude has the right-of-way. However, pilots should not take advantage of this rule to cut in front of another aircraft that is on final approach to land or to overtake that aircraft. If the turn to final would create a collision hazard, a go-around or avoidance maneuver is in order. A pilot trying to overtake another aircraft might be tempted to make an overly steep turn to final. If rushing the turn to increase the distance from another aircraft, there is good reason to abandon the approach and go around.

The final approach leg is a descending flightpath starting from the completion of the base-to-final turn and extending to the point of touchdown. This is probably the most important leg of the entire pattern, because of the sound judgment and precision needed to accurately control the airspeed and descent angle while approaching the intended touchdown point.

The pilot on final approach focuses on making a safe approach. If there is traffic on the runway, there should be sufficient time for that traffic to clear. If it appears that there may be a conflict, an early go-around may be in order. A pilot may go around and inform the controller. This is also a good time to verify the correct landing surface and avoid lining up with the wrong runway, an airport road, or a taxiway.

The upwind leg is a course flown parallel to the landing runway in the same direction as landing traffic. The upwind leg is flown at controlled airports and after go-arounds. When necessary, the upwind leg is the part of the traffic pattern in which the pilot will transition from the final approach to the climb altitude to initiate a go-around. When a safe altitude is attained, the pilot should commence a shallow-bank turn to the upwind side of the airport. This allows better visibility of the runway for departing aircraft.

The departure leg of the rectangular pattern is a straight course aligned with, and leading from, the takeoff runway. This leg begins at the point the airplane leaves the ground and continues until the pilot begins the 90° turn onto the crosswind leg.

On the departure leg after takeoff, the pilot should continue climbing straight ahead and, if remaining in the traffic pattern, commence a turn to the crosswind leg beyond the departure end of the runway within 300 feet of the traffic pattern altitude. If departing the traffic pattern, the pilot should continue straight out or exit with a 45° turn (to the left when in a left-hand traffic pattern or to the right when in a right-hand traffic pattern) beyond the departure end of the runway after reaching the traffic pattern altitude.

The crosswind leg is the part of the rectangular pattern that is horizontally perpendicular to the extended centerline of the takeoff runway. The pilot should enter the crosswind leg by making approximately a 90° turn from the upwind leg. The pilot should continue on the crosswind leg, to the downwind leg position.

If the takeoff is made into the wind, the wind will now be approximately perpendicular to the airplane's flightpath. As a result, the pilot should turn or head the airplane slightly into the wind while on the crosswind leg to maintain a ground track that is perpendicular to the runway centerline extension.

Non-Towered Airports

Non-towered airports traffic patterns are always entered at pattern altitude. How a pilot enters the pattern depends upon the direction of arrival. The preferred method for entering from the downwind leg side of the pattern is to approach the pattern on a course 45° to the downwind leg and join the pattern at midfield.

There are several ways to enter the pattern if the arrival occurs on the upwind leg side of the airport. One method of entry from the opposite side of the pattern is to announce intentions and cross over midfield at least 500 feet above pattern altitude (normally 1,500 feet AGL). However, if large or turbine aircraft operate at the airport, it is best to remain 2,000 feet AGL so as not to conflict with their traffic pattern. When well clear of the pattern—approximately 2 miles—the pilot should scan carefully for traffic, descend to pattern altitude, then turn right to enter at 45° to the downwind leg at midfield. *[Figure 8-3A]* An alternate method is to enter on a midfield crosswind at pattern altitude, carefully scan for traffic, announce intentions, and then turn downwind. *[Figure 8-3B]* This technique should not be used if the pattern is busy.

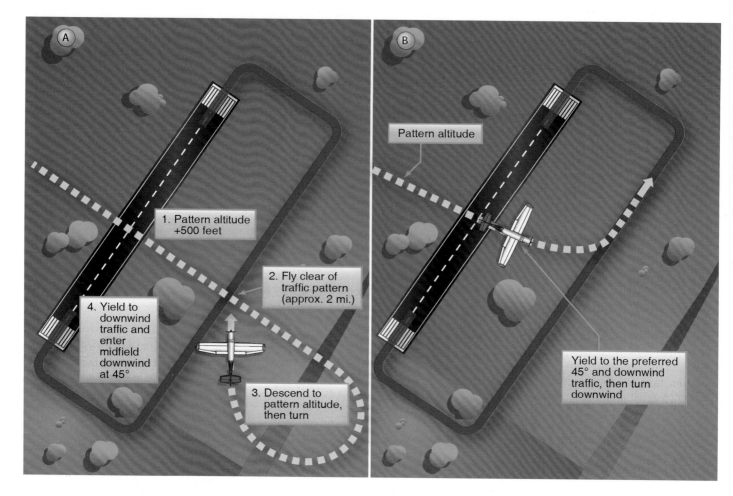

Figure 8-3. Preferred *entry from upwind leg side of airport (A). Alternate midfield entry from upwind leg side of airport (B).*

In either case, it is vital to announce intentions and to scan outside. Make course and speed adjustments that will lead to a successful pattern entry and give way to other aircraft on the preferred 45° entry or to aircraft already established on downwind.

Why is it advantageous to use the preferred 45° entry? If it is not possible to enter the pattern due to conflicting traffic, the pilot on a 45° entry can continue to turn away from the downwind, fly a safe distance away, and return for another attempt to join on the 45° entry—all while scanning for traffic.

Before joining the downwind leg, adjust course or speed to fit the traffic. Once fitting into the flow of traffic, adjust power on the downwind leg to avoid flying too fast or too slow. Speeds recommended by the airplane manufacturer should be used. They will generally fall between 70 to 90 knots for typical piston single-engine airplanes.

Safety Considerations

According to the National Transportation Safety Board (NTSB), the most probable cause of mid-air collisions is the pilot failing to see and avoid other aircraft. When near an airport, pilots should continue to scan for other aircraft and check blind spots caused by fixed aircraft structures, such as doorposts and wings. High-wing airplanes have restricted visibility above while low-wing airplanes have limited visibility below. The worst-case scenario is a low-wing airplane flying above a high-wing airplane. Banking from time to time can uncover blind spots. The pilot should also occasionally look to the rear of the airplane to check for other aircraft. *Figure 8-4* depicts the greatest threat area for mid-air collisions in the traffic pattern. Listed below are important facts regarding mid-air collisions:

- Mid-air collisions generally occur during daylight hours—56 percent occur in the afternoon, 32 percent occur in the morning, and 2 percent occur at night, dusk, or dawn.

- Most mid-air collisions occur under good visibility.

- A mid-air collision is most likely to occur between two aircraft going in the same direction.

- The majority of pilots involved in mid-air collisions are not on a flight plan.

- Nearly all accidents occur at or near uncontrolled airports and at altitudes below 1,000 feet.

- Pilots of all experience levels can be involved in mid-air collisions.

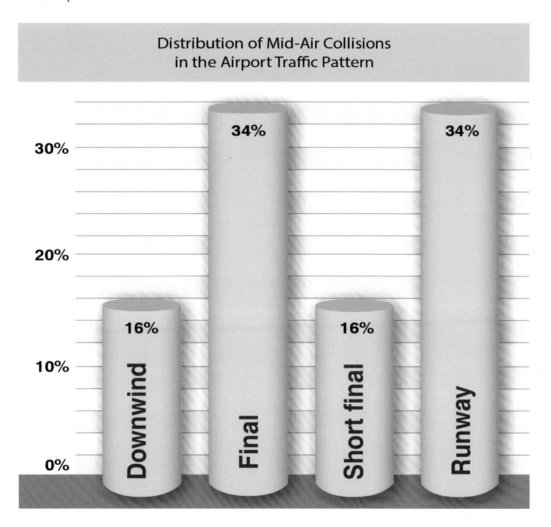

Figure 8-4. *Location distribution of mid-air collisions in the airport traffic pattern.*

The following are some important procedures that all pilots should follow when flying in a traffic pattern or in the vicinity of an airport.

1. Tune and verify radio frequencies before entering the airport traffic area.
2. Monitor the correct Common Traffic Advisory Frequency (CTAF).
3. Report position 10 miles out and listen for reports from other inbound traffic.
4. At a non-towered airport, report entering downwind, turning downwind to base, and base to final.
5. Descend to traffic pattern altitude before entering the pattern.
6. Maintain a constant visual scan for other aircraft.
7. Be aware that there may be aircraft in the pattern without radios.
8. Use exterior lights to improve the chances of being seen.

Chapter Summary

The volume of traffic at an airport can create a hazardous environment. Airport traffic patterns are procedures that improve the flow of traffic at an airport and enhance safety when properly executed. Most reported mid-air collisions occur during the final or short-final approach leg of the airport traffic pattern.

Introduction

There is an old saying that while takeoff is optional, landing is mandatory. In consideration of that adage, this chapter focuses on the approach to landing, factors that affect landings, types of landings, and aspects of faulty landings. A careful pilot knows that the safe outcome of a landing should never be in doubt. Pilots who respect their own limitations are able to approach each landing with confidence and achieve the satisfaction that comes from successful aircraft control. After any landing, a pilot performs a self-evaluation. If there is a question, a read of the relevant section in this chapter may help. When needed, additional flight instruction enhances safety.

The manufacturer's recommended procedures, including airplane configuration and airspeeds, and other information relevant to approaches and landings in a specific make and model airplane are contained in the Federal Aviation Administration (FAA)-approved Airplane Flight Manual and/or Pilot's Operating Handbook (AFM/POH) for that airplane. If any of the information in this chapter differs from the airplane manufacturer's recommendations as contained in the AFM/POH, the airplane manufacturer's recommendations take precedence.

Use of Flaps

The following general discussion applies to airplanes equipped with flaps. The pilot may use landing flaps during the descent to adjust lift and drag. Flap settings help determine the landing spot and the descent angle to that spot. [*Figure 9-1* and *Figure 9-2*] Flap extension during approaches and landings provides several advantages by:

1. Producing greater lift and permitting lower approach and landing speeds,

2. Producing greater drag and permitting a steeper descent angle,

3. Increasing forward visibility by allowing a lower pitch, and

4. Reducing the length of the landing roll.

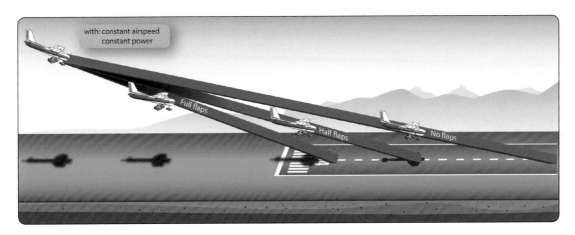

Figure 9-1. *Effect of flaps on the landing point.*

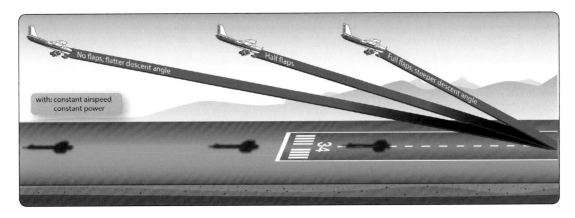

Figure 9-2. *Effect of flaps on the approach angle.*

The increased camber from flap deflection increases lift, primarily on the rear portion of the wing. This produces a nose-down pitching moment which may cause the airplane to pitch down. Flap deployment may also affect wing downwash on the horizontal tail and alter the tail-down force. Consequently, pitch behavior from flap extension depends on the design of the particular airplane.

Flap deflection of up to 15° primarily produces lift with minimal drag. The airplane has a tendency to balloon up with initial flap deflection because of the lift increase. The nose-down pitching moment, however, tends to offset the balloon. Flap deflection beyond 15° produces a large increase in drag. Deflection beyond 15° also produces a significant nose-up pitching moment in certain high-wing airplanes because the resulting downwash changes the airflow over the horizontal tail.

The time of flap extension and the degree of deflection are related. Large changes in flap deflection at one single point in the landing pattern can produce large lift changes that require significant pitch and power changes in order to maintain airspeed and descent angle. Consequently, there is an advantage to extending flaps in increments while in the landing pattern. Incremental deflection of flaps on downwind, base leg, and final approach allow smaller adjustments of pitch and power and support a stabilized approach.

Whenever the flap setting is changed, the pilot should be prepared to re-trim the airplane as needed to compensate for the change in aerodynamic forces. Throughout this chapter, more detail is provided on the use of flaps during specific approach and landing situations, as appropriate.

Normal Approach and Landing

Normal approach and landing procedures are used when the engine power is available, the wind is light or the final approach is made directly into the wind, the final approach path has no obstacles, and the landing surface is firm and of ample length to gradually bring the airplane to a stop. The selected landing point is normally beyond the runway approach threshold but within the first 1/3 of the runway.

The factors involved and the procedures described for the normal approach and landing also have applications to the other-than-normal approaches and landings discussed later in this chapter. The principles of normal operations are explained first and need to be understood before proceeding to the more complex operations. To better understand the factors that influence judgment and procedures, the last part of the approach pattern and the actual landing are divided into five phases:

1. the base leg

2. the final approach

3. the round out (flare)

4. the touchdown

5. the after-landing roll

Base Leg

The placement of the base leg is one of the important judgments made by the pilot to set up for a good landing. [*Figure 9-3*] The pilot accurately judges the height, distance from the approach end of the runway, and rate of descent to allow a stabilized approach, round out, and touchdown at the desired spot. The distance depends on the altitude of the base leg, the current winds, and the amount of wing flaps used. When there is a strong wind on final approach or the flaps are used to produce a steep angle of descent, the base leg should be positioned closer to the approach end of the runway than would be required with normal winds or flap settings. Normally, the landing gear is extended and the before-landing check completed prior to reaching the base leg.

After turning onto the base leg, the pilot starts or continues the descent with reduced power and a target airspeed of approximately 1.4 V_{SO}—the stalling speed in the landing configuration. For example, if V_{SO} is 60 knots, 1.4 V_{SO} is 84 knots (84 = 1.4 x 60). Landing flaps should be deployed as recommended. Full flaps are not recommended until the final approach is established. Since the final approach and landing are normally made into the wind, there is usually a crosswind during the base leg. A drift correction is established and maintained to follow a ground track perpendicular to the extension of the landing runway centerline. This requires that the airplane be angled sufficiently into the wind to prevent drifting farther away from the intended landing spot.

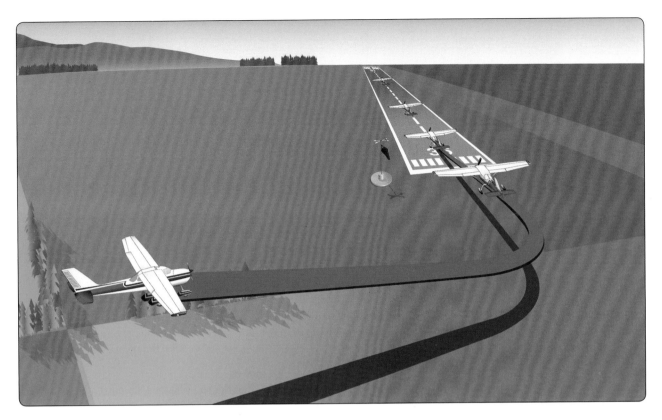

Figure 9-3. *Base leg and final approach.*

Final Approach

After the base-to-final approach turn is completed, the pilot aligns the longitudinal axis of the airplane with the centerline of the runway or landing surface. On a final approach, with no crosswind, the longitudinal axis is kept aligned with the runway centerline throughout the final approach and landing. (Methods to correct for a crosswind are explained in the "Crosswind Approach and Landing" section of this chapter. For now, only approaches and landings where the wind is straight down the runway are discussed.)

After aligning the airplane with the runway centerline, the final flap setting is completed and the pitch attitude adjusted as required. Some adjustment of pitch and power may be necessary to maintain the desired rate of descent and approach airspeed. The pilot should use the manufacturer's recommended airspeed or 1.3 V_{SO} if there is no manufacturer's recommendation. As the pitch attitude and airspeed stabilize, the airplane is re-trimmed to relieve any control pressure.

The descent angle is controlled throughout the approach so that the airplane lands in the center of the first third of the runway. The descent angle is affected by all four fundamental forces that act on an airplane (lift, drag, thrust, and weight). If all the forces are balanced out such that the net force on the airplane is zero, the descent angle remains constant in a steady state wind condition. The pilot controls these forces by adjusting the airspeed, attitude, power, and drag (flaps or forward slip). However, wind may affect the gliding distance over the ground [*Figure 9-4*]; the pilot does not have control over the wind, but corrects for its effect on the airplane's descent by adjusting pitch and power appropriately.

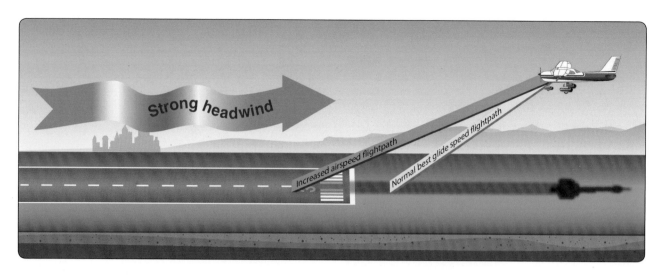

Figure 9-4. *Effect of headwind on final approach.*

A well-executed final approach includes reaching the desired touchdown point at an airspeed that results in minimum floating just before touchdown. To accomplish this, both the descent angle and the airspeed need to be controlled. This is one reason for performing approaches with partial power; if the approach is too high, the pilot can lower the nose and reduce the power to maintain the correct airspeed. When the approach is too low, the pilot can add power and raise the nose.

While the proper angle of descent and airspeed are maintained by integrating pitch and power changes, an untrained or inexperienced pilot may try to reach a landing spot by applying back-elevator pressure without adding power. However, attempting to stretch the final approach by raising the pitch attitude alone is almost always a bad idea. Using pitch alone causes a significant increase in AOA and decay in airspeed that leads to an excessive rate of descent or a low altitude stall. It is possible for either or both to occur.

Wrong Surface Landing Avoidance

A wrong surface landing occurs when an aircraft lands or tries to land on the wrong runway, on a taxiway in error, or at the wrong airport. The pilot should take a moment on every final approach to verify the correctness of the landing zone ahead.

Lack of familiarity with a particular airport, complacency, or fatigue may lead to pilot confusion, and occasionally a pilot will line up with the wrong surface while perceiving the situation as normal. A pilot may compensate for any lack of destination airport familiarity by studying an airport diagram and lighting ahead of time and noting key features and geometry. On final approach, the pilot should verify correct runway alignment and runway number. Pilots often refer to moving map displays driven by GPS, and these devices should increase situational awareness and safety. If there is a doubt over the landing surface, the pilot should go around and consider the situation further.

Stabilized Approach Concept

A stabilized approach is one in which the pilot establishes and maintains a constant-angle glide path towards a predetermined point on the landing runway. It is based on the pilot's judgment of certain visual clues and depends on maintaining a constant final descent airspeed and configuration.

An airplane descending on final approach at a constant rate and airspeed travels in a straight line towards a spot on the ground ahead, commonly called the aiming point. If the airplane maintains a constant glide path without a round out for landing, it will strike the ground at the aiming point. [*Figure 9-5*]

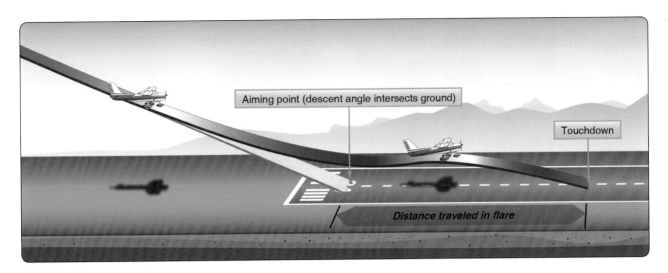

Figure 9-5. *Stabilized approach.*

To the pilot, the aiming point appears to be stationary. It does not appear to move under the nose of the aircraft and does not appear to move forward away from the aircraft. This feature identifies the aiming point—it does not move. However, objects in front of and beyond the aiming point do appear to move as the distance is closed, and they appear to move in opposite directions! For a constant angle glide path, the distance between the horizon and the aiming point remains constant. If descending at a constant angle and the distance between the perceived aiming point and the horizon appears to increase (aiming point moving down away from the horizon), then the true aiming point is farther down the runway. If the distance between the perceived aiming point and the horizon decreases, meaning that the aiming point is moving up toward the horizon, the true aiming point is closer than perceived.

During instruction in landings, one of the important skills a pilot acquires is how to use visual cues to discern the true aiming point from any distance out on final approach. From this, the pilot determines if the current glide path will result in either an under or overshoot. Note that the aiming point is not where the airplane actually touches down. Since the pilot reduces the rate of descent during the round out (flare), the actual touchdown occurs farther down the runway. Considering float during round out, the pilot is also able to predict the point of touchdown with some accuracy.

When the airplane is established on final approach, the shape of the runway image also presents clues as to what should be done to maintain a stabilized approach to a safe landing.

Obviously, a runway is normally shaped in the form of an elongated rectangle. When viewed from the air during the approach, the phenomenon, known as perspective, causes the runway to assume the shape of a trapezoid with the far end looking narrower than the approach end and the edge lines converging ahead.

As an airplane continues down the glide path at a constant angle (stabilized), the image the pilot sees is still trapezoidal, but of proportionately larger dimensions. In other words, during a stabilized approach, the runway shape does not change. [*Figure 9-6*]

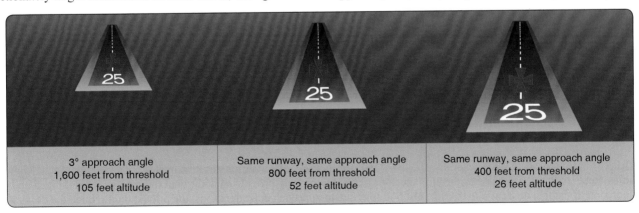

Figure 9-6. *Runway shape during stabilized approach.*

If the approach becomes shallow, the runway appears to shorten and become wider. Conversely, if the approach is steepened, the runway appears to become longer and narrower. [*Figure 9-7*]

Figure 9-7. *Change in runway shape if approach becomes narrow or steep.*

Immediately after rolling out on final approach, the pilot adjusts the pitch attitude, power, and trim so that the airplane is descending directly toward the aiming point at the appropriate airspeed in the landing configuration. If it appears that the airplane is going to overshoot the desired landing spot, a steeper approach results by reducing power and lowering the pitch attitude to maintain airspeed. If available and not fully extended, the pilot may further extend the flaps. If the desired landing spot is being undershot and a shallower approach is needed, the pilot increases both power and pitch attitude to reduce the descent angle. Once the approach is set up and control pressures removed with trim, the pilot is free to devote significant attention toward outside references and use the available visual cues to fine tune the approach. The pilot should not stare at any one place, but rather scan from one point to another, such as from the aiming point to the horizon, to any objects along the runway, to an area well short of the runway, and back to the aiming point. This makes it easier to perceive any deviation from the desired glide path and determine if the airplane is proceeding directly toward the aiming point. The pilot should also glance at the airspeed indicator periodically and correct for any airspeed deviation.

Pilots normally establish a stabilized approach before short final. The round out, touchdown, and landing roll are much easier to accomplish when preceded by a stabilized final approach, which reduces the chance of a landing mishap. Therefore, deviations from the desired glide path should be detected and corrected early so that the magnitude of corrections during the later portion of the approach is small. If the approach is very high or very low, it may not be possible to establish a stabilized approach, and the pilot normally executes a go-around. If the airplane is initially low and undershooting the aiming point, the pilot may intercept the desired glide path by increasing pitch attitude and adding power to level off while maintaining the correct airspeed. This may necessitate a substantial increase in power if the aircraft is operating on the backside of the power curve. As the airplane intercepts the desired glide path, the pilot reduces power and pitches down to remain on the glide path. Retracting the flaps to correct for an undershoot creates an unnecessary risk. It may cause a sudden decrease in lift, an excessive sink rate, and an aggravated unstable condition.

If the approach is too high or too low, it may not be possible to establish a stabilized approach, and the pilot should execute a go-around. Typically, pilots go-around if unable to establish a stabilized approach by 500 ft above airport elevation in visual meteorological conditions (VMC) or 1,000 ft above airport elevation in instrument conditions (IMC). For a typical GA piston aircraft in a traffic pattern, an immediate go-around should be initiated if the approach becomes unstabilized below 300 ft AGL.

Pilots may consider the following elements when attempting to set up and fly a stabilized approach to landing. The pilot should focus on the elements that lead to a stabilized approach rather than the order of the elements or the insistence on meeting all of the approach criteria. For a typical piston aircraft, an approach is stabilized when the following criteria are met:

1. **Glide path.** Typically a constant 3 degrees to the touchdown zone on the runway (obstructions permitting).

2. **Heading.** The aircraft tracks the centerline to the runway with only minor heading/pitch changes necessary to correct for wind or turbulence to maintain alignment. Bank angle normally limited to 15 degrees once established on final.

3. **Airspeed.** The aircraft speed is within +10 /-5 KIAS of the recommended landing speed specified in the AFM, $1.3V_{SO}$, or on approved placards/markings. If the pilot applies a gust factor, indicated airspeed should not decay below the recommended landing speed.

4. **Configuration.** The aircraft is in the correct landing configuration with flaps as required; landing gear extended, and is in trim.

5. **Descent rate.** A descent rate (generally 500-1000 fpm for light general aviation aircraft) makes for a safe approach. Minimal adjustments to the descent rate as the airplane approaches the runway provide an additional indication of a stabilized and safe approach. If using a descent rate in excess of 500 fpm due to approach considerations, the pilot should reduce the descent rate prior to 300 ft AGL.

6. **Power setting.** The pilot should use a power setting appropriate for the aircraft configuration and not below the minimum power for approach as defined by the AFM.

7. **Briefings and checklists.** Completing all briefings and checklists prior to initiating the approach (except the landing checklist), ensures the pilot can focus on the elements listed above.

Estimating Airplane Movement and Height

During short final, round out, and touchdown, vision is of prime importance. To provide a wide scope of vision and to foster good judgment of height and movement, the pilot's head should assume a natural, straight-ahead position. Visual focus is not fixed on any one side or any one spot ahead of the airplane. Instead, it is changed slowly from a point just over the airplane's nose to the desired touchdown zone and back again. This is done while maintaining a deliberate awareness of distance from either side of the runway using peripheral vision.

Accurate estimation of distance, besides being a matter of practice, depends upon how clearly objects are seen. It requires that vision be focused properly so that the important objects stand out as clearly as possible.

Speed blurs objects at close range. For example, most everyone has noted this in an automobile moving at high speed. Nearby objects seem to merge together in a blur, while objects farther away stand out clearly. The driver subconsciously focuses the eyes sufficiently far ahead of the automobile to see objects distinctly.

The distance at which the pilot's vision is focused should be proportional to the speed at which the airplane is traveling over the ground. Thus, as speed is reduced during the round out, the distance ahead of the airplane at which it is possible to focus is brought closer accordingly.

If the pilot attempts to focus on a reference that is too close or looks directly down, the reference becomes blurred, [*Figure 9-8*] and the reaction is either too abrupt or too late. In this case, the pilot's tendency is to over-control, round out high, and make full-stall, drop-in landings. If the pilot focuses too far ahead, accuracy in judging the closeness of the ground is lost and the consequent reaction is too slow, since there does not appear to be a necessity for action. This sometimes results in the airplane flying into the ground nose first. The change of visual focus from a long distance to a short distance requires a definite time interval, and even though the time is brief, the airplane's speed during this interval is such that the airplane travels an appreciable distance, both forward and downward toward the ground.

Figure 9-8. *Focusing too close blurs vision.*

Visual cues are important in flaring at the proper height and maintaining the wheels a few inches above the runway until eventual touchdown. Flare cues are primarily dependent on the angle at which the pilot's central vision intersects the ground (or runway) ahead and slightly to the side. Proper depth perception is a factor in a successful flare, but the visual cues used most are those related to changes in runway or terrain perspective and to changes in the size and texture of familiar objects near the landing area. The pilot should focus direct central vision at a shallow downward angle from 10° to 15° relative to the runway as the round out/flare is initiated. [*Figure 9-9*] When using this steady viewing angle, the point of visual interception with the runway appears progressively closer as the airplane loses altitude. This rate of closure is an important visual cue in assessing the rate of altitude loss. Conversely, movement of the visual interception point further down the runway indicates an increase in altitude and means that the pitch angle was increased too rapidly during the flare. Location of the visual interception point in conjunction with assessment of flow velocity of nearby off-runway terrain, as well as the similarity of appearance of height above the runway ahead of the airplane (in comparison to the way it looked when the airplane was taxied prior to takeoff), is also used to judge when the wheels are just a few inches above the runway.

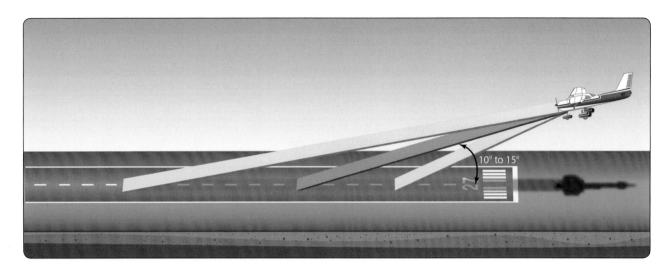

Figure 9-9. *To obtain necessary visual cues, the pilot should look toward the runway at a shallow angle.*

Round Out (Flare)

The round out is a slow, smooth transition from a normal approach attitude to a landing attitude, gradually rounding out the flightpath to one that is parallel to and a few inches above the runway. When the airplane approaches 10 to 20 feet above the ground in a normal descent, the round out or flare is started. Back-elevator pressure is gradually applied to slowly increase the pitch attitude and AOA. [*Figure 9-10*] The AOA is increased at a rate that allows the airplane to continue settling slowly as forward speed decreases. This is a continuous process until the airplane touches down on the ground.

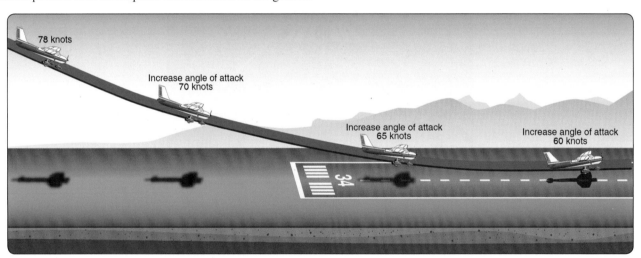

Figure 9-10. *Changing angle of attack during round out.*

When the AOA is increased, the lift is momentarily increased and this decreases the rate of descent. Since power normally is reduced to idle during the round out, the airspeed also gradually decreases. This causes lift to decrease again and necessitates raising the nose and further increasing the AOA. During the round out, the airspeed is decreased to touchdown speed while the lift is controlled so the airplane settles gently onto the landing surface. The round out is executed at a rate such that the proper landing attitude and the proper touchdown airspeed are attained simultaneously just as the wheels contact the landing surface.

The rate at which the round out is executed depends on the airplane's height above the ground, the rate of descent, and the pitch attitude. A round out started excessively high needs to be executed more slowly than one started from a lower height. The round out rate should also be proportional to the rate of closure with the ground. When the airplane appears to be descending very slowly, the increase in pitch attitude should be made at a correspondingly slow rate.

The pitch attitude of the airplane in a full-flap approach is considerably lower than in a no-flap approach. To attain the proper landing attitude before touching down, the nose needs to travel through a greater pitch change when flaps are fully extended. Since the round out is usually started at approximately the same height above the ground regardless of the degree of flaps used, the pitch attitude should be increased at a faster rate when full flaps are used. However, the round out should still be executed at a rate that takes the airplane's downward motion into account.

Once the actual process of rounding out is started, the pilot should not push the elevator control forward. If too much back-elevator pressure was exerted, this pressure is either slightly relaxed or held constant, depending on the degree of the error. In some cases, it may be necessary to advance the throttle slightly to prevent an excessive rate of sink or a stall, either of which results in a hard, drop-in type landing.

It is recommended that a pilot form the habit of keeping one hand on the throttle throughout the approach and landing should a sudden and unexpected hazardous situation require an immediate application of power.

Touchdown

The touchdown is the gentle settling of the airplane onto the landing surface. The round out and touchdown are normally made with the engine idling. During the round out, the airspeed decays such that the airplane touches down on the main gear at or just above the approximate stalling speed. As the airplane settles, proper landing attitude is attained by application of whatever back-elevator pressure is necessary.

Some pilots try to force or fly the airplane onto the ground without establishing proper landing attitude. The airplane should never be flown onto the runway with excessive speed. A common technique to making a smooth touchdown is to actually focus on holding the wheels of the aircraft a few inches off the ground as long as possible using the elevators while the power is smoothly reduced to idle. In most cases, when the wheels are within 2 or 3 feet of the ground, the airplane is still settling too fast for a gentle touchdown. Therefore, this descent is retarded by increasing back-elevator pressure. Since the airplane is already close to its stalling speed and is settling, this added back-elevator pressure only slows the settling instead of stopping it. At the same time, it results in the airplane touching the ground in the proper landing attitude and the main wheels touching down first so that little or no weight is on the nose-wheel. [*Figure 9-11*]

Figure 9-11. *A well-executed round out results in attaining the proper landing attitude.*

After the main wheels make initial contact with the ground, back-elevator pressure is held to maintain a positive AOA for aerodynamic braking and to hold the nose-wheel off the ground as the airplane decelerates. The pilot should be certain not to inadvertently have brake pressure engaged as touchdown occurs. Early use of brakes can result in a sudden drop in the nose and a loss of aerodynamic braking. As the airplane's momentum decreases, back-elevator pressure is gradually relaxed to allow the nose-wheel to gently settle onto the runway. This permits steering if the airplane has a steerable nose-wheel. At the same time, it decreases the AOA and reduces lift on the wings to prevent floating or skipping and allows the full weight of the airplane to rest on the wheels for better mechanical braking action. As the airplane slows, the mechanical braking becomes more effective.

It is extremely important that the touchdown occur with the airplane's longitudinal axis exactly parallel to the direction in which the airplane is moving along the runway. Failure to accomplish this imposes severe side loads on the landing gear. To avoid these side stresses, the pilot should not allow the airplane to touch down while turned into the wind or drifting.

After-Landing Roll

The landing process should never be considered complete until the airplane decelerates to the normal taxi speed during the landing roll or has been brought to a complete stop when clear of the landing area. Accidents may occur as a result of pilots abandoning their vigilance and failing to maintain positive control after getting the airplane on the ground.

A pilot should be alert for directional control difficulties immediately upon and after touchdown due to the ground friction on the wheels. Loss of directional control may lead to an aggravated, uncontrolled, tight turn on the ground, or a ground loop. The combination of centrifugal force acting on the center of gravity (CG) and ground friction of the main wheels resisting it during the ground loop may cause the airplane to tip or lean enough for the outside wingtip to contact the ground. This imposes a sideward force that could collapse the landing gear.

The rudder serves the same purpose on the ground as it does in the air—it controls the yawing of the airplane. The effectiveness of the rudder is dependent on the airflow, which depends on the speed of the airplane. As the speed decreases and the nose-wheel has been lowered to the ground, the steerable nose provides more positive directional control.

The brakes of an airplane serve the same primary purpose as the brakes of an automobile—to reduce speed on the ground. In airplanes, they are also used as an aid in directional control when more positive control is required than could be obtained with rudder or nose-wheel steering alone.

To use brakes, on an airplane equipped with toe brakes, the pilot slides the toes or feet up from the rudder pedals to the brake pedals. If rudder pressure is being held at the time braking action is needed, the pilot should not release that pressure as the feet or toes are being slid up to the brake pedals because control may be lost before brakes can be applied.

Putting maximum weight on the wheels after touchdown is an important factor in obtaining optimum braking performance. During the early part of rollout, some lift continues to be generated by the wing. After touchdown, the nose-wheel is lowered to the runway to maintain directional control. During deceleration, applying brakes may cause the nose to pitch down and some weight to transfer to the nose-wheel from the main wheels. This does not aid in braking action, so back pressure is applied to the controls without lifting the nose-wheel off the runway. This enables directional control while keeping weight on the main wheels.

Careful application of the brakes is initiated after the nose-wheel is on the ground and directional control is established. Maximum brake effectiveness is just short of the point where skidding occurs. If the brakes are applied so hard that skidding takes place, braking becomes ineffective. Skidding is stopped by releasing the brake pressure. Braking effectiveness is not enhanced by alternately applying, releasing, and reapplying brake pressure. The brakes are applied firmly and smoothly as necessary.

During the ground roll, the airplane's direction of movement can be changed by carefully applying pressure on one brake or uneven pressures on each brake in the desired direction. Caution must be exercised when applying brakes to avoid over-controlling.

The ailerons serve the same purpose on the ground as they do in the air—they change the lift and drag components of the wings. During the after-landing roll, they are used to keep the wings level in much the same way they are used in flight. If a wing starts to rise, aileron control is applied toward that wing to lower it. The amount required depends on speed because as the forward speed of the airplane decreases, the ailerons become less effective. Procedures for using ailerons in crosswind conditions are explained in the "Crosswind Approach and Landing" section of this chapter.

Once the airplane has slowed sufficiently and has turned onto the taxiway and stopped, the pilot performs the after-landing checklist. Many accidents have occurred as a result of the pilot unintentionally operating the landing gear control and retracting the gear instead of the flap control when the airplane was still rolling. The habit of positively identifying both of these controls, before actuating them, should be formed from the very beginning of flight training and continued in all future flying activities. If available runway permits, the speed of the airplane is allowed to dissipate in a normal manner.

Common Errors

Common errors in the performance of normal approaches and landings are:

1. Failure to complete the landing checklist in a timely manner.
2. Inadequate wind drift correction on the base leg.
3. An overshooting, undershooting, too steep, or too shallow a turn onto final approach.
4. A skidding turn from base leg to final approach as a result of overshooting/inadequate wind drift correction.
5. Poor coordination during turn from base to final approach.
6. Unstable approach.
7. Failure to adequately compensate for flap extension.
8. Poor trim technique on final approach.
9. Attempting to maintain altitude or reach the runway using elevator alone.
10. Focusing too close to the airplane resulting in a too high round out.
11. Focusing too far from the airplane resulting in a too low round out.
12. Touching down prior to attaining proper landing attitude.
13. Failure to hold sufficient back-elevator pressure after touchdown.
14. Excessive braking after touchdown.
15. Loss of aircraft control during touchdown and rollout.

Go-Arounds (Rejected Landings)

A go-around is a normal maneuver that is used when approach and landing parameters deviate from expectations or when it is hazardous to continue. Situations such as air traffic control (ATC) requirements, unexpected appearance of hazards on the runway, overtaking another airplane, wind shear, wake turbulence, mechanical failure, or an unstable approach are all reasons to discontinue a landing approach. Like any other normal maneuver, the go-around should be practiced and perfected. The flight instructor should emphasize early on, and the pilot should understand, that any approach or landing may result in a go-around. The assumption that an aborted landing is invariably the consequence of a poor approach, which in turn is due to insufficient experience or skill, is a fallacy.

Although the need to discontinue a landing may arise at any point in the landing process, the most critical go-around is one started when very close to the ground. The go-around maneuver is not inherently dangerous in itself. It becomes dangerous only when delayed unduly or executed improperly. Delay in initiating the go-around normally stems from two sources:

1. Landing expectancy or set—the anticipatory belief that conditions are not as threatening as they are and that the approach is sure to terminate with a safe landing.

2. Pride—the mistaken belief that the act of going around is an admission of failure—failure to execute the approach properly.

The proper execution of a go-around maneuver includes three cardinal principles:

1. Power

2. Attitude

3. Configuration

Power

Power is the pilot's first concern. The instant a pilot decides to go around, full or maximum allowable takeoff power should be applied smoothly, without hesitation, and held until flying speed and controllability are restored. An airplane that is settling toward the ground has inertia that needs to be overcome, and sufficient power is needed to stop the descent. The application of power is smooth, as well as positive. Abrupt movements of the throttle in some airplanes cause the engine to falter. Carburetor heat is turned off to obtain maximum power, as applicable.

Attitude

A pilot executing a go-around needs to accept the fact that an airplane cannot fly below stall speed, and it cannot climb below minimum power required speed. The pilot should resist any impulse to pitch-up for a climb if airspeed is insufficient. In some circumstances, it may be desirable to lower the nose briefly to gain airspeed and not be on the backside of the power curve.

At the time a pilot decides to go around, a trim setting for low airspeed is in place. The sudden addition of power tends to raise the airplane's nose and causes left yaw. Allowing the nose to rise too early could result in an unrecoverable stall when the go-around occurs at a low altitude. The pilot should anticipate the need for considerable forward elevator pressure to hold the nose level or in a safe climb attitude. The pilot applies sufficient right rudder pressure to counteract torque and P-factor. Trim helps to relieve adverse control pressures and assists in maintaining a proper pitch attitude. After attaining the appropriate airspeed and adjusting pitch attitude for a climb, the pilot should "rough trim" the airplane to relieve any adverse control pressures. More precise trim adjustments can be made when flight conditions have stabilized. On airplanes that produce high control pressures when using maximum power on go-arounds, the pilot should use caution when reaching for the flap handle. Airplane control is the first consideration during this high-workload phase.

Configuration

After establishing the proper climb attitude and power settings, the pilot's next concern is flap retraction. After the descent has been stopped, the landing flaps are partially retracted or placed in the takeoff position as recommended by the manufacturer. Depending on the airplane's altitude and airspeed, it is wise to retract the flaps intermittently in small increments to allow time for the airplane to accelerate progressively as they are being raised. A sudden and complete retraction of the flaps could cause a loss of lift resulting in the airplane settling into the ground. [*Figure 9-12*]

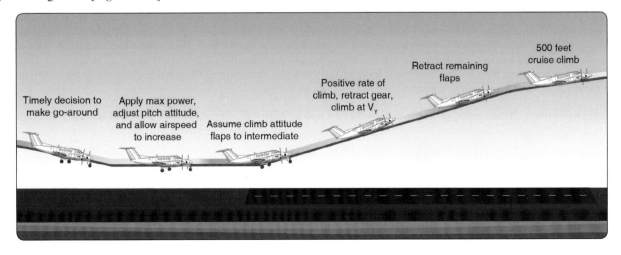

Figure 9-12. *Go-around procedure.*

Unless otherwise specified in the AFM/POH, it is generally recommended that the flaps be retracted (at least partially) before retracting the landing gear for two reasons. First, on most airplanes full flaps produce more drag than the landing gear; and second, in case the airplane inadvertently touches down as the go-around is initiated, it is desirable to have the landing gear in the down-and-locked position. After a positive rate of climb is established, the landing gear is retracted.

The landing gear is retracted only after the initial or rough trim is accomplished and when it is certain the airplane will remain airborne. During the initial part of an extremely low go-around, it is possible for the airplane to settle onto the runway and bounce. This situation is not particularly dangerous provided the airplane is kept straight and a constant, safe pitch attitude is maintained. With the application of power, the airplane attains a safe flying speed rapidly and the advanced power cushions any secondary touchdown.

Ground Effect

Ground effect is a factor in every landing and every takeoff in fixed-wing airplanes. Ground effect can also be an important factor in go-arounds. If the go-around is made close to the ground, the airplane may be in the ground effect area. Pilots are often lulled into a sense of false security by the apparent "cushion of air" under the wings that initially assists in the transition from an approach descent to a climb. This "cushion of air," however, is imaginary. The apparent increase in airplane performance is, in fact, due to a reduction in induced drag in the ground effect area. It is "borrowed" performance that is repaid when the airplane climbs out of the ground effect area. The pilot needs to factor in ground effect when initiating a go-around close to the ground. An attempt to climb prematurely may result in the airplane not being able to climb or even maintain altitude at full power.

Common Errors

Common errors in the performance of go-arounds (rejected landings) are:

1. Failure to recognize a condition that warrants a rejected landing.

2. Indecision.

3. Delay in initiating a go-around.

4. Failure to apply maximum allowable power in a timely manner.

5. Abrupt power application.

6. Improper pitch attitude.

7. Failure to configure the airplane appropriately.

8. Attempting to climb out of ground effect prematurely.

9. Failure to adequately compensate for torque/P factor.

10. Loss of aircraft control.

Intentional Slips

A slip occurs when the bank angle of an airplane is too steep for the existing rate of turn. Unintentional slips are most often the result of uncoordinated rudder/aileron application. Intentional slips, however, are used to dissipate altitude without increasing airspeed and/or to adjust airplane ground track during a crosswind. Intentional slips are especially useful in forced landings and in situations where obstacles need to be cleared during approaches to confined areas. A slip can also be used as a means of rapidly reducing airspeed in situations where wing flaps are inoperative or not installed.

A slip is a combination of forward movement and sideward (with respect to the longitudinal axis of the airplane) movement, the lateral axis being inclined and the sideward movement being toward the low end of this axis (low wing). An airplane in a slip is in fact flying sideways through the air even though it may appear to be going straight over the ground. This results in a change in the direction that the relative wind strikes the airplane. Slips are characterized by a marked increase in drag and corresponding decrease in airplane climb, cruise, and glide performance. Because the airplane is banked, the vertical component of lift is reduced allowing for an airplane in a slip to descend rapidly without an increase in airspeed.

Most airplanes exhibit the characteristic of positive static directional stability and, therefore, have a natural tendency to compensate for slipping. An intentional slip usually requires deliberate cross-controlling of ailerons and rudder throughout the maneuver.

There are two types of intentional slips: sideslip and forward slips. Sideslips are frequently used when landing with a crosswind to keep the aircraft aligned with the runway centerline. A sideslip is entered by lowering a wing and applying just enough opposite rudder to prevent a turn. In a sideslip, the airplane's longitudinal axis remains parallel to the original flightpath, but the airplane no longer flies straight ahead. Instead, the horizontal component of lift forces the airplane also to move somewhat sideways toward the low wing. [*Figure 9-13*] The amount of slip, and therefore the rate of sideward movement, is determined by the bank angle. The steeper the bank, the greater the degree of slip. As bank angle is increased, additional opposite rudder is required to prevent turning.

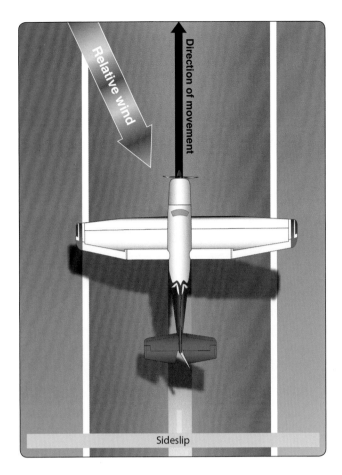

Figure 9-13. *Sideslip.*

A forward slip is used to dissipate altitude and increase descent rate without increasing airspeed. In a forward slip, the airplane's direction of motion continues the same as before the slip was begun. Assuming the airplane is originally in straight coordinated flight, the wing on one side is lowered by use of the ailerons. Simultaneously, sufficient opposite rudder is used to yaw the airplane's nose in the opposite direction such that the airplane remains on its original flightpath. However, the nose of the airplane will no longer point in the direction of flightpath. [*Figure 9-14*] In a forward slip, the amount of slip, and therefore the sink rate, is determined by the bank angle. The steeper the bank, the steeper the descent. In order to use the maneuver to lose altitude, power is normally reduced to idle. The pilot controls airspeed using elevator control. When a crosswind is present, the pilot should lower the upwind wing such that the airplane is banked into the crosswind since slipping into the wind makes it easier to remain on the original flightpath.

In most light airplanes, the steepness of a slip is limited by the amount of rudder travel available. In both sideslips and forward slips, the point may be reached where full rudder is required to maintain heading even though the ailerons are capable of further steepening the bank angle. This is the practical slip limit because any additional bank would cause the airplane to turn even though full opposite rudder is being applied. If there is a need to descend more rapidly, even though the practical slip limit has been reached, lowering the nose not only increases the sink rate but also increases airspeed. The increase in airspeed increases rudder effectiveness permitting a steeper slip. Conversely, when the nose is raised, rudder effectiveness decreases and the bank angle should be reduced.

Discontinuing a slip is accomplished by leveling the wings and simultaneously releasing the rudder pressure while readjusting the pitch attitude to the normal glide attitude. If the pressure on the rudder is released abruptly, the nose swings too quickly into line and the airplane tends to acquire excess speed. Because of the location of the pitot tube and static vents, airspeed indicators in some airplanes may have considerable error when the airplane is in a slip. The pilot needs to be aware of this possibility and recognize a properly performed slip by the attitude of the airplane, the sound of the airflow, and the feel of the flight controls. Unlike skids, however, if an airplane in a slip is made to stall, it displays very little of the yawing tendency that causes a skidding stall to develop into a spin. The airplane in a slip may do little more than tend to roll into a wings-level attitude.

Note that some airplanes have limitations regarding slips. In some cases slips are limited in duration or by fuel quantity. These limitations are meant to preclude fuel starvation caused when fuel is forced to one side of a tank in uncoordinated flight. If a forward slip is being used to reach a landing area in an actual engine-out emergency, the time limitation or fuel limitation is irrelevant (unless a prolonged slip caused the engine issue). For aerodynamic reasons, there may also be recommendations or limitations related to slips with flaps extended. Consult the manufacturer's AFM/POH for specific airplane information.

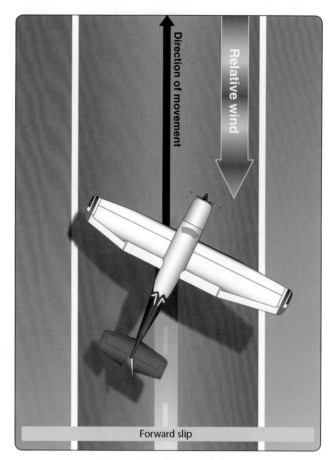

Figure 9-14. *Forward slip.*

Some pilots try to avoid using forward slips. An approach with flaps allows for coordinated and more familiar flight orientation, while the sideways force on the occupants of the aircraft during a forward slip may seem uncomfortable. However, in a real emergency that involves engine failure, the ability to use a forward slip provides a pilot with a technique contributing to a better outcome. In that situation, a pilot may initiate a descent using a forward slip much more quickly than by deploying flaps. To reduce the descent, the pilot can remove the slip without penalty. On the other hand, retracting flaps on an approach could lead to an unwanted loss of altitude. Even with full rudder displacement during a forward slip, the pilot can adjust to the left and right of the intended ground track by increasing and decreasing aileron deflection. The value of the maneuver explains its inclusion as a task in the Private Pilot Airman Certification Standards (ACS).

Forward Slip to a Landing

When demonstrating a forward slip to a landing in an airport traffic pattern, the pilot plans the descent such that a forward slip may be used on final approach. Flaps usually remain retracted, and using a forward slip on downwind or base may be a necessary part of the maneuver. When abeam the landing point on the downwind leg, the pilot initiates a descent by reducing power to idle. If an insufficient rate of descent occurs on downwind, the pilot uses a forward slip to increase the rate of descent. The pilot should make a coordinated turn to base. At this point, ongoing evaluation of height takes place. If the airplane is high on base, continued forward slip should occur. However, the pilot should make a coordinated turn to line up with the final approach course. Once established on a final approach, the height above ground should be sufficient to allow the pilot to use a forward slip and establish a suitable approach path to the runway aiming point. At the appropriate time, when the round out begins, the pilot removes the forward slip and transitions to a normal landing.

Common Errors

Common errors with forward slips to a landing:

1. Incorrect pitch adjustments that result in poor airspeed control.

2. Reacting to erroneous airspeed indications.

3. Using excess power while trying to lose altitude.

4. A slip in the same direction as any crosswind.

5. Poor glidepath control.

6. Late transition to a sideslip during landing with crosswinds.

7. Landing without the longitudinal axis parallel to runway.

8. Landing off the centerline.

Crosswind Approach and Landing

Most runways or landing areas are such that landings need to be made while the wind is blowing at an angle to the runway rather than parallel to the landing direction. All pilots should be prepared to manage a crosswind situation when it arises. Many of the same basic principles and factors involved in a normal approach and landing apply to a crosswind approach and landing; therefore, only the additional procedures required for correcting for wind drift are discussed here.

Crosswind landings are a little more difficult to perform than crosswind takeoffs, mainly due to different inputs involved in maintaining accurate control of the airplane while its speed is decreasing rather than increasing as on takeoff.

There are two usual methods of accomplishing a crosswind approach and landing—the crab method and the wing-low (sideslip) method. Although the crab method may be easier for the pilot to maintain during final approach, it requires judgment and precise timing when removing the crab immediately prior to touchdown. The wing-low method is recommended in most cases, although a combination of both methods may be used. While current testing standards allow for either method, pilots should learn to do both.

Crosswind Final Approach

When using the crab method, the pilot makes a coordinated turn to establish a heading (crab) toward the wind. The selected heading should align the airplane's wings-level ground track with the centerline of the runway. The pilot makes small heading corrections, if needed, to maintain alignment with the runway. [*Figure 9-15*] The appropriate crab angle is maintained until just prior to touchdown, when the pilot uses rudder control to align the longitudinal axis of the airplane with the runway to avoid sideward contact of the wheels with the runway. A change in alignment made too early or too late results in a side load. If a long final approach is being flown, one option is to use the crab method initially and smoothly transition to the wing-low method before the round out is started.

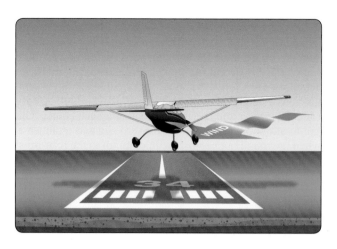

Figure 9-15. *Crabbed approach.*

While the wing-low (sideslip) method also compensates for a crosswind from any angle, it keeps the airplane's ground track and longitudinal axis aligned with the runway centerline throughout the final approach, round out, touchdown, and after-landing roll. This prevents the airplane from touching down in a sideward motion and imposing damaging side loads on the landing gear. When first experienced, it may seem odd to land while holding a bank angle. Although some pilots state that it appears the upwind wingtip will strike the ground, this is not the case. This method sets up the crosswind correction well before touchdown, does not require a heading change at the moment before touchdown, and allows the pilot to exercise smooth and continuous control. Pilots using this technique use precise airplane control as changes in control pressure occur near the ground, on short final, and while over the runway.

To use the wing-low method, the pilot first uses rudder to align and maintain the airplane's heading with the runway direction. Since the airplane is now exposed to an uncorrected crosswind, the airplane will begin to drift. Note the rate and direction of drift, and oppose it using ailerons resulting in just enough bank to cancel the drift. [*Figure 9-16*] Varying the amount of bank allows the pilot to drift either to the left or to the right, and the pilot adjusts control pressures as needed to intercept and maintain the runway centerline. If the crosswind changes, the sideslip is adjusted to keep the airplane in line with the center of the runway. [*Figure 9-17*]

Figure 9-16. *Sideslip approach.*

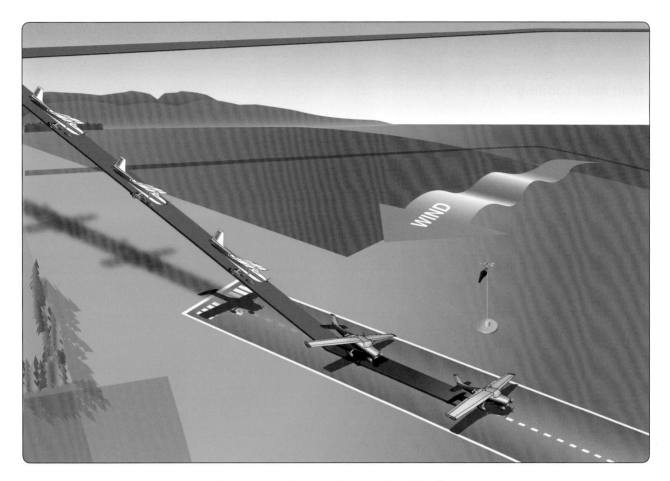

Figure 9-17. *Crosswind approach and landing.*

To correct for strong crosswind, the slip into the wind is increased by lowering the upwind wing as needed. As a consequence, this results in a greater tendency of the airplane to turn. Since turning is not desired, considerable opposite rudder is applied to keep the airplane's longitudinal axis aligned with the runway. In some airplanes, there may not be sufficient rudder travel available to compensate for the strong turning tendency caused by the steep bank. If the required bank is such that full opposite rudder does not prevent a turn, the wind is too strong to safely land the airplane on that particular runway with those wind conditions. Since the airplane's capability is exceeded, it is imperative that the landing be made on a more favorable runway either at that airport or at an alternate airport.

Flaps are used during most approaches since they tend to have a stabilizing effect on the airplane. The degree to which flaps are extended vary with the airplane's handling characteristics, as well as the wind velocity.

Crosswind Round Out (Flare)

Generally, the round out is made like a normal landing approach, but the application of a crosswind correction is continued as necessary to prevent drifting.

Since the airspeed decreases as the round out progresses, the flight controls gradually become less effective. As a result, the crosswind correction being held becomes inadequate. When using the wing-low method, it is necessary to gradually increase the deflection of the rudder and ailerons to maintain the proper amount of drift correction.

Keep the upwind wing down throughout the round out. If the wings are leveled, the airplane begins drifting and the touchdown occurs while drifting. Remember, the primary objective is to land the airplane without subjecting it to any side loads that result from touching down while drifting.

Crosswind Touchdown

If the crab method of drift correction is used throughout the final approach and round out, the crab needs to be removed the instant before touchdown by applying rudder to align the airplane's longitudinal axis with its direction of movement.

If the wing-low method is used, the crosswind correction is maintained throughout the round out, and the initial touchdown occurs on the upwind main wheel. During gusty or high wind conditions, prompt adjustments are made in the crosswind correction to assure that the airplane does not drift as the airplane touches down. As the forward momentum decreases after initial contact, the weight of the airplane causes the downwind main wheel to gradually settle onto the runway.

In those airplanes having nose-wheel steering interconnected with the rudder, the nose-wheel is not aligned with the runway as the main wheels touch down because opposite rudder is being held for the crosswind correction. To prevent swerving in the direction the nose-wheel is offset, the corrective rudder pressure needs to be relaxed as the nose-wheel touches down.

Crosswind After-Landing Roll

Particularly during the after-landing roll, special attention should be given to maintaining directional control by the use of rudder or nose-wheel steering, while keeping the upwind wing from rising by the use of aileron. When an airplane is airborne, it moves with the air mass in which it is flying regardless of the airplane's heading and speed. When an airplane is on the ground, it is unable to move with the air mass (crosswind) because of the resistance created by ground friction on the wheels.

Characteristically, an airplane has a greater profile or side area behind the main landing gear than forward of the gear. With the main wheels acting as a pivot point and the greater surface area exposed to the crosswind behind that pivot point, the airplane tends to turn or weathervane into the wind.

The relative wind acting on an airplane during the after-landing roll is the result of two factors. One is the natural wind, which acts in the direction the air mass is traveling. It has a headwind component acting along the airplane's ground track and a crosswind component acting 90° to its track. The other factor is the wind induced by the forward movement of the airplane, which acts parallel and opposite to the direction of movement. The relative wind is the resultant of these two factors and acts from a direction somewhere between the two components. The faster the airplane's groundspeed, the more the relative wind aligns towards the nose of the aircraft. As the airplane's forward speed decreases during the after-landing roll, the forward component of the relative wind decreases, causing the relative wind to act in a direction more aligned with the crosswind component. The greater the crosswind component, the more difficult it is to prevent weathervaning, especially with a conventional-gear airplane.

Maintaining control on the ground is a critical part of the after-landing roll because of the weathervaning effect of the wind on the airplane. Additionally, tire side load from runway contact while drifting may generate a "roll-over" in a tricycle-geared airplane. This occurs when one main wheel lifts up off the ground and the airplane tips forward along the axis between the nose-wheel and the main wheel still on the ground. A roll-over could cause one wingtip or the prop to contact the ground. The basic factors involved are cornering angle and side load.

Cornering angle is the angular difference between the heading of a tire and its path. Whenever a load-bearing tire's path and heading diverge, a side load is created. It is accompanied by tire distortion. Although side load differs in varying tires and air pressures, it is completely independent of speed, and through a considerable range, is directly proportional to the cornering angle and the weight supported by the tire. As little as 10° of cornering angle creates a side load equal to half the supported weight; after 20°, the side load does not increase with increasing cornering angle. For each high-wing, tricycle-geared airplane, there is a cornering angle at which roll-over is inevitable. At lesser angles, the roll-over may be avoided by use of ailerons, rudder, or steerable nose-wheel, but not brakes.

While the airplane is decelerating during the after-landing roll, more and more aileron is applied to keep the upwind wing from rising. Since the airplane is slowing down, there is less airflow around the ailerons and they become less effective. At the same time, the relative wind becomes more of a crosswind and exerts a greater lifting force on the upwind wing. When the airplane is coming to a stop, the aileron control should be held fully toward the wind.

Maximum Safe Crosswind Velocities

Takeoffs and landings in certain crosswind conditions are inadvisable or even dangerous. [*Figure 9-18*] If the crosswind is great enough to warrant an extreme drift correction, a hazardous landing condition may result. Therefore, the takeoff and landing capabilities with respect to the reported surface wind conditions and the available landing directions should be considered.

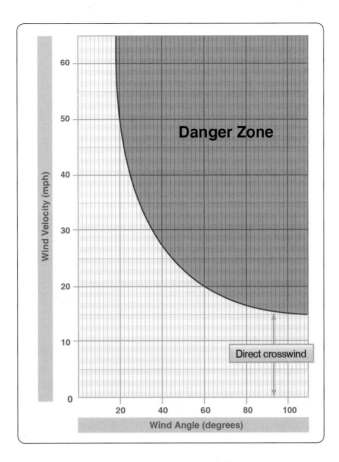

Figure 9-18. *Crosswind chart.*

Before an airplane is type certificated by the Federal Aviation Administration (FAA), it is flight tested to ensure it meets certain requirements. Among these is the demonstration of being satisfactorily controllable with no exceptional degree of skill or alertness on the part of the pilot in 90° crosswinds up to a velocity equal to 0.2 V_{SO}. This means a wind speed of two-tenths of the airplane's stalling speed with power off and in landing configuration. The demonstrated crosswind velocity is included on a placard in airplanes certificated after May 3, 1962.

The headwind component and the crosswind component for a given situation is determined by reference to a crosswind component chart. [*Figure 9-19*] It is imperative that pilots determine the maximum crosswind component of each airplane they fly and avoid operations in wind conditions that exceed the capability of the airplane.

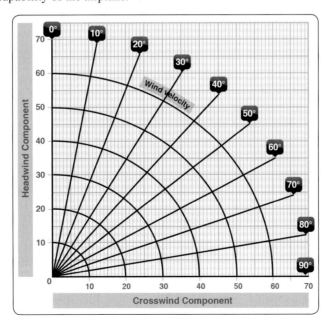

Figure 9-19. *Crosswind component chart.*

Common Errors

Common errors in the performance of crosswind approaches and landings are:

1. Attempted landing in crosswinds that exceed the airplane's maximum demonstrated crosswind component.
2. Undershooting or overshooting the turn from base leg to final approach.
3. Inadequate compensation for wind drift on final approach.
4. Unstable approach.
5. Excessive sink rate or too low an airspeed from increased drag and reduced vertical lift during sideslip.
6. Failure to touch down with the longitudinal axis aligned with the runway.
7. Touching down while drifting.
8. Excessive airspeed on touchdown.
9. Failure to apply appropriate flight control inputs during rollout.
10. Failure to maintain direction control on rollout.
11. Excessive braking.
12. Loss of aircraft control.

Turbulent Air Approach and Landing

For landing in turbulent conditions, the pilot should use a power-on approach at an airspeed slightly above the normal approach speed. This provides for more positive control of the airplane when strong horizontal wind gusts, or up and down drafts, are experienced. Like other power-on approaches, a coordinated combination of both pitch and power adjustments is usually required. The proper approach attitude and airspeed require a minimum round out and should result in little or no floating during the landing.

To maintain control during an approach in turbulent air with gusty crosswind, the pilot should use partial wing flaps. With less than full flaps, the airplane is in a higher pitch attitude. Thus, it requires less of a pitch change to establish the landing attitude and touchdown at a higher airspeed to ensure more positive control.

Pilots often use the normal approach speed plus one-half of the wind gust factors in turbulent conditions. If the normal speed is 70 knots, and the wind gusts are 15 knots, an increase of airspeed to 77 knots is appropriate. In any case, the airspeed and the flap setting should conform to airplane manufacturer's recommendations in the AFM/POH.

Use an adequate amount of power to maintain the proper airspeed and descent path throughout the approach, and retard the throttle to idling position only after the main wheels contact the landing surface. Care should be exercised in closing the throttle before the pilot is ready for touchdown. In turbulent conditions, the sudden or premature closing of the throttle may cause a sudden increase in the descent rate, resulting in a hard landing.

When landing from power approaches in turbulence, the touchdown is made with the airplane in approximately level flight attitude. The pitch attitude at touchdown would be only enough to prevent the nose-wheel from contacting the surface before the main wheels have touched the surface. After touchdown, the pilot should avoid the tendency to apply forward pressure on the yoke, as this may result in wheelbarrowing and possible loss of control. The pilot should allow the airplane to decelerate normally, assisted by careful use of wheel brakes and avoid heavy braking until the wings are devoid of lift and the airplane's full weight is resting on the landing gear.

Short-Field Approach and Landing

Short-field approaches and landings require the use of procedures for approaches and landings at fields with a relatively short landing area or where an approach is made over obstacles that limit the available landing area. [*Figure 9-20* and *Figure 9-21*] This low-speed type of power-on approach is closely related to the performance of flight near minimum controllable airspeeds.

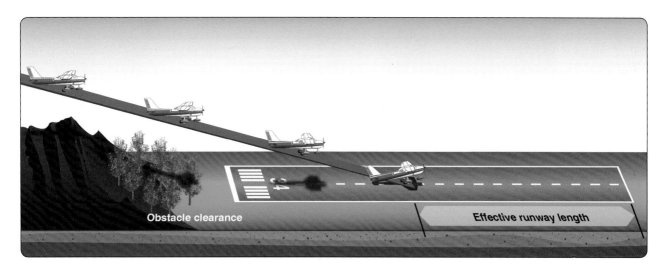

Figure 9-20. *Landing over an obstacle.*

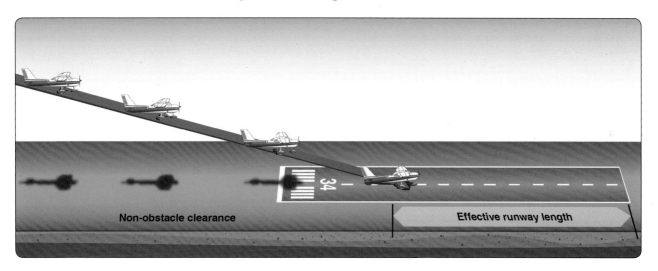

Figure 9-21. *Landing on a short field.*

To land within a short field or a confined area, the pilot needs to have precise, positive control of the rate of descent and airspeed, and fly an approach that clears any obstacles, results in little or no floating during the round out, and permits the airplane to be stopped in the shortest possible distance. When safety and conditions permit, a wider-than-normal pattern with a longer final approach may be used. This allows the pilot ample opportunity to adjust and stabilize the descent angle after the airplane is configured and trimmed. A stabilized approach is essential.

The procedures for landing on a short field or for landing approaches over obstacles as recommended in the AFM/POH should be used. [*Figure 9-22* and *Figure 9-23*] These procedures generally involve a final approach started from an altitude of at least 500 feet higher than the touchdown area and the use of full flaps at an appropriate point during the final approach. For many general aviation airplanes this means flying a stabilized final approach with the flap setting that precedes full flaps. When the field is made, the pilot should extend full flaps and lower the nose in order to maintain airspeed and keep the aiming point stationary in the windscreen. When over the obstacle, the pilot may reduce power slightly. Ideally, if full flaps are extended at the correct point, the pilot will be in a position to slowly reduce power. When no manufacturer's recommended approach speed is available, a speed of not more than 1.3 V_{SO} is used. In gusty air, no more than one-half the gust factor is added. An excessive amount of airspeed could result in a touchdown too far from the runway threshold or an after-landing roll that exceeds the available landing area. When obstacles are present, a slightly steeper approach angle places the touchdown closer to the obstacle, which gives the pilot more room to stop.

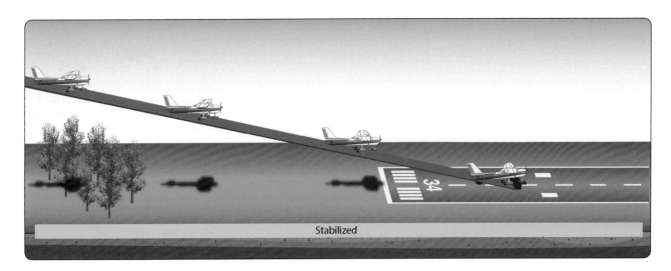

Figure 9-22. *Stabilized approach.*

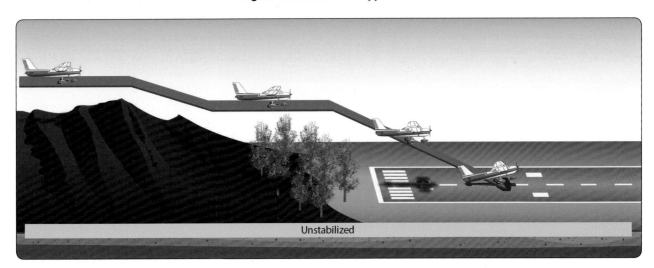

Figure 9-23. *Unstabilized approach.*

After the landing gear has been extended, if applicable, or when beginning a suitable final approach, the pilot simultaneously adjusts the power and the pitch attitude to establish and maintain the proper descent angle and airspeed. During a stabilized approach, small changes in the airplane's pitch attitude and power setting are needed when making corrections to the angle of descent and airspeed.

The short-field approach and landing is an accuracy approach to an aiming point. The procedures previously outlined in the section on the stabilized approach concept are used. If it appears that the obstacle clearance is excessive and touchdown occurs well beyond the desired aiming point, leaving insufficient room to stop, power is reduced while lowering the pitch attitude to steepen the descent path and increase the rate of descent. If it appears that the descent angle does not ensure safe clearance of obstacles, power is increased while simultaneously raising the pitch attitude to shallow the descent path and decrease the rate of descent. Care should be taken to avoid excessively low airspeeds. When operating at high AOAs and low airspeeds, an increase in pitch attitude increases the rate of descent. When there is doubt regarding the outcome of the approach, the pilot should execute a go-around, evaluate the situation, and decide whether to make another approach or divert to a more suitable landing area.

Because the final approach over obstacles is made at a relatively steep approach angle and close to the airplane's stalling speed, the initiation of the round out or flare needs to be judged accurately to avoid flying into the ground or stalling prematurely and sinking rapidly. A lack of floating during the flare with sufficient control to touch down properly is verification that the approach speed was correct.

Touchdown should occur at the minimum controllable airspeed with the airplane in approximately the pitch attitude that results in a power-off stall when the throttle is closed. Care should be exercised to avoid closing the throttle too rapidly, as closing the throttle may result in an immediate increase in the rate of descent and a hard landing. Note that a small amount of power provides more airflow over the elevator giving it more authority at low airspeeds to enable the pilot to flare. There is a risk that low airspeed and a windmilling propeller blocking airflow over the elevator may make it difficult to flare.

Upon touchdown, the airplane is held in this positive pitch attitude as long as the elevators remain effective and if recommended by the manufacturer. This provides aerodynamic braking to assist in deceleration. However, immediately upon touchdown of the nose-wheel, maximum braking is applied to minimize the after-landing roll. For most airplanes, aerodynamic drag is the single biggest factor in slowing the aircraft in the first quarter of its speed decay. Brakes become increasingly effective as airspeed and lift decrease. The pilot increases braking effectiveness by holding the wheel or stick full back while smoothly applying brakes. Back pressure is needed because the airplane tends to lean forward with heavy braking. Best braking results are always achieved with the wheels in an "incipient skid condition." That means a little more brake pressure would lock up the wheels entirely. In an incipient skid, the wheels are turning, but with great reluctance. If the wheels lock, braking effectiveness drops dramatically in a skid and the tires could be damaged. The airplane is normally stopped within the shortest possible distance consistent with safety and controllability. If the proper approach speed has been maintained, resulting in minimum float during the round out and the touchdown made at minimum control speed, excessive braking should not be needed.

Common Errors

Common errors in the performance of short-field approaches and landings are:

1. A final approach that necessitates an overly steep approach and high sink rate.

2. Unstable approach.

3. Undue delay in initiating glide path corrections.

4. Too low an airspeed on final resulting in inability to flare properly and landing hard.

5. Too high an airspeed resulting in floating on round out.

6. Prematurely reducing power to idle on round out resulting in hard landing.

7. Touchdown with excessive airspeed.

8. Excessive and/or unnecessary braking after touchdown.

9. Failure to maintain directional control.

10. Failure to recognize and abort a poor approach that cannot be completed safely.

Soft-Field Approach and Landing

Landing on fields that are rough or have soft surfaces, such as snow, sand, mud, or tall grass, requires unique procedures. When landing on such surfaces, the objective is to touch down as smoothly as possible and at the slowest possible landing speed. A pilot needs to control the airplane in a manner that the wings support the weight of the airplane as long as practical to minimize stresses imposed on the landing gear by a rough surface or to prevent sinking into a soft surface.

The approach for the soft-field landing is similar to the normal approach used for operating into long, firm landing areas. The major difference between the two is that a degree of power is used throughout the level-off and touchdown for the soft-field landing. This allows the airspeed to slowly dissipate while the airplane is flown 1 to 2 feet off the surface in ground effect. When the wheels first touch the ground, the wings continue to support much of the weight of the airplane. [*Figure 9-24*] This technique minimizes the nose-over forces that suddenly affect the airplane at the moment of touchdown.

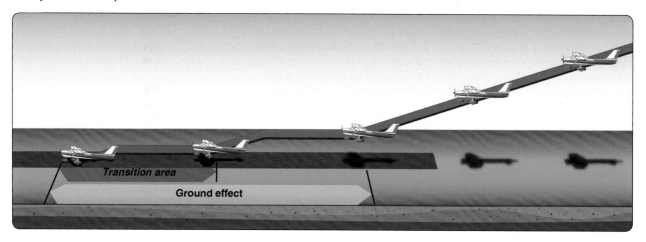

Figure 9-24. *Soft/rough field approach and landing.*

The use of flaps during soft-field landings aids in touching down at minimum speed and is recommended whenever practical. In low-wing airplanes, the flaps may suffer damage from mud, stones, or slush thrown up by the wheels. If flaps are used, it is generally inadvisable to retract them during the after-landing roll because the need for flap retraction is less important than the need for total concentration on maintaining full control of the airplane.

The final-approach airspeed used for short-field landings is equally appropriate to soft-field landings. The use of higher approach speeds may result in excessive float in ground effect, and floating makes a smooth, controlled touchdown even more difficult. There is no reason for a steep angle of descent unless obstacles are present in the approach path.

Touchdown on a soft or rough field is made at the lowest possible airspeed with the airplane in a nose-high pitch attitude. In nose-wheel type airplanes, after the main wheels touch the surface, the pilot should hold sufficient back-elevator pressure to keep the nose-wheel off the surface. Using back-elevator pressure and engine power, the pilot can control the rate at which the weight of the airplane is transferred from the wings to the wheels.

Field conditions may warrant that the pilot maintain a flight condition in which the main wheels are just touching the surface but the weight of the airplane is still being supported by the wings until a suitable taxi surface is reached. At any time during this transition phase, before the weight of the airplane is being supported by the wheels, and before the nose-wheel is on the surface, the ability is retained to apply full power and perform a safe takeoff (obstacle clearance and field length permitting) should the pilot elect to abandon the landing. Once committed to a landing, the pilot should gently lower the nose-wheel to the surface. A slight addition of power usually aids in easing the nose-wheel down.

The use of brakes on a soft field is not needed and should be avoided as this may tend to impose a heavy load on the nose-gear due to premature or hard contact with the landing surface, causing the nose-wheel to dig in. The soft or rough surface itself provides sufficient reduction in the airplane's forward speed. Often upon landing on a very soft field, an increase in power may be needed to keep the airplane moving and from becoming stuck in the soft surface.

Common Errors

Common errors in the performance of soft-field approaches and landings are:

1. Excessive descent rate on final approach.

2. Excessive airspeed on final approach.

3. Unstable approach.

4. Round out too high above the runway surface.

5. Poor power management during round out and touchdown.

6. Hard touchdown.

7. Inadequate control of the airplane weight transfer from wings to wheels after touchdown.

8. Allowing the nose-wheel to "fall" to the runway after touchdown rather than controlling its descent.

Power-Off Accuracy Approaches

Power-off accuracy approaches and landings involve gliding to a touchdown at a given point (or within a specified distance beyond that point), while using a specific pattern and with the engine idling. The objective is to instill in the pilot the judgment and procedures necessary for accurately flying the airplane, without power, to a safe landing.

The ability to estimate the distance an airplane glides to a landing is the real basis of all power-off accuracy approaches and landings. The distance to be covered largely determines the amount of maneuvering needed to complete an approach from a given altitude. While developing the pilot's ability to estimate gliding distance, power-off accuracy approaches call upon the pilot to use a variety of techniques to set and maintain an appropriate glide angle and airspeed to the aiming point.

With experience and practice, altitudes up to approximately 1,000 feet can be estimated with fair accuracy; while above this level the accuracy in judgment of height above the ground decreases, since all features tend to merge. The best aid in perfecting the ability to judge height above this altitude is through the indications of the altimeter and associating them with the general appearance of the earth.

The judgment of altitude in feet, hundreds of feet, or thousands of feet is not as important as the ability to estimate gliding angle and its resultant distance. Regardless of altitude, a pilot who knows the normal glide angle of the airplane can estimate, with reasonable accuracy, the approximate spot along a given ground path at which the airplane will land. A pilot who has the ability to accurately estimate altitude, can also judge how much maneuvering is possible and safe during the glide, which is important to the choice of landing areas in an actual emergency.

The objective of a good final approach is to descend at an angle that permits the airplane to reach the desired aiming point at an airspeed that results in a predictable float where touchdown occurs on or within a specified distance beyond a designated point. To accomplish this, it is essential that both the descent angle and the airspeed be accurately controlled.

Unlike a normal approach when the power setting is variable, on a power-off approach the power is fixed at the idle setting. Pitch attitude is adjusted to control the airspeed. This also changes the glide or descent angle. If an airplane is on approach with an airspeed higher than best glide, pitching down will increase the airspeed and steepen the descent angle, while pitching up will reduce the airspeed and shallow the descent angle. Conversely, if the airspeed is below best glide, then pitching down will increase the airspeed and shallow the descent angle, while pitching up will reduce the airspeed and will greatly steepen the descent angle. If the airspeed is too high, the pilot should raise the nose; and when the airspeed is too low, lower the nose. If the pitch attitude is raised too high, the airplane settles rapidly due to a slow airspeed and insufficient lift. For this reason, the pilot should never try to stretch a glide to reach the desired landing spot.

Note that certain single-engine turboprop airplanes experience an excessive rate of descent if the power is set to flight idle. In some cases, if the powerplant failed, the manufacturer's checklist calls for feathering the propeller during a power-off glide. During flight training in these airplanes, the propeller is not feathered as would be the case in an emergency or true power-off glide. During training and pilot certification, where the manufacturer's checklist calls for propeller feathering in a power-off situation, the pilot should set sufficient power to provide the performance that would be expected with the propeller feathered.

Uniform approach patterns, such as the 90° or 180° power-off approaches, are described further in this chapter. Practicing these approaches provides a pilot with a basis on which to develop judgment in gliding distance and in planning an approach. While square patterns demonstrate good planning, they are not required and may not be appropriate for every approach. For example, when conditions are not as expected, pilots may need to dog-leg away from the runway on base or dog-leg toward the runway on base. Pilots may use S-turns, slips, early or late extension of flaps, reduce airspeed below best glide, or increase airspeed slightly above best glide in a headwind in order to stabilize the remaining approach, to reach the desired aiming point at an appropriate speed, and to touch down where planned. Note that selection of the runway numbers as the touchdown point does not provide a safety cushion in case of a mechanical problem or misjudgment. Selecting a point farther down the runway establishes an increased safety margin.

The basic procedure in these approaches involves closing the throttle at a given altitude and gliding to a key position. Starting with the same energy (airspeed and height) each time the throttle is closed, makes the maneuver more predictable. The key position, like the pattern itself, is not the primary objective; it is merely a convenient point in the air from which the pilot can judge what to do such that the landing occurs at or just beyond the desired point. The selected key position should be one that is appropriate for the available altitude and the wind condition. From the key position, the pilot should constantly evaluate the situation.

It should be emphasized that, although accurate spot touchdowns are important, safe and properly executed approaches and landings are vital. A pilot should never sacrifice a good approach or landing just to land on the desired spot.

90° Power-Off Approach

The 90° power-off approach is made from a base leg and requires an approximate 90° turn onto the final approach. The approach path may be varied by positioning the base leg closer to or farther out from the approach end of the runway according to wind conditions. [*Figure 9-25*] The glide from the key position on the base leg through the 90° turn to the final approach is the final part of all accuracy landing maneuvers. The 90° power-off approach usually begins from a rectangular pattern at approximately 1,000 feet above the ground or at normal traffic pattern altitude. The airplane is flown on a downwind leg at the same distance from the landing surface as in a normal traffic pattern. The before-landing checklist should be completed on the downwind leg, including extension of the landing gear if the airplane is equipped with retractable gear.

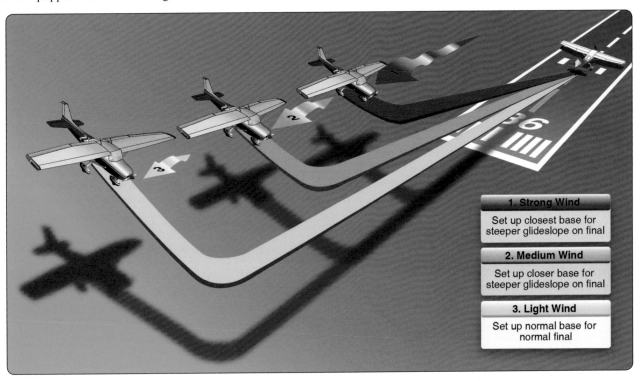

| 1. Strong Wind |
| Set up closest base for steeper glideslope on final |

| 2. Medium Wind |
| Set up closer base for steeper glideslope on final |

| 3. Light Wind |
| Set up normal base for normal final |

Figure 9-25. *Plan the base leg for wind conditions.*

After a medium-banked turn onto the base leg is completed, the throttle is retarded slightly and the airspeed allowed to decrease to the normal base-leg speed. [*Figure 9-26*] On the base leg, the airspeed, wind drift correction, and altitude are maintained while proceeding to the 45° key position. At this position, the intended landing spot appears to be on a 45° angle from the airplane's nose.

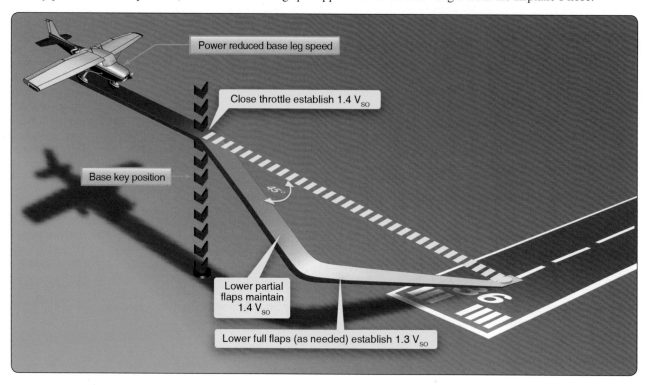

Figure 9-26. *90° power-off approach.*

The pilot can determine the strength and direction of the wind from the amount of crab necessary to hold the desired ground track on the base leg. This helps in planning the turn onto the final approach and provides some indication of when to lower the flaps.

At the 45° key position, the throttle is closed completely, the propeller control (if equipped) advanced to the full increase revolution per minute (rpm) position, and altitude maintained until the airspeed decreases to the manufacturer's recommended glide speed. In the absence of a recommended speed, the pilot should use 1.4 V_{SO}. When this airspeed is attained, the nose is lowered to maintain the gliding speed and the controls trimmed. The wing flaps may be gradually lowered and the pitch attitude adjusted, as needed, to establish the proper descent angle. The base-to-final turn is planned and accomplished so that upon rolling out of the turn, the airplane is aligned with the runway centerline. If the approach is planned to be slightly high in the current configuration, the pilot will be assured of making the aiming point. The wing flaps may be lowered, as needed, and the pitch attitude adjusted, as needed, to establish the proper descent angle and airspeed (1.3 V_{SO}), and the controls trimmed. Slight adjustments in pitch attitude and slips are used as necessary to control the glide angle and airspeed. A crab or side slip can be used to maintain the desired flight path. A forward slip may be used momentarily to steepen the descent without changing the airspeed. Full flaps should be delayed until it is clear that adding them will not cause the landing to be short of the point. The pilot should never try to stretch the glide or retract the flaps to reach the desired landing spot.

On short final, full attention is given to making a good, safe landing rather than concentrating on the selected landing spot. The approach angle used and final approach airspeed determine the probability of landing on the spot, and late adjustments to these parameters are not appropriate. It is always better to execute a good landing away from the spot than to make a poor landing precisely on or just past the spot.

180° Power-Off Approach

The 180° power-off approach is executed by gliding with idle power from a given point on a downwind leg to a preselected landing spot. [*Figure 9-27*] It is an extension of the principles involved in the 90° power-off approach just described. The objective is to further develop judgment in estimating distances and glide ratios, in that the airplane is flown without power from a higher altitude and through a 90° turn to reach the base-leg position at a proper altitude for executing the 90° approach.

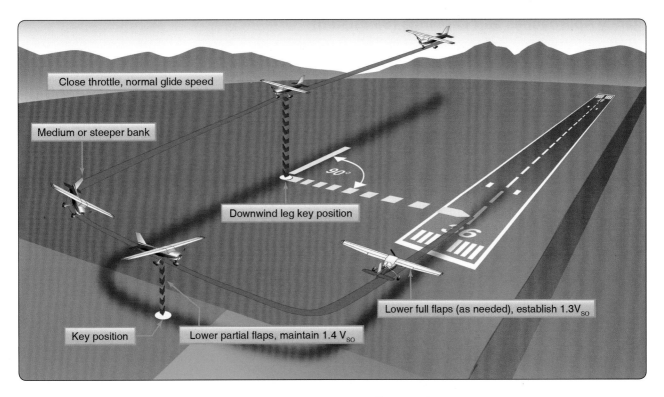

Figure 9-27. *180° power-off approach.*

The 180° power-off approach requires more planning and judgment than the 90° power-off approach. In the execution of 180° power-off approaches, the airplane is flown on a downwind heading parallel to the landing runway. The altitude from which this type of approach is started varies with the type of airplane, but should usually not exceed 1,000 feet above the ground, except with large airplanes. Greater accuracy in judgment and maneuvering is required at higher altitudes.

When abreast of or opposite the desired landing spot, the throttle is closed and altitude maintained while decelerating to the manufacturer's recommended glide speed or 1.4 V_{SO}. The point at which the throttle is closed is the downwind key position.

The turn from the downwind leg to the base leg is a uniform turn with a medium or slightly steeper bank. The degree of bank and amount of this initial turn depend upon the glide angle of the airplane and the velocity and direction of the wind. Again, the base leg is positioned as needed for the altitude or wind condition. Position the base leg to conserve or dissipate altitude so as to reach the desired landing spot.

The turn onto the base leg is made at an altitude high enough and close enough to permit the airplane to glide to what would normally be the base key position in a 90° power-off approach. Initial flaps may be extended prior to the base key position if needed.

Although the base key position is important, it should not be overemphasized nor considered as a fixed point on the ground. Many inexperienced pilots may gain a conception of it as a particular landmark, such as a tree, crossroad, or other visual reference, to be reached at a certain altitude. This misconception leaves the pilot at a total loss any time such objects are not present. Both altitude and geographical location should be varied as much as is practical to eliminate any such misconceptions. After reaching the base key position, the approach and landing are the same as in the 90° power-off approach.

Common Errors

Common errors in the performance of power-off accuracy approaches are:

1. Downwind leg is too far from the runway/landing area.

2. Overextension of downwind leg resulting from a tailwind.

3. Inadequate compensation for wind drift on base leg.

4. Skidding turns in an effort to increase gliding distance.

5. Failure to lower landing gear in retractable gear airplanes.

6. Attempting to "stretch" the glide during an undershoot.

7. Premature flap extension/landing gear extension.

8. Use of throttle to increase the glide instead of merely clearing the engine.

9. Forcing the airplane onto the runway in order to avoid overshooting the designated landing spot.

Emergency Approaches and Landings (Simulated)

During dual training flights, the instructor should give simulated emergency landings by retarding the throttle and calling "simulated emergency landing." The objective of these simulated emergency landings is to develop a pilot's accuracy, judgment, planning, procedures, and confidence when little or no power is available. A simulated emergency landing may be given with the airplane in any configuration. If the simulated power failure occurs while above best glide speed, the pilot allows the airplane to slow (or may even bleed off speed by climbing) until reaching best glide speed. When reaching that speed, the nose can be lowered and the airplane trimmed to maintain that speed. If the failure occurs at or below best glide speed, the nose should be lowered immediately to maintain or accelerate to best glide speed. The pilot should ensure that the flaps and landing gear are in the proper configuration for the existing situation.

A constant gliding speed is usually maintained because variations of gliding speed nullify all attempts at accuracy in judgment of gliding distance and the landing spot. The many variables, such as altitude, obstruction, wind direction, landing direction, landing surface and gradient, and landing distance requirements of the airplane, determine the pattern and approach procedures to use.

The pilot may use any combination of normal gliding maneuvers, from wings level to spirals to eventually arrive at the normal key position at a normal traffic pattern altitude for the selected landing area. From the key point on, the approach is a normal power-off approach. [*Figure 9-28*]

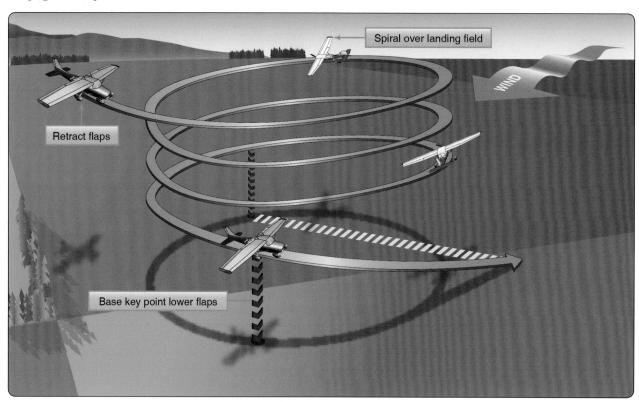

Figure 9-28. *Remain over intended landing area.*

With the greater choice of fields afforded by higher altitudes, the inexperienced pilot may be inclined to delay making a decision, and with considerable altitude in which to maneuver, errors in maneuvering and estimation of glide distance may develop.

All pilots should learn to determine the wind direction and estimate its speed from the windsock at the airport, smoke from factories or houses, dust, brush fires, wind farms, or patterns displayed on nearby bodies of water .

Once a field has been selected, a pilot should indicate the proposed landing area to the instructor. Normally, the pilot should plan and fly a pattern for landing on the field first elected until the instructor terminates the simulated emergency landing. This provides the instructor an opportunity to explain and correct any errors; it also gives the pilot an opportunity to see the results of the errors. However, if the pilot realizes during the approach that a poor field has been selected—one that would obviously result in disaster if a landing were to be made—and there is a more advantageous field within gliding distance, a change to the better field should be permitted. The instructor should thoroughly explain the hazards involved in these last-minute decisions, such as excessive maneuvering at very low altitudes.

Instructors should stress slipping the airplane, using flaps, varying the position of the base leg, and varying the turn onto final approach as ways of correcting for misjudgment of altitude and glide angle.

Eagerness to get down is one of the most common faults of inexperienced pilots during simulated emergency landings. They forget about speed and arrive at the edge of the field with too much speed to permit a safe landing. Too much speed is just as dangerous as too little;

it results in excessive floating and overshooting the desired landing spot. Instructors need to stress during their instruction that pilots cannot dive at a field and expect to land on it.

During all simulated emergency landings, keep the engine warm and cleared. During a simulated emergency landing, either the instructor or the pilot should have complete control of the throttle. There should be no doubt as to who has control since many near accidents have occurred from such misunderstandings.

Every simulated emergency landing approach is terminated as soon as it can be determined whether or not a safe landing is assured. In no case should it be continued to a point where it creates an undue hazard or an annoyance to persons or property on the ground.

In addition to flying the airplane from the point of simulated engine failure to where it is known that a reasonable safe landing could be made (or to where it is known that the approach cannot be salvaged), a pilot should also receive instruction on certain emergency flight deck procedures. The habit of performing these procedures should be developed to such an extent that, if an engine failure actually occurs, a pilot checks the critical items that might get the engine operating again while selecting a field and planning an approach. Combining the two operations—accomplishing emergency procedures and planning and flying the approach—is difficult during the early training in emergency landings.

There are steps and procedures pilots should follow in a simulated emergency landing. Although they may differ somewhat from the procedures used in an actual emergency, they should be learned thoroughly and each step called out to the instructor. The use of a checklist is strongly recommended. Most airplane manufacturers provide a checklist of the appropriate items. [*Figure 9-29*]

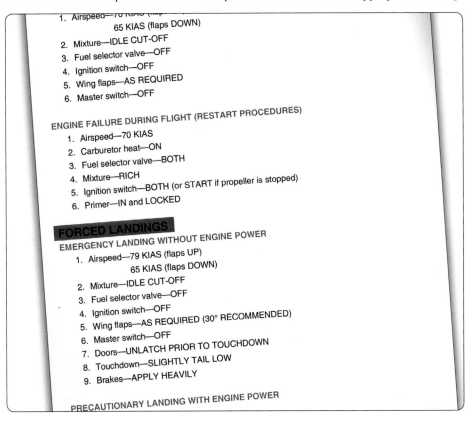

Figure 9-29. *Sample emergency checklist.*

Critical items to be checked include the position of the fuel tank selector, the quantity of fuel in the tank selected, the fuel pressure gauge to see if the electric fuel pump is needed, the position of the mixture control, the position of the magneto switch, and the use of carburetor heat. Many actual emergency landings have been made and later found to be the result of the fuel selector valve being positioned to an empty tank while the other tank had plenty of fuel. It may be wise to change the position of the fuel selector valve even though the fuel gauge indicates fuel in all tanks because fuel gauges can be inaccurate. Many actual emergency landings could have been prevented if the pilots had developed the habit of checking these critical items during flight training.

Instruction in emergency procedures is not limited to simulated emergency landings caused by power failures. Other emergencies associated with the operation of the airplane should be explained, demonstrated, and practiced if practicable. Among these emergencies are fire in flight, electrical or hydraulic system malfunctions, unexpected severe weather conditions, engine overheating, imminent fuel exhaustion, and the emergency operation of airplane systems and equipment.

Faulty Approaches and Landings

Landing involves many precise, time-sensitive, and sequential control inputs. When corrected early, small errors are often not noticeable. On the other hand, uncorrected errors may place the airplane and occupants in an undesirable state. Since pilot training normally includes exposure to landing deviations and their appropriate remedies, this section covers several common landing imperfections.

Low Final Approach

When the base leg is too low, insufficient power is used, landing flaps are extended prematurely, or the velocity of the wind is misjudged, the airplane may be well below the proper final approach path. In such a situation, the pilot would have to apply considerable power to fly the airplane (at an excessively low altitude) up to the runway threshold. When it is realized the runway cannot be reached unless appropriate action is taken, power should be applied immediately to maintain the airspeed while the pitch attitude is raised to increase lift and stop the descent. When the proper approach path has been intercepted, the correct approach attitude is reestablished and the power reduced and a stabilized approach maintained. [*Figure 9-30*] The pilot should not increase the pitch attitude without increasing the power because the airplane decelerates rapidly and may approach the critical AOA and stall. In addition, the pilot should not retract the flaps since this causes a sudden decrease in lift and causes the airplane to sink more rapidly. If there is any doubt about the approach being safely completed, it is advisable to execute an immediate go-around.

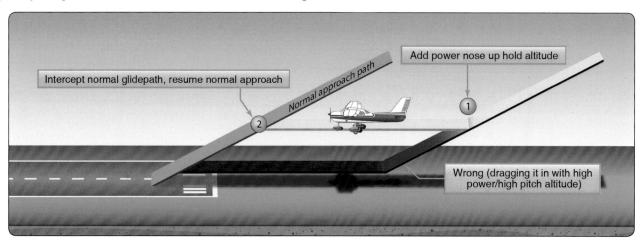

Figure 9-30. *Right and wrong methods of correction for low final approach.*

High Final Approach

When the final approach is too high, the pilot may lower the flaps as required. Further reduction in power may be necessary, while lowering the nose simultaneously to maintain approach airspeed and steepen the approach path. [*Figure 9-31*] Alternatively, the pilot could use a forward slip to increase the descent angle and rate of descent while maintaining proper approach speed. Since a sink rate in excess of 800–1,000 feet per minute (fpm) is considered excessive, either technique avoids the high sink rates that would occur if the pilot dives the airplane toward the aiming point. Since a high sink rate continued close to the surface makes it be difficult to slow to a proper rate prior to ground contact, it is not a good idea to dive toward the aiming point. Therefore, when intercepting the proper approach path from above, the pilot adjusts the power as required to maintain a stabilized approach. A go-around should be initiated if the sink rate becomes excessive.

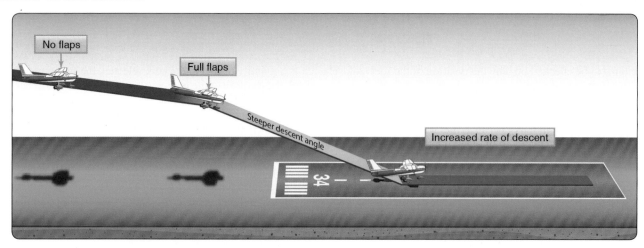

Figure 9-31. *Change in glidepath and increase in descent rate for high final approach.*

Slow Final Approach

On the final approach, when the airplane is flown at a slower than normal airspeed, the pilot's judgment of the rate of sink (descent) and the height of round out is difficult. During an excessively slow approach, the wing is operating near the critical AOA and, depending on the pitch attitude changes and control usage, the airplane may stall or sink rapidly, contacting the ground with a hard impact.

Whenever a slow speed approach is noted, the pilot should apply power to accelerate the airplane and increase the lift to reduce the sink rate and to prevent a stall. This is done while still at a high enough altitude to reestablish the correct approach airspeed and attitude. If too slow and too low, it is best to execute a go-around.

Use of Power

Power can be used effectively during the approach and round out to compensate for errors in judgment. Power may be added to accelerate the airplane, to increase lift without increasing the AOA, and to slow the descent to an acceptable rate. The increased propwash over the wing behind the propeller(s) also provides an immediate boost in lift that also helps slow the descent rate. If the proper landing attitude is attained and the airplane is only slightly high, the landing attitude is held constant and sufficient power applied to help ease the airplane onto the ground. After the airplane has touched down, the pilot closes the throttle so the additional thrust and lift are removed and the airplane remains on the ground.

High Round Out

Sometimes when the airplane appears to temporarily stop moving downward, the round out has been made too rapidly and the airplane is flying level, too high above the runway. Continuing the round out further reduces the airspeed and increases the AOA to the critical angle. This results in the airplane stalling and dropping hard onto the runway. To prevent this, the pitch attitude is held constant until the airplane decelerates enough to again start descending. Then the round out is continued to establish the proper landing attitude. This procedure is only used when there is adequate airspeed. It may be necessary to add a slight amount of power to keep the airspeed from decreasing excessively and to avoid losing lift too rapidly.

When the proper landing attitude is attained, the airplane is approaching a stall because the airspeed is decreasing and the critical AOA is being approached, even though the pitch attitude is no longer being increased. [*Figure 9-32*]

Figure 9-32. *Rounding out too high.*

Although back-elevator pressure may be relaxed slightly, the nose should not be lowered to make the airplane descend when fairly close to the runway unless some power is added momentarily. The momentary decrease in lift that results from lowering the nose and decreasing the AOA might cause the airplane to contact the ground with the nose-wheel first and may result in nose gear damage or collapse.

It is recommended that a go-around be executed any time it appears the nose needs to be lowered significantly or that the landing is in any other way uncertain.

Late or Rapid Round Out

Starting the round out too late or pulling the elevator control back too rapidly to prevent the airplane from touching down prematurely can impose a significant load on the wings and cause an accelerated stall.

Suddenly increasing the AOA and stalling the airplane during a round out is a dangerous situation since it may cause the airplane to land extremely hard on the main landing gear and then bounce back into the air. As the airplane contacts the ground, the tail is forced down very rapidly by the back-elevator pressure and by inertia acting downward on the tail.

Recovery from this situation requires prompt and positive application of power prior to occurrence of the stall. This may be followed by a normal landing if sufficient runway is available—otherwise the pilot should execute a go-around immediately.

If the round out is late and uncorrected, the nose-wheel may strike the runway first, causing the nose to bounce upward. Do not attempt to force the airplane back onto the ground; execute a go-around immediately.

Floating During Round Out

If the airspeed on final approach is excessive, it usually results in the airplane floating. [*Figure 9-33*] Before touchdown can be made, the airplane may be well past the desired landing point and the available runway may be insufficient. When diving the airplane on final approach to land at the proper point, there is an appreciable increase in airspeed. The proper touchdown attitude cannot be established without producing an excessive AOA and lift. This causes the airplane to gain altitude or balloon.

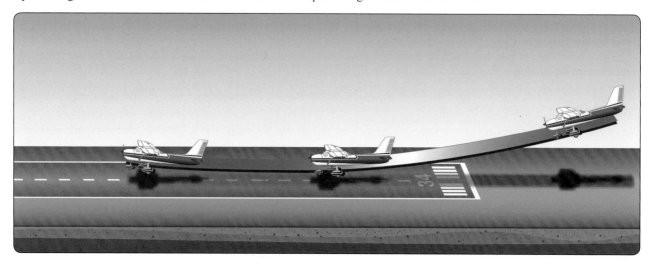

Figure 9-33. *Floating during round out.*

Any time the airplane floats, judgment of speed, height, and rate of sink needs to be especially acute. The pilot should smoothly and gradually adjust the pitch attitude as the airplane decelerates to touchdown speed and starts to settle, so the proper landing attitude is attained at the moment of touchdown. The slightest error in judgment and timing results in either ballooning or bouncing.

The recovery from floating is dependent upon the amount of floating and the effect of any crosswind, as well as the amount of runway remaining. Since prolonged floating utilizes considerable runway length, it should be avoided especially on short runways or in strong crosswinds. If a landing cannot be made on the first third of the runway, or the airplane drifts sideways, execute a go-around.

Ballooning During Round Out

If the pilot misjudges the rate of sink during a landing and thinks the airplane is descending faster than it should, there is a tendency to increase the pitch attitude and AOA too rapidly. This not only stops the descent, but actually starts the airplane climbing. This climbing during the round out is known as ballooning. [*Figure 9-34*] Ballooning is dangerous because the height above the ground is increasing and the airplane is rapidly approaching a stalled condition. The altitude gained in each instance depends on the airspeed or the speed with which the pitch attitude is increased.

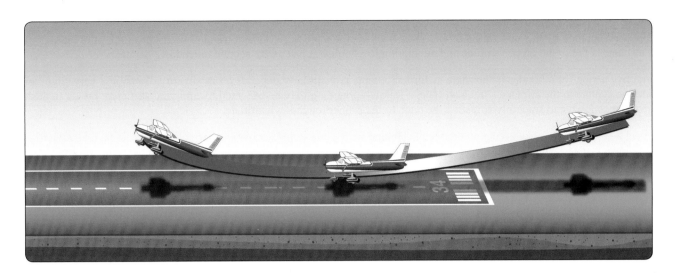

Figure 9-34. *Ballooning during roundout.*

Depending on the severity of ballooning, the use of throttle is helpful in cushioning the landing. By adding power, thrust is increased to keep the airspeed from decelerating too rapidly and the wings from suddenly losing lift, but throttle should be closed immediately after touchdown. Torque effects vary as power is changed, and it is necessary to use rudder pressure to keep the airplane straight as it settles onto the runway.

The pilot needs to be extremely cautious of ballooning when there is a crosswind present because the crosswind correction may be inadvertently released or it may become inadequate. Because of the lower airspeed after ballooning, the crosswind affects the airplane more. Consequently, the wing has to be lowered even further to compensate for the increased drift. It is imperative that the pilot makes certain that the appropriate wing is down and that directional control is maintained with opposite rudder. If there is any doubt, or the airplane starts to drift, the pilot should execute a go-around.

When ballooning is excessive, it is best to execute a go-around immediately and not attempt to salvage the landing. Power should be applied before the airplane enters a stalled condition.

Bouncing During Touchdown

When the airplane contacts the ground with a sharp impact as the result of an improper attitude or an excessive rate of sink, it tends to bounce back into the air. Though the airplane's tires and shock struts provide some springing action, the airplane does not bounce like a rubber ball. Instead, it rebounds into the air because the wing's AOA was abruptly increased, producing a sudden addition of lift. [*Figure 9-35*]

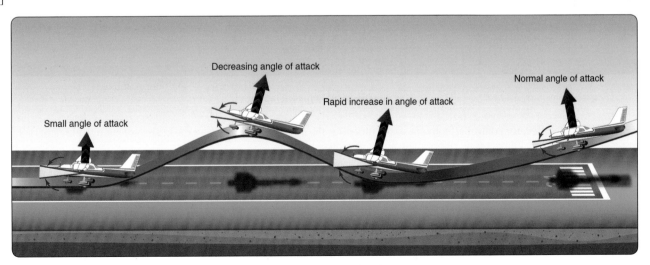

Figure 9-35. *Bouncing during touchdown.*

The abrupt change in AOA is the result of inertia instantly forcing the airplane's tail downward when the main wheels contact the ground sharply. The severity of the bounce depends on the airspeed at the moment of contact and the degree to which the AOA or pitch attitude was increased.

Since a bounce occurs when the airplane makes contact with the ground before the proper touchdown attitude is attained, it is almost invariably accompanied by the application of excessive back-elevator pressure. This is usually the result of the pilot realizing too late that the airplane is not in the proper attitude and attempting to establish it just as the second touchdown occurs.

The corrective action for a bounce is the same as for ballooning and similarly depends on its severity. When it is very slight and there is no extreme change in the airplane's pitch attitude, a follow-up landing may be executed by applying sufficient power to cushion the subsequent touchdown and smoothly adjusting the pitch to the proper touchdown attitude.

In the event a very slight bounce is encountered while landing with a crosswind, crosswind correction needs to be maintained while the next touchdown is made. Since the subsequent touchdown is made at a slower airspeed, the upwind wing has to be lowered even further to compensate for drift.

Extreme caution and alertness should be exercised any time a bounce occurs, but particularly when there is a crosswind. Pilots should not release the crosswind correction. When one main wheel of the airplane strikes the runway, the other wheel touches down immediately afterwards, and the wings become level. Then, with no crosswind correction as the airplane bounces, the wind causes the airplane to roll with the wind, thus exposing even more surface to the crosswind and increasing any drift.

When a bounce is severe, the safest procedure is to execute a go-around immediately. The pilot should not attempt to salvage the landing. Apply full power while simultaneously maintaining directional control and lowering the nose to a safe climb attitude. The go-around procedure should be continued even though the airplane may descend and another bounce may be encountered. Landing from a bad bounce should not be attempted, since airspeed diminishes very rapidly in the nose-high attitude, and a stall may occur before a subsequent touchdown can be made.

Porpoising

In a bounced landing that is improperly recovered, the airplane comes in nose first, initiating a series of motions imitating the jumps and dives of a porpoise. [*Figure 9-36*] The improper airplane attitude at touchdown may be caused by inattention, not knowing where the ground is, miss-trimming, or forcing the airplane onto the runway.

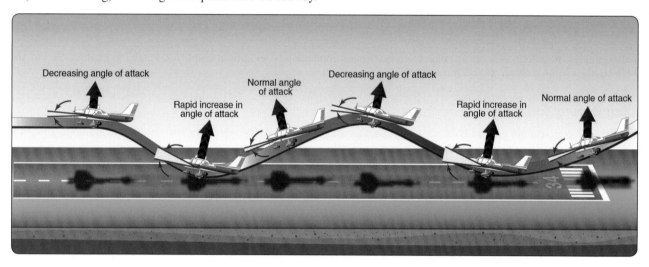

Figure 9-36. *Porpoising.*

Ground effect decreases elevator control effectiveness and increases the effort required to raise the nose. Not enough elevator or stabilator trim can result in a nose low contact with the runway and a porpoise develops.

Porpoising can also be caused by improper airspeed control. Usually, if an approach is too fast, the airplane floats and the pilot tries to force it on the runway when the airplane still wants to fly. A gust of wind, a bump in the runway, or even a slight tug on the control wheel sends the airplane aloft again.

The corrective action for a porpoise is the same as for a bounce and similarly depends on its severity. When it is very slight and there is no extreme change in the airplane's pitch attitude, a follow-up landing may be executed by applying sufficient power to cushion the subsequent touchdown and smoothly adjusting the pitch to the proper touchdown attitude.

When pilots attempt to correct a severe porpoise with flight control and power inputs, the inputs are often untimely may increase the severity of each successive contact with the surface. These unintentional and increasing pilot-induced oscillations may lead to damage or collapse of the nose gear. When porpoising is severe or seems to be getting worse, the safest procedure is to execute a go-around immediately by applying full power while simultaneously maintaining directional control and lowering the nose to a safe climb attitude.

Wheelbarrowing

When a pilot permits the airplane weight to become concentrated about the nose-wheel during the takeoff or landing roll, a condition known as wheelbarrowing occurs. Wheelbarrowing may cause loss of directional control during the landing roll because braking action is ineffective, and the airplane tends to swerve or pivot on the nose-wheel, particularly in crosswind conditions. One of the most common causes of wheelbarrowing during the landing roll is a simultaneous touchdown of the main and nose-wheel with excessive speed, followed by application of forward pressure on the elevator control. Usually, the situation can be corrected by smoothly applying back-elevator pressure.

Wheelbarrowing does not occur if the pilot achieves and maintains the correct landing attitude, touches down at the proper speed, and gently lowers the nose-wheel while losing speed on rollout. However, if wheelbarrowing is encountered and runway and other conditions permit, it is advisable to promptly initiate a go-around. If the pilot decides it's safer to stay on the ground rather than attempt a go-around when directional control is lost, close the throttle and adjust the pitch attitude smoothly but firmly to the proper landing attitude.

Hard Landing

When the airplane contacts the ground during landings, its vertical speed is instantly reduced to zero. Unless provisions are made to slow this vertical speed and cushion the impact of touchdown, the force of contact with the ground could cause structural damage to the airplane.

The purpose of pneumatic tires, shock absorbing landing gear, and other devices is to cushion the impact and to increase the time in which the airplane's vertical descent is stopped. The importance of this cushion may be understood from the computation that a 6-inch free fall on landing is roughly equal to a 340 fpm descent. Within a fraction of a second, the airplane gets slowed from this rate of vertical descent to zero without damage.

During this time, the landing gear, together with some aid from the lift of the wings, supplies whatever force is needed to counteract the force of the airplane's inertia and weight. However, the lift decreases rapidly as the airplane's forward speed is decreased, and the force on the landing gear increases by the impact of touchdown. When the descent stops, the lift is practically zero, leaving the landing gear alone to carry both the airplane's weight and inertia force. The load imposed at the instant of touchdown may easily be three or four times the actual weight of the airplane depending on the severity of contact.

Touchdown in a Drift or Crab

At times, it is necessary to correct for wind drift by crabbing on the final approach. If the round out and touchdown are made while the airplane is drifting or in a crab, it contacts the ground while moving sideways. This imposes extreme side loads on the landing gear and, if severe enough, may cause structural failure.

The most effective method to prevent drift is the wing-low method. This technique keeps the longitudinal axis of the airplane aligned with both the runway and the direction of motion throughout the approach and touchdown. There are three factors that cause the longitudinal axis and the direction of motion to be misaligned during touchdown: drifting, crabbing, or a combination of both.

If the pilot does not take adequate corrective action to avoid drift during a crosswind landing, the main wheels' tire tread offers resistance to the airplane's sideward movement with respect to the ground. Consequently, any sideward velocity of the airplane is abruptly decelerated, as shown in *Figure 9-37*. This creates a moment around the main wheel when it contacts the ground, tending to overturn or tip the airplane. If the upwind wingtip is raised by the action of this moment, all the weight and shock of landing is borne by one main wheel. This concentration of forces may cause tire failure or structural damage.

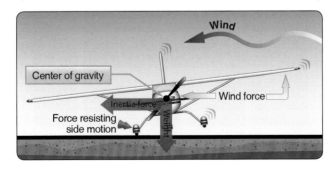

Figure 9-37. *Drifting during touchdown.*

Not only are the same factors present that are attempting to raise a wing, but the crosswind is also acting on the fuselage surface behind the main wheels, tending to yaw (weathervane) the airplane into the wind. This often results in a ground loop.

Ground Loop

A ground loop is an uncontrolled turn during ground operation that may occur while taxiing or taking off. However, an airplane is especially vulnerable to this occurrence during the after-landing roll. A ground loop may result if the pilot fails to control an initial swerve. Drift or weathervaning may cause the initial swerve. Careless use of the rudder, an uneven ground surface, or a soft spot that retards one main wheel of the airplane may also cause a swerve. In any case, the initial swerve tends to make the airplane ground loop, whether it is a tailwheel-type or nose-wheel type. [*Figure 9-38*]

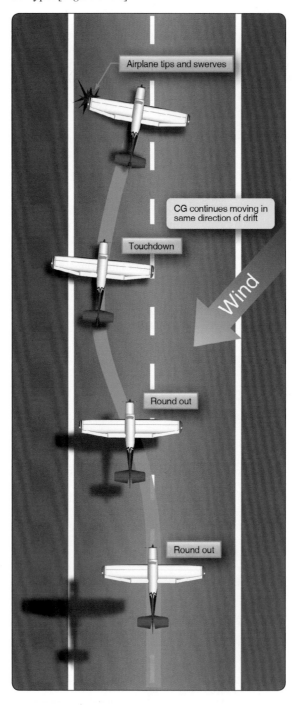

Figure 9-38. *Start of a ground loop.*

Nose-wheel type airplanes are somewhat less prone to ground loop than tailwheel-type airplanes. Since the center of gravity (CG) is located forward of the main landing gear on these airplanes, any time a swerve develops, centrifugal force acting on the CG tends to stop the swerving action.

If the airplane touches down while drifting or in a crab, apply aileron toward the high wing and stop the swerve with the rudder. Brakes are used to correct for turns or swerves only when the rudder is inadequate. Exercise caution when applying corrective brake action because it is very easy to over control and aggravate the situation.

If brakes are used, sufficient brake is applied on the low-wing wheel (outside of the turn) to stop the swerve. When the wings are approximately level, the new direction should be maintained until the airplane has slowed to taxi speed or has stopped.

In nose-wheel airplanes, a ground loop is almost always a result of wheelbarrowing. A pilot should be aware that even though the nose-wheel type airplane is less prone than the tailwheel-type airplane, virtually every type of airplane, including large multiengine airplanes, can be made to ground loop when sufficiently mishandled.

Wing Rising After Touchdown

When landing in a crosswind, there may be instances when a wing rises during the after-landing roll. This may occur whether or not there is a loss of directional control, depending on the amount of crosswind and the degree of corrective action.

Any time an airplane is rolling on the ground in a crosswind condition, the upwind wing is receiving a greater force from the wind than the downwind wing. This causes a lift differential. Also, as the upwind wing rises, there is an increase in the AOA, which increases lift on the upwind wing, rolling the airplane downwind.

When the effects of these two factors are great enough, the upwind wing may rise even though directional control is maintained. If no correction is applied, it is possible that the upwind wing rises sufficiently to cause the downwind wing to strike the ground.

In the event a wing starts to rise during the landing roll, the pilot should immediately apply more aileron pressure toward the high wing and continue to maintain direction. The sooner the aileron control is applied, the more effective it is. The further a wing is allowed to rise before taking corrective action, the more airplane surface is exposed to the force of the crosswind. This diminishes the effectiveness of the aileron.

Hydroplaning

Hydroplaning is a condition that can exist when an airplane has landed on a runway surface contaminated with standing water, slush, or wet snow. Hydroplaning can have serious adverse effects on ground controllability and braking efficiency. The three basic types of hydroplaning are dynamic hydroplaning, reverted rubber hydroplaning, and viscous hydroplaning. Any one of the three can render an airplane partially or totally uncontrollable anytime during the landing roll.

Dynamic Hydroplaning

Dynamic hydroplaning is a relatively high-speed phenomenon that occurs when there is a film of water on the runway that is at least one-tenth of an inch deep. As the speed of the airplane and the depth of the water increase, the water layer builds up an increasing resistance to displacement, resulting in the formation of a wedge of water beneath the tire. At some speed, termed the hydroplaning speed (V_p), the water pressure equals the weight of the airplane, and the tire is lifted off the runway surface. In this condition, the tires no longer contribute to directional control and braking action is nil.

Dynamic hydroplaning is related to tire inflation pressure. Data obtained during hydroplaning tests have shown the minimum dynamic hydroplaning speed (V_p) of a tire to be 8.6 times the square root of the tire pressure in pounds per square inch (PSI). For an airplane with a main tire pressure of 24 PSI, the calculated hydroplaning speed would be approximately 42 knots. It is important to note that the calculated speed referred to above is for the start of dynamic hydroplaning. Once hydroplaning has started, it may persist to a significantly slower speed depending on the type being experienced.

Reverted Rubber Hydroplaning

Reverted rubber (steam) hydroplaning occurs during heavy braking that results in a prolonged locked-wheel skid. Only a thin film of water on the runway is required to facilitate this type of hydroplaning. The tire skidding generates enough heat to cause the rubber in contact with the runway to revert to its original uncured state. The reverted rubber acts as a seal between the tire and the runway and delays water exit from the tire footprint area. The water heats and is converted to steam, which supports the tire off the runway.

Reverted rubber hydroplaning frequently follows an encounter with dynamic hydroplaning, during which time the pilot may have the brakes locked in an attempt to slow the airplane. Eventually the airplane slows enough to where the tires make contact with the runway surface and the airplane begins to skid. The remedy for this type of hydroplaning is to release the brakes and allow the wheels to spin up and apply moderate braking. Reverted rubber hydroplaning is insidious in that the pilot may not know when it begins, and it can persist to very slow groundspeeds (20 knots or less).

Viscous Hydroplaning

Viscous hydroplaning is due to the viscous properties of water. A thin film of fluid no more than one-thousandth of an inch in depth is all that is needed. The tire cannot penetrate the fluid and the tire rolls on top of the film. This can occur at a much lower speed than dynamic hydroplaning, but requires a smooth or smooth acting surface, such as asphalt or a touchdown area coated with the accumulated rubber from previous landings. Such a surface can have the same friction coefficient as wet ice.

When confronted with the possibility of hydroplaning, it is best to land on a grooved runway (if available). Touchdown speed should be as slow as possible consistent with safety. After the nose-wheel is lowered to the runway, moderate braking is applied. If deceleration is not detected and hydroplaning is suspected, raise the nose and use aerodynamic drag to decelerate to a point where the brakes become effective.

Proper braking technique is essential. The brakes are applied firmly until reaching a point just short of a skid. At the first sign of a skid, release brake pressure and allow the wheels to spin up. Directional control is maintained as far as possible with the rudder. Remember that in a crosswind, if hydroplaning occurs, the crosswind causes the airplane to simultaneously weathervane into the wind, as well as slide downwind.

Chapter Summary

Accident statistics show that a pilot is more at risk during the approach and landing than during any other phase of a flight. There are many factors that contribute to accidents in this phase, but an overwhelming percentage of these accidents result from a lack of pilot proficiency. This chapter presents procedures that, when learned and practiced correctly, are key to attaining proficiency. Additional information on aerodynamics, aircraft performance, and other aspects affecting approaches and landings can be found in the Pilot's Handbook of Aeronautical Knowledge (FAA-H-8083-25, as revised). For information concerning risk assessment as a means of preventing accidents, refer to the Risk Management Handbook (FAA-H-8083-2, as revised). Both of these publications are available at www.faa.gov/regulations_policies/handbooks_manuals/aviation/.

Airplane Flying Handbook (FAA-H-8083-3C)
Chapter 10: Performance Maneuvers

Introduction

Basic flight maneuvers taught to pilots include: straight-and-level, turns, climbs, and descents. As training advances, other performance maneuvers serve to further develop piloting skills. Performance maneuvers enhance a pilot's proficiency in flight control application, maneuver planning, situational awareness, and division of attention. To further that intent, performance maneuver design allows for the application of flight control pressures, attitudes, airspeeds, and orientations that constantly change throughout the maneuver.

Deficiencies during execution of performance maneuvers often occur when a pilot lacks an understanding of fundamental skills or never mastered them. Performance maneuver training should not take place until the pilot demonstrates consistent competency in the fundamentals. Further, initial training for performance maneuvers should always begin with a detailed ground lesson for each maneuver, so that the learner understands the technicalities prior to flight. In addition, performance maneuver training should use segmented building blocks of instruction so as to allow the pilot an appropriate level of repetition necessary to develop the required skills.

Performance maneuvers, once grasped by the pilot, are very satisfying and rewarding. As the pilot develops skills in executing performance maneuvers, they may likely see an increased smoothness in their flight control application and an increased ability to sense the airplane's attitude and orientation without significant conscious effort.

Steep Turns

Steep turns consist of single to multiple 360° and 720° turns, in either or both directions, using a bank angle between 45° and 60°. The objective of the steep turn is to develop a pilot's skill in flight control smoothness and coordination, an awareness of the airplane's orientation to outside references, division of attention between flight control applications, and the constant need to scan for hazards and other traffic in the area. [Figure 10-1]

Figure 10-1. *Steep turns.*

While the fundamental concepts of all turns are the same, when steep turns are first demonstrated and performed, the pilot will be exposed to:

1. Higher G-forces
2. The airplane's inherent overbanking tendency
3. Significant loss of the vertical component of lift when the wings are steeply banked
4. Substantial pitch control pressures
5. The need for increased additional power to maintain altitude at a constant airspeed during the turn

As discussed in previous chapters, when banking an airplane for a level turn, the total lift divides into vertical and horizontal components of lift. In order to maintain altitude at a constant airspeed, the pilot increases the angle of attack (AOA) to ensure that the vertical component of lift is sufficient to maintain altitude. The pilot adds power as needed to maintain airspeed. For a steep turn, as in any level turn, the horizontal component of lift provides the necessary force to turn the airplane. Regardless of the airspeed or airplane, for a given bank angle in a level altitude turn, the same load factor will always be produced. The load factor is the vector addition of gravity and centrifugal force. When the bank becomes steep as in a level altitude 45° banked turn, the resulting load factor is 1.41. In a level altitude 60° banked turn, the resulting load factor is 2.0. To put this in perspective, with a load factor of 2.0, the effective weight of the aircraft (and its occupants) doubles. Pilots may have difficulty with orientation and movement when first experiencing these forces. Pilots should also understand that load factors increase dramatically during a level turn beyond 60° of bank. Note that the design of a standard category general aviation airplane accommodates a load factor up to 3.8. A level turn using 75° of bank exceeds that limit.

Because of higher load factors, steep turns should be performed at an airspeed that does not exceed the airplane's design maneuvering speed (V_A) or operating maneuvering speed (V_O). Maximum turning performance for a given speed is accomplished when an airplane has a high angle of bank. Each airplane's level turning performance is limited by structural and aerodynamic design, as well as available power. The airplane's limiting load factor determines the maximum bank angle that can be maintained in level flight without exceeding the airplane's structural limitations or stalling. As the load factor increases, so does the stalling speed. For example, if an airplane stalls in level flight at 50 knots, it will stall at 60 knots in a 45° steep turn while maintaining altitude. It will stall at 70 knots if the bank is increased to 60°. Stalling speed increases at the square root of the load factor. As the bank angle increases in level flight, the margin between stalling speed and maneuvering speed decreases. At speeds at or below V_A or V_O, the airplane will stall before exceeding the design load limit.

In addition to the increased load factors, the airplane will exhibit what is called "overbanking tendency" as previously discussed in Chapter 3, Basic Flight Maneuvers. In most flight maneuvers, bank angles are shallow enough that the airplane exhibits positive or neutral stability about the longitudinal axis. However, as bank angles steepen, the airplane will continue rolling in the direction of the bank unless deliberate and opposite aileron pressure is held. Pilots should also be mindful of the various left-turning tendencies, such as P-factor, which require effective rudder/aileron coordination. While performing a steep turn, a significant component of yaw is experienced as motion away from and toward the earth's surface, which may seem confusing when first experienced.

Before starting any practice maneuver, the pilot ensures that the area is clear of air traffic and other hazards. Further, distant references should be chosen to allow the pilot to assess when to begin rollout from the turn. After establishing the manufacturer's recommended entry speed, V_A, or V_O, as applicable, the airplane should be smoothly rolled into a predetermined bank angle between 45° and 60°. As the bank angle is being established, generally prior to 30° of bank, elevator back pressure should be smoothly applied to increase the AOA and power should be added. Pilots should keep in mind that as the AOA increases, so does drag, and additional power allows the airplane to maintain airspeed. After the selected bank angle has been reached, the pilot will find that considerable force is required on the elevator control to hold the airplane in level flight.

The certification testing standards do not specify trim requirements for a steep turn. The decision whether to use trim depends on the airplane characteristics, speed of the trim system, and preference of the instructor and learner. As the bank angle transitions from medium to steep, increasing elevator-up trim and smoothly increasing engine power to that required for the turn removes some or all of the control forces required to maintain a higher angle of attack. However, if trim is used, pilots should not forget to remove both the trim and power inputs as the maneuver is completed.

Maintaining bank angle, altitude, and orientation requires an awareness of the relative position of the horizon to the nose and the wings. The pilot who references the aircraft's attitude by observing only the nose will have difficulty maintaining altitude. A pilot who observes both the nose and the wings relative to the horizon is likely able to maintain altitude within performance standards. Altitude deviations are primary errors exhibited in the execution of steep turns. Minor corrections for pitch attitude are accomplished with proportional elevator back pressure while the bank angle is held constant with the ailerons. However, during steep turns, it is not uncommon for a pilot to allow the nose to get excessively low resulting in a significant loss in altitude in a very short period of time. The pilot can recover from such an altitude loss by first reducing the angle of bank with coordinated use of opposite aileron and rudder and then increasing the pitch attitude by increasing elevator back pressure. Attempting to recover from an excessively nose-low, steep bank condition by using only the elevator causes a steepening of the bank and puts unnecessary stress on the airplane.

The rollout from the steep turn should be timed so that the wings reach level flight when the airplane is on the heading from which the maneuver was started. A good rule of thumb is to begin the rollout at ½ the number of degrees of bank prior to reaching the terminating heading. For example, if a right steep turn was begun on a heading of 270° and if the bank angle is 60°, the pilot should begin the rollout 30° prior or at a heading of 240°. While the rollout is being made, elevator back pressure, trim (if used), and power should be gradually reduced, as necessary, to maintain the altitude and airspeed.

Common errors when performing steep turns are:

1. Not clearing the area
2. Inadequate pitch control on entry or rollout
3. Gaining or losing altitude
4. Failure to maintain constant bank angle
5. Poor flight control coordination
6. Ineffective use of trim
7. Ineffective use of power
8. Inadequate airspeed control
9. Becoming disoriented
10. Performing by reference to the flight instruments rather than visual references
11. Failure to scan for other traffic during the maneuver
12. Attempting to start recovery prematurely
13. Failure to stop the turn on the designated heading

Steep Spiral

The objective of the steep spiral is to provide a flight maneuver for rapidly dissipating substantial amounts of altitude while remaining over a selected spot. This maneuver may be useful during an emergency landing. A steep spiral is a gliding turn wherein the pilot maintains a constant radius around a surface-based reference point—similar to the turns around a point maneuver, but in this case the airplane is rapidly descending. The maneuver consists of the completion of at least three 360° turns *[Figure 10-2]*, and should begin at sufficient altitude such that the maneuver concludes no lower than 1,500 feet above ground level (AGL). Note that while there are similarities between a steep spiral and an emergency descent, the reasons for using the two maneuvers may differ, and the airspeed and configuration are usually different.

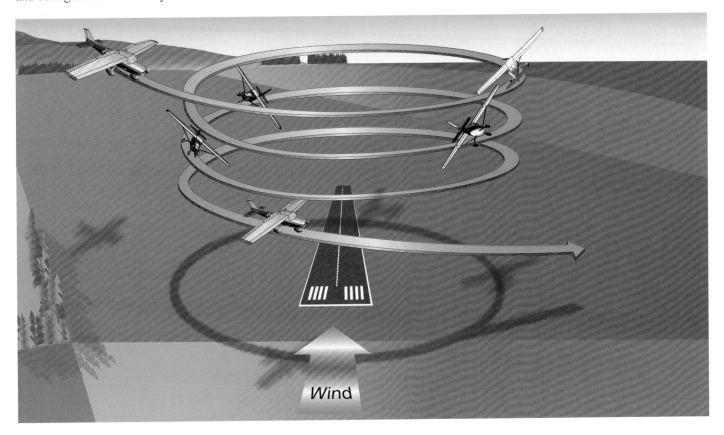

Figure 10-2. *Steep spiral.*

The steep spiral is initiated by properly clearing the airspace for air traffic and hazards. In general, the throttle is closed to idle, carburetor heat is applied if equipped, and gliding speed is established. Once the proper airspeed is attained, the pitch should be lowered and the airplane rolled to the desired bank angle as the reference point is reached. The pilot should consider the distance from the reference point since that establishes the turning radius, and the steepest bank should not exceed 60°. The gliding spiral should be a turn of constant radius while maintaining the airplane's position relative to the reference. This can only be accomplished by proper correction for wind drift by steepening the bank on downwind headings and shallowing the bank on upwind headings. During the steep spiral, the pilot should continually correct for any changes in wind direction and velocity to maintain a constant radius.

Operating the engine at idle speed for any prolonged period during the glide may result in excessive engine cooling, spark plug fouling, or carburetor ice. To assist in avoiding these issues, the throttle should be periodically advanced and sustained for a few seconds. Monitoring cylinder head temperature gauges, if available, provides a pilot with additional information on engine cooling. When advancing the throttle, the pitch attitude should be adjusted to maintain a constant airspeed and, preferably, this should be done when headed into the wind.

Maintaining a constant airspeed throughout the maneuver is an important skill for a pilot to develop. This is necessary because the airspeed tends to fluctuate as the bank angle is changed throughout the maneuver. The pilot should anticipate pitch corrections as the bank angle is varied throughout the maneuver. During practice of the maneuver, the pilot should execute at least three turns and roll out toward a definite object or on a specific heading. To make the exercise more challenging, the pilot rolls out on a heading perpendicular to or directly into the wind rather toward a specific object. This ability would be a particularly useful skill in the event of an actual emergency. In addition, noting the altitude lost during each revolution would help the pilot determine when to roll out in an actual emergency so as not to be too high or too low to make a safe approach. During rollout, the smooth and accurate application of the flight controls allow the airplane to recover to a wings-level glide with no change in airspeed. Recovering to normal cruise flight would proceed after the establishment of a wings-level glide.

Common errors when performing steep spirals are:

1. Not clearing the area
2. Inadequate pitch control on entry or rollout
3. Not correcting the bank angle to compensate for wind
4. Poor flight control coordination
5. Ineffective use of trim
6. Inadequate airspeed control
7. Becoming disoriented
8. Performing by reference to the flight instruments rather than visual references
9. Not scanning for other traffic during the maneuver
10. Not completing the turn on the designated heading or reference

Chandelle

A chandelle is a maximum performance, 180° climbing turn that begins from approximately straight-and-level flight and concludes with the airplane in a wings-level, nose-high attitude just above stall speed. *[Figure 10-3]* The goal is to gain the most altitude possible for a given bank angle and power setting; however, the standard used to judge the maneuver is not the amount of altitude gained, but rather the pilot's proficiency as it pertains to maximizing climb performance for the power and bank selected, as well as the skill demonstrated.

A chandelle is best described in two specific phases: the first 90° of turn and the second 90° of turn. The first 90° of turn is described as constant bank and continuously increasing pitch; and the second 90° as constant pitch and continuously decreasing bank. During the first 90°, the pilot will set the bank angle, increase power, and increase pitch attitude at a rate such that maximum pitch-up occurs at the completion of the first 90°. The maximum pitch-up attitude achieved at the 90° mark is held for the remainder of the maneuver. If the pitch attitude is set too low, the airplane's airspeed will never decrease to just above stall speed. If the pitch attitude is set too high, the airplane may aerodynamically stall prior to completion of the maneuver. Starting at the 90° point, and while maintaining the pitch attitude set at the end of the first 90°, the pilot begins a slow and coordinated constant rate rollout so as to have the wings level when the airplane is at the 180° point. If the rate of rollout is too rapid or sluggish, the airplane either exceeds the 180° turn or does not complete the turn as the wings come level to the horizon.

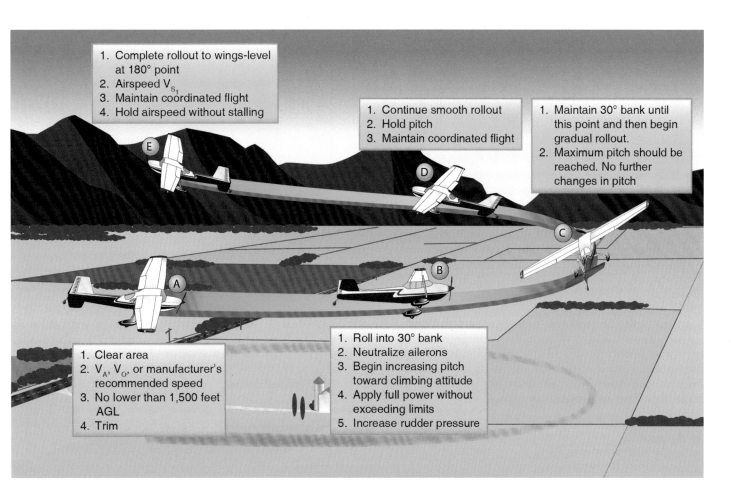

Figure 10-3. *Chandelle.*

Prior to starting the chandelle, the flaps and landing gear (if retractable) should be in the UP position. The chandelle is initiated by properly clearing the airspace for air traffic and hazards. The maneuver should be entered from straight-and-level flight or a shallow dive at an airspeed recommended by the manufacturer—in many cases this is the airplane's design maneuvering speed (V_A) or operating maneuvering speed (V_O). *[Figure 10-3A]* After the appropriate entry airspeed has been established, the chandelle is started by smoothly entering a coordinated turn to the desired angle of bank. Once the bank angle is established, which is generally 30°, a climbing turn should be started by smoothly applying elevator back pressure at a constant rate while simultaneously increasing engine power to the recommended setting. In airplanes with a fixed-pitch propeller, the throttle should be set so as to not exceed rotations per minute (rpm) limitations. In airplanes with constant-speed propellers, power may be set at the normal cruise or climb setting as appropriate. *[Figure 10-3B]*

As airspeed decreases during the chandelle, left-turning tendencies, such as P-factor, have greater effect. As airspeed decreases, right rudder pressure is progressively increased to ensure that the airplane remains in coordinated flight. The pilot maintains coordinated flight by sensing physical slipping or skidding, by glancing at the ball in the turn-and-slip or turn coordinator, and by using appropriate control pressures.

At the 90° point, the pilot should begin to smoothly roll out of the bank at a constant rate while maintaining the pitch attitude attained at the end of the first 90°. While the angle of bank is fixed during the first 90°, recall that as airspeed decreases, the overbanking tendency increases. *[Figure 10-3C]* As a result, proper use of the ailerons allows the bank to remain at a fixed angle until rollout is begun at the start of the final 90°. As the rollout continues, the vertical component of lift increases. However, as speed continues to decrease, a slight increase of elevator back pressure is required to keep the pitch attitude from decreasing.

When the airspeed is slowest, near the completion of the chandelle, right rudder pressure is significant, especially when rolling out from a left chandelle due to left adverse yaw and left-turning tendencies, such as P-factor. *[Figure 10-3D]* When rolling out from a right chandelle, the yawing moment is to the right, which partially cancels some of the left-turning tendency's effect. Depending on the airplane, either very little left rudder or a reduction in right rudder pressure is required during the rollout from a right chandelle. At the completion of 180° of turn, the wings should be level to the horizon, the airspeed should be just above the power-on stall speed, and the airplane's pitch-high attitude should be held momentarily. *[Figure 10-3E]*

Once the airplane is in controlled flight, the pitch attitude may be reduced and the airplane returned to straight-and-level cruise flight.

Common errors when performing chandelles are:

1. Not clearing the area
2. Initial bank is too shallow resulting in a stall
3. Initial bank is too steep resulting in failure to gain maximum performance
4. Allowing the bank angle to increase after initial establishment
5. Not starting the recovery at the 90° point in the turn
6. Allowing the pitch attitude to increase as the bank is rolled out during the second 90° of turn
7. Leveling the wings prior to the 180° point being reached
8. Pitch attitude is low on recovery resulting in airspeed well above stall speed
9. Application of flight control pressures is not smooth
10. Poor flight control coordination
11. Stalling at any point during the maneuver
12. Execution of a steep turn instead of a climbing maneuver
13. Not scanning for other traffic during the maneuver
14. Performing by reference to the instruments rather than visual references

Lazy Eight

The lazy eight is a maneuver that is designed to develop the proper coordination of the flight controls across a wide range of airspeeds and attitudes. It is the only standard flight training maneuver in which flight control pressures are constantly changing. In an attempt to simplify the discussion about this maneuver, the lazy eight can be loosely compared to the ground reference maneuver, S-turns across the road. Recall that S-turns across the road are made of opposing 180° turns. For example, first a 180° turn to the right, followed immediately by a 180° turn to the left. The lazy eight adds both a climb and descent to each 180° segment. The first 90° is a climb; the second 90° is a descent. *[Figure 10-4]*

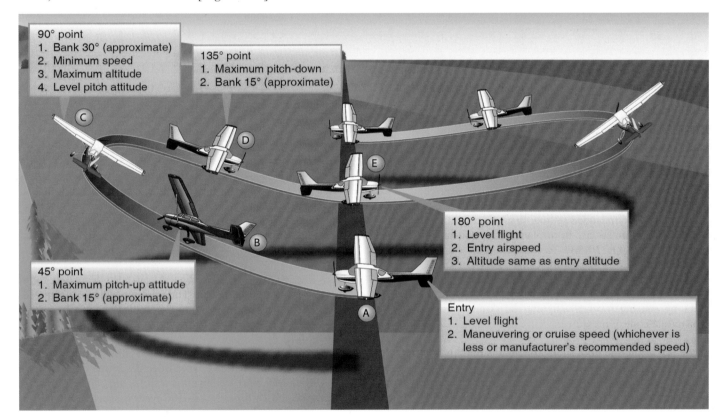

Figure 10-4. *Lazy eight.*

The previous description of a lazy eight and *figure 10-4* describe how a lazy eight looks from outside the flight deck and describes it as two 180° turns with altitude changes. How does it look from the pilot's perspective? Think of the longitudinal axis of the airplane as a pencil, which draws on whatever it points to. During this maneuver, the longitudinal axis of the airplane traces a symmetrical eight on its side with segments of the eight above and below the horizon, and it takes both 180° turns to form both loops of an eight. The first 90° of the first 180° turn traces the upper portion of one of the loops. The second 90° portion of the second 180° turn traces the lower portion of that loop at the end of the maneuvuer. The second 90° of the first 180° turn and the first 90° of the second 180° turn complete the other loop of the eight. The sensation of using the airplane to slowly draw this symbol gives the maneuver its name.

To aid in the performance of the lazy eight's symmetrical climbing/descending turns, the pilot selects prominent reference points on the natural horizon. The reference points selected should be at 45°, 90°, and 135° from the direction in which the maneuver is started for each 180° turn. With the general concept of climbing and descending turns grasped, specifics of the lazy eight can then be discussed.

Shown in *Figure 10-4A*, from level flight a gradual climbing turn is begun in the direction of the 45° reference point. The climbing turn should be planned and controlled so that the maximum pitch-up attitude is reached at the 45° point with an approximate bank angle of 15°. *[Figure 10-4B]* As the pitch attitude is raised, the airspeed decreases, which causes the rate of turn to increase. As such, the lazy eight should begin with a slow rate of roll as the combination of increasing pitch and increasing bank may cause the rate of turn to be so rapid that the 45° reference point will be reached before the highest pitch attitude is attained. At the 45° reference point, the pitch attitude should be at the maximum pitch-up selected for the maneuver while the bank angle is slowly increasing. Beyond the 45° reference point, the pitch-up attitude should begin to decrease slowly toward the horizon until the 90° reference point is reached where the pitch attitude passes through level.

The lazy eight requires substantial skill in coordinating the aileron and rudder; therefore, some discussion about coordination is warranted. As pilots understand, the purpose of the rudder is to maintain coordination; slipping or skidding is to be avoided. Pilots should remember that since the airspeed is still decreasing as the airplane is climbing; additional right rudder pressure should be applied to counteract left-turning tendencies, such as P-factor. As the airspeed decreases, right rudder pressure should be gradually applied to counteract yaw at the apex of the lazy eight in both the right and left turns; however, additional right rudder pressure is required when using right aileron control pressure. When displacing the ailerons for more lift on the left wing, left adverse yaw augments with the left-yawing P-factor in an attempt to yaw the nose to the left. In contrast, in left climbing turns or rolling to the left, the left yawing P-factor tends to cancel the effects of adverse yaw to the right; consequently, less right rudder pressure is required. These concepts can be difficult to remember; however, to simplify, rolling right at low airspeeds and high-power settings requires substantial right rudder pressures.

At the lazy eight's 90° reference point, the bank angle should also have reached its maximum angle of approximately 30°. *[Figure 10-4C]* The airspeed should be at its minimum, just about 5 to 10 knots above stall speed, with the airplane's pitch attitude passing through level flight. Coordinated flight at this point requires that, in some flight conditions, a slight amount of opposite aileron pressure may be required to prevent the wings from overbanking while maintaining rudder pressure to cancel the effects of left-turning tendencies.

The pilot should not hesitate at the 90° point but should continue to maneuver the airplane into a descending turn. The rollout from the bank should proceed slowly while the airplane's pitch attitude is allowed to decrease. When the airplane has turned 135°, the airplane should be in its lowest pitch attitude. *[Figure 10-4D]* Pilots should remember that the airplane's airspeed is increasing as the airplane's pitch attitude decreases; therefore, maintaining proper coordination will require a decrease in right rudder pressure. As the airplane approaches the 180° point, it is necessary to progressively relax rudder and aileron pressure while simultaneously raising pitch and roll to level flight. As the rollout is being accomplished, the pilot should note the amount of turn remaining and adjust the rate of rollout and pitch change so that the wings and nose are level at the original airspeed just as the 180° point is reached.

Upon arriving at 180° point, a climbing turn should be started immediately in the opposite direction toward the preselected reference points to complete the second half of the lazy eight in the same manner as the first half. *[Figure 10-4E]*

Power should be set so as not to enter the maneuver at an airspeed that would exceed manufacturer's recommendations, which is generally no greater than V_A or V_O. Power and bank angle have significant effect on the altitude gained or lost; if excess power is used for a given bank angle, altitude is gained at the completion of the maneuver; however, if insufficient power is used for a given bank angle, altitude is lost.

Common errors when performing lazy eights are:

1. Not clearing the area
2. Maneuver is not symmetrical across each 180°
3. Inadequate or improper selection or use of 45°, 90°, 135° references
4. Ineffective planning
5. Gain or loss of altitude at each 180° point
6. Poor control at the top of each climb segment resulting in the pitch rapidly falling through the horizon
7. Airspeed or bank angle standards not met
8. Control roughness
9. Poor flight control coordination
10. Stalling at any point during the maneuver
11. Execution of a steep turn instead of a climbing maneuver
12. Not scanning for other traffic during the maneuver
13. Performing by reference to the flight instruments rather than visual references

Chapter Summary

Performance maneuvers are used to develop a pilot's skills in coordinating the flight control's use and effect while enhancing the pilot's ability to divide attention across the various demands of flight. Performance maneuvers are also designed to further develop a pilot's application and correlation of the fundamentals of flight and integrate developing skills into advanced maneuvers. Developing highly-honed skills in performance maneuvers allows the pilot to effectively progress toward the mastery of flight. Mastery is developed as the mechanics of flight become a subconscious, rather than a conscious, application of the flight controls to maneuver the airplane in attitude, orientation, and position.

Airplane Flying Handbook (FAA-H-8083-3C)
Chapter 11: Night Operations

Introduction

The mechanical operation of an airplane at night is no different than operating the same airplane during the day. The airplane does not know if it is being operated in the dark or in bright sunlight. It performs and responds to control inputs by the pilot. The pilot, however, is affected by various aspects of night operations and should take them into consideration during night flight operations. Some are actual physical limitations affecting all pilots. Others, such as equipment requirements, procedures, and emergency situations, should also be considered.

According to 14 CFR part 1, section 1.1, Definitions and Abbreviations, "night" means the time between the end of evening civil twilight and the beginning of morning civil twilight, as published in the Air Almanac, converted to local time. To explain further, the National Weather Service defines evening civil twilight as the time that begins in the morning, or ends in the evening, when the geometric center of the sun is 6 degrees below the horizon. Therefore, morning civil twilight begins when the geometric center of the sun is 6 degrees below the horizon, and ends at sunrise. Evening civil twilight begins at sunset and ends when the geometric center of the sun is 6 degrees below the horizon. The FAA has an online tool to calculate sunrise, sunset, and civil twilight for any given location.

For 14 CFR part 61, section 61.57(b)(1) night operations that meet recent flight experience requirements, the term "night" refers to the time period beginning 1 hour after sunset and ending 1 hour before sunrise. The same regulation requires that during those hours, no person may act as pilot-in-command (PIC) of an aircraft carrying passengers unless within the preceding 90 days and during those specified hours, that person has made 3 takeoffs and landings to a full stop. 14 CFR part 61, sections 61.57(b)(1)(i) and (ii) require the pilot to have made the required takeoffs and landings acting as the sole manipulator of the controls, and to have performed the takeoffs and landings in an aircraft of the same category, class, and type (if a type rating is required). Other conditions apply if using a full flight simulator to meet the requirement as described in 14 CFR part 61 (section 61.57(b)(2)) or if seeking to use another alternative provided in the regulation.

Night flying operations should not be encouraged or attempted except by certificated pilots with knowledge of and experience in the topics discussed in this chapter.

Night Vision

Due to the physiology of the eye *[Figure 11-1]*, humans experience diminished vision in low-light conditions. Because vision involves the eyes and brain working together, understanding eye function leads to pilot behaviors that can improve night vision significantly.

Anatomy of the Eye

- Light from an object enters the eye through the cornea and then continues through the pupil.

- The opening (dilation) and closing (constriction) of the pupil is controlled by the iris, which is the colored part of the eye. The function of the pupil is similar to that of the diaphragm of a photographic camera: to control the amount of light.

- The lens is located behind the pupil and its function is to focus light on the surface of the retina.

- The retina is the inner layer of the eyeball that contains photosensitive cells called rods and cones. The function of the retina is similar to that of the film in a photographic camera: to record an image.

- The cones are located in higher concentrations than rods in the central area of the retina known as the macula, which measures about 4.5 mm in diameter. The exact center of the macula has a very small depression called the fovea, which contains cones only. The cones are used for day or high-intensity light vision. They are involved with central vision to detect detail, perceive color, and identify far-away objects.

- The rods are located mainly in the periphery of the retina—an area that is about 10,000 times more sensitive to light than the fovea. Rods are used for low light intensity or night vision and are involved with peripheral vision to detect position references, including objects (fixed and moving) in shades of gray, but cannot be used to detect detail or to perceive color.

- Although there is not a clear-cut division of function, the rods make night vision possible. The rods and cones function in daylight and in moonlight, but in the absence of normal light, the process of night vision is placed almost entirely on the rods.

- Light energy (an image) enters the eyes and is transformed by the cones and rods into electrical signals that are carried by the optic nerve to the posterior area of the brain (occipital lobes). This part of the brain interprets the electrical signals and creates a mental image of the actual object that was seen by the person.

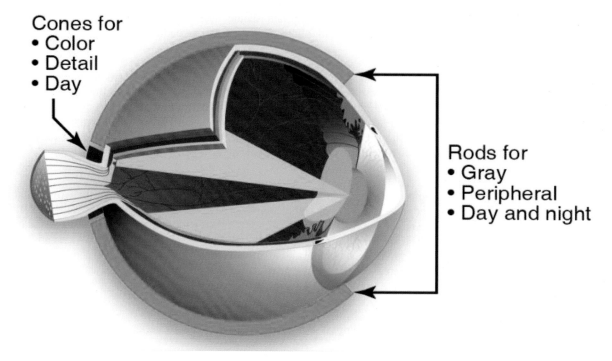

Figure 11-1. *Rods and cones.*

Types of Vision

Photopic Vision. During daytime or high-intensity artificial illumination conditions, the eyes rely on central vision (foveal cones) to perceive and interpret sharp images and color of objects. *[Figure 11-2]*

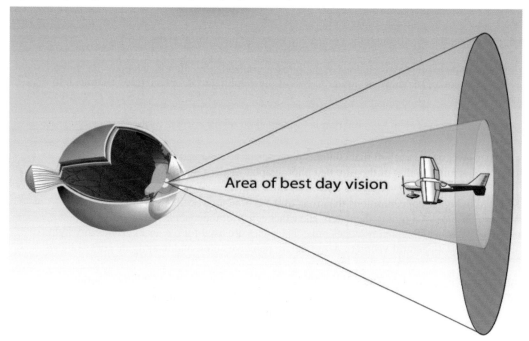

Figure 11-2. *Central Vision.*

Mesopic Vision. Occurs at dawn, dusk, or under full moonlight levels and is characterized by decreasing visual acuity and color vision. Under these conditions, a combination of central (foveal cones) and peripheral (rods)vision is required to maintain appropriate visual performance.

Scotopic Vision. During nighttime, partial moonlight, or low intensity artificial illumination conditions, central vision (foveal cones) becomes ineffective to maintain visual acuity and color perception. Under these conditions, if looking directly at an object for more than a few seconds, the image of the object fades away completely (night blind spot). Peripheral vision (off center scanning) provides the only means of seeing very dim objects in the dark.

Night Blind Spot

The "Night Blind Spot" appears under conditions of low ambient illumination due to the absence of rods in the fovea. *[Figure 11-3]* This absence of rods affects the central 5 to 10 degrees of the visual field. If an object is viewed directly at night, it may go undetected or it may fade away after initial detection. The night blind spot can hide larger objects as the distance between the pilot and an object increases.

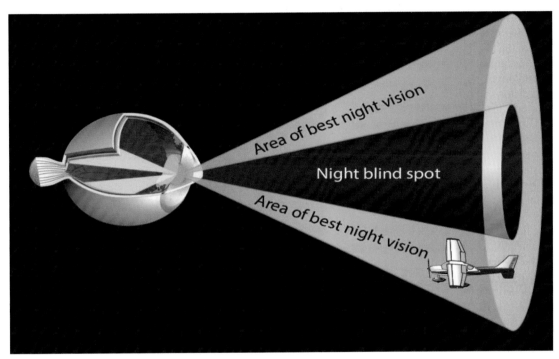

Figure 11-3. *The night blind spot.*

Vision Under Dim and Bright Illumination

The eye's adaptation to darkness is another important aspect of night vision. When a dark room is entered, it is difficult to see anything until the eyes become adjusted to the darkness. Almost everyone experiences this when entering a darkened movie theater.

In darkness, vision gradually becomes more sensitive to light. Maximum dark adaptation can take up to 30 minutes. Exposure to aircraft anti-collision lights does not impair night vision adaptation because the intermittent flashes have a very short duration (less than 1 second). However, if dark-adapted eyes are exposed to a bright light source (searchlights, landing lights, flares, etc.) for a period of 1 second or more, night vision is temporarily impaired. If it is safe to do so, pilots may close one eye when bright exposure begins in order to preserve dark adaptation for that eye.

Factors Affecting Vision

• During the day, identification of objects at a distance is aided by good resolution. At night, the identification range of dim objects is limited and the detail resolution is poor.

• Surface references or the horizon may become obscured by smoke, fog, smog, haze, dust, ice particles, or other phenomena, even when visibility meets Visual Flight Rule (VFR) minimums. This is especially true at airports located adjacent to large bodies of water or sparsely populated areas where few, if any, surface references are available. Lack of horizon or surface reference is common on over-water flights, at night, and in low-visibility conditions.

- Presence of uncorrected refractive eye disorders such as myopia (nearsightedness–impaired focusing of distant objects), hyperopia (farsightedness–impaired focusing of near objects), astigmatism (impaired focusing of objects in different meridians), or presbyopia (impaired focusing of near objects) affect day and night vision.

- Self-imposed stresses such as self-medication, alcohol consumption (including hangover effects), tobacco use (including withdrawal), hypoglycemia, sleep deprivation/fatigue, and extreme emotional upset can seriously impair vision.

- Inflight exposure to low barometric pressure without the use of supplemental oxygen (above 10,000 feet during the day and above 5,000 feet at night) can result in hypoxia, which impairs visual performance.

- Due to the effects of carbon monoxide on the blood, smokers may experience a physiological altitude that is much higher than actual altitude. The smoker is thus more susceptible to hypoxia at lower altitudes than the nonsmoker.

- Other factors that may have an adverse effect on visual performance include windscreen haze, improper illumination of the flight deck and/or instruments, scratched and/or dirty instrumentation, use of flight deck red lighting, inadequate flight deck environmental control (temperature and humidity), inappropriate sunglasses and/or prescription glasses/contact lenses, and sustained visual workload during flight. Red light illumination distorts colors (magenta and yellow pigments both appear as red, and cyan pigment appears black) on aeronautical charts. Pilots should use it only where optimum outside night vision capability is necessary. Dim white flight deck lighting should be available when needed for map and instrument reading.

- Monovision contact lenses (one contact lens for distant vision and the other lens for near vision) make the pilot alternate his/her vision; that is, a person uses one eye at a time, suppressing the other, and consequently impairs binocular vision and depth perception. The FAA recommends not using these lenses when piloting an aircraft..

- A flickering light in the flight deck, anti-collision lights, or other aircraft lights, may cause interference with brain function. Although rare, this may occur at a frequencies from 1 to 20 hertz. If continuous, the possible physical reactions can be nausea, dizziness, grogginess, unconsciousness, headaches, or confusion. Pilots should try to eliminate or screen out any light source that might cause an unwanted reaction to blinking or flickering lights.

- Sunglasses can aid the dark adaptation process, which is delayed by prolonged exposure to bright sunlight.

Night Illusions

Visual illusions are especially hazardous because pilots rely on their eyes for correct information. Darkness or low visibility increases pilot susceptibility to error. Two illusions that lead to spatial disorientation, false horizon and autokinesis, concern the visual system only.

False Horizon

Flying at night under clear skies with ground lights below can result in situations where it is difficult to distinguish the ground lights from the stars. A dark scene spread with ground lights and stars, and certain geometric patterns of ground lights can provide inaccurate visual information, making it difficult to align the aircraft correctly with the actual horizon. An aurora borealis display at night or a visible sloping cloud formation can also affect a pilot's sense of the horizon. A similar problem is encountered during certain daylight operations over large bodies of water. Various atmospheric and water conditions can create a visual scene without a discernible horizon.

Autokinesis

In the dark, a stationary light will appear to move about when stared at for many seconds. The disoriented pilot could lose control of the aircraft in attempting to align it with the false movements of this light.

Featureless Terrain Illusion

A black-hole approach occurs when the landing is made from over water or non-lighted terrain where the runway lights are the only source of light. Without peripheral visual cues to help, orientation is difficult. The runway can seem out of position (down-sloping or up-sloping) and in the worst case, results in landing short of the runway. If an electronic glide slope or visual approach slope indicator (VASI) is available, it should be used. If navigation aids (NAVAIDs) are unavailable, the flight instruments assist in maintaining orientation and a normal approach. Anytime position in relation to the runway or altitude is in doubt, the pilot should execute a go-around.

Bright runway and approach lighting systems, especially where few lights illuminate the surrounding terrain, may create the illusion of being lower or having less distance to the runway. In this situation, the tendency is to fly a higher approach. Also, flying over terrain with only a few lights makes the runway recede or appear farther away. With this situation, the tendency is to fly a lower-than-normal approach. If the runway has a city in the distance on higher terrain, the tendency is to fly a lower-than-normal approach. A good review of the airfield layout and boundaries before initiating any approach helps maintain a safe approach angle.

Ground Lighting Illusions

Lights along a straight path, such as a road or lights on moving trains, can be mistaken for runway and approach lights. Bright runway and approach lighting systems, especially where few lights illuminate the surrounding terrain, may create the illusion of less distance to the runway. The pilot who does not recognize this illusion will often fly a higher approach.

Illusions created by runway lights result in a variety of problems. Bright lights or bold colors advance the runway, making it appear closer. Night landings are further complicated by the difficulty of judging distance and the possibility of confusing approach and runway lights. For example, when a double row of approach lights joins the boundary lights of the runway, there can be confusion as to where the approach lights terminate and runway lights begin. Under certain conditions, approach lights can make the aircraft seem higher in a turn to final, than when its wings are level.

Pilot Equipment

As part of preflight preparation, pilots should carefully consider the personal equipment that should be readily available during the flight to include a flashlight, aeronautical charts, pertinent data for the flight, and a flight deck checklist containing procedures for the following tasks:

1. Before starting engines
2. Before takeoff
3. Cruise
4. Before landing
5. After landing
6. Stopping engines
7. Emergencies

At least one reliable flashlight is recommended as standard equipment on all night flights. A reliable incandescent or light-emitting diode (LED) dimmable flashlight able to produce white/red light is preferable. The flashlight should be large enough to be easily located in the event it is needed. It is also recommended to have a spare set of batteries for the flashlight readily available. The white light is used while performing the preflight visual inspection of the airplane, the red light is used when performing flight deck operations, and the dim white light may be used for chart reading. Many charts can be displayed on a EFB, which does not require a flashlight. However, its brightness should be set so as not to seriously impair night vision.

Since the red light is non-glaring, it will not impair night vision. Some pilots prefer two flashlights, one with a white light for preflight and the other a penlight type with a red light. The latter can be suspended by a string from around the neck to ensure the light is always readily available. As mentioned earlier, red light distorts color perception of pigments other than red on charts.

Aeronautical charts are essential for night cross-country flight and, if the intended course is near the edge of the chart, the adjacent chart should also be available. The lights of cities and towns can be seen at surprising distances at night, and if this adjacent chart is not available to identify those landmarks, confusion could result. Regardless of the equipment used, organization of the flight deck eases the burden and enhances safety. Organize equipment and charts and place them within easy reach prior to taxiing.

Airplane Equipment and Lighting

14 CFR part 91, section 91.205(c) specifies the basic minimum airplane equipment that is required for VFR flight at night. This equipment includes basic instruments, lights, electrical energy source, and spare fuses if applicable.

The standard instruments required by 14 CFR part 91, section 91.205(d) for IFR flight are valuable assets for aircraft control at night. 14 CFR part 91, section 91.205(c)(3) specifies that during VFR flight at night, operating aircraft are required to have an approved anti-collision light system, which can include a flashing or rotating beacon and position lights. However, 14 CFR part 91, section 91.209(b) gives the pilot-in-command leeway to turn off the anti-collision lights in the interest of safety. Airplane position lights are arranged similar to those of boats and ships. A red light is positioned on the left wingtip, a green light on the right wingtip, and a white light on the tail. *[Figure 11-4]*

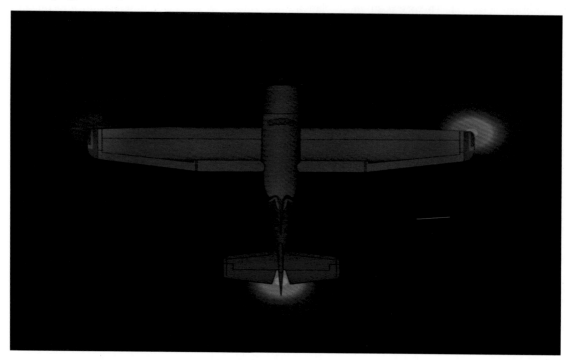

Figure 11-4. *Position lights.*

This arrangement provides a means to determine the general direction of movement of other airplanes in flight. If both a red and green light of another aircraft are observed, and the red light is on the left and the green to the right, the airplane is flying the same direction. Care must be taken to maintain clearance. If red were on the right and green to the left, the airplane could be on a collision course.

Landing lights are not only useful for taxi, takeoffs, and landings, but also provide a means by which airplanes can be seen at night by other pilots. Pilots are encouraged to turn on their landing lights when operating within 10 miles of an airport and below 10,000 feet. Operation with landing lights on applies to both day and night or in conditions of reduced visibility. This should also be done in areas where flocks of birds may be expected.

Although turning on aircraft lights supports the "see and be seen" concept, pilots should continue to keep a sharp lookout for other aircraft. Aircraft lights may blend in with the stars or the lights of the cities at night and go unnoticed unless a conscious effort is made to distinguish them from other lights.

Airport and Navigation Lighting Aids

The lighting systems used for airports, runways, obstructions, and other visual aids at night are other important aspects of night flying. Lighted airports located away from congested areas are identified readily at night by the lights outlining the runways. Airports located near or within large cities are often difficult to identify as the airport lights tend to blend with the city lights. It is important to not only know the exact location of an airport relative to the city, but also to be able to identify these airports by the characteristics of their lighting patterns.

Aeronautical lights are designed and installed in a variety of colors and configurations, each having its own purpose. Although some lights are used only during low ceiling and visibility conditions, this discussion includes only the lights that are fundamental to visual flight rules (VFR) night operation.

It is recommended that prior to a night flight, and particularly a cross-country night flight, that a check of the availability and status of lighting systems at the destination airport is made. This information can be found on aeronautical charts and in the Chart Supplements. The status of each facility can be determined by reviewing pertinent Notices to Airmen (NOTAMs).

Most airports have rotating beacons. The beacon rotates at a constant speed, thus producing a series of light flashes at regular intervals. These flashes may consist of a white flash and one or two different colors that are used to identify various types of landing areas. For example:

- Lighted civilian land airports—alternating white and green lights

- Lighted civilian water airports—alternating white and yellow lights

- Lighted military airports—alternating white and green lights, but are differentiated from civil airports by dual peaked (two quick) white flashes, then green

Beacons producing red flashes indicate obstructions or areas considered hazardous to aerial navigation. Steady-burning red lights are used to mark obstructions on or near airports and sometimes to supplement flashing lights on en route obstructions. High-intensity, flashing white lights are used to mark some supporting structures of overhead transmission lines that stretch across rivers, chasms, and gorges. These high-intensity lights are also used to identify tall structures, such as chimneys and towers.

As a result of technological advancements, runway lighting systems have become quite sophisticated to accommodate takeoffs and landings in various weather conditions. However, if flying is limited to VFR only, it is important to be familiar with the basic lighting of runways and taxiways.

The basic runway lighting system consists of two straight parallel lines of runway edge lights defining the lateral limits of the runway. These lights are aviation white, although aviation yellow may be substituted for a distance of 2,000 feet from the far end of the runway to indicate a caution zone. At some airports, the intensity of the runway edge lights can be activated and adjusted by radio control. The control system consists of a 3-step control responsive to 7, 5, and/or 3 microphone clicks. This 3-step control turns on lighting facilities capable of either 3-step, 2-step, or 1-step operation. The 3-step and 2-step lighting facilities can be altered in intensity, while the 1-step cannot. All lighting is illuminated for a period of 15 minutes from the most recent time of activation and may not be extinguished prior to end of the 15-minute period. Suggested use is to always initially key the mike 7 times; this assures that all controlled lights are turned on to the maximum available intensity. If desired, adjustment can then be made, where the capability is provided, to a lower intensity by keying 5 and/or 3 times. Due to the close proximity of airports using the same frequency, radio-controlled lighting receivers may be set at a low sensitivity requiring the aircraft to be relatively close to activate the system. Consequently, even when lights are on, the pilot should always key the mike as directed when overflying an airport of intended landing or just prior to entering the final segment of an approach. This assures the aircraft is close enough to activate the system and a full 15-minute lighting duration is available.

The length limits of the runway are defined by straight lines of lights across the runway ends. At some airports, the runway threshold lights are aviation green, and the runway end lights are aviation red. At many airports, the taxiways are also lighted. A taxiway edge lighting system consists of blue lights that outline the usable limits of taxi paths.

Training for Night Flight

Learning to fly safely at night takes time and experience. Pilot's should practice maneuvers at night including straight-and-level flight, climbs and descents, level turns, climbing and descending turns, and steep turns. Practicing recovery from unusual attitudes should only be done with a flight instructor. Pilots may practice these maneuvers with all the flight deck lights turned OFF, as well as ON. This blackout training simulates an electrical or instrument light failure. Pilots should also use the navigation equipment and local NAVAIDs during the training. In spite of fewer references or checkpoints, night cross-country flights do not present particular problems if pre-planning is adequate. Just as during the day, the pilot continuously monitors position, time estimates, fuel consumed, and uses NAVAIDs, if available, to assist in monitoring en route progress.

Preparation and Preflight

Night flying requires that pilots are aware of, and operate within, their abilities and limitations. Although careful planning of any flight is essential, night flying demands more attention to the details of preflight preparation and planning.

Preparation for a night flight includes a thorough review of the available weather reports and forecasts with particular attention given to temperature/dew point spread. A narrow temperature/dew point spread may indicate the possibility of fog. Emphasis should also be placed on wind direction and speed, since its effect on the airplane cannot be as easily detected at night as during the day.

On night cross-country flights, pilots should select and use appropriate aeronautical charts to include the appropriate adjacent charts. Course lines should be drawn in black to be more distinguishable in low-light conditions. Rotating beacons at airports, lighted obstructions, lights of cities or towns, and lights from major highway traffic all provide excellent visual checkpoints. If using a global positioning system (GPS) for navigation, the pilot should ensure that it works properly. All necessary waypoints should be loaded before the flight, and the database should be checked for accuracy prior to taking off and again once in flight. The use of radio navigation aids and communication facilities add significantly to the safety and efficiency of night flying.

Check all personal equipment prior to flight to ensure proper functioning and operation. All airplane lights should be checked for operation by turning them on momentarily during the preflight inspection. Position lights can be checked for loose connections by tapping the light fixture. If the lights blink while being tapped, determine the cause prior to flight. Parking ramps should be checked with a flashlight prior to entering the airplane. During the day, it is quite easy to see stepladders, chuckholes, wheel chocks, and other obstructions, but at night, it is more difficult and a check of the area can prevent taxiing mishaps.

Starting, Taxiing, and Run-up

Once seated in the airplane and prior to starting the engine, a careful pilot will organize and arrange all items and materials to be used during the flight. The pilot should also take extra care at night to clear the propeller area. While turning the rotating beacon ON or flashing the airplane position lights helps alert persons nearby to remain clear of the propeller, the pilot should carefully and methodically scan the area around the aircraft. To avoid excessive drain of electrical current from the battery, the pilot may turn off unnecessary electrical equipment until after the engine has been started.

After starting the engine and when ready to taxi, the pilot turns the taxi or landing light ON. In some airplanes, continuous use of the landing light while taxiing may place an excessive drain on the airplane's electrical system. Also, overheating of some types of landing lights is possible because of inadequate airflow to carry the heat away. If overheating or electrical power is an issue, the landing light may be used only if necessary. When using lights, consideration should be given to not blinding other pilots. Pilots should taxi slowly, particularly in congested areas. If taxi lines are painted on the ramp or taxiway, following the lines ensures a proper path along the route. An instrument check should be done while taxiing to check for proper and correct operation prior to takeoff.

While taxiing for any takeoff, the pilot should verify that the aircraft position, taxi route, and runway for the departure all appear as expected. The taxi diagram, signage, pavement markings, and instruments should all reinforce the pilot's situational awareness. If any conflicting information or doubt exists, the pilot should not proceed with the taxi or the takeoff. A wrong turn, wrong-surface takeoff, or takeoff on a closed runway can have catastrophic results, and preventing any of these depends on maintaining situational awareness while the aircraft moves on the ground.

When using the checklist for the before-takeoff and run-up checks during the day, any forward movement of the airplane can be detected easily. However, at night, the airplane could creep forward without being noticed unless the pilot takes steps to prevent this possibility. Pilots should hold or lock the brakes during the run-up and be alert for any forward movement.

Takeoff and Climb

The most noticeable difference between daylight and nighttime flying is the limited availability of outside visual references at night. Therefore, flight instruments should be used to a greater degree in controlling the airplane. This is particularly true on night takeoffs and climbs. The pilot should adjust the flight deck lights to a minimum brightness that will allow for reading the instruments and switches but not hinder outside vision. Dimming the lights also eliminates reflections on the windshield and windows.

After ensuring that the final approach and runway are clear of other air traffic, or when cleared for takeoff by the air traffic controller, the pilot turns the landing and taxi lights ON and lines the airplane up with the centerline of the runway. If the runway does not have centerline lighting, the painted centerline and the distance from the runway edge lights on each side indicate the center. The heading indicator should be noted and correspond or set to the known runway direction. To begin the takeoff, the pilot releases the brakes and advances the throttle smoothly to maximum allowable power. As it accelerates, the airplane should be kept moving straight ahead between and parallel to the runway edge lights.

The procedure for night takeoffs is the same as for normal daytime takeoffs except that many of the runway visual cues are not available. The pilot should check the flight instruments frequently during the takeoff to ensure proper airspeed, attitude, and heading. As the airspeed reaches the normal lift-off speed, the pilot adjusts the pitch attitude to establish a normal climb by referring to both outside visual references, such as lights, and to the flight instruments. *[Figure 11-5]* Without visual references ahead, inexperienced pilots may relax right rudder pressure after takeoff and veer off to the left.

After becoming airborne, the darkness of night often makes it difficult to note whether the airplane is getting closer to or farther from the surface. The attitude indicator, vertical speed indicator (VSI), and altimeter should all indicate a positive climb. It is also important to ensure the airspeed is at best climb speed.

Figure 11-5. *Establish a positive climb.*

The pilot makes necessary pitch and bank adjustments by referencing the attitude and heading indicators. It is recommended that turns not be made until reaching a safe maneuvering altitude. Although the use of the landing lights is helpful during the takeoff, they become ineffective after the airplane has climbed to an altitude where the light beam no longer extends to the surface. The light can cause distortion when it is reflected by haze, smoke, or clouds that might exist in the climb. Therefore, when the landing light is used for the takeoff, it should be turned off after the climb is well established provided it is not being used for collision avoidance.

Orientation and Navigation

Generally, at night, it is difficult to see clouds and restrictions to visibility, particularly on dark nights or under an overcast. When flying under VFR, pilots should exercise caution to avoid flying into clouds. Usually, the first indication of flying into restricted visibility conditions is the gradual disappearance of lights on the ground. If the lights begin to appear surrounded by a halo or glow, further flight in the same direction calls for caution. Such a halo or glow around lights on the ground is indicative of ground fog. If a descent occurs through clouds, smoke, or haze in order to land, the horizontal visibility is considerably less when looking through the restriction than it is when looking straight down through it from above. Pilots should avoid a VFR night flight if expecting conditions below VFR minimums. If encountering IMC, risk increases dramatically unless both the pilot and aircraft are equipped for flight under IFR, and the pilot has prepared and filed an IFR flight plan that can be activated, if needed.

Crossing large bodies of water at night in single-engine airplanes could be potentially hazardous, because in the event of an engine failure, the pilot may be forced to land (ditch) the airplane in the water. Another hazard faced by pilots of all aircraft, due to limited or no lighting, is that the horizon blends with the water. During poor visibility conditions over water, the horizon becomes obscure and may result in a loss of orientation. Even on clear nights, the stars may be reflected on the water surface, which could appear as a continuous array of lights, thus making the horizon difficult to identify.

Lighted runways, buildings, or other objects may cause illusions when seen from different altitudes. At an altitude of 2,000 feet, a group of lights on an object may be seen individually, while at 5,000 feet or higher, the same lights could appear to be one solid light mass. These illusions may become quite acute with altitude changes and, if not overcome, could present problems when making approaches to lighted runways.

Approaches and Landings

When approaching the airport to enter the traffic pattern and land, it is important that the runway lights and other airport lighting be identified as early as possible. If the airport layout is unfamiliar, sighting of the runway may be difficult until very close-in due to the maze of lights observed in the area. *[Figure 11-6]* A pilot should normally fly toward the rotating beacon until the lights outlining the runway are distinguishable. To fly a traffic pattern of proper size and direction, the runway threshold and runway-edge lights need to be positively identified. Once the airport lights are seen, these lights should be kept in sight throughout the approach.

Figure 11-6. *Use light patterns for orientation.*

Distance may be deceptive at night due to limited lighting conditions. A lack of intervening references on the ground and the inability to compare the size and location of different ground objects cause this. This also applies to the estimation of altitude and speed. Consequently, more dependence should be placed on flight instruments, particularly the altimeter and the airspeed indicator. Monitoring the altimeter prevents flying too low for the distance from the airport. When entering the traffic pattern, the pilot should allow adequate time to complete the before-landing checklist. If the heading indicator contains a heading bug, setting it to the runway heading is an excellent reference for the pattern legs.

The pilot maintains the recommended airspeeds and executes the approach and landing in the same manner as during the day. A low, shallow approach is definitely inappropriate during a night operation. The altimeter and VSI should be constantly cross-checked against the airplane's position along the base leg and final approach. A visual approach slope indicator (VASI) is an indispensable aid in establishing and maintaining a proper glide path. *[Figure 11-7]*

After turning onto the final approach and aligning the airplane midway between the two rows of runway-edge lights, the pilot should note and correct for any wind drift. Throughout the final approach, proper use of pitch and power helps to maintain a stabilized approach. Flaps are used as in a normal approach. Usually, halfway through the final approach, the landing light is turned on. The landing light is sometimes ineffective since the light beam will usually not reach the ground from higher altitudes. The light may even be reflected back into the pilot's eyes by any existing haze, smoke, or fog. Safety considerations regarding local traffic and collision avoidance may overshadow these disadvantages.

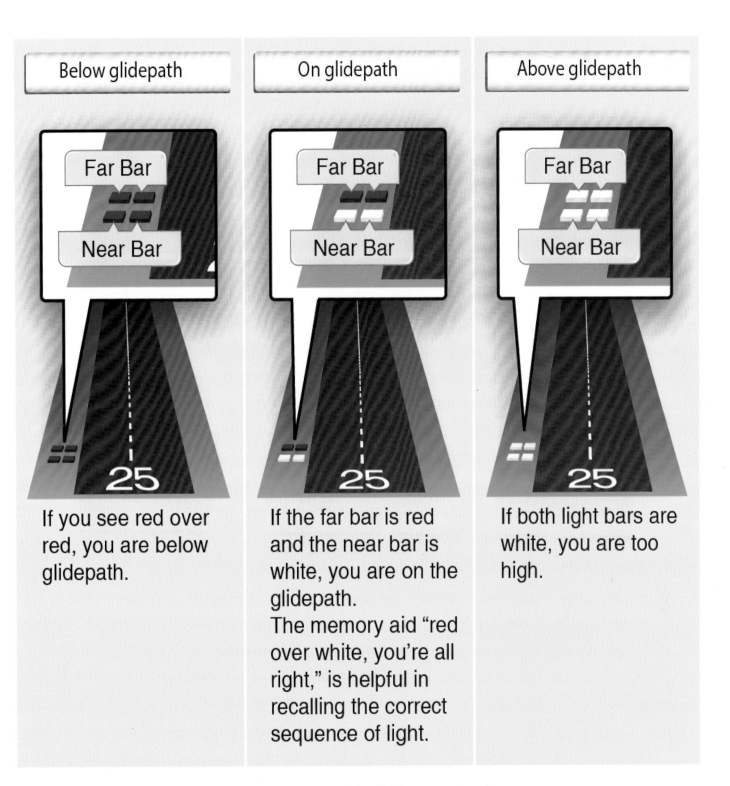

Below glidepath	On glidepath	Above glidepath

Far Bar **Near Bar** — 25

If you see red over red, you are below glidepath.

If the far bar is red and the near bar is white, you are on the glidepath.
The memory aid "red over white, you're all right," is helpful in recalling the correct sequence of light.

If both light bars are white, you are too high.

Figure 11-7. *VASI.*

The round out and touchdown is made in the same manner as in day landings. At night, the judgment of height, speed, and sink rate is impaired by the scarcity of observable objects in the landing area. An inexperienced pilot may have a tendency to round out too high. Continuing a constant approach descent until the landing lights reflect on the runway and tire marks on the runway can be seen clearly helps identify the point to begin the round out. At this point, the round out is started smoothly and the throttle gradually reduced to idle as the airplane is touching down. *[Figure 11-8]* During landings without the use of landing lights, the round out may be started when the runway lights at the far end of the runway first appear to be rising higher than the nose of the airplane. This demands a smooth and very timely round out and requires that the pilot feel for the runway surface using power and pitch changes, as necessary, for the airplane to settle slowly to the runway. Blackout landings should always be included in night pilot training as an emergency procedure.

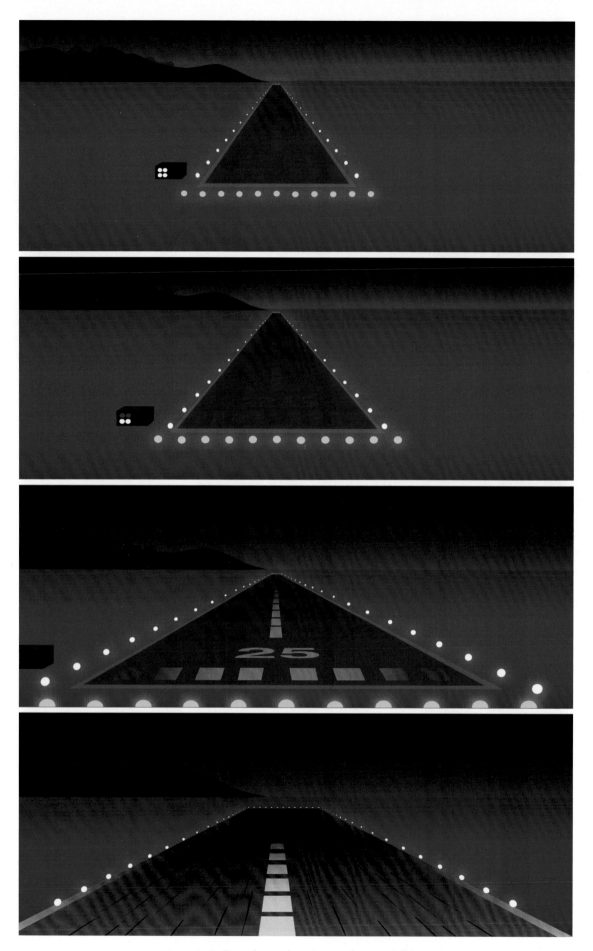

Figure 11-8. *Round out when tire marks are visible.*

How to Prevent Landing Errors Due to Optical Illusions

To prevent these illusions and their potentially hazardous consequences, pilots can:

1. Anticipate the possibility of visual illusions during approaches to unfamiliar airports, particularly at night or in adverse weather conditions.

2. Consult airport diagrams and the Chart Supplements for information on runway slope, terrain, and lighting.

3. Make frequent reference to the altimeter, especially during all approaches, day and night.

4. If possible, conduct aerial visual inspection of unfamiliar airports before landing.

5. Use Visual Approach Slope Indicator (VASI) or Precision Approach Path Indicator (PAPI) systems for a visual reference or an electronic glideslope, whenever they are available.

6. Utilize the visual descent point (VDP) found on many nonprecision instrument approach procedure charts.

7. Recognize that the chances of being involved in an approach accident increase when some emergency or other activity distracts from usual procedures.

8. Maintain optimum proficiency in landing procedures.

Night Emergencies

Perhaps the greatest concern about flying a single-engine airplane at night is the possibility of a complete engine failure and the subsequent emergency landing. This is a legitimate concern, even though continuing flight into adverse weather and poor pilot judgment account for most serious accidents.

If the engine fails at night, there are several important procedures and considerations to keep in mind. They are as follows:

- Maintain positive control of the airplane and establish the best glide configuration and airspeed. Turn the airplane towards an airport or away from congested areas.

- Check to determine the cause of the engine malfunction, such as the position of fuel selectors, magneto switch, or primer. If possible, the cause of the malfunction should be corrected immediately and the engine restarted.

- Announce the emergency situation to air traffic control (ATC) or Universal Communications (UNICOM). If already in radio contact with a facility, do not change frequencies unless instructed to change.

- If the condition of the nearby terrain is known and is suitable for a forced landing, turn towards an unlighted portion of the area and plan an emergency forced landing to an unlighted portion.

- Consider an emergency landing area close to public access if possible. This may facilitate rescue or help, if needed.

- Maintain orientation with the wind to avoid a downwind landing.

- Complete the before-landing checklist, and check the landing lights for operation at altitude and turn ON in sufficient time to illuminate the terrain or obstacles along the flightpath. The landing should be completed in the normal landing attitude at the slowest possible airspeed. If the landing lights are unusable and outside visual references are not available, the airplane should be held in level-landing attitude until the ground is contacted.

- After landing, turn off all switches and evacuate the airplane as quickly as possible.

Chapter Summary

Night operations present additional risks that pilots should identify and assess. Night flying operations should not be encouraged or attempted, except by pilots that are certificated, current, and proficient in night flying. Prior to attempting night operations, pilots should receive training and be familiar with the risks associated with night flight and how they differ from daylight operations. Even for experienced pilots, night VFR operations should only be conducted in unrestricted visibility, favorable winds, both on the surface and aloft, and no turbulence. Additional information on pilot vision and illusions can be found in FAA brochure AM-400-98/2 at www.faa.gov/pilots/safety/pilotsafetybrochures/ and also in Chapters 2 and 17 of the Pilot's Handbook of Aeronautical Knowledge (FAA-H-8083-25) at www.faa.gov. Additional information on lighting aids can be found in Chapter 2 of the Aeronautical Information Manual (AIM), which can be accessed at www.faa.gov.

Airplane Flying Handbook (FAA-H-8083-3C)
Chapter 12: Transition to Complex Airplanes

Introduction

A high-performance airplane is defined as an airplane with an engine capable of developing more than 200 horsepower (see 14 CFR part 61, section 61.31(f)(1)). A complex airplane (see 14 CFR part 61, section 61.1) means an airplane that has a retractable landing gear, flaps, and a controllable pitch propeller, including airplanes equipped with an engine control system consisting of a digital computer and associated accessories for controlling the engine and propeller, such as a full authority digital engine control; or, in the case of a seaplane, flaps and a controllable pitch propeller, including seaplanes equipped with an engine control system consisting of a digital computer and associated accessories for controlling the engine and propeller, such as a full authority digital engine control.

Transition to a complex airplane, or a high-performance airplane, can be demanding for many pilots. Both increased performance and complexity require additional planning, judgment, and piloting skills. Transition to these types of airplanes, therefore, should be accomplished in a systematic manner through a structured course of training administered by a qualified flight instructor.

Airplanes can be designed to fly through a wide range of airspeeds. High speed flight requires smaller wing areas and moderately cambered airfoils whereas low speed flight is obtained with airfoils with a greater camber and larger wing area. *[Figure 12-1]* Many compromises are often made by designers to provide for higher speed cruise flight and low speeds for landing. Flaps are a common design effort to increase an airfoil's camber and surface area for lower-speed flight. *[Figure 12-2]*

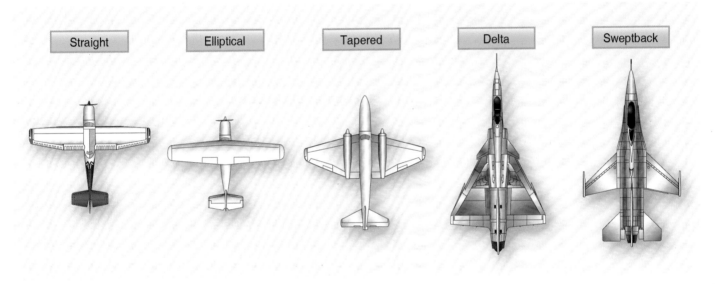

| Straight | Elliptical | Tapered | Delta | Sweptback |

Figure 12-1. *Airfoil types.*

Since an airfoil cannot have two different cambers at the same time, designers and engineers deliver the desired performance characteristics using two different methods. Either the airfoil can be a compromise, or a cruise airfoil can be combined with a device for increasing the camber of the airfoil for low-speed flight. Camber is the asymmetry between the top and the bottom surfaces of an airfoil. One method for varying an airfoil's camber is the addition of trailing-edge flaps. Engineers call these devices a high-lift system.

Function of Flaps

Flaps work primarily by changing the camber of the airfoil, which increases the wing's lift coefficient. With some flap designs, the surface area of the wing is also increased. Flap deflection does not increase the critical (stall) angle of attack (AOA). In some cases, flap deflection actually decreases the critical AOA. Deflection of a wing's control surfaces, such as ailerons and flaps, alters both lift and drag. With aileron deflection, there is asymmetrical lift which imparts a rolling moment about the airplane's longitudinal axis. Wing flaps act symmetrically about the longitudinal axis producing no rolling moment; however, both lift and drag increase as well as a pitching moment about the lateral axis. Lift is a function of several variables including air density, velocity, surface area, and lift coefficient. Since flaps increase an airfoil's lift coefficient, lift is increased. *[Figure 12-3]*

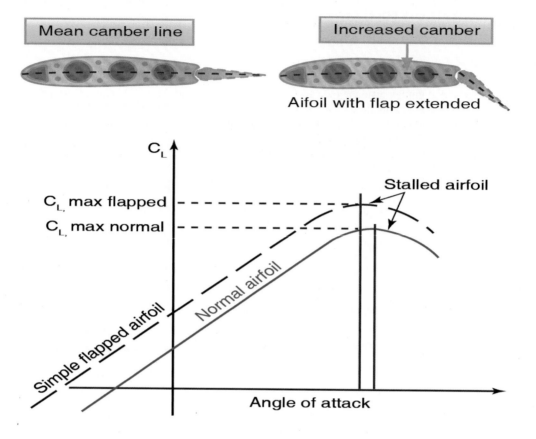

Figure 12-2. *Coefficient of lift comparison for flap extended and retracted positions.*

$$L = \frac{1}{2}pV^2SC_L$$

L = Lift produced

P = Air density

V = Velocity relative to the air

S = Surface area of the wing

C_L = lift coefficient which is determined by the camber of the airfoil used, the chord of the wing and AOA

Figure 12-3. *Lift equation.*

As flaps are deflected, the aircraft may pitch nose-up, nose-down, or have minimal changes in pitch attitude. Pitching moment is caused by the rearward movement of the wing's center of pressure; however, that pitching behavior depends on several variables including flap type, wing position, downwash behavior, and horizontal tail location. Consequently, pitch behavior depends on the design features of the particular airplane.

Flap deflection of up to 15° primarily produces lift with minimal increases in drag. Deflection beyond 15° produces a large increase in drag. Drag from flap deflection is parasite drag and, as such, is proportional to the square of the speed. Also, deflection beyond 15° produces a significant nose-up pitching moment in most high-wing airplanes because the resulting downwash changes the airflow over the horizontal tail.

Flap Effectiveness

Flap effectiveness depends on a number of factors, but the most noticeable are size and type. For the purpose of this chapter, trailing edge flaps are classified as four basic types: plain (hinge), split, slotted, and Fowler. *[Figure 12-4]*

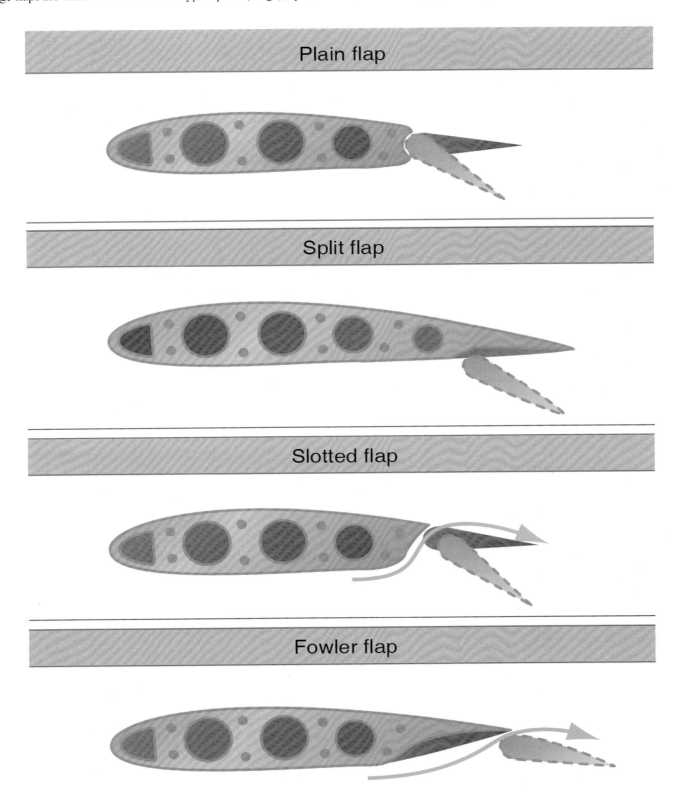

Figure 12-4. *Four basic types of flaps.*

The plain or hinge flap is a hinged section of the wing. The structure and function are comparable to the other control surfaces—ailerons, rudder, and elevator. The split flap is more complex. It is the lower or underside portion of the wing; deflection of the flap leaves the upper trailing edge of the wing undisturbed. It is, however, more effective than the hinge flap because of greater lift and less pitching moment, but there is more drag. Split flaps are more useful for landing, but the partially deflected hinge flaps have the advantage in takeoff. The split flap has significant drag at small deflections, whereas the hinge flap does not because airflow remains "attached" to the flap.

The slotted flap has a gap between the wing and the leading edge of the flap. The slot allows high-pressure airflow on the wing undersurface to energize the lower pressure over the top, thereby delaying flow separation. The slotted flap has greater lift than the hinge flap but less than the split flap; but, because of a higher lift-drag ratio, it gives better takeoff and climb performance. Small deflections of the slotted flap give a higher drag than the hinge flap but less than the split. This allows the slotted flap to be used for takeoff.

The Fowler flap deflects down and aft to increase the wing area. This flap can be multi-slotted making it the most complex of the trailing-edge systems. This system does, however, give the maximum lift coefficient. Drag characteristics at small deflections are much like the slotted flap. Fowler flaps are most commonly used on larger airplanes because of their structural complexity and difficulty in sealing the slots.

Operational Procedures

It would be impossible to discuss all the many airplane design and flap combinations. Pilots should refer to the Federal Aviation Administration (FAA) approved Airplane Flight Manual and/or Pilot's Operating Handbook (AFM/POH) for a given airplane. However, while some AFM/POHs are specific as to operational use of flaps, others leave the use of flaps to pilot discretion. Since flaps are often used for landings and takeoffs, when the airplane is close to the ground, pilot judgment and error avoidance are of critical importance.

Since the recommendations given in the AFM/POH are based on the airplane and the flap design, the pilot should relate the manufacturer's recommendation to aerodynamic effects of flaps. This requires basic background knowledge of flap aerodynamics and geometry. With this information, a decision as to the degree of flap deflection and time of deflection based on runway and approach conditions relative to the wind conditions can be made.

The time of flap extension and the degree of deflection are related. Large changes in flap deflection at one single point in the landing pattern can produce large lift changes that require significant pitch and power changes in order to maintain airspeed and descent angle. Consequently, there is an advantage to extending flaps in increments while in the landing pattern. Incremental deflection of flaps on downwind, base leg, and final approach allow smaller adjustments of pitch and power and support a stabilized approach.

While normal, soft-field, or short-field landings require minimal speed at touchdown, a short-field obstacle approach requires minimum speed and a steep approach angle. Flap extension, particularly beyond 30°, results in significant levels of drag. The drag can produce a high sink rate that the pilot needs to control with power. When a pilot uses power during a steep approach or short-field approach to offset the drag produced by the flaps, the landing flare becomes critical. A reduction in power too early can result in a hard landing, airplane damage, or loss of control. A reduction in power too late causes the airplane to float down the runway.

Crosswind component is another factor to be considered in the degree of flap extension. The deflected flap presents a surface area for the wind to act on. With flaps extended in a crosswind, the wing on the upwind side is more affected than the downwind wing. The effect is reduced to a slight extent in the crabbed approach since the airplane is more nearly aligned with the wind. When using a wing-low approach, the lowered wing partially blocks the upwind flap. The dihedral of the wing combined with the flap and wind make lateral control more difficult. Lateral control becomes more difficult as flap extension reaches maximum and the crosswind becomes perpendicular to the runway.

With flaps extended, the crosswind effects on the wing become more pronounced as the airplane reaches the ground. The wing, flap, and ground on the upwind side of the airplane form a "container" that is filled with air by the crosswind. Since the flap is located behind the main landing gear, wind striking the deflected flap tends to yaw the airplane into the wind and raise the upwind wing. The raised wing reduces the tire forces and further increases the tendency to turn into the wind. Proper control position (ailerons into the wind) is essential for maintaining runway alignment. Depending on the amount of crosswind, it may be necessary to retract the flaps soon after touchdown in order to maintain control of the airplane.

The go-around is another factor to consider when making a decision about degree of flap deflection and about where in the landing pattern to extend flaps. Because of the nose-down pitching moment produced with flap extension, trim is used to offset this pitching moment. Application of full power in the go-around increases the airflow over the wing. This produces additional lift causing significant changes in pitch. The pitch-up tendency does not diminish completely with flap retraction because of the trim setting. Expedient retraction of flaps is desirable to eliminate drag; however, the pilot should be prepared for rapid changes in pitch forces as the result of trim and the increase in airflow over the control surfaces. *[Figure 12-5]*

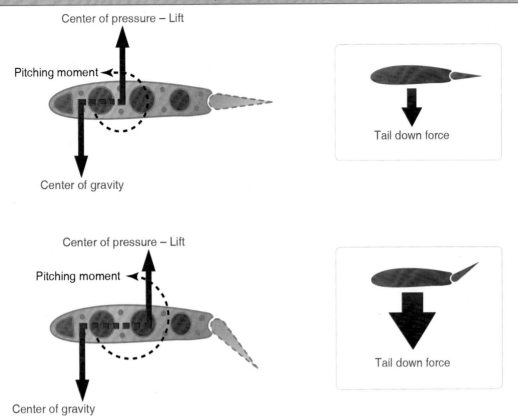

Figure 12-5. *Flaps extended pitching moment.*

During a go-around, the pilot should carefully monitor pitch and airspeed and expect that the degree of flap deflection and the design configuration of the horizontal tail relative to the wing will affect go-around characteristics. The pilot should carefully monitor pitch and airspeed, control flap retraction to minimize altitude loss, and use rudder for coordination. Considering these factors, it is good practice to extend the same degree of flaps at the same point in the landing pattern for each landing. Consistent use of flaps in the traffic pattern allows for a preplanned and familiar go-around sequence based on the airplane's position in the landing pattern.

There is no single formula to determine the degree of flap deflection to be used on landing because a landing involves variables that are dependent on each other. The AFM/POH for the particular airplane contains the manufacturer's recommendations for some landing situations. On the other hand, AFM/POH information on flap usage for takeoff is more precise. The manufacturer's requirements are based on the climb performance produced by a given flap design. Under no circumstances should a flap setting given in the AFM/POH be exceeded for takeoff.

Controllable-Pitch Propeller

Fixed-pitch propellers are designed for best efficiency at one particular revolutions per minute (rpm) setting and one airspeed. A fixed-pitch propeller provides suitable performance in a narrow range of airspeeds. However, fixed-pitch efficiency suffers considerably when operating outside of this range. To provide improved propeller efficiency through a wide range of operation, the propeller blade angle needs to be controllable.

Constant-Speed Propeller

A constant-speed propeller keeps the blade angle adjusted for maximum efficiency during most flight conditions. The pilot controls the engine rpm indirectly by means of a propeller control, which is connected to the propeller governor. For maximum takeoff power, the propeller control is moved all the way forward to the low pitch/high rpm position, and the throttle is moved forward to the maximum allowable manifold pressure position. *[Figure 12- 6]* To reduce power for climb or cruise, the pilot reduces manifold pressure to the desired value with the throttle, and then reduces engine rpm by moving the propeller control back toward the high pitch/low rpm position. The pilot sets the rpm accurately using the tachometer.

High pitch – Low RPM   Low pitch – High RPM

Figure 12-6. *Controllable-pitch propeller pitch angles.*

When an airplane engine runs at a constant governed speed, the torque (force) exerted by the engine at the propeller shaft equals the force resisting the moving blades. The pilot uses the propeller control to change engine rpm by adjusting the propeller blade pitch, which increases or decreases the air resistance on the rotating propeller. For example, pulling back on the propeller control moves the propeller blades to a higher pitch. This increases the air resistance exerted on the spinning propeller and puts an additional load on the engine, which causes it to slow down until the forces reach equilibrium. Advancing the propeller control reduces the propeller blade pitch. This reduces the resistance of the air against the propeller. In response, the engine rpm increases until the opposing forces balance. In order for this system to function, a constant-speed propeller governor needs the means to sense engine rpm and a means to control the propeller AOA. In most cases, the governor is geared to the engine crankshaft giving it a means to sense engine rpm. The "Blade Angle Control" section of this chapter discusses the ways a propeller governor adjusts propeller blade angle.

Other factors affect constant-speed propeller blade pitch. When an airplane is nosed up into a climb from level flight, the engine tends to slow down. Since the governor is sensitive to small changes in engine rpm, it decreases the blade angle just enough to keep the engine speed constant. If the airplane is nosed down into a dive, the governor increases the blade angle just enough to keep the engine speed constant. This allows the engine to maintain a constant rpm and power output. The pilot can also set engine power output by changing rpm at a constant manifold pressure; by changing the manifold pressure at a constant rpm; or by changing both rpm and manifold pressure. The constant-speed propeller makes it possible to obtain an infinite number of power settings.

Takeoff, Climb, and Cruise

During takeoff, when the forward motion of the airplane is at a low speed and when maximum power and thrust are required, the constant-speed propeller sets up a low propeller blade pitch. The low blade angle keeps the blade angle of attack, with respect to the relative wind, small and efficient at the low speed. *[Figure 12-7]*

At the same time, low blade pitch allows the propeller to handle a smaller mass of air per revolution. This light propeller load allows the engine to turn at maximum rpm and develop maximum engine power. Although the mass of air per revolution is small, the number of rpm is high, and propeller thrust is maximized until brake release. Thrust is maximum at the beginning of the takeoff roll and then decreases as the airplane gains speed.

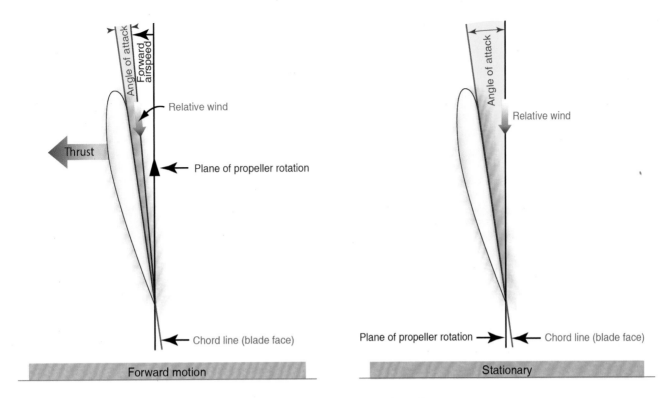

Figure 12-7. *Propeller blade angle.*

As the airspeed increases after lift-off, the load on the engine is lightened because of the small blade angle. The governor senses this and increases the blade angle slightly. Again, the higher blade angle, with the higher speed, keeps the blade AOA with respect to the relative wind small and efficient.

For climb after takeoff, the power output of the engine is reduced to climb power by decreasing the manifold pressure and increasing the blade angle to lower engine rpm. At the higher (climb) airspeed and the higher blade angle, the propeller is handling a greater mass of air per second at a lower slipstream velocity. This reduction in power is offset by the increase in propeller efficiency. The blade AOA is again kept small by the increase in the blade angle with an increase in airspeed.

At cruising altitude, when the airplane is in level flight, airspeed increases, and less power is required. Consequently, the pilot uses the throttle to reduce manifold pressure and uses the propeller control to reduce engine rpm. The higher airspeed and higher blade angle enable the propeller to handle a still greater mass of air per second at still smaller slipstream velocity. At normal cruising speeds, propeller efficiency is at or near maximum efficiency.

Blade Angle Control

Once the rpm settings for the propeller are selected, the propeller governor automatically adjusts the blade angle to maintain the selected rpm. It does this by using oil pressure. Generally, the oil pressure used for pitch change comes directly from the engine lubricating system. When a governor is employed, engine oil is used and the oil pressure is usually boosted by a pump that is integrated with the governor. The higher pressure provides a quicker blade angle change. The rpm at which the propeller is to operate is adjusted in the governor head. The pilot changes this setting by changing the position of the governor rack through the flight deck propeller control.

On some constant-speed propellers, changes in pitch are obtained by the use of an inherent centrifugal twisting moment of the blades that tends to flatten the blades toward low pitch and oil pressure applied to a hydraulic piston connected to the propeller blades which moves them toward high pitch. Another type of constant-speed propeller uses counterweights attached to the blade shanks in the hub. Governor oil pressure and the blade twisting moment move the blades toward the low pitch position, and centrifugal force acting on the counterweights moves them (and the blades) toward the high pitch position. In the first case above, governor oil pressure moves the blades towards high pitch and in the second case, governor oil pressure and the blade twisting moment move the blades toward low pitch. A loss of governor oil pressure, therefore, affects each differently.

Governing Range

The blade angle range for constant-speed propellers varies from about 11.5° to 40°. The higher the speed of the airplane, the greater the blade angle range. *[Figure 12-8]*

Aircraft Type	Design Speed (mph)	Blade Angle Range	Pitch	
			Low	High
Fixed gear	160	11$^1/_2$°	10$^1/_2$°	22°
Retractable	180	15°	11°	26°
Turbo retractable	225/240	20°	14°	34°
Turbine retractable	250/300	30°	10°	40°
Transport retractable	325	40°	10/15°	50/55°

Figure 12-8. *Blade angle range (values are approximate).*

The range of possible blade angles between high and low blade angle pitch stops define the propeller's governing range. As long as the propeller's blades operate within the governing range and not against either pitch stop, a constant engine rpm is maintained. However, once the propeller blades reach their pitch-stop limit, the engine rpm increases or decreases with changes in airspeed and propeller load similar to a fixed-pitch propeller. For example, once a specific rpm is selected, if the airspeed decreases enough, the propeller blades reduce pitch in an attempt to maintain the selected rpm until they contact their low pitch stops. From that point, any further reduction in airspeed causes the engine rpm to decrease. Conversely, if the airspeed increases, the pitch angle of the propeller blades increase until the high pitch stop is reached. The engine rpm then begins to increase.

Constant-Speed Propeller Operation

The engine is started with the propeller control in the low pitch/high rpm position. This position reduces the load or drag of the propeller and the result is easier starting and warm-up of the engine. During warm-up, the propeller blade changing mechanism is operated slowly and smoothly through a full cycle. This is done by moving the propeller control (with the manifold pressure set to produce about 1,600 rpm) to the high pitch/low rpm position, allowing the rpm to stabilize, and then moving the propeller control back to the low pitch takeoff position. This is done for two reasons: to determine whether the system is operating correctly and to circulate fresh warm oil through the propeller governor system. Remember the oil has been trapped in the propeller cylinder since the last time the engine was shut down. There is a certain amount of leakage from the propeller cylinder, and the oil tends to congeal, especially if the outside air temperature is low. Consequently, if the propeller is not exercised before takeoff, there is a possibility that the engine may over-speed on takeoff.

An airplane equipped with a constant-speed propeller has better takeoff performance than a similarly powered airplane equipped with a fixed-pitch propeller. This is because with a constant-speed propeller, an airplane can develop its maximum rated horsepower (red line on the tachometer) while motionless. An airplane with a fixed-pitch propeller, on the other hand, needs to accelerate down the runway to increase airspeed and aerodynamically unload the propeller so that rpm and horsepower can steadily build up to their maximum. With a constant-speed propeller, the tachometer reading should come up to within 40 rpm of the red line as soon as full power is applied and remain there for the entire takeoff. Excessive manifold pressure raises the cylinder combustion pressures, resulting in high stresses within the engine. Excessive pressure also produces high-engine temperatures. A combination of high manifold pressure and low rpm can induce damaging detonation. In order to avoid these situations, the following sequence should be followed when making power changes.

- When increasing power, increase the rpm first and then the manifold pressure
- When decreasing power, decrease the manifold pressure first and then decrease the rpm

The cruise power charts in the AFM/POH should be consulted when selecting cruise power settings. Whatever the combinations of rpm and manifold pressure listed in these charts—they have been flight tested and approved by engineers for the respective airframe and engine manufacturer. Therefore, if there are power settings, such as 2,100 rpm and 24 inches manifold pressure in the power chart, they are approved for use. With a constant-speed propeller, a power descent can be made without over-speeding the engine. The system compensates for the increased airspeed of the descent by increasing the propeller blade angles. If the descent is too rapid or is being made from a high altitude, the maximum blade angle limit of the blades is not sufficient to hold the rpm constant. When this occurs, the rpm is responsive to any change in throttle setting.

Although the governor responds quickly to any change in throttle setting, a sudden and large increase in the throttle setting causes a momentary over-speeding of the engine until the blades become adjusted to absorb the increased power. If an emergency demanding full power should arise during approach, the sudden advancing of the throttle causes momentary over-speeding of the engine beyond the rpm for which the governor is adjusted.

Some important points to remember concerning constant speed propeller operation are:

- The red line on the tachometer not only indicates maximum allowable rpm; it also indicates the rpm required to obtain the engine's rated horsepower.

- A momentary propeller overs-peed may occur when the throttle is advanced rapidly for takeoff. This is usually not serious if the rated rpm is not exceeded by 10 percent for more than 3 seconds.

- The green arc on the tachometer indicates the normal operating range. When developing power in this range, the engine drives the propeller. Below the green arc, however, it is usually the windmilling propeller that powers the engine. Prolonged operation below the green arc can be detrimental to the engine. On takeoffs from low elevation airports, the manifold pressure in inches of mercury may exceed the rpm. This is normal in most cases, but the pilot should always consult the AFM/POH for limitations.

- All power changes should be made smoothly and slowly to avoid over-boosting and/or over-speeding.

Turbocharging

The turbocharged engine allows the pilot to maintain sufficient cruise power at high altitudes where there is less drag, which means faster true airspeeds and increased range with fuel economy. At the same time, the powerplant has flexibility and can be flown at a low altitude without the increased fuel consumption of a turbine engine. When attached to the standard powerplant, the turbocharger does not take any horsepower from the engine to operate; it is relatively simple mechanically, and some models can pressurize the cabin as well.

The turbocharger is an exhaust-driven device that raises the pressure and density of the induction air delivered to the engine. It consists of two separate components: a compressor and a turbine connected by a common shaft. The compressor supplies pressurized air to the engine for high-altitude operation. The compressor and its housing are between the ambient air intake and the induction air manifold. The turbine and its housing are part of the exhaust system and utilize the flow of exhaust gases to drive the compressor. *[Figure 12-9]*

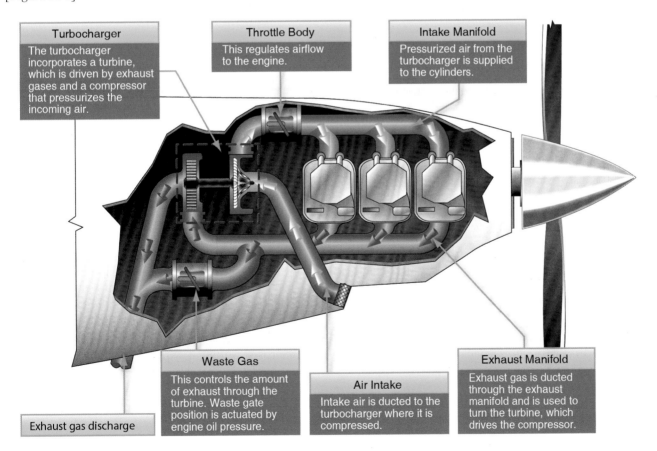

Figure 12-9. *Turbocharging system.*

The turbine has the capability of producing manifold pressure in excess of the maximum allowable for the particular engine. In order not to exceed the maximum allowable manifold pressure, a bypass or waste gate is used so that some of the exhaust is diverted overboard before it passes through the turbine.

The position of the waste gate regulates the output of the turbine and therefore, the compressed air available to the engine. When the waste gate is closed, all of the exhaust gases pass through and drive the turbine. As the waste gate opens, some of the exhaust gases are routed around the turbine through the exhaust bypass and overboard through the exhaust pipe.

The waste gate actuator is a spring-loaded piston operated by engine oil pressure. The actuator, which adjusts the waste gate position, is connected to the waste gate by a mechanical linkage.

The control center of the turbocharger system is the pressure controller. This device simplifies turbocharging to one control: the throttle. Once the desired manifold pressure is set, virtually no throttle adjustment is required with changes in altitude. The controller senses compressor discharge requirements for various altitudes and controls the oil pressure to the waste gate actuator, which adjusts the waste gate accordingly. Thus the turbocharger will maintain the manifold pressure called for by the throttle setting.

Ground Boosting Versus Altitude Turbocharging

Altitude turbocharging (sometimes called "normalizing") is accomplished by using a turbocharger that maintains maximum allowable sea level manifold pressure (normally 29–30 "Hg) up to a certain altitude. This altitude is specified by the airplane manufacturer and is referred to as the airplane's critical altitude. Above the critical altitude, the manifold pressure decreases as additional altitude is gained. Ground boosting, on the other hand, is an application of turbocharging where more than the standard 29 inches of manifold pressure is used in flight. In various airplanes using ground boosting, takeoff manifold pressures may go as high as 45 "Hg.

Although a sea-level manifold pressure setting and maximum rpm can be maintained up to the critical altitude, the engine may not be developing sea-level power. Because the turbocharged induction air is heated by compression, lower induction air density causes a loss of engine power. Maintaining the equivalent horsepower output requires a somewhat higher manifold pressure at a given altitude than if the induction air were not compressed and heated by turbocharging. If, on the other hand, the system incorporates an automatic density controller, which automatically positions the waste gate so as to maintain constant air density to the engine, a near equivalent to sea-level horsepower output results.

Operating Characteristics

First and foremost, all movements of the power controls on turbocharged engines should be slow and smooth. Aggressive or abrupt throttle movements increase the possibility of over-boosting. Carefully monitor engine indications when making power changes.

When the waste gate is open, the turbocharged engine reacts the same as a normally aspirated engine when the rpm is varied. That is, when the rpm is increased, the manifold pressure decreases slightly. When the engine rpm is decreased, the manifold pressure increases slightly. However, when the waste gate is closed, manifold pressure variation with engine rpm is just the opposite of the normally aspirated engine. An increase in engine rpm results in an increase in manifold pressure, and a decrease in engine rpm results in a decrease in manifold pressure.

Above the critical altitude, where the waste gate is closed, any change in airspeed results in a corresponding change in manifold pressure. This is true because the increase in ram air pressure with an increase in airspeed is magnified by the compressor resulting in an increase in manifold pressure. The increase in manifold pressure creates a higher mass flow through the engine, causing higher turbine speeds and thus further increasing manifold pressure.

When running at high altitudes, aviation gasoline tends to vaporize prior to reaching the cylinder. If this occurs in the portion of the fuel system between the fuel tank and the engine-driven fuel pump, an auxiliary positive pressure pump may be needed in the tank. Since engine-driven pumps pull fuel, they are easily vapor locked. A boost pump provides positive pressure, which pushes the fuel and reduces the tendency to vaporize.

Heat Management

Turbocharged engines should be thoughtfully and carefully operated with continuous monitoring of pressures and temperatures. There are two temperatures that are especially important—turbine inlet temperature (TIT) or, in some installations, exhaust gas temperature (EGT) and cylinder head temperature. TIT or EGT limits are set to protect the elements in the hot section of the turbocharger, while cylinder head temperature limits protect the engine's internal parts.

Due to the heat of compression of the induction air, a turbocharged engine runs at higher operating temperatures than a non-turbocharged engine. Because turbocharged engines operate at high altitudes, their environment is less efficient for cooling. At altitude, the air is less dense and, therefore, cools less efficiently. Also, the less dense air causes the compressor to work harder. Compressor turbine speeds can reach 80,000–100,000 rpm, adding to the overall engine operating temperatures. Turbocharged engines are also operated at higher power settings a greater portion of the time.

High heat is detrimental to piston engine operation. Its cumulative effects can lead to piston, ring, and cylinder head failure and place thermal stress on other operating components. Excessive cylinder head temperature can lead to detonation, which in turn can cause catastrophic engine failure. Turbocharged engines are especially heat sensitive. The key to turbocharger operation is effective heat management.

Monitor the condition of a turbocharged engine with manifold pressure gauge, tachometer, exhaust gas temperature/turbine inlet temperature gauge, and cylinder head temperature gauge. Manage the "heat system" with the throttle, propeller rpm, mixture, and cowl flaps. At any given cruise power, the mixture is the most influential control over the exhaust gas/TIT. The throttle regulates total fuel flow, but the mixture governs the fuel-to-air ratio. The mixture, therefore, controls temperature.

Exceeding temperature limits in an after-takeoff climb is usually not a problem since a full rich mixture cools with excess fuel. At cruise, power is normally reduced and mixture adjusted accordingly. Under cruise conditions, monitor temperature limits closely because that is when the temperatures are most likely to reach the maximum, even though the engine is producing less power. Overheating in an en route climb, however, may require fully open cowl flaps and a higher airspeed.

Since turbocharged engines operate hotter at altitude than normally aspirated engines, they are more prone to damage from cooling stress. Gradual reductions in power and careful monitoring of temperatures are essential in the descent phase. Extending the landing gear during the descent may help control the airspeed while maintaining a higher engine power setting. This allows the pilot to reduce power in small increments which allows the engine to cool slowly. It may also be necessary to lean the mixture slightly to eliminate roughness at the lower power settings.

Turbocharger Failure

Because of the high temperatures and pressures produced in the turbine exhaust system, any malfunction of the turbocharger should be treated with extreme caution. In all cases of turbocharger operation, the manufacturer's recommended procedures should be followed. This is especially so in the case of turbocharger malfunction. However, in those instances where the manufacturer's procedures do not adequately describe the actions to be taken in the event of a turbocharger failure, the following procedures should be used.

Over-Boost Condition

If an excessive rise in manifold pressure occurs during normal advancement of the throttle (possibly owing to faulty operation of the waste gate):

- Immediately retard the throttle smoothly to limit the manifold pressure below the maximum for the rpm
  and mixture setting.
- Operate the engine in such a manner as to avoid a further over-boost condition.

Low Manifold Pressure

Although this condition may be caused by a minor fault, it is quite possible that a serious exhaust leak has occurred creating a potentially hazardous situation:

- Shut down the engine in accordance with the recommended engine failure procedures, unless a greater
  emergency exists that warrants continued engine operation.

- If continuing to operate the engine, use the lowest power setting demanded by the situation and land as
  soon as practicable.

It is very important to ensure that corrective maintenance is undertaken following any turbocharger malfunction.

Retractable Landing Gear

The primary benefits of being able to retract the landing gear are increased climb performance and higher cruise airspeeds due to a decrease in drag after gear retraction. Retractable landing gear systems may be operated either hydraulically or electrically or may employ a combination of the two systems. Warning indicators are provided in the flight deck to show the pilot when the wheels are down and locked and when they are up and locked or if they are in intermediate positions. Systems for emergency operation are also provided. Due to the complexity of a retractable landing gear system, the pilot should adhere to specific operating procedures and should not exceed any operating limitations.

Landing Gear Systems

An electrical landing gear retraction system utilizes an electrically-driven motor for gear operation. The system is basically an electrically-driven jack for raising and lowering the gear. When a switch in the flight deck is moved to the UP position, the electric motor operates. Through a system of shafts, gears, adapters, an actuator screw, and a torque tube, a force is transmitted to the drag strut linkages. Thus, the gear retracts and locks. Struts are also activated that open and close the gear doors. If the switch is moved to the DOWN position, the motor reverses and the gear moves down and locks. Once activated, the gear motor continues to operate until an up or down limit switch on the motor's gearbox is tripped.

A hydraulic landing gear retraction system utilizes pressurized hydraulic fluid to actuate linkages to raise and lower the gear. When a switch in the flight deck is moved to the UP position, hydraulic fluid is directed into the gear up line. The fluid flows through sequenced valves and downlocks to the gear actuating cylinders. A similar process occurs during gear extension. The pump that pressurizes the fluid in the system can be either engine-driven or electrically-powered. If an electrically-powered pump is used to pressurize the fluid, the system is referred to as an electrohydraulic system. The system also incorporates a hydraulic reservoir to contain excess fluid and to provide a means of determining system fluid level.

Regardless of its power source, the hydraulic pump is designed to operate within a specific range. When a sensor detects excessive pressure, a relief valve within the pump opens, and hydraulic pressure is routed back to the reservoir. Another type of relief valve prevents excessive pressure that may result from thermal expansion. Hydraulic pressure is also regulated by limit switches. Each gear has two limits switches—one dedicated to extension and one dedicated to retraction. These switches de-energize the hydraulic pump after the landing gear has completed its gear cycle. In the event of limit switch failure, a backup pressure relief valve activates to relieve excess system pressure.

Controls and Position Indicators

Landing gear position is controlled by a switch on the flight deck panel. In most airplanes, the gear switch is shaped like a wheel in order to facilitate positive identification and to differentiate it from other flight deck controls.

Landing gear position indicators vary with different make and model airplanes. Some types of landing gear position indicators utilize a group of lights. One type consists of one green light to indicate when the landing gear is down and an amber light to indicate when the gear is up. *[Figure 12-10]* Another type consists of a group of three green lights, which illuminate when the landing gear is down and locked. *[Figure 12-10]* Still other systems incorporate a red or amber light to indicate when the gear is in transit or unsafe for landing. *[Figure 12-11]* When the lights use a "press to test" feature, the bulbs are often interchangeable. Integrated electronic displays may also indicate gear position on a portion of the screen without any dedicated lights.

Other types of landing gear position indicators consist of tab-type indicators with markings "UP" to indicate the gear is up and locked, a display of red and white diagonal stripes to show when the gear is unlocked, or a silhouette of each gear to indicate when it locks in the DOWN position.

Landing Gear Safety Devices

Most airplanes with a retractable landing gear have a gear warning horn that sounds when the airplane is configured for landing and the landing gear is not down and locked. Normally, the horn is linked to the throttle or flap position and/or the airspeed indicator so that when the airplane is below a certain airspeed, configuration, or power setting with the gear retracted, the warning horn sounds.

Accidental retraction of a landing gear may be prevented by such devices as mechanical downlocks, safety switches, and ground locks. Mechanical downlocks are built-in components of a gear retraction system and are operated automatically by the gear retraction system. To prevent accidental operation of the downlocks and inadvertent landing gear retraction while the airplane is on the ground, electrically-operated safety switches are installed.

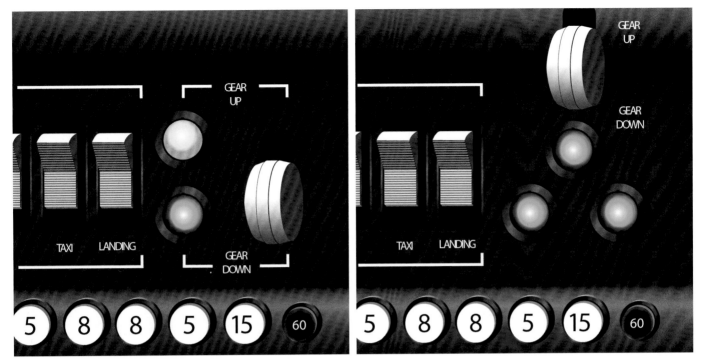

Figure 12-10. *Typical landing gear switch with combination amber and green. Another combination has a three light indicator.*

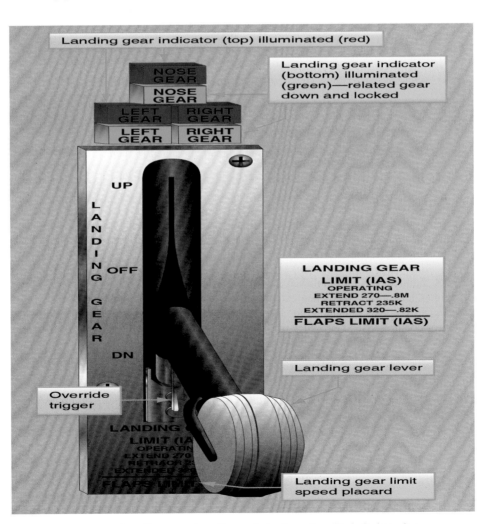

Figure 12-11. *Landing gear handles and single and multiple light indictor.*

A landing gear safety switch, sometimes referred to as a squat switch, is usually mounted in a bracket on one of the main gear shock struts. *[Figure 12-12]* When the strut is compressed by the weight of the airplane, the switch opens the electrical circuit to the motor or mechanism that powers retraction. In this way, if the landing gear switch in the flight deck is placed in the RETRACT position when weight is on the gear, the gear remains extended, and the warning horn may sound as an alert to the unsafe condition. Once the weight is off the gear, however, such as on takeoff, the safety switch releases and the gear retracts.

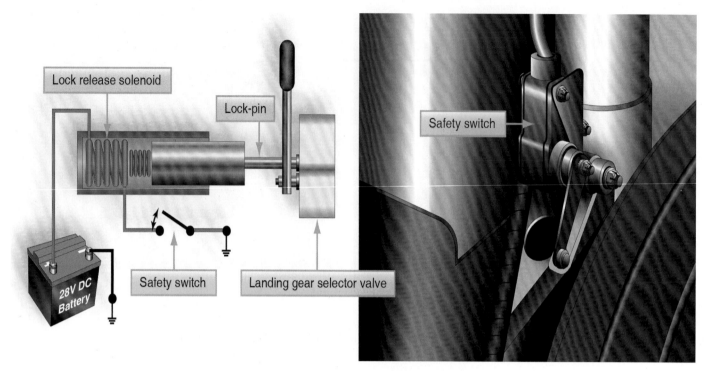

Figure 12-12. *Landing gear safety switch.*

Many airplanes are equipped with removable safety devices to prevent collapse of the gear when the airplane is on the ground. These devices are called ground locks. One common type is a pin installed in aligned holes drilled in two or more units of the landing gear support structure. Another type is a spring-loaded clip designed to fit around and hold two or more units of the support structure together. All types of ground locks usually have red streamers permanently attached to them to readily indicate whether or not they are installed.

Emergency Gear Extension Systems

The emergency gear extension system lowers the landing gear if the main power system fails. Some airplanes have an emergency release handle in the flight deck, which is connected through a mechanical linkage to the gear uplocks. When the handle is operated, it releases the uplocks and allows the gear to free fall or extend under their own weight. *[Figure 12-13]*

On other airplanes, release of the uplock is accomplished using compressed gas, which is directed to uplock release cylinders. In some airplanes, design configurations make emergency extension of the landing gear by gravity and air loads alone impossible or impractical. In these airplanes, provisions are included for forceful gear extension in an emergency. Some installations are designed so that either hydraulic fluid or compressed gas provides the necessary pressure, while others use a manual system, such as a hand crank for emergency gear extension. *[Figure 12-14]* Hydraulic pressure for emergency operation of the landing gear may be provided by an auxiliary hand pump, an accumulator, or an electrically-powered hydraulic pump depending on the design of the airplane.

Operational Procedures
Preflight

Because of their complexity, retractable landing gear demands a close inspection prior to every flight. The inspection should begin inside the flight deck. First, make certain that the landing gear selector switch is in the GEAR DOWN position. Then, turn on the battery master switch and ensure that the landing gear position indicators show that the gear is down and locked.

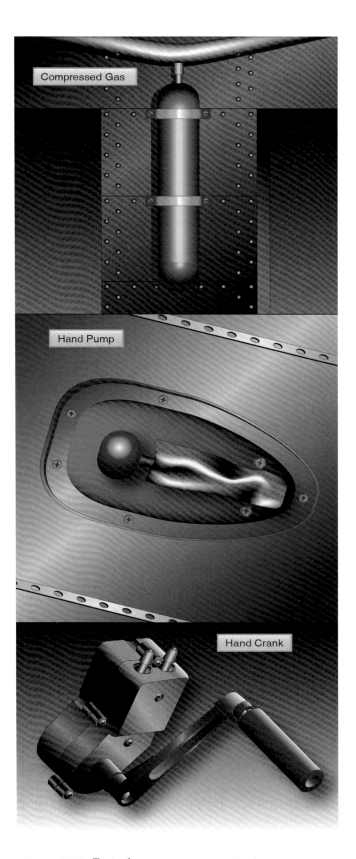

Figure 12-13. *Typical emergency gear extension systems.*

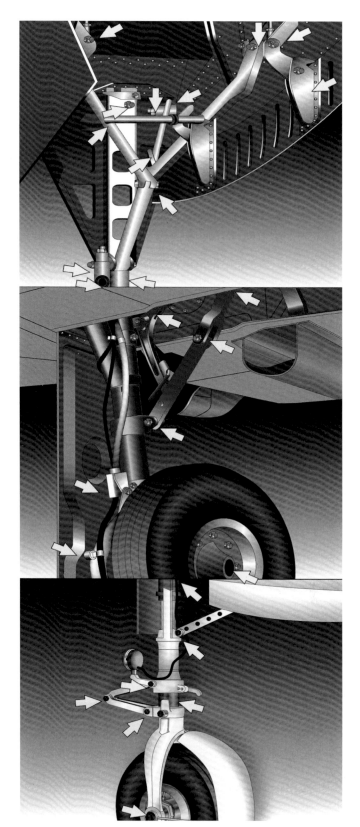

Figure 12-14. *Retractable landing gear inspection checkpoints.*

External inspection of the landing gear consists of checking individual system components. *[Figure 12-14]* The landing gear, wheel well, and adjacent areas should be clean and free of mud and debris. Dirty switches and valves may cause false safe light indications or interrupt the extension cycle before the landing gear is completely down and locked. The wheel wells should be clear of any obstructions, as foreign objects may damage the gear or interfere with its operation. Bent gear doors may be an indication of possible problems with normal gear operation.

Ensure shock struts are properly inflated and that the pistons are clean. Check main gear and nose gear uplock and downlock mechanisms for general condition. Power sources and retracting mechanisms are checked for general condition, obvious defects, and security of attachment. Check hydraulic lines for signs of chafing and leakage at attach points. Warning system micro switches (squat switches) are checked for cleanliness and security of attachment. Actuating cylinders, sprockets, universal joints, drive gears, linkages, and any other accessible components are checked for condition and obvious defects. The airplane structure to which the landing gear is attached is checked for distortion, cracks, and general condition. All bolts and rivets should be intact and secure.

Takeoff and Climb

Normally, the landing gear is retracted after lift-off when the airplane has reached an altitude where, in the event of an engine failure or other emergency requiring an aborted takeoff, the airplane could no longer be landed on the runway. This procedure, however, may not apply to all situations. Preplan landing gear retraction taking into account the following:

- Length of the runway
- Climb gradient
- Obstacle clearance requirements
- The characteristics of the terrain beyond the departure end of the runway
- The climb characteristics of the particular airplane

For example, in some situations it may be preferable, in the event of an engine failure, to make an off airport forced landing with the gear extended in order to take advantage of the energy absorbing qualities of the terrain (see Chapter 18, "Emergency Procedures"). In which case, a delay in retracting the landing gear after takeoff from a short runway may be warranted. In other situations, obstacles in the climb path may warrant a timely gear retraction after takeoff. Also, in some airplanes the initial climb pitch attitude is such that any view of the runway remaining is blocked, making an assessment of the feasibility of touching down on the remaining runway difficult.

Avoid premature landing gear retraction and do not retract the landing gear until a positive rate of climb is indicated on the flight instruments. If the airplane has not attained a positive rate of climb, there is always the chance it may settle back onto the runway with the gear retracted. This is especially so in cases of premature lift-off. Remember that leaning forward to reach the landing gear selector may result in inadvertent forward pressure on the yoke, which causes the airplane to descend.

As the landing gear retracts, airspeed increases and the airplane's pitch attitude may change. The gear may take several seconds to retract. Gear retraction and locking (and gear extension and locking) is accompanied by sound and feel that are unique to the specific make and model airplane. Become familiar with the sound and feel of normal gear retraction so that any abnormal gear operation can be readily recognized. Abnormal landing gear retraction is most often a clear sign that the gear extension cycle will also be abnormal.

Approach and Landing

The operating loads placed on the landing gear at higher airspeeds may cause structural damage due to the forces of the airstream. Limiting speeds, therefore, are established for gear operation to protect the gear components from becoming overstressed during flight. These speeds may not be found on the airspeed indicator. They are published in the AFM/POH for the particular airplane and are usually listed on placards in the flight deck. [*Figure 12-15*] The maximum landing extended speed (V_{LE}) is the maximum speed at which the airplane can be flown with the landing gear extended. The maximum landing gear operating speed (V_{LO}) is the maximum speed at which the landing gear may be operated through its cycle.

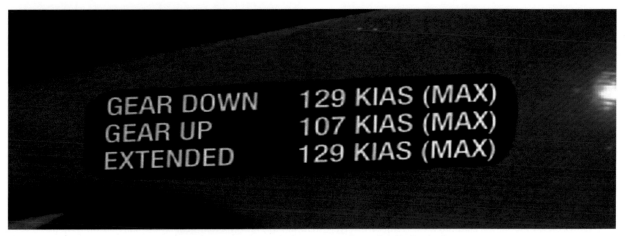

Figure 12-15. *Placarded gear speeds in the flight deck.*

The landing gear is extended by placing the gear selector switch in the GEAR DOWN position. As the landing gear extends, the airspeed decreases and the pitch attitude may change. During the several seconds it takes for the gear to extend, be attentive to any abnormal sounds or feel. Confirm that the landing gear has extended and locked by the normal sound and feel of the system operation, as well as by the gear position indicators in the flight deck. Unless the landing gear has been previously extended to aid in a descent to traffic pattern altitude, the landing gear should be extended by the time the airplane reaches a point on the downwind leg that is opposite the point of intended landing. Establish a standard procedure consisting of a specific position on the downwind leg at which to lower the landing gear. Strict adherence to this procedure aids in avoiding unintentional gear up landings.

Operation of an airplane equipped with a retractable landing gear requires the deliberate, careful, and continued use of an appropriate checklist. When on the downwind leg, make it a habit to complete the before-landing checklist for that airplane. This accomplishes two purposes—it ensures that action has been taken to lower the gear and establishes awareness so that the gear down indicators can be rechecked prior to landing.

Unless good operating practices dictate otherwise, the landing roll should be completed and the airplane should be clear of the runway before any levers or switches are operated. This technique greatly reduces the chance of inadvertently retracting the landing gear while on the ground. Wait until after rollout and clearing the runway to focus attention on the after-landing checklist. This practice allows for positive identification of the proper controls.

When transitioning to retractable gear airplanes, it is important to consider some frequent pilot errors. These include pilots that have:

- Neglected to extend landing gear
- Inadvertently retracted landing gear
- Activated gear but failed to check gear position
- Misused emergency gear system
- Retracted gear prematurely on takeoff
- Extended gear too late

These mistakes are not only committed by pilots who have just transitioned to complex aircraft, but also by pilots who have developed a sense of complacency over time. In order to minimize the chances of a landing gear-related mishap:

- Use an appropriate checklist. (A condensed checklist mounted in view is a reminder for its use and easy reference can be especially helpful.)

- Be familiar with, and periodically review, the landing gear emergency extension procedures for the particular airplane.

- Be familiar with the landing gear warning horn and warning light systems for the particular airplane. Use the horn system to cross-check the warning light system when an unsafe condition is noted.

- Review the procedure for replacing light bulbs in the landing gear warning light displays for the particular airplane, if applicable, so that you can properly replace a bulb to determine if the bulb(s) in the display is good. If bulbs are replaceable, check to see if spare bulbs are available in the airplane spare bulb supply as part of the preflight inspection.

- Be familiar with and aware of the sounds and feel of a properly operating landing gear system.

Transition Training

Transition to a complex airplane or a high-performance airplane should be accomplished through a structured course of training administered by a competent and qualified flight instructor. The training should be accomplished in accordance with a ground and flight training syllabus. *[Figure 12-16]*

This sample syllabus for transition training is an example. The arrangement of the subject matter may be changed and the emphasis shifted to fit the qualifications of the transitioning pilot, the airplane involved, and the circumstances of the training situation. The goal is to ensure proficiency standards are achieved. These standards are contained in the Airman Certification Standards for the certificate that the transitioning pilot holds or is working toward.

Ground Instruction	Flight Instruction
One hour	One hour
1. Operations sections of flight manual	1. Flight training maneuvers
2. Line inspection	2. Takeoffs, landings and go-arounds
3. Flight deck familiarization	
One hour	One hour
1. Aircraft loading, limitations and servicing	1. Emergency operations
2. Instruments, radio and special equipment	2. Control by reference to instruments
3. Aircraft systems	3. Use of radio and autopilot
One hour	One hour
1. Performance section of flight manual	1. Short and soft-field takeoffs and landings
2. Cruise control	2. Maximum performance operations
3. Review	

Figure 12-16. *Sample transition training syllabus.*

The training times indicated in the syllabus are for illustration purposes. Actual times should be based on the capabilities of the pilot. The time periods may be minimal for pilots with higher qualifications or increased for pilots who do not meet certification requirements or have had little recent flight experience.

Chapter Summary

Flying a complex or high-performance airplane requires a pilot to further divide his or her attention during the most critical phases of flight: takeoff and landing. The knowledge, judgment, and piloting skills required to fly these airplanes needs to be developed. It is essential that adequate training is received to ensure a complete understanding of the systems, their operation (both normal and emergency), and operating limitations.

Airplane Flying Handbook (FAA-H-8083-3C)
Chapter 13: Transition to Multiengine Airplanes

Introduction

This chapter is devoted to the factors associated with the operation of small multiengine airplanes. For the purpose of this handbook, a "small" multiengine airplane is a reciprocating or turbopropeller-powered airplane with a maximum certificated takeoff weight of 12,500 pounds or less. This discussion assumes a conventional design with two engines—one mounted on each wing. Reciprocating engines are assumed unless otherwise noted. The term "light-twin," although not formally defined in the regulations, is used herein as a small multiengine airplane with a maximum certificated takeoff weight of 6,000 pounds or less.

There are several unique characteristics of multiengine airplanes that make them worthy of a separate class rating. The one engine inoperative (OEI) flight information presented in this chapter emphasizes the significant difference between flying a multiengine and a single-engine airplane. However, all pilots need appropriate knowledge, risk management strategies, and skills to fly safely in any airplane they fly, and mastery of OEI flight is only one aspect of safe multiengine flying. The modern, well-equipped multiengine airplane can be remarkably capable under many circumstances, but, the performance and system redundancy of a multiengine airplane only increase safety if the pilot is trained and proficient.

The airplane manufacturer is the final authority on the operation of a particular make and model airplane. Flight instructors and learners should use the Federal Aviation Administration's Approved Flight Manual (AFM) and/or the Pilot's Operating Handbook (POH). The airplane manufacturer's guidance and procedures take precedence over any general recommendations made in this handbook.

General

Multiengine and single-engine airplanes operate differently during an engine failure. In a multiengine airplane, loss of thrust from one engine affects both *performance and control*. The most obvious problem is the loss of 50 percent of power, which reduces climb performance 80 to 90 percent. In some cases after an engine failure, the ability to climb or maintain altitude in a light-twin may not exist. After an engine failure, asymmetrical thrust also creates control issues for the pilot. Attention to both these factors is crucial to safe OEI flight.

Terms and Definitions

Pilots of single-engine airplanes are already familiar with many performance "V" speeds and their definitions. Twin-engine airplanes have several additional V-speeds unique to OEI operation. These speeds are differentiated by the notation "SE" for single engine. A review of some key V-speeds and several new V-speeds unique to twin-engine airplanes are listed below.

- V_R—rotation speed—speed at which back pressure is applied to rotate the airplane to a takeoff attitude.

- V_{LOF}—lift-off speed—speed at which the airplane leaves the surface. (Note: Some manufacturers reference takeoff performance data to V_R, others to V_{LOF}.)

- V_X—best angle of climb speed—speed at which the airplane gains the greatest altitude for a given distance of forward travel.

- V_{XSE}—best angle-of-climb speed with OEI.

- V_Y—best rate of climb speed—speed at which the airplane gains the most altitude for a given unit of time.

- V_{YSE}—best rate of climb speed with OEI. Marked with a blue radial line on most airspeed indicators. Above the single-engine absolute ceiling, V_{YSE} yields the minimum rate of sink.

- V_{SSE}—safe, intentional OEI speed—originally known as safe single-engine speed. It is the minimum speed to intentionally render the critical engine inoperative.

- V_{REF}—reference landing speed—an airspeed used for final approach, which is normally 1.3 times V_{SO}, the stall speed in the landing configuration. The pilot may adjust the approach speed for winds and gusty conditions by using V_{REF} plus an additional number of units (e.g., V_{REF}+5).

- V_{MC}—currently defined in 14 CFR part 23, section 23.2135(c) as the calibrated airspeed at which, following the sudden critical loss of thrust, it is possible to maintain control of the airplane. V_{MC} is typically marked with a red radial line on most airspeed indicators [*Figure 13-1*]. V_{MC} was previously defined in 14 CFR part 23, section 23.149 as the calibrated airspeed at which, when the critical engine is suddenly made inoperative, it is possible to maintain control of the airplane with that engine still inoperative, and thereafter maintain straight flight at the same speed with an angle of bank of not more than 5 degrees. This definition still applies to airplanes certified under that regulation. There is no requirement under either determination that the airplane be capable of climbing at this airspeed. V_{MC} only addresses directional control. Further discussion of V_{MC} as determined during airplane certification and demonstrated in pilot training follows later in this chapter.

Figure 13-1. *Airspeed indicator markings for a multiengine airplane*

Unless otherwise noted, when V-speeds are given in the AFM/POH, they apply to sea level, standard day conditions at maximum takeoff weight. Performance speeds vary with aircraft weight, configuration, and atmospheric conditions. The speeds may be stated in statute miles per hour (mph) or knots (kt), and they may be given as calibrated airspeeds (CAS) or indicated airspeeds (IAS). As a general rule, the newer AFM/POHs show V-speeds in knots indicated airspeed (KIAS). Some V-speeds are also stated in knots calibrated airspeed (KCAS) to meet certain regulatory requirements. Whenever available, pilots should operate the airplane from published indicated airspeeds.

Rate of climb is the altitude gain per unit of time, while climb gradient is the actual measure of altitude gained per 100 feet of horizontal travel, expressed as a percentage. An altitude gain of 1.5 feet per 100 feet of travel (or 15 feet per 1,000 or 150 feet per 10,000) is a climb gradient of 1.5 percent.

There is a dramatic performance loss associated with the loss of an engine, particularly just after takeoff. Any airplane's climb performance is a function of thrust horsepower, which is in excess of that required for level flight. In a hypothetical twin with each engine producing 200 thrust horsepower, assume that the total level flight thrust horsepower required is 175. In this situation, the airplane would ordinarily have a reserve of 225 thrust horsepower available for climb. Loss of one engine would leave only 25 (200 minus 175) thrust horsepower available for climb, a drastic reduction.

The performance characteristics of an airplane depend upon the rules in effect during type certification and do not depend on the production year after certification. The current amendment to 14 CFR part 23, 81 FR 96689, went into effect on December 30, 2016. This includes certification of normal category airplanes with passenger seating configuration of 19 or less and a maximum certificated takeoff weight of 19,000 pounds or less (section 23.2005(a)). Current 14 CFR part 23 certification rules (section 23.2005(b)) classify airplanes into certification levels 1 through 4 based on maximum passenger seating configuration. For example, a level 2 airplane has a passenger seating configuration between two and six passengers. The rule further divides airplanes into two different performance levels based on speed (section 23.2005(c)). After a critical loss of thrust, a level 2 low speed airplane (V_{NO} or V_{MO} less than or equal to 250 knots calibrated airspeed and M_{MO} less than or equal to 0.6) that does not meet single-engine crashworthiness requirements requires a climb gradient of at least 1.5 percent at a pressure altitude of 5,000 feet in the cruise configuration for certification (section 23.2120(b)(1)).

While, the various subsets of airplanes receiving certification under the current part 23 meet specific single-engine climb performance criteria as listed in 14 CFR part 23, section 23.2120(b), the historical 14 CFR part 23 single-engine climb performance requirements for reciprocating engine-powered multiengine airplanes are broken down as follows:

- More than 6,000 pounds maximum weight and/or V_{SO} more than 61 knots: the single-engine rate of climb in feet per minute (fpm) at 5,000 feet mean sea level (MSL) must be equal to at least $0.027 V_{SO} 2$. For airplanes type certificated February 4, 1991, or thereafter, the climb requirement is expressed in terms of a climb gradient, 1.5 percent. The climb gradient is not a direct equivalent of the $.027 V_{SO} 2$ formula. Do not confuse the date of type certification with the airplane's model year. The type certification basis of many multiengine airplanes dates back to the Civil Aviation Regulations (CAR) 3.

- 6,000 pounds or less maximum weight and V_{SO} 61 knots or less: the single-engine rate of climb at 5,000 feet MSL must simply be determined. The rate of climb could be a negative number. There is no requirement for a single-engine positive rate of climb at 5,000 feet or any other altitude. For light-twins type certificated February 4, 1991, or thereafter, the single-engine climb gradient (positive or negative) is simply determined.

Operation of Systems

This section deals with systems and equipment that are generally installed in multiengine airplanes. Multiengine airplanes share many features with complex single-engine airplanes. However, there are certain features that are found more often in airplanes with two or more engines.

Feathering Propellers

Although the propellers of a multiengine airplane may appear identical to a constant-speed propeller used in many single-engine airplanes, this is usually not the case. The pilot of a typical multiengine airplane can feather the propeller of an inoperative engine. Since it stops engine rotation with the propeller blade streamlined with the airplane's relative wind, feathering the propeller of an inoperative engine minimizes propeller drag. *[Figure 13-2]* Depending upon single-engine performance, this feature often permits continued flight to a suitable airport following an engine failure.

Feathering is important because of the change in parasite drag with propeller blade angle. *[Figure 13-3]* When the propeller blade angle is in the feathered position, parasite drag from the propeller is at a minimum. In a typical multiengine airplane, the parasite drag from a single, feathered propeller is a small part the airplane's total drag.

At the smaller blade angles near the flat pitch position, the drag added by the propeller is large. At these small blade angles, the propeller windmilling at high revolutions per minute (rpm) can create enough drag to make the airplane difficult or impossible to control. A propeller windmilling at high speed in the low range of blade angles can produce parasite drag as great as the parasite drag of the entire airframe.

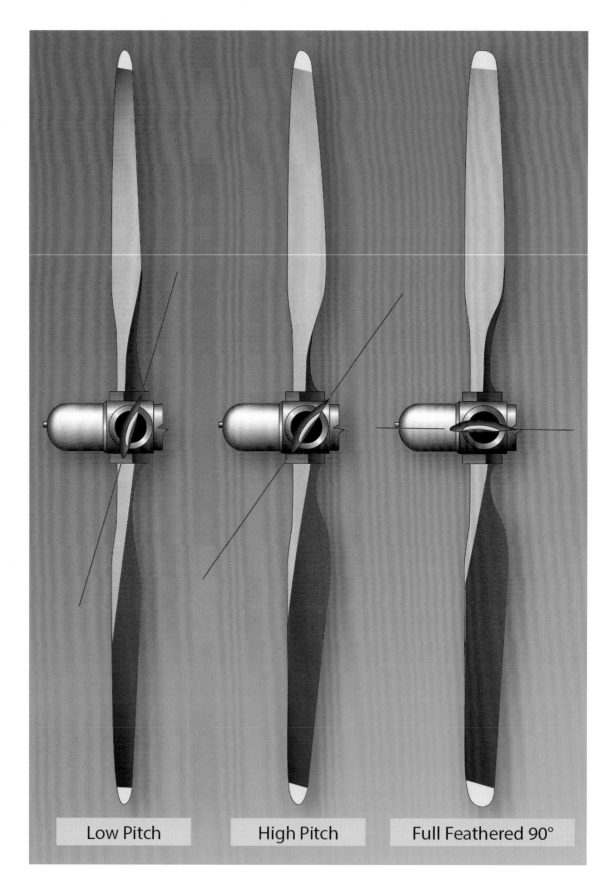

Figure 13-2. *Feathered propeller.*

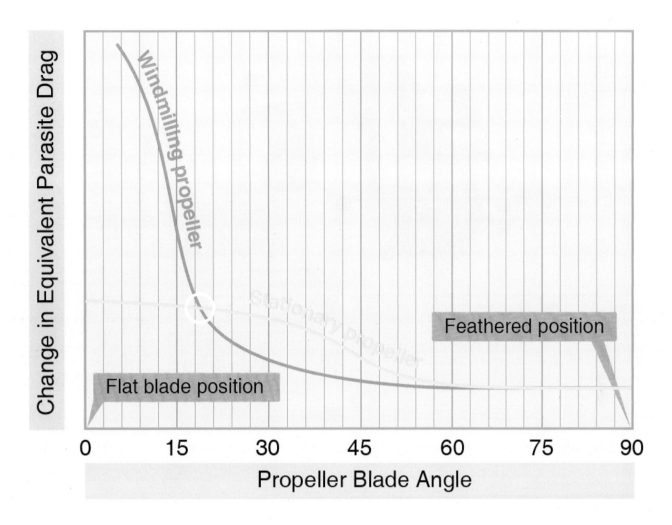

Figure 13-3. *Propeller drag contribution.*

As a review, the constant-speed propellers on almost all single-engine airplanes are of the non-feathering, oil-pressure-to-increase-pitch design. In this design, increased oil pressure from the propeller governor drives the blade angle towards high pitch, low rpm.

In contrast, the constant-speed propellers installed on most multiengine airplanes are full feathering, counterweighted, oil-pressure-to-decrease-pitch designs. In this design, increased oil pressure from the propeller governor drives the blade angle toward low pitch, high rpm—away from the feather blade angle. In effect, the only thing that keeps these propellers from feathering is a constant supply of high-pressure engine oil. This is a necessity to enable propeller feathering in the event of a loss of oil pressure or a propeller governor failure.

Aerodynamic forces acting upon a windmilling propeller tend to drive the blades to low pitch, high rpm. Counterweights attached to the shank of each blade tend to force the blades to high pitch, low rpm. Inertia, or the apparent force (called centrifugal force) acting through the counterweights, is generally slightly greater than the aerodynamic forces. Therefore, centrifugal force would drive the blades to high pitch and low rpm were it not for an additional force acting through the propeller governor. A controlling force generated from high pressure oil from the propeller governor pushes the propeller blade angles toward low pitch and high rpm. Thus, a reduction in oil pressure allows the counterweights to drive the blades to a higher pitch and decreases engine rpm. *[Figure 13-4]*

To feather the propeller, the propeller control is brought fully aft. All oil pressure is dumped from the governor, and the counterweights drive the propeller blades toward feather. As centrifugal force acting on the counterweights decays from decreasing rpm, additional forces are needed to completely feather the blades. This additional force comes from either a spring or high-pressure air stored in the propeller dome, which forces the blades into the feathered position. The entire process may take up to 10 seconds.

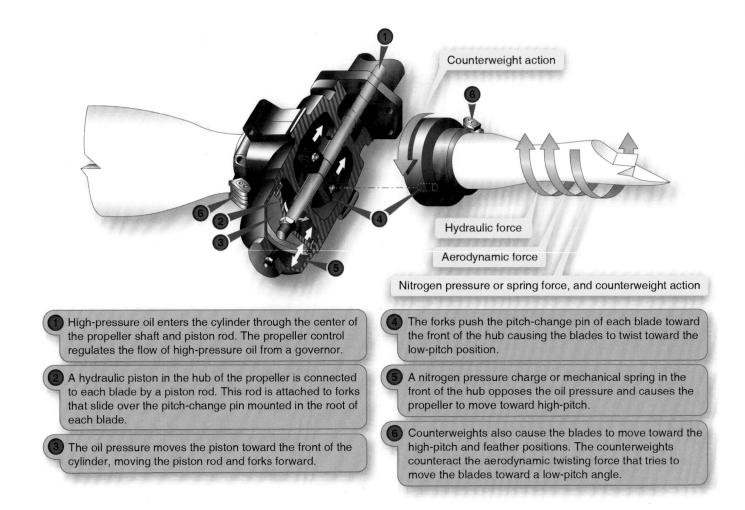

Counterweight action

Hydraulic force

Aerodynamic force

Nitrogen pressure or spring force, and counterweight action

① High-pressure oil enters the cylinder through the center of the propeller shaft and piston rod. The propeller control regulates the flow of high-pressure oil from a governor.

② A hydraulic piston in the hub of the propeller is connected to each blade by a piston rod. This rod is attached to forks that slide over the pitch-change pin mounted in the root of each blade.

③ The oil pressure moves the piston toward the front of the cylinder, moving the piston rod and forks forward.

④ The forks push the pitch-change pin of each blade toward the front of the hub causing the blades to twist toward the low-pitch position.

⑤ A nitrogen pressure charge or mechanical spring in the front of the hub opposes the oil pressure and causes the propeller to move toward high-pitch.

⑥ Counterweights also cause the blades to move toward the high-pitch and feather positions. The counterweights counteract the aerodynamic twisting force that tries to move the blades toward a low-pitch angle.

Figure 13-4. *Pitch change forces.*

Feathering a propeller only alters blade angle and stops engine rotation. To completely secure the engine, the pilot turns off the fuel (mixture, electric boost pump, and fuel selector), ignition, alternator/generator, and closes the cowl flaps. If the airplane is pressurized, there may also be an air bleed to close for the failed engine. Some airplanes are equipped with firewall shutoff valves that secure several of these systems with a single switch.

Completely securing a failed engine may not be necessary or even desirable depending upon the failure mode, altitude, and time available. The position of the fuel controls, ignition, and alternator/generator switches of the failed engine has no effect on aircraft performance, and the pilot might manipulate the incorrect switch under conditions of haste or pressure.

To unfeather a propeller, the engine should be rotated so that oil pressure can be generated to move the propeller blades from the feathered position. The ignition is turned on prior to engine rotation with the throttle at low idle and the mixture rich. With the propeller control in a high rpm position, the starter is engaged. The engine begins to windmill, start, and run as oil pressure moves the blades out of feather. As the engine starts, the propeller rpm should be immediately reduced until the engine has had several minutes to warm up; the pilot should monitor cylinder head and oil temperatures.

An unfeathering accumulator is a device that permits starting a feathered engine in-flight without the use of the electric starter. An accumulator is any device that stores a reserve of high pressure. On multiengine airplanes, the unfeathering accumulator stores a small reserve of engine oil under pressure from compressed air or nitrogen. To start a feathered engine in-flight, the pilot moves the propeller control out of the feather position to release the accumulator pressure. The oil flows under pressure to the propeller hub and drives the blades toward the high rpm, low pitch position, whereupon the propeller usually begins to windmill. If fuel and ignition are present, the engine starts and runs. High oil pressure from the propeller governor recharges the accumulator just moments after engine rotation begins making it available for another unfeathering cycle, if needed. For airplanes used in training, an unfeathering accumulator may prolong the life of the electric starter and battery. If the accumulator fails to bring the propeller out of feather, the electric starter may be engaged.

In any event, the AFM/POH procedures should be followed for the exact unfeathering procedure. Both feathering and starting a feathered reciprocating engine on the ground are strongly discouraged by manufacturers due to the excessive stress and vibrations generated.

As just described, a loss of oil pressure from the propeller governor allows the counterweights, spring, and/or dome charge to drive the blades to feather. Logically then, the propeller blades should feather every time an engine is shut down as oil pressure falls to zero. However, below approximately 800 rpm, a reduction in centrifugal force allows small anti-feathering lock pins in the pitch changing mechanism of the propeller hub to move into place and block feathering. Therefore, if a propeller is to be feathered, it needs to be done before engine rpm decays below approximately 800. On one popular model of turboprop engine, the propeller blades do, in fact, feather with each shutdown. This propeller is not equipped with such centrifugally-operated pins due to a unique engine design.

Propeller Synchronization

Many multiengine airplanes have a propeller synchronizer (prop sync) installed to eliminate the annoying "drumming" or "beat" of propellers whose rpm are close, but not precisely the same. To use prop sync, the propeller rpms are coarsely matched by the pilot and the system is engaged. The prop sync adjusts the rpm of the "slave" engine to precisely match the rpm of the "master" engine and then maintains that relationship.

The prop sync should be disengaged when the pilot selects a new propeller rpm and then re-engaged after the new rpm is set. The prop sync should always be off for takeoff, landing, and single-engine operation. The AFM/POH should be consulted for system description and limitations.

A variation on the propeller synchronizer is the propeller synchrophaser. A propeller synchrophaser acts much like a synchronizer to precisely match rpm, but the synchrophaser goes one step further. It not only matches rpm but actually compares and adjusts the positions of the individual blades of the propellers in their arcs. There can be significant propeller noise and vibration reductions with a propeller synchrophaser. From the pilot's perspective, operation of a propeller synchronizer and a propeller synchrophaser are very similar. A synchrophaser is also commonly referred to as prop sync, although that is not entirely correct nomenclature from a technical standpoint.

As a pilot aid to manually synchronizing the propellers, some twins have a small gauge mounted in or by the tachometer(s) with a propeller symbol on a disk that spins. The pilot manually fine tunes the engine rpm so as to stop disk rotation, thereby synchronizing the propellers. This is a useful backup to synchronizing engine rpm using the audible propeller beat. This gauge is also found installed with most propeller synchronizer and synchrophase systems. Some synchrophase systems use a knob for the pilot to control the phase angle.

Fuel Crossfeed

Fuel crossfeed systems are also unique to multiengine airplanes. Using crossfeed, an engine can draw fuel from a fuel tank located in the opposite wing.

On most multiengine airplanes, operation in the crossfeed mode is an emergency procedure used to extend airplane range and endurance in OEI flight. There are a few models that permit crossfeed as a normal, fuel balancing technique in normal operation, but these are not common. The AFM/POH describes crossfeed limitations and procedures that vary significantly among multiengine airplanes.

Checking crossfeed operation on the ground with a quick repositioning of the fuel selectors does nothing more than ensure freedom of motion of the handle. To actually check crossfeed operation, a complete, functional crossfeed system check should be accomplished. To do this, each engine should be operated from its crossfeed position during the run-up. The engines should be checked individually and allowed to run at moderate power (1,500 rpm minimum) for at least 1 minute to ensure that fuel flow can be established from the crossfeed source. Upon completion of the check, each engine should be operated for at least 1 minute at moderate power from the main (takeoff) fuel tanks to reconfirm fuel flow prior to takeoff.

This suggested check is not required prior to every flight. Crossfeed lines are ideal places for water and debris to accumulate unless they are used from time to time and drained using their external drains during preflight. Crossfeed is ordinarily not used for completing a flight with one engine inoperative when an alternate airport is nearby. Pilots should never use crossfeed during takeoff or for normal landing operations with both engines operating. A landing with one engine inoperative using crossfeed may be necessary if setting normal fuel flow would cause the operative engine to fail.

Combustion Heater

Combustion heaters are another common item on multiengine airplanes not found on single-engine airplanes. A combustion heater is best described as a small furnace that burns gasoline to produce heated air for occupant comfort and windshield defogging. Most are thermostatically operated and have a separate hour meter to record time in service for maintenance purposes. Automatic over-temperature protection is provided by a thermal switch mounted on the unit that cannot be accessed in flight. This requires the pilot or mechanic to visually inspect the unit for possible heat damage in order to reset the switch.

Manufacturers often suggest a cool-down period when shutting down a combustion heater. Most heater instructions recommend that outside air be permitted to circulate through the unit for at least 15 seconds in flight or that the ventilation fan can be operated for at least 2 minutes on the ground. Failure to provide an adequate cool down usually trips the thermal switch and renders the heater inoperative until the switch is reset.

Flight Director/Autopilot

Multiengine airplanes are often equipped with flight director/autopilot (FD/AP) systems. The system integrates pitch, roll, heading, altitude, and radio navigation signals in a computer. The outputs, called computed commands, are displayed on a flight command indicator (FCI). The FCI replaces the conventional attitude indicator on the instrument panel. The FCI is occasionally referred to as a flight director indicator (FDI) or as an attitude director indicator (ADI).

The entire flight director/autopilot system is called an integrated flight control system (IFCS) by some manufacturers. Others may use the term automatic flight control system (AFCS).

The FD/AP system may be employed at the following different levels:

- Off (raw data)
- Flight director (computed commands)
- Autopilot

With the system off, the FCI operates as an ordinary attitude indicator. On most FCIs, the command bars are biased out of view when the FD is off. The pilot maneuvers the airplane as though the system were not installed.

To maneuver the airplane using the FD, the pilot enters the desired modes of operation (heading, altitude, navigation (NAV) intercept, and tracking) on the FD/AP mode controller. The computed flight commands are then displayed to the pilot through either a single-cue or dual-cue system in the FCI. On a single-cue system, the commands are indicated by "V" bars. On a dual-cue system, the commands are displayed on two separate command bars, one for pitch and one for roll. To maneuver the airplane using computed commands, the pilot "flies" the symbolic airplane of the FCI to match the steering cues presented.

On most systems, the FD needs to be operating to engage the autopilot. At any time thereafter, the pilot may engage the autopilot through the mode controller. The autopilot then maneuvers the airplane to satisfy the computed commands of the FD.

Like any computer, the FD/AP system only does what it is told. The pilot should ensure that it has been programmed properly for the particular phase of flight desired. The armed and/or engaged modes are usually displayed on the mode controller or separate annunciator lights. When the airplane is being hand-flown, if the FD is not being used at any particular moment, it should be off so that the command bars are pulled from view.

Prior to system engagement, all FD/AP computer and trim checks should be accomplished. Many newer systems cannot be engaged without the completion of a self-test. The pilot should also be familiar with various methods of disengagement, both normal and emergency. System details, including approvals and limitations, can be found in the supplements section of the AFM/POH. Additionally, many avionics manufacturers can provide informative pilot operating guides upon request.

Yaw Damper

The yaw damper is a servo that moves the rudder in response to inputs from a gyroscope or accelerometer that detects yaw rate or lateral Gs, respectively. The yaw damper reduces motion about the vertical axis caused by turbulence. (Yaw dampers on swept wing airplanes provide another, more vital function of damping Dutch roll characteristics.) Occupants feel a smoother ride, particularly if seated in the rear of the airplane, when the yaw damper is engaged. The yaw damper should be off for takeoff and landing. There may be additional restrictions against its use with one engine inoperative. Most yaw dampers can be engaged independently of the autopilot.

Alternator/Generator

On a multiengine aircraft, each engine has an alternator or generator installed. Alternator or generator paralleling circuitry matches the output of each engine's alternator/generator so that the electrical system load is shared equally between them. In the event of an alternator/generator failure, the inoperative unit can be isolated and the entire electrical system powered from the remaining one. Depending upon the electrical capacity of the alternator/generator, the pilot may need to reduce the electrical load (referred to as load shedding) when operating on a single unit. The AFM/POH contains system description and limitations.

Nose Baggage Compartment

Nose baggage compartments are common on multiengine airplanes (and are even found on a few single-engine airplanes). There is nothing strange or exotic about a nose baggage compartment, and the usual guidance concerning observation of load limits applies. Pilots occasionally neglect to secure the latches properly. When improperly secured, the door may open and the contents may be drawn out, usually into the propeller arc and just after takeoff. Even when the nose baggage compartment is empty, airplanes have been lost when the pilot became distracted by the open door. Security of the nose baggage compartment latches and locks is a vital preflight item.

Most airplanes continue to fly with a nose baggage door open. There may be some buffeting from the disturbed airflow, and there is an increase in noise. Pilots should never become so preoccupied with an open door (of any kind) that they fail to fly the airplane.

Inspection of the compartment interior is another important preflight item. More than one pilot has been surprised to find a supposedly empty compartment packed to capacity or loaded with ballast. The tow bars, engine inlet covers, windshield sun screens, oil containers, spare chocks, and miscellaneous small hand tools that find their way into baggage compartments should be secured to prevent damage from shifting in flight.

Anti-Icing/Deicing Equipment

Anti-icing/deicing equipment is frequently installed on multiengine airplanes and may consist of a combination of different systems. These may be classified as either anti-icing or deicing, depending upon function. The presence of anti-icing and deicing equipment, even though it may appear elaborate and complete, does not necessarily mean that the airplane is approved for flight in icing conditions. The AFM/POH, placards, and even the manufacturer should be consulted for specific determination of approvals and limitations. Anti-icing equipment is provided to prevent ice from forming on certain protected surfaces. Examples of anti-icing equipment include heated pitot tubes, heated or non-icing static ports and fuel vents, propeller blades with electrothermal boots or alcohol slingers, windshields with alcohol spray or electrical resistance heating, windshield defoggers, and heated stall warning lift detectors. On many turboprop engines, the "lip" surrounding the air intake is heated either electrically or with bleed air. In the absence of AFM/POH guidance to the contrary, anti-icing equipment should be actuated prior to flight into known or suspected icing conditions.

Deicing equipment is generally limited to pneumatic boots on wing and tail leading edges. Deicing equipment is installed to remove ice that has already formed on protected surfaces. Upon pilot actuation, the boots inflate with air from the pneumatic pumps to break off accumulated ice. After a few seconds of inflation, they are deflated back to their normal position with the assistance of a vacuum. The pilot monitors the buildup of ice and cycles the boots as directed in the AFM/POH. An ice light on the left engine nacelle allows the pilot to monitor wing ice accumulation at night.

Other airframe equipment necessary for flight in icing conditions includes an alternate induction air source and an alternate static system source. Ice tolerant antennas are also installed.

In the event of impact ice accumulating over normal engine air induction sources, carburetor heat (carbureted engines) or alternate air (fuel-injected engines) should be selected. Ice buildup on normal induction sources can be detected by a loss of engine rpm with fixed-pitch propellers and a loss of manifold pressure with constant-speed propellers. On some fuel-injected engines, an alternate air source is automatically activated with blockage of the normal air source.

An alternate static system provides an alternate source of static air for the pitot-static system in the unlikely event that the primary static source becomes blocked. In non-pressurized airplanes, most alternate static sources are plumbed to the cabin. On pressurized airplanes, they are usually plumbed to a non-pressurized baggage compartment. The pilot may activate the alternate static source by opening a valve or a fitting in the flight deck. Activation may create airspeed indicator, altimeter, or vertical speed indicator (VSI) errors. A correction table is frequently provided in the AFM/POH.

Anti-icing/deicing equipment only eliminates ice from the protected surfaces. Significant ice accumulations may form on unprotected areas, even with proper use of anti-ice and deice systems. Flight at high angles of attack (AOA) or even normal climb speeds permit significant ice accumulations on lower wing surfaces, which are unprotected. Many AFM/POHs provide minimum speeds to be maintained in icing conditions. Degradation of all flight characteristics and large performance losses can be expected with ice accumulations. Pilots should not rely upon the stall warning devices for adequate stall warning with ice accumulations.

Ice accumulates unevenly on the airplane. It adds weight and drag (primarily drag) and decreases thrust and lift. Even wing shape affects ice accumulation; thin airfoil sections are more prone to ice accumulation than thick, highly-cambered sections. For this reason, certain surfaces, such as the horizontal stabilizer, are more prone to icing than the wing. With ice accumulations, landing approaches should be made with a minimum wing flap setting (flap extension increases the AOA of the horizontal stabilizer) and with an added margin of airspeed. Sudden and large configuration and airspeed changes should be avoided.

Unless otherwise recommended in the AFM/POH, the autopilot should not be used in icing conditions. Continuous use of the autopilot masks trim and handling changes that occur with ice accumulation. Without this control feedback, the pilot may not be aware of ice accumulation building to hazardous levels. The autopilot suddenly disconnects when it reaches design limits, and the pilot may find the airplane has assumed unsatisfactory handling characteristics.

The installation of anti-ice/deice equipment on airplanes without AFM/POH approval for flight into icing conditions is to facilitate escape when such conditions are inadvertently encountered. Even with AFM/POH approval, the prudent pilot avoids icing conditions to the maximum extent practicable and avoids extended flight in any icing conditions. No multiengine airplane is approved for flight into severe icing conditions and none are intended for indefinite flight in continuous icing conditions.

Performance and Limitations

Discussion of performance and limitations requires the definition of the following terms.

- Accelerate-stop distance is the runway length required to accelerate to a specified speed (either V_R or V_{LOF}, as specified by the manufacturer), experience an engine failure, and bring the airplane to a complete stop. *[Figure 13-5A]*

- Accelerate-go distance is the horizontal distance required to continue the takeoff and climb to 50 feet, assuming an engine failure at V_R or V_{LOF}, as specified by the manufacturer. *[Figure 13-5A]*

- Climb gradient is a slope most frequently expressed in terms of altitude gain per 100 feet of horizontal distance, whereupon it is stated as a percentage. A 1.5 percent climb gradient is an altitude gain of one and one-half feet per 100 feet of horizontal travel. Climb gradient may also be expressed as a function of altitude gain per nautical mile (NM), or as a ratio of the horizontal distance to the vertical distance (10:1, for example). *[Figure 13-5B]* Unlike rate of climb, climb gradient is affected by wind. Climb gradient is improved with a headwind component and reduced with a tailwind component.

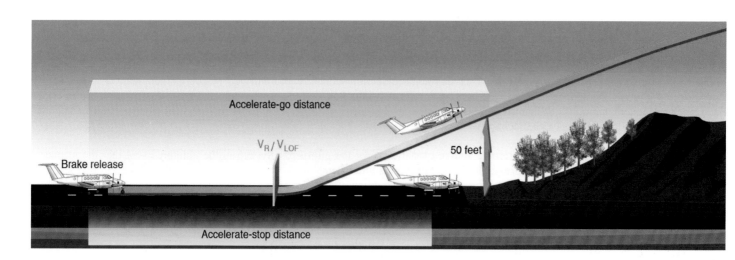

Figure 13-5A. *Accelerate-stop distance and accelerate-go distance.*

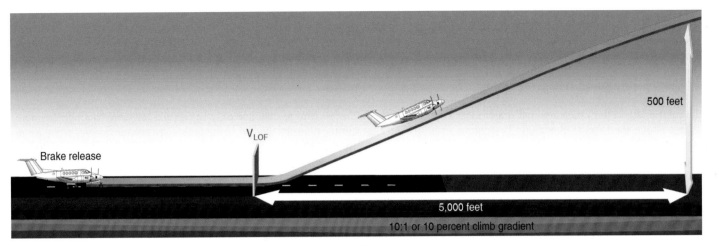

Figure 13-5B. *Climb gradient.*

- The all-engine service ceiling of multiengine airplanes is the highest altitude at which the airplane can maintain a steady rate of climb of 100 fpm with both engines operating. The airplane has reached its absolute ceiling when climb is no longer possible.

- The single-engine service ceiling is reached when the multiengine airplane can no longer maintain a 50 fpm rate of climb with OEI, and its single-engine absolute ceiling when climb is no longer possible.

The takeoff in a multiengine airplane should be planned in sufficient detail so that the appropriate action is taken in the event of an engine failure. The pilot should be thoroughly familiar with the airplane's performance capabilities and limitations in order to make an informed takeoff decision as part of the preflight planning. That decision should be reviewed as the last item of the "before takeoff" checklist.

In the event of an engine failure shortly after takeoff, the decision is basically one of continuing flight or landing, even off-airport. If single-engine climb performance is adequate for continued flight, and the airplane has been promptly and correctly configured, the climb after takeoff may be continued. If single-engine climb performance is such that climb is unlikely or impossible, a landing has to be made in the most suitable area. To be avoided above all is attempting to continue flight when it is not within the airplane's performance capability to do so. *[Figure 13-6]*

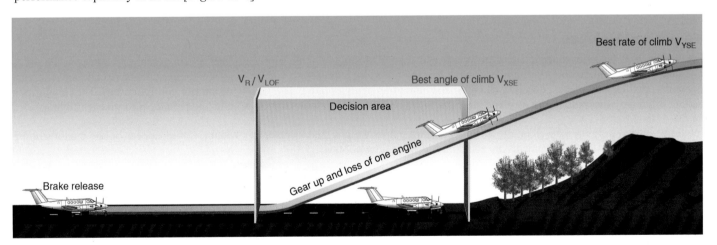

Figure 13-6. *Area of decision for engine failure after lift-off.*

Takeoff planning factors include weight and balance, airplane performance (both single and multiengine), runway length, slope and contamination, terrain and obstacles in the area, weather conditions, and pilot proficiency. Most multiengine airplanes have AFM/POH performance charts and the pilot should be proficient in their use. Prior to takeoff, the multiengine pilot should ensure that the weight and balance limitations have been observed, the runway length is adequate, and the normal flightpath clears obstacles and terrain. The pilot should also consider the appropriate actions expected in the event of an engine failure at any point during the takeoff.

The regulations do not specifically require that the runway length be equal to or greater than the accelerate-stop distance. Most AFM/POHs publish accelerate-stop distances only as an advisory. It becomes a limitation only when published in the limitations section of the AFM/POH. Experienced multiengine pilots, however, recognize the safety margin of runway lengths in excess of the bare minimum required for normal takeoff, and they insist on runway lengths of at least accelerate-stop distance as a matter of safety and good operating practice.

The multiengine pilot considers that under ideal circumstances, the accelerate-go distance only brings the airplane to a point a mere 50 feet above the takeoff elevation. To achieve even this meager climb, the pilot had to instantaneously recognize and react to an unanticipated engine failure, retract the landing gear, identify and feather the correct engine, all the while maintaining precise airspeed control and bank angle as the airspeed is nursed to V_{YSE}. Assuming flawless airmanship thus far, the airplane has now arrived at a point little more than one wingspan above the terrain, assuming it was absolutely level and without obstructions.

For the purpose of illustration, with a near 150 fpm rate of climb at a 90-knot V_{YSE}, it takes approximately 3 minutes to climb an additional 450 feet to reach 500 feet AGL. In doing so, the airplane has traveled an additional 5 NM beyond the original accelerate-go distance, with a climb gradient of about 1.6 percent. Any turn, such as to return to the airport, seriously degrades the already marginal climb performance of the airplane.

Not all multiengine airplanes have published accelerate-go distances in their AFM/POH and fewer still publish climb gradients. When such information is published, the figures have been determined under ideal flight testing conditions. It is unlikely that this performance is duplicated in service conditions.

The point of the previous discussion is to illustrate the marginal climb performance of a multiengine airplane that suffers an engine failure shortly after takeoff, even under ideal conditions. The prudent multiengine pilot should pick a decision point in the takeoff and climb sequence in advance. If an engine fails before this point, the takeoff should be rejected, even if airborne, for a landing on whatever runway or surface lies essentially ahead. If an engine fails after this point, the pilot should promptly execute the appropriate engine failure procedure and continue the climb, assuming the performance capability exists. As a general recommendation, if the landing gear has not been selected up, the takeoff should be rejected, even if airborne.

As a practical matter for planning purposes, the option of continuing the takeoff probably does not exist unless the published single-engine rate-of-climb performance is at least 100 to 200 fpm. Thermal turbulence, wind gusts, engine and propeller wear, or poor technique in airspeed, bank angle, and rudder control can easily negate even a 200 fpm rate of climb.

A pre-takeoff safety brief clearly defines all pre-planned emergency actions to all crewmembers. Even if operating the aircraft alone, the pilot should review and be familiar with takeoff emergency considerations. Indecision at the moment an emergency occurs degrades reaction time and the ability to make a proper response.

Weight and Balance

The weight and balance concept is no different than that of a single-engine airplane. The actual execution, however, is almost invariably more complex due to a number of new loading areas, including nose and aft baggage compartments, nacelle lockers, main fuel tanks, auxiliary fuel tanks, nacelle fuel tanks, and numerous seating options in a variety of interior configurations. The flexibility in loading offered by the multiengine airplane places a responsibility on the pilot to address weight and balance prior to each flight.

The terms empty weight, licensed empty weight, standard empty weight, and basic empty weight as they appear on the manufacturer's original weight and balance documents are sometimes confused by pilots.

In 1975, the General Aviation Manufacturers Association (GAMA) adopted a standardized format for AFM/POHs. It was implemented by most manufacturers in model year 1976. Airplanes whose manufacturers conform to the GAMA standards utilize the following terminology for weight and balance:

standard empty weight + optional equipment = basic empty weight

Standard empty weight is the weight of the standard airplane, full hydraulic fluid, unusable fuel, and full oil. Optional equipment includes the weight of all equipment installed beyond standard. Basic empty weight is the standard empty weight plus optional equipment. Note that basic empty weight includes no usable fuel, but full oil.

Airplanes manufactured prior to the GAMA format generally utilize the following terminology for weight and balance, although the exact terms may vary somewhat:

empty weight + unusable fuel = standard empty weight

standard empty weight + optional equipment = licensed empty weight

Empty weight is the weight of the standard airplane, full hydraulic fluid, and undrainable oil. Unusable fuel is the fuel remaining in the airplane not available to the engines. Standard empty weight is the empty weight plus unusable fuel. When optional equipment is added to the standard empty weight, the result is licensed empty weight. Licensed empty weight, therefore, includes the standard airplane, optional equipment, full hydraulic fluid, unusable fuel, and undrainable oil.

The major difference between the two formats (GAMA and the old) is that basic empty weight includes full oil and licensed empty weight does not. Oil should always be added to any weight and balance utilizing a licensed empty weight.

When the airplane is placed in service, amended weight and balance documents are prepared by appropriately-rated maintenance personnel to reflect changes in installed equipment. The old weight and balance documents are customarily marked "superseded" and retained in the AFM/POH. Maintenance personnel are under no regulatory obligation to utilize the GAMA terminology, so weight and balance documents subsequent to the original may use a variety of terms. Pilots should use care to determine whether or not oil has to be added to the weight and balance calculations or if it is already included in the figures provided.

The multiengine airplane is where most pilots encounter the term "zero fuel weight" for the first time. Not all multiengine airplanes have a zero fuel weight limitation published in their AFM/POH, but many do. Zero fuel weight is simply the maximum allowable weight of the airplane and payload, assuming there is no usable fuel on board. The actual airplane is not devoid of fuel at the time of loading, of course. This is merely a calculation that assumes it was. If a zero fuel weight limitation is published, then all weight in excess of that figure should consist of usable fuel. The purpose of a zero fuel weight is to limit load forces on the wing spars with heavy fuselage loads.

Assume a hypothetical multiengine airplane with the following weights and capacities:

    Basic empty weight 3,200 lbs
    Zero fuel weight 4,400 lbs
    Maximum takeoff weight 5,200 lbs
    Maximum usable fuel 180 gal

1. Calculate the useful load:

    Maximum takeoff weight 5,200 lbs
    Basic empty weight –3,200 lbs
    Useful load 2,000 lbs

The useful load is the maximum combination of usable fuel, passengers, baggage, and cargo that the airplane is capable of carrying.

2. Calculate the payload:

    Zero fuel weight 4,400 lbs
    Basic empty weight –3,200 lbs
    Payload 1,200 lbs

The payload is the maximum combination of passengers, baggage, and cargo that the airplane is capable of carrying. A zero fuel weight, if published, is the limiting weight.

3. Calculate the fuel capacity at maximum payload (1,200 lb):

    Maximum takeoff weight 5,200 lbs
    Zero fuel weight –4,400 lbs
    Fuel allowed 800 lbs

Assuming maximum payload, the only weight permitted in excess of the zero fuel weight should consist of usable fuel. In this case, 133.3 gallons (gal).

4. Calculate the payload at maximum fuel capacity (180 gal):

Basic empty weight 3,200 lbs
Maximum usable fuel +1,080 lbs
Weight with max. fuel 4,280 lbs
Maximum takeoff weight 5,200 lbs
Weight with max. fuel –4,280 lbs
Payload allowed 920 lbs

Assuming maximum fuel, the payload is the difference between the weight of the fueled airplane and the maximum takeoff weight.

Some multiengine airplanes have a ramp weight, which is in excess of the maximum takeoff weight. The ramp weight allows for fuel that would be burned during taxi and run-up, permitting a takeoff at full maximum takeoff weight. The airplane should weigh no more than maximum takeoff weight at the beginning of the takeoff roll.

A maximum landing weight is a limitation against landing at a weight in excess of the published value. This requires preflight planning of fuel burn to ensure that the airplane weight upon arrival at destination is at or below the maximum landing weight. In the event of an emergency requiring an immediate landing, the pilot should recognize that the structural margins designed into the airplane are not fully available when over landing weight. An overweight landing inspection may be advisable—the service manual or manufacturer should be consulted.

Although the foregoing problems only dealt with weight, the balance portion of weight and balance is equally vital. The flight characteristics of the multiengine airplane vary significantly with shifts of the center of gravity (CG) within the approved envelope.

At forward CG, the airplane is more stable, with a slightly higher stalling speed, a slightly slower cruising speed, and favorable stall characteristics. At aft CG, the airplane is less stable, with a slightly lower stalling speed, a slightly faster cruising speed, and less desirable stall characteristics. Forward CG limits are usually determined in certification by elevator/stabilator authority in the landing round out. Aft CG limits are determined by the minimum acceptable longitudinal stability. It is contrary to the airplane's operating limitations and 14 CFR to exceed any weight and balance parameter.

Some multiengine airplanes may require ballast to remain within CG limits under certain loading conditions. Several models require ballast in the aft baggage compartment with only a learner and instructor on board to avoid exceeding the forward CG limit. When passengers are seated in the aft-most seats of some models, ballast or baggage may be required in the nose baggage compartment to avoid exceeding the aft CG limit. The pilot should direct the seating of passengers and placement of baggage and cargo to achieve a CG within the approved envelope. Most multiengine airplanes have general loading recommendations in the weight and balance section of the AFM/POH. When ballast is added, it should be securely tied down, and it should not exceed the maximum allowable floor loading.

Some airplanes make use of a special weight and balance plotter. It consists of several movable parts that can be adjusted over a plotting board on which the CG envelope is printed. The reverse side of the typical plotter contains general loading recommendations for the particular airplane. A pencil line plot can be made directly on the CG envelope imprinted on the working side of the plotting board. This plot can easily be erased and recalculated anew for each flight. This plotter is to be used only for the make and model airplane for which it was designed.

Ground Operation

Good habits learned with single-engine airplanes are directly applicable to multiengine airplanes for preflight and engine start. Upon placing the airplane in motion to taxi, the new multiengine pilot may notice several differences. The most obvious is the increased wingspan and the need for even greater vigilance while taxiing in close quarters. Ground handling may seem somewhat ponderous and the multiengine airplane is not as nimble as the typical two- or four-place single-engine airplane. As always, the pilot should use care not to ride the brakes by keeping engine power to a minimum. One ground handling advantage of the multiengine airplane over single-engine airplanes is the differential power capability. Turning with an assist from differential power minimizes both the need for brakes during turns and the turning radius.

The pilot should be aware, however, that making a sharp turn assisted by brakes and differential power can cause the airplane to pivot about a stationary inboard wheel and landing gear. The airplane was not designed for this action, and the pilot should not allow it to occur. Unless otherwise directed by the AFM/POH, all ground operations should be conducted with the cowl flaps fully open. The use of strobe lights is normally deferred until taxiing onto the active runway.

Normal and Crosswind Takeoff and Climb

After completing the before takeoff checklist and pre-takeoff safety brief, and after receiving an air traffic control (ATC) clearance (if applicable), the pilot should check for approaching aircraft and line up on the runway centerline. If departing from an airport without an operating control tower, the pilot should listen on the appropriate frequency, make a careful check for traffic, and transmit a radio advisory before entering the runway. Sharp turns onto the runway combined with a rolling takeoff are not a good operating practice and may be prohibited by the AFM/POH due to the possibility of "unporting" a fuel tank pickup. The takeoff itself may be prohibited by the AFM/POH under any circumstances below certain fuel levels. The flight controls should be positioned for a crosswind, if present. Exterior lights, such as landing and taxi lights, and wingtip strobes should be illuminated immediately prior to initiating the takeoff roll, day or night. If holding in takeoff position for any length of time, particularly at night, the pilot should activate all exterior lights upon taxiing into position.

Takeoff power should be set as recommended in the AFM/POH. With normally aspirated (non-turbocharged) engines, this is full throttle. Full throttle is also used in most turbocharged engines. There are some turbocharged engines, however, that require the pilot to set a specific power setting, usually just below red line manifold pressure. This yields takeoff power with less than full throttle travel. Turbocharged engines often require special consideration. Throttle motion with turbocharged engines should be exceptionally smooth and deliberate. It is acceptable, and may even be desirable, to hold the airplane in position with brakes as the throttles are advanced. Brake release customarily occurs after significant boost from the turbocharger is established. This prevents utilizing the available runway with slow, partial throttle acceleration as the engine power is increased. If runway length or obstacle clearance is critical, full power should be set before brake release as specified in the performance charts. Note that for all airplanes equipped with constant speed propellers, the engines can turn at maximum rpm and can develop maximum engine power before brake release. Although the mass of air per revolution is small, the number of rpm is high and propeller thrust is maximized. Thrust is at a maximum at the beginning of the takeoff roll and then decreases as the airplane gains speed. The high slipstream velocity during takeoff increases the effective lift of the wing behind the propeller(s).

As takeoff power is established, initial attention should be divided between tracking the runway centerline and monitoring the engine gauges. Many novice multiengine pilots tend to fixate on the airspeed indicator just as soon as the airplane begins its takeoff roll. Instead, the pilot should confirm that both engines are developing full-rated manifold pressure and rpm, and that as the fuel flows, fuel pressures, exhaust gas temperatures (EGTs), and oil pressures are matched in their normal ranges. A directed and purposeful scan of the engine gauges can be accomplished well before the airplane approaches rotation speed. If a crosswind is present, the aileron displacement in the direction of the crosswind may be reduced as the airplane accelerates. The elevator/stabilator control should be held neutral throughout.

Full rated takeoff power should be used for every takeoff. Partial power takeoffs are not recommended. There is no evidence to suggest that the life of modern reciprocating engines is prolonged by partial power takeoffs. In actuality, excessive heat and engine wear can occur with partial power as the fuel metering system fails to deliver the slightly over-rich mixture vital for engine cooling during takeoff.

There are several key airspeeds to be noted during the takeoff and climb sequence in any twin. The first speed to consider is V_{MC}. If an engine fails below V_{MC} while the airplane is on the ground, the takeoff needs to be rejected. Directional control can only be maintained by promptly closing both throttles and using rudder and brakes as required. If an engine fails below V_{MC} while airborne, directional control is not possible with the remaining engine producing takeoff power. On takeoffs, therefore, the airplane should never be airborne before the airspeed exceeds V_{MC}. Pilots should use the manufacturer's recommended rotation speed (V_R) or lift-off speed (V_{LOF}). If no such speeds are published, a minimum of V_{MC} plus 5 knots should be used for V_R.

The rotation to a takeoff pitch attitude is performed with smooth control inputs. With a crosswind, the pilot should ensure that the landing gear does not momentarily touch the runway after the airplane has lifted off, as a side drift is present. The rotation may be accomplished more positively and/or at a higher speed under these conditions. However, the pilot should keep in mind that the AFM/POH performance figures for accelerate-stop distance, takeoff ground roll, and distance to clear an obstacle were calculated at the recommended V_R and/or V_{LOF} speed.

After lift-off, the next consideration is to gain altitude as rapidly as possible. To assist the pilot in takeoff and initial climb profile, some AFM/POHs give a "50-foot" or "50-foot barrier" speed to use as a target during rotation, lift-off, and acceleration to V_Y. Prior to takeoff, pilots should review the takeoff distance to 50 feet above ground level (AGL) and the stopping distance from 50 feet AGL and add the distance together. If the runway is no longer than the total value, the odds are very good that if anything fails, it will be an off-runway landing at the least. After leaving the ground, altitude gain is more important than achieving an excess of airspeed. Experience has shown that excessive speed cannot be effectively converted into altitude in the event of an engine failure. Additional altitude increases the time available to recognize and respond to any aircraft abnormality or emergency during the climb segment.

Excessive climb attitudes can be just as dangerous as excessive airspeed. Steep climb attitudes limit forward visibility and impede the pilot's ability to detect and avoid other traffic. The airplane should be allowed to accelerate in a shallow climb to attain V_Y, the best all-engine rate-of-climb speed. V_Y should then be maintained until achieving a safe single-engine maneuvering altitude, which considers terrain and obstructions. Any speed above or below V_Y reduces the performance of the airplane. Even with all engines operating normally, terrain and obstruction clearance during the initial climb after takeoff is an important preflight consideration. Most airliners and most turbine-powered airplanes climb out at an attitude that yields best rate of climb (V_Y) usually utilizing a flight management system (FMS).

When to raise the landing gear after takeoff depends on several factors. Normally, the gear should be retracted when there is insufficient runway available for landing and after a positive rate of climb is established as indicated on the altimeter. If an excessive amount of runway is available, it would not be prudent to leave the landing gear down for an extended period of time and sacrifice climb performance and acceleration. Leaving the gear extended after the point at which a landing cannot be accomplished on the runway is a hazard. In some multiengine airplanes, operating in a high-density altitude environment, a positive rate of climb with the landing gear down is not possible. Waiting for a positive rate of climb under these conditions is not practicable. An important point to remember is that raising the landing gear as early as possible after liftoff drastically decreases the drag profile and significantly increases climb performance should an engine failure occur. An equally important point to remember is that leaving the gear down to land on sufficient runway or overrun is a much better option than landing with the gear retracted. A general recommendation is to raise the landing gear not later than V_{YSE} airspeed, and once the gear is up, consider it a GO commitment if climb performance is available. Some AFM/POHs direct the pilot to apply the wheel brakes momentarily after lift-off to stop wheel rotation prior to landing gear retraction. If flaps were extended for takeoff, they should be retracted as recommended in the AFM/POH.

Once a safe, single-engine maneuvering altitude has been reached, typically a minimum of 400−500 feet AGL, the transition to an en route climb speed should be made. This speed is higher than V_Y and is usually maintained to cruising altitude. En route climb speed gives better visibility, increased engine cooling, and a higher groundspeed. Takeoff power can be reduced, if desired, as the transition to en route climb speed is made.

Some airplanes have a climb power setting published in the AFM/POH as a recommendation (or sometimes as a limitation), which should then be set for en route climb. If there is no climb power setting published, it is customary, but not a requirement, to reduce manifold pressure and rpm somewhat for en route climb. The propellers are usually synchronized after the first power reduction and the yaw damper, if installed, engaged. The AFM/POH may also recommend leaning the mixtures during climb. The climb checklist should be accomplished as traffic and work load allow. *[Figure 13-7]*

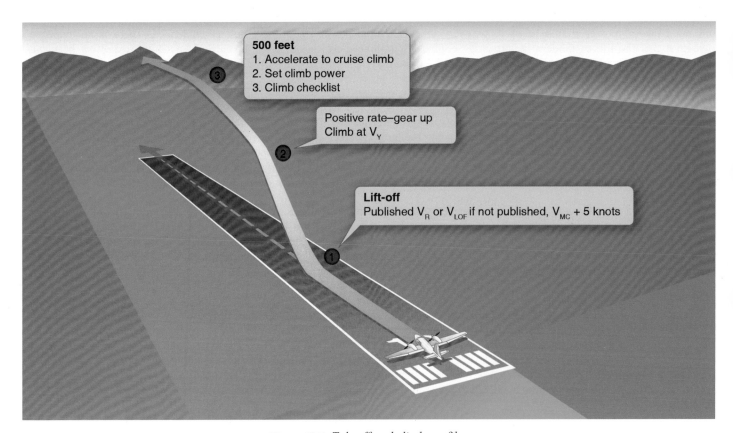

Figure 13-7. *Takeoff and climb profile.*

Short-Field Takeoff and Climb

The short-field takeoff and climb differs from the normal takeoff and climb in the airspeeds and initial climb profile. Some AFM/POHs give separate short-field takeoff procedures and performance charts that recommend specific flap settings and airspeeds. Other AFM/POHs do not provide separate short-field procedures. In the absence of such specific procedures, the airplane should be operated only as recommended in the AFM/POH. No operations should be conducted contrary to the recommendations in the AFM/POH.

On short-field takeoffs in general, just after rotation and lift-off, the airplane should be allowed to accelerate to V_X, making the initial climb over obstacles at V_X and transitioning to V_Y as obstacles are cleared. *[Figure 13-8]*

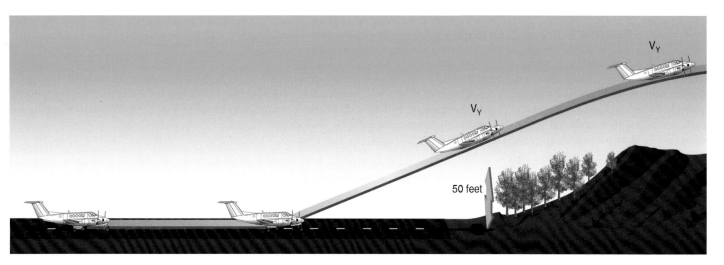

Figure 13-8. *Short-field takeoff and climb*

When partial flaps are recommended for short-field takeoffs, many light-twins have a strong tendency to become airborne prior to V_{MC} plus 5 knots. Attempting to prevent premature lift-off with forward elevator pressure results in wheel barrowing. To prevent this, allow the airplane to become airborne, but only a few inches above the runway. The pilot should be prepared to promptly abort the takeoff and land in the event of engine failure on takeoff with landing gear and flaps extended at airspeeds below V_X.

Engine failure on takeoff, particularly with obstructions, is compounded by the low airspeeds and steep climb attitudes utilized in short-field takeoffs. V_X and V_{XSE} are often perilously close to V_{MC}, leaving scant margin for error in the event of engine failure as V_{XSE} is assumed. If flaps were used for takeoff, the engine failure situation becomes even more critical due to the additional drag incurred. If V_X is less than 5 knots higher than V_{MC}, give strong consideration to reducing useful load or using another runway in order to increase the takeoff margins so that a short-field technique is not required.

Rejected Takeoff

A takeoff can be rejected for the same reasons a takeoff in a single-engine airplane would be rejected. Once the decision to reject a takeoff is made, the pilot should promptly close both throttles and maintain directional control with the rudder, nose-wheel steering, and brakes. Aggressive use of rudder, nose-wheel steering, and brakes may be required to keep the airplane on the runway, particularly if an engine failure is not immediately recognized and accompanied by prompt closure of both throttles. However, the primary objective is not necessarily to stop the airplane in the shortest distance, but to maintain control of the airplane as it decelerates. In some situations, it may be preferable to continue into the overrun area under control, rather than risk directional control loss, landing gear collapse, or tire/brake failure in an attempt to stop the airplane in the shortest possible distance.

Level Off and Cruise

Upon leveling off at cruising altitude, the pilot should allow the airplane to accelerate at climb power until cruising airspeed is achieved, and then cruise power and rpm should be set. To extract the maximum cruise performance from any airplane, the power setting tables provided by the manufacturer should be closely followed. If the cylinder head and oil temperatures are within their normal ranges, the cowl flaps may be closed. When the engine temperatures have stabilized, the mixtures may be leaned per AFM/POH recommendations. The remainder of the cruise checklist should be completed by this point.

Fuel management in multiengine airplanes is often more complex than in single-engine airplanes. Depending upon system design, the pilot may need to select between main tanks and auxiliary tanks or even employ fuel transfer from one tank to another. In complex fuel systems, limitations are often found restricting the use of some tanks to level flight only or requiring a reserve of fuel in the main tanks for descent and landing. Electric fuel pump operation can also vary widely among different models, particularly during tank switching or fuel transfer. Some fuel pumps are to be on for takeoff and landing; others are to be off. There is simply no substitute for thorough systems and AFM/POH knowledge when operating complex aircraft.

Slow Flight

There is nothing unusual about maneuvering during slow flight in a multiengine airplane. Slow flight may be conducted in straight-and-level flight, turns, climbs, or descents. It can also be conducted in the clean configuration, landing configuration, or at any other combination of landing gear and flaps. Slow flight in a multiengine airplane should be conducted so the maneuver can be completed no lower than 3,000 feet AGL or higher if recommended by the manufacturer. In all cases, practicing slow flight should be conducted at an adequate height above the ground for recovery should the airplane inadvertently stall.

Pilots should closely monitor cylinder head and oil temperatures during slow flight. Some high performance multiengine airplanes tend to heat up fairly quickly under some conditions of slow flight, particularly in the landing configuration. Simulated engine failures should not be conducted during slow flight. The airplane will be well below V_{SSE} and very close to V_{MC}. Stability, stall warning, or stall avoidance devices should not be disabled while maneuvering during slow flight.

Spin Awareness and Stalls

No multiengine airplane is approved for spins, and their spin recovery characteristics are generally very poor. It is therefore prudent to practice spin avoidance and maintain a high awareness of situations that can result in an inadvertent spin.

Spin Awareness

In order to spin any airplane, a stalled condition needs to exist. At the stall, the presence or introduction of a yawing moment can initiate spin entry. In a multiengine airplane, the yawing moment may be generated by rudder input or asymmetrical thrust. It follows, then, that spin awareness be at its greatest during V_{MC} demonstrations, stall practice, slow flight, or any condition of high asymmetrical thrust, particularly at low speed/high AOA. Single-engine stalls are not part of any multiengine training curriculum.

No engine failure should ever be introduced below safe, intentional one-engine inoperative speed (V_{SSE}). If no V_{SSE} is published, use V_{YSE}. Other than training situations, the multiengine airplane is only operated below V_{SSE} for mere seconds just after lift-off or during the last few dozen feet of altitude in preparation for landing.

For spin avoidance when practicing engine failures, the flight instructor should pay strict attention to the maintenance of proper airspeed and bank angle as the learner executes the appropriate procedure. The instructor should also be particularly alert during stall and slow flight practice. While flying with a center-of-gravity closer to the forward limit provides better stall and spin avoidance characteristics, it does not eliminate the hazard.

When performing a V_{MC} demonstration, the instructor should also be alert for any sign of an impending stall. The learner may be highly focused on the directional control aspect of the maneuver to the extent that impending stall indications go unnoticed. If a V_{MC} demonstration cannot be accomplished under existing conditions of density altitude, the instructor may, for training purposes, utilize a rudder blocking technique.

As very few twins have ever been spin-tested (none are required to), the recommended spin recovery techniques are based only on the best information available. The departure from controlled flight may be quite abrupt and possibly disorienting. The direction of an upright spin can be confirmed from the turn needle or the symbolic airplane of the turn coordinator, if necessary. Do not rely on the ball position or other instruments.

If a spin is entered, most manufacturers recommend immediately retarding both throttles to idle, applying full rudder opposite the direction of rotation, and applying full forward elevator/stabilator pressure (with ailerons neutral). These actions should be taken as near simultaneously as possible. The controls should then be held in that position until the spin has stopped. At that point adjust rudder pressure, back elevator pressure, and power as necessary to return to the desired flight path. Pilots should be aware that a spin recovery will take considerable altitude; therefore, it is critical that corrective action be taken immediately.

Stall Training

It is recommended that stalls be practiced at an altitude that allows recovery no lower than 3,000 feet AGL for multiengine airplanes, or higher if recommended by the AFM/POH. Losing altitude during recovery from a stall is to be expected.

Stall characteristics vary among multiengine airplanes just as they do with single-engine airplanes, and therefore, a pilot should be familiar with them. Yet, the most important stall recovery step in a multiengine airplane is the same as it is in all airplanes: reduce the angle of attack (AOA). For reference, the stall recovery procedure described in Chapter 5 is included in *Figure 13-9*. Following a reduction in the AOA and the stall warning being eliminated, the wings should be rolled level and power added as needed. Immediate full application of power in a stalled condition has an associated risk due to the possibility of asymmetric thrust. In addition, single-engine stalls, or stalls with significantly more power on one engine than the other, should not be attempted due to the likelihood of a departure from controlled flight and possible spin entry. Similarly, simulated engine failures should not be performed during stall entry and recovery.

Stall Recovery Template	
1. Wing leveler or autopilot	1. Disconnect
2. a) Nose-down pitch control	2. a) Apply until impending stall indications are eliminated
b) Nose-down pitch trim	b) As needed
3. Bank	3. Wings Level
4. Thrust/Power	4. As needed
5. Speed brakes/spoilers	5. Retract
6. Return to the desired flight path	

Figure 13-9. *Stall recovery procedure.*

Power-Off Approach to Stall (Approach and Landing)

A power-off approach to stall is trained and checked to simulate problematic approach and landing scenarios. A power-off approach to stall may be performed with wings level, or from shallow turns (up to 20 degrees of bank). To initiate a power-off approach to stall maneuver, the area surrounding the airplane should first be cleared for possible traffic. The airplane should then be slowed and configured for an approach and landing. A stabilized descent should be established (approximately 500 fpm) and trim adjusted. A turn should be initiated at this point, if desired. The pilot should then smoothly increase the AOA to induce a stall warning. Power is reduced further during this phase, and trimming should cease at speeds slower than takeoff.

When the airplane reaches the stall warning (e.g., aural alert, buffet, etc.), the recovery is accomplished by first reducing the AOA until the stall warning is eliminated. The pilot then rolls the wings level with coordinated use of the rudder and smoothly applies power as required. The airplane should be accelerated to V_X (if simulated obstacles are present) or V_Y during recovery and climb. Considerable forward elevator/stabilator pressure will be required after the stall recovery as the airplane accelerates to V_X or V_Y. Appropriate trim input should be anticipated. The flap setting should be reduced from full to approach, or as recommended by the manufacturer. Then, with a positive rate of climb, the landing gear is selected up. The remaining flaps are then retracted as a positive rate-of-climb continues.

Power-On Approach to Stall (Takeoff and Departure)

A power-on approach to stall is trained and checked to simulate problematic takeoff scenarios. A power-on approach to stall may be performed from straight-and-level flight or from shallow and medium banked turns (up to 20 degrees of bank). To initiate a power-on approach to stall maneuver, the area surrounding the airplane should always be cleared to look for potential traffic. The airplane is slowed to the manufacturer's recommended lift-off speed. The airplane should be configured in the takeoff configuration. Trim should be adjusted for this speed. Engine power is then increased to that recommended in the AFM/POH for the practice of power-on approach to stall. In the absence of a recommended setting, use approximately 65 percent of maximum available power. Begin a turn, if desired, while increasing AOA to induce a stall warning (e.g., aural alert, buffet, etc.). Other specified (reduced) power settings may be used to simulate performance at higher gross weights and density altitudes.

When the airplane reaches the stall warning, the recovery is made first by reducing the AOA until the stall warning is eliminated. The pilot then rolls the wings level with coordinated use of the rudder and applying power as needed. However, if simulating limited power available for high gross weight and density altitude situations, the power during the recovery should be limited to that specified. The landing gear should be retracted when a positive rate of climb is attained, and flaps retracted, if flaps were set for takeoff. The target airspeed on recovery is V_X if (simulated) obstructions are present, or V_Y. The pilot should anticipate the need for nose-down trim as the airplane accelerates to V_X or V_Y after recovery.

Full Stall

It is not recommended that full stalls be practiced unless a qualified flight instructor is present. A power-off or power-on full stall should only be practiced in a structured lesson with clear learning objectives and cautions discussed. The goals of the training are (a) to provide the pilots the experience of the handling characteristics and dynamic cues (e.g., buffet, roll off) near and at full stall and (b) to reinforce the proper application of the stall recovery procedures. Given the associated risk of asymmetric thrust at high angles of attack and low rudder effectiveness due to low airspeeds, this reinforces the primary step of first lowering the AOA, which allows all control surfaces to become more effective and allows for roll to be better controlled. Thrust should only be used as needed in the recovery.

Accelerated Approach to Stall

Accelerated approach to stall should be performed with a bank of approximately 45°, and in no case at a speed greater than the airplane manufacturer's recommended airspeed, the specified design maneuvering speed (V_A), or operating maneuvering speed (V_O). The pilot should select an entry altitude that will allow completion of the maneuver no lower than 3,000 feet AGL.

The entry method for the maneuver is no different than for a single-engine airplane. Once at an appropriate speed, begin increasing the back pressure on the elevator while maintaining a coordinated 45° turn. A good speed reduction rate is approximately 3 to 5 knots per second. Once a stall warning occurs, recover promptly by reducing the AOA until the stall warning stops. Then, roll the wings level with coordinated rudder and add power as necessary to return to the desired flightpath.

Normal Approach and Landing

Given the higher cruising speed (and frequently altitude) of multiengine airplanes over most single-engine airplanes, the descent needs to be planned in advance. A hurried, last minute descent with power at or near idle is inefficient and can cause excessive engine cooling. It may also lead to passenger discomfort, particularly if the airplane is unpressurized. As a rule of thumb, if terrain and passenger conditions permit, a maximum of a 500 fpm rate of descent should be planned. Pressurized airplanes can plan for higher descent rates, if desired.

In a descent, some airplanes require a minimum EGT or may have a minimum power setting or cylinder head temperature to maintain. In any case, combinations of very low manifold pressure and high rpm settings are strongly discouraged by engine manufacturers. If higher descent rates are necessary, the pilot should consider extending partial flaps or lowering the landing gear before retarding the power excessively. The descent checklist should be initiated upon leaving cruising altitude and completed before arrival in the terminal area. Upon arrival in the terminal area, pilots are encouraged to turn on their landing and recognition lights when operating below 10,000 feet, day or night, and especially when operating within 10 miles of any airport or in conditions of reduced visibility.

The traffic pattern and approach are typically flown at somewhat higher indicated airspeeds in a multiengine airplane contrasted to most single-engine airplanes. The pilot may allow for this through an early start on the before-landing checklist. This provides time for proper planning, spacing, and thinking well ahead of the airplane. Many multiengine airplanes have partial flap extension speeds above V_{FE}, and partial flaps can be deployed prior to traffic pattern entry. Normally, the landing gear should be selected and confirmed down when abeam the intended point of landing as the downwind leg is flown. *[Figure 13-10]*

The FAA recommends a stabilized approach concept. To the greatest extent practical, on final approach and within 500 feet AGL, the airplane should be on speed, in trim, configured for landing, tracking the extended centerline of the runway, and established in a constant angle of descent toward an aim point in the touchdown zone. Absent unusual flight conditions, only minor corrections are required to maintain this approach to the round out and touchdown.

The final approach should be made with power and at a speed recommended by the manufacturer; if a recommended speed is not furnished, the speed should be no slower than the single-engine best rate-of-climb speed (V_{YSE}) until short final with the landing assured, but in no case less than critical engine-out minimum control speed (V_{MC}). Some multiengine pilots prefer to delay full flap extension to short final with the landing assured. This is an acceptable technique with appropriate experience and familiarity with the airplane.

In the round out for landing, residual power is gradually reduced to idle. With the higher wing loading of multiengine airplanes and with the drag from two windmilling propellers, there is minimal float. Full stall landings are generally undesirable in twins. The airplane should be held off as with a high performance single-engine model, allowing touchdown of the main wheels prior to a full stall.

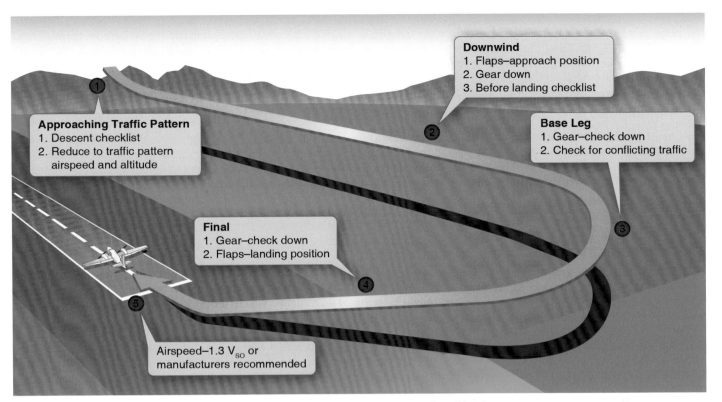

Downwind
1. Flaps–approach position
2. Gear down
3. Before landing checklist

Approaching Traffic Pattern
1. Descent checklist
2. Reduce to traffic pattern airspeed and altitude

Base Leg
1. Gear–check down
2. Check for conflicting traffic

Final
1. Gear–check down
2. Flaps–landing position

Airspeed–1.3 V_{SO} or manufacturers recommended

Figure 13-10. *Normal two-engine approach and landing.*

Under favorable wind and runway conditions, the nose-wheel can be held off for best aerodynamic braking. Even as the nose-wheel is gently lowered to the runway centerline, continued elevator back pressure greatly assists the wheel brakes in stopping the airplane.

If runway length is critical, or with a strong crosswind, or if the surface is contaminated with water, ice, or snow, it is undesirable to rely solely on aerodynamic braking after touchdown. The full weight of the airplane should be placed on the wheels as soon as practicable. The wheel brakes are more effective than aerodynamic braking alone in decelerating the airplane.

Once on the ground, elevator back pressure should be used to place additional weight on the main wheels. When necessary, wing flap retraction also adds additional weight to the wheels and improves braking effectivity. Flap retraction during the landing rollout is discouraged, however, unless there is a clear, operational need. It should not be accomplished as routine with each landing.

Some multiengine airplanes, particularly those of the cabin class variety, can be flown through the round out and touchdown with a small amount of power. This is an acceptable technique to prevent high sink rates and to cushion the touchdown. The pilot should keep in mind, however, that the primary purpose in landing is to get the airplane down and stopped. This technique should only be attempted when there is a generous margin of runway length. As propeller blast flows directly over the wings, lift as well as thrust is produced. The pilot should taxi clear of the runway as soon as speed and safety permit, and then accomplish the after-landing checklist. Ordinarily, no attempt should be made to retract the wing flaps or perform other checklist duties until the airplane has been brought to a halt when clear of the active runway. Exceptions to this would be the rare operational needs discussed above, to relieve the weight from the wings and place it on the wheels. In these cases, AFM/POH guidance should be followed. The pilot should not indiscriminately reach out for any switch or control on landing rollout. An inadvertent landing gear retraction while meaning to retract the wing flaps may result.

Crosswind Approach and Landing

The multiengine airplane is often easier to land in a crosswind than a single-engine airplane due to its higher approach and landing speed. In any event, the principles are no different between singles and twins. Prior to touchdown, the longitudinal axis should be aligned with the runway centerline to avoid landing gear side loads.

The two primary methods, crab and wing-low, are typically used in conjunction with each other. As soon as the airplane rolls out onto final approach, the crab angle to track the extended runway centerline is established. This is coordinated flight with adjustments to heading to compensate for wind drift either left or right. Prior to touchdown, the transition to a sideslip is made with the upwind wing lowered and opposite rudder applied to prevent a turn. The airplane touches down on the landing gear of the upwind wing first, followed by that of the downwind wing, and then the nose gear. Follow-through with the flight controls involves an increasing application of aileron into the wind until full control deflection is reached.

The point at which the transition from the crab to the sideslip is made is dependent upon pilot familiarity with the airplane and experience. With high skill and experience levels, the transition can be made during the round out just before touchdown. With lesser skill and experience levels, the transition is made at increasing distances from the runway. Some multiengine airplanes (as some single-engine airplanes) have AFM/POH limitations against slips in excess of a certain time period; 30 seconds, for example. This is prevent engine power loss from fuel starvation as the fuel in the tank of the lowered wing flows toward the wingtip, away from the fuel pickup point. This time limit should be observed if the wing-low method is utilized.

Some multiengine pilots prefer to use differential power to assist in crosswind landings. The asymmetrical thrust produces a yawing moment little different from that produced by the rudder. When the upwind wing is lowered, power on the upwind engine is increased to prevent the airplane from turning. This alternate technique is completely acceptable, but most pilots feel they can react to changing wind conditions quicker with rudder and aileron than throttle movement. This is especially true with turbocharged engines where the throttle response may lag momentarily. The differential power technique should be practiced with an instructor before being attempted alone.

Short-Field Approach and Landing

The primary elements of a short-field approach and landing do not differ significantly from a normal approach and landing. Many manufacturers do not publish short-field landing techniques or performance charts in the AFM/POH. In the absence of specific short-field approach and landing procedures, the airplane should be operated as recommended in the AFM/POH. No operations should be conducted contrary to the AFM/POH recommendations.

The emphasis in a short-field approach is on configuration (full flaps), a stabilized approach with a constant angle of descent, and precise airspeed control. As part of a short-field approach and landing procedure, some AFM/POHs recommend a slightly slower than normal approach airspeed. If no such slower speed is published, use the AFM/POH-recommended normal approach speed.

Full flaps are used to provide the steepest approach angle. If obstacles are present, the approach should be planned so that no drastic power reductions are required after they are cleared. The power should be smoothly reduced to idle in the round out prior to touchdown. Pilots should keep in mind that the propeller blast blows over the wings providing some lift in addition to thrust. Reducing power significantly, just after obstacle clearance, usually results in a sudden, high sink rate that may lead to a hard landing. After the short-field touchdown, maximum stopping effort is achieved by retracting the wing flaps, adding back pressure to the elevator/stabilator, and applying heavy braking. However, if the runway length permits, the wing flaps should be left in the extended position until the airplane has been stopped clear of the runway. There is always a significant risk of retracting the landing gear instead of the wing flaps when flap retraction is attempted on the landing rollout.

Landing conditions that involve a short field, high winds, or strong crosswinds are just about the only situations where flap retraction on the landing rollout should be considered. When there is an operational need to retract the flaps just after touchdown, it needs to be done deliberately with the flap handle positively identified before it is moved.

Go-Around

When the decision to go around is made, the throttles should be advanced to takeoff power and pitch adjusted to arrest the sink rate. With adequate airspeed, the airplane should be placed in a climb pitch attitude. These actions, which are accomplished sequentially, arrest the sink rate and place the airplane in the proper attitude for transition to a climb. The initial target airspeed is V_Y or V_X if obstructions are present. With sufficient airspeed, the flaps should be retracted from full to an intermediate position and the landing gear retracted when there is a positive rate of climb and no chance of runway contact. The remaining flaps should then be retracted. [Figure 13-11]

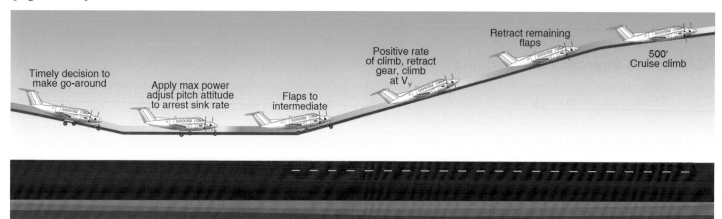

Figure 13-11. *Go-around procedure.*

If the go-around was initiated due to conflicting traffic on the ground or aloft, the pilot should consider maneuvering to the side to keep the conflicting traffic in sight. This may involve a slight turn to offset from the runway/landing area.

If the airplane was in trim for the landing approach when the go-around was commenced, it soon requires a great deal of forward elevator/stabilator pressure as the airplane accelerates away in a climb. The pilot should apply appropriate forward pressure to maintain the desired pitch attitude. Trim should be commenced immediately. The balked landing checklist should be reviewed as work load permits.

Flaps should be retracted before the landing gear for two reasons. First, on most airplanes, full flaps produce more drag than the extended landing gear. Secondly, the airplane tends to settle somewhat with flap retraction, and the landing gear should be down in the event of an inadvertent, momentary touchdown.

Many multiengine airplanes have a landing gear retraction speed significantly less than the extension speed. Care should be exercised during the go-around not to exceed the retraction speed. If the pilot desires to return for a landing, it is essential to re-accomplish the entire before-landing checklist. An interruption to a pilot's habit patterns, such as a go-around, is a classic scenario for a subsequent gear-up landing.

The preceding discussion about performing a go-around assumes that the maneuver was initiated from normal approach speeds or faster. If the go-around was initiated from a low airspeed, the initial pitch up to a climb attitude should be tempered with the necessity to maintain adequate flying speed throughout the maneuver. Examples of where this applies include a go-around initiated from the landing round out or recovery from a bad bounce, as well as a go-around initiated due to an inadvertent approach to a stall. The first priority is always to maintain control and obtain adequate flying speed. A few moments of level or near level flight may be required as the airplane accelerates up to climb speed.

Engine Inoperative Flight Principles

There are two main considerations for OEI operations—*performance and control*. Multiengine pilots learn to operate the airplane for maximum *rate of climb performance* at the blue radial indicated airspeed by training to fly without sideslip. Pilots also learn to recognize and recover from loss of *directional control* associated with the red radial indicated airspeed by performing a V_{MC} demonstration. Since the object of a V_{MC} demonstration is not performance, sideslip occurs during the maneuver. Detailed discussion on both the loss of directional control and maximum OEI climb performance follows.

Derivation of V_{MC}

V_{MC} is a speed established by the manufacturer, published in the AFM/POH, and marked on most airspeed indicators with a red radial line. A knowledgeable and competent multiengine pilot understands that V_{MC} is **not** a fixed airspeed under all conditions. V_{MC} is a fixed airspeed only for the very specific set of circumstances under which it was determined during aircraft certification. In reality, V_{MC} varies with a variety of factors as outlined below. The V_{MC} noted in practice and demonstration, or in actual OEI operation, could be less or even greater than the published value, depending on conditions and pilot technique.

Historically, in aircraft certification, V_{MC} is the sea level calibrated airspeed at which, when the critical engine is suddenly made inoperative, it is possible to maintain control of the airplane with that engine still inoperative and then maintain straight flight at the same speed with an angle of bank not more than 5°.

The foregoing refers to the determination of V_{MC} under *dynamic* conditions. This technique is only used by highly experienced test pilots during aircraft certification. It is unsafe to be attempted outside of these circumstances.

In aircraft certification, there is also a determination of V_{MC} under *static*, or steady-state conditions. If there is a difference between the dynamic and static speeds, the higher of the two is published as V_{MC}. The *static* determination is simply the ability to maintain straight flight at V_{MC} with a bank angle of not more than 5°. This more closely resembles the V_{MC} demonstration task in the practical test for a multiengine rating.

The AFM/POH-published V_{MC} is determined with the *critical* engine inoperative. The critical engine is the engine whose failure had the most adverse effect on directional control. On twins with each engine rotating in conventional, clockwise rotation as viewed from the pilot's seat, the critical engine will be the left engine.

Multiengine airplanes are subject to P-factor just as single-engine airplanes are. The descending propeller blade of each engine will produce greater thrust than the ascending blade when the airplane is operated under power and at positive angles of attack. The descending propeller blade of the right engine is also a greater distance from the center of gravity, and therefore has a longer moment arm than the descending propeller blade of the left engine. As a result, failure of the left engine will result in the most asymmetrical thrust (adverse yaw) as the right engine will be providing the remaining thrust. *[Figure 13-12]*

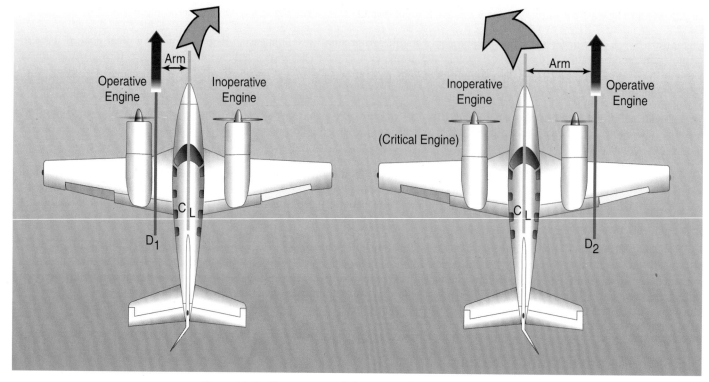

Figure 13-12. *Forces created during single-engine operation.*

Many twins are designed with a counter-rotating right engine. With this design, the degree of asymmetrical thrust is the same with either engine inoperative. No engine is more critical than the other, and a V$_{MC}$ demonstration may be performed with either engine windmilling.

The following bullets describe the way several factors affect V$_{MC}$ speed for those multiengine airplanes often used during training, which were certified in accordance with historical 14 CFR part 23, section 23.149. They also describe the conditions used to determine the manufacturer's published speed. Historically, in aircraft certification, *dynamic* V$_{MC}$ has been determined under the following conditions outlined in historical 14 CFR part 23, section 23.149:

- **Maximum available takeoff power initially on each engine (section 23.149(b)(1)).** V$_{MC}$ increases as power is increased on the operating engine. With normally aspirated engines, V$_{MC}$ is highest at takeoff power and sea level, and decreases with altitude. With turbocharged engines, takeoff power, and therefore V$_{MC}$, remains constant with increases in altitude up to the engine's critical altitude (the altitude where the engine can no longer maintain 100 percent power). Above the critical altitude, V$_{MC}$ decreases just as it would with a normally aspirated engine whose critical altitude is sea level. In order to avoid accidents, test pilots conduct V$_{MC}$ tests at a variety of altitudes, and the results of those tests are then extrapolated to a single, sea level value.

- **All propeller controls in the recommended takeoff position throughout V$_{MC}$ determination (section 23.149(b)(5)).** V$_{MC}$ increases with increased drag on the inoperative engine. V$_{MC}$ is highest, therefore, when the critical engine propeller is windmilling at the low pitch, high rpm blade angle. V$_{MC}$ is normally determined with the critical engine propeller windmilling in the takeoff position, unless the engine is equipped with an autofeather system.

- **Most unfavorable weight and center-of-gravity position (section 23.149(b)).** V$_{MC}$ increases as the center-of-gravity (CG) is moved aft. The moment arm of the rudder is reduced, and therefore its effectivity is reduced, as the CG is moved aft. For a typical light twin, the aft-most CG limit is the most unfavorable CG position. Historically, 14 CFR part 23 calls for V$_{MC}$ to be determined at the most unfavorable weight. For twins certified under CAR 3 or early 14 CFR part 23, the weight at which V$_{MC}$ was determined was not specified. V$_{MC}$ increases as weight is reduced. [*Figure 13-13*]

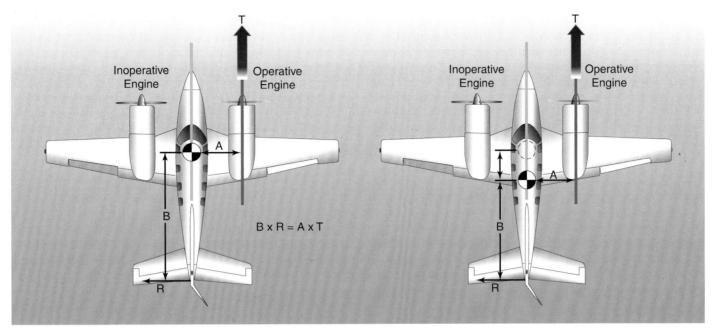

Figure 13-13. *Effect of CG location on yaw.*

- **Landing gear retracted (section 23.149(b)(4)).** V_{MC} increases when the landing gear is retracted. Extended landing gear aids directional stability, which tends to decrease V_{MC}.

- **Flaps in the takeoff position (section 23.149(b)(3)).** This normally includes wing flaps and cowl flaps. For most twins, this will be 0° of flaps.

- **Airplane trimmed for takeoff (section 23.149(b)(2)).**

- **Airplane airborne and the ground effect negligible (section 23.149(b)).**

- **Maximum of 5° angle of bank (section 23.149(a)).** V_{MC} is highly sensitive to bank angle. To prevent claims of an unrealistically low V_{MC} speed in aircraft certification, the manufacturer is permitted to use a maximum of a 5° bank angle toward the operative engine. The horizontal component of lift generated by the bank balances the side force from the rudder, rather than using sideslip to do so. Sideslip requires more rudder deflection, which in turn increases V_{MC}. The bank angle works in the manufacturer's favor in lowering V_{MC} since using high bank angles reduces required rudder deflection. However, this method may result in unsafe flight from both the large sideslip and the need to increase the angle of attack in order to maintain the vertical component of lift.

V_{MC} increases as bank angle decreases. In fact, V_{MC} may increase more than 3 knots for each degree of bank reduction between 5° and wings-level. Since V_{MC} was determined with up to 5° of bank, loss of directional control may be experienced at speeds almost 20 knots above published V_{MC} when the wings are held level.

The 5° bank angle maximum is a historical limit imposed upon manufacturers in aircraft certification. The 5° bank does not inherently establish zero sideslip or best single-engine climb performance. Zero sideslip, and therefore best single-engine climb performance, may occur at bank angles less than 5°. The determination of V_{MC} in certification is solely concerned with the minimum speed for directional control under a very specific set of circumstances, and not the optimum airplane attitude or configuration for climb performance.

During *dynamic* V_{MC} determination in aircraft certification, cuts of the critical engine using the mixture control are performed by flight test pilots while gradually reducing the speed with each attempt. V_{MC} is the minimum speed at which directional control could be maintained within 20° of the original entry heading when a cut of the critical engine was made. During such tests, the climb angle with both engines operating was high, and the pitch attitude following the engine cut had to be quickly lowered to regain the initial speed. Transitioning pilots should understand that attempting to demonstrate V_{MC} with an engine cut from high power, or intentionally failing an engine at speeds less than V_{SSE} creates a high likelihood for loss of control and an accident.

V$_{MC}$ Demo

The actual demonstration of V$_{MC}$ and recovery in flight training more closely resembles *static* V$_{MC}$ determination in aircraft certification. For a demonstration that avoids the hazard of unintended contact with the ground, the pilot selects an altitude that will allow performance of the maneuver at least 3,000 feet AGL. The following description assumes a twin with non-counter-rotating engines, where the left engine is critical.

With the landing gear retracted and the flaps set to the takeoff position, the pilot slows the airplane to approximately 10 knots above V$_{SSE}$ or V$_{YSE}$ (whichever is higher) and trims for takeoff. For the remainder of the maneuver, the trim setting remains unaltered. The pilot selects an entry heading and sets high rpm on both propeller controls. Power on the left engine is throttled back to idle as the right engine power is advanced to the takeoff setting. The landing gear warning horn will sound as long as a throttle is retarded, however the pilot listens carefully for the stall warning horn or watches for the stall warning light. The left yawing and rolling moment of the asymmetrical thrust is counteracted primarily with right rudder. A bank angle of up to 5° (a right bank in this case) may be established as appropriate for the airplane make and model.

While maintaining entry heading, the pitch attitude is slowly increased to decelerate at a rate of 1 knot per second (no faster). As the airplane slows and control effectivity decays, the pilot counteracts the increasing yawing tendency with additional rudder pressure. Aileron displacement will also increase in order to maintain the established bank. An airspeed is soon reached where full right rudder travel and up to a 5° right bank can no longer counteract the asymmetrical thrust, and the airplane will begin to yaw uncontrollably to the left.

The moment the pilot first recognizes the uncontrollable yaw, or experiences any symptom associated with a stall, the pilot simultaneously retards the throttle for the operating engine to stop the yaw and lowers the pitch attitude to regain speed. Recovery is made to straight flight on the entry heading at V$_{SSE}$ or V$_{YSE}$. The pilot increases power to the operating engine, and demonstrates controlled flight before restoring symmetrical power.

To keep the foregoing description simple, there were several important background details that were not covered. The rudder pressure during the demonstration can be quite high. During certification under historical 14 CFR part 23, section 23.149(e), 150 pounds of force was permitted. Most twins will run out of rudder travel long before 150 pounds of pressure is required. Still, the rudder pressure used during any V$_{MC}$ demonstration may seem considerable.

Maintaining altitude is not a criterion in accomplishing this maneuver. This is a demonstration of controllability, not performance. Many airplanes will lose (or gain) altitude during the demonstration. Remaining at or above a minimum of 3,000 feet AGL throughout the maneuver is considered to be effective risk mitigation of certain hazards.

V$_{MC}$ Demo Stall Avoidance

As discussed earlier, with normally aspirated engines, V$_{MC}$ decreases with altitude. Stalling speed (V$_S$), however, remains the same. Except for a few models, published V$_{MC}$ is almost always higher than V$_S$. At sea level there is usually a margin of several knots between V$_{MC}$ and V$_S$, but the margin decreases with altitude, and at some altitude, V$_{MC}$ and V$_S$ are the same. *[Figure 13-14]*

Should a stall occur while the airplane is under asymmetrical power, a spin entry is likely. The yawing moment induced from asymmetrical thrust is little different from that induced by full rudder in an intentional spin in the appropriate model of single-engine airplane. In this case, however, the airplane will depart controlled flight in the direction of the idle engine, not in the direction of applied rudder. Twins are not required to demonstrate recoveries from spins, and their spin recovery characteristics are generally very poor.

Where V$_S$ is encountered before V$_{MC}$, the departure from controlled flight might be quite sudden, with strong yawing and rolling tendencies to the inverted orientation and a spin entry. Therefore, during a V$_{MC}$ demonstration, if there are any symptoms of an impending stall such as a stall warning light or horn, airframe or elevator buffet, or sudden loss of control effectiveness; the pilot should terminate the maneuver immediately by reducing the angle of attack as the throttle is retarded and return the airplane to the entry airspeed. Note that noise within the flight deck may mask the sound of the stall warning horn.

While the V$_{MC}$ demonstration shows the earliest onset of a loss of directional control when performed in accordance with the foregoing procedures, avoid a stalled condition. Avoid stalls with asymmetrical thrust, such that the V$_{MC}$ demonstration does not degrade into a single-engine stall. A V$_{MC}$ demonstration that is allowed to degrade into a single-engine stall with high asymmetrical thrust may result in an unrecoverable loss of control and a fatal accident.

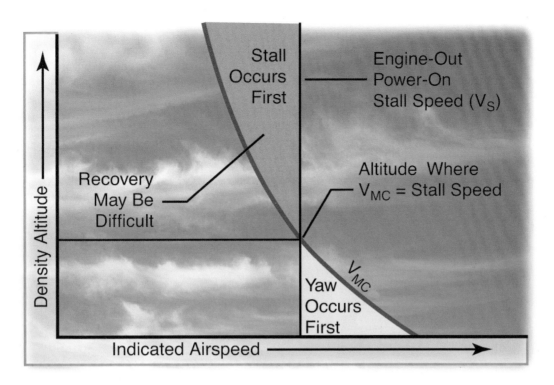

Figure 13-14. *Graph depicting relationship of V$_{MC}$ to V$_S$.*

An actual demonstration of V$_{MC}$ may not be possible under certain conditions of density altitude, or with airplanes whose V$_{MC}$ is equal to or less than V$_S$. Under those circumstances, as a training technique, a demonstration of V$_{MC}$ may safely be conducted by artificially limiting rudder travel to simulate maximum available rudder. A speed well above V$_S$ (approximately 20 knots) is recommended when limiting rudder travel.

The rudder limiting technique avoids the hazards of spinning as a result of stalling with high asymmetrical power, yet is effective in demonstrating the loss of directional control.

To reduce the risk of a loss of control, avoid performing any V$_{MC}$ demonstration from a high pitch attitude with both engines operating and then reducing power on one engine.

OEI Climb Performance

Best OEI climb performance is obtained at V$_{YSE}$ with maximum available power and minimum drag. After the flaps and landing gear have been retracted and the propeller of the failed engine feathered, a key element in best climb performance is minimizing sideslip.

For any airplane, sideslip can be confirmed through the use of a yaw string. A yaw string is a piece of string or yarn approximately 18 to 36 inches in length taped to the base of the windshield or to the nose near the windshield along the airplane centerline. In two-engine coordinated flight, the relative wind causes the string to align itself with the longitudinal axis of the airplane, and it positions itself straight up the center of the windshield. This is zero sideslip. Experimentation with slips and skids vividly displays the location of the relative wind. A particular combination of aileron and rudder also establishes zero sideslip during OEI flight. Adequate altitude, flying speed, and caution should be maintained if attempting these maneuvers.

With a single-engine airplane or a multiengine airplane with both engines operative, sideslip is eliminated when the ball of the turn and bank instrument is centered. This is a condition of zero sideslip, and the airplane is presenting its smallest possible profile to the relative wind. As a result, drag is at its minimum. Pilots know this as coordinated flight.

In a multiengine airplane with an inoperative engine, the centered ball is no longer the indicator of zero sideslip due to asymmetric thrust. In fact, there is no flight deck instrument that directly indicates conditions for zero sideslip. In the absence of a yaw string, the pilot needs to place the airplane at a predetermined bank angle and ball position. Since the AFM/POH performance charts for one engine inoperative flight were determined at zero sideslip, this technique should be used to obtain the charted OEI performance. There are two different control inputs that can be used to counteract the asymmetric thrust of a failed engine:

1. Yaw from the rudder
2. The horizontal component of lift that results from bank with the ailerons

Used individually, neither is correct. Used together in the proper combination, zero sideslip and best climb performance are achieved.

Three different scenarios of airplane control inputs are presented below. The first two are not correct and can increase the risk of a loss of control. They are presented to illustrate the reasons for the zero sideslip approach to best climb performance.

1. Engine inoperative flight with wings level and ball centered requires large rudder input toward the operative engine. *[Figure 13-15]* The result is a moderate sideslip toward the inoperative engine. Climb performance is reduced by the moderate sideslip. With wings level, V_{MC} is significantly higher than published as there is no horizontal component of lift available to help the rudder combat asymmetrical thrust.

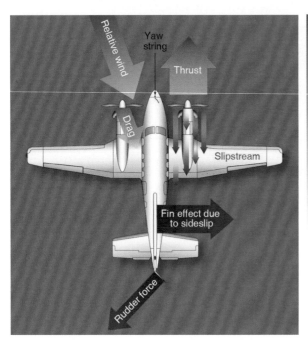
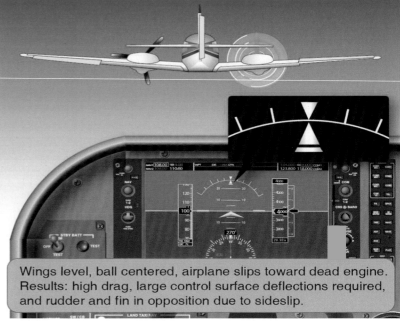

Wings level, ball centered, airplane slips toward dead engine. Results: high drag, large control surface deflections required, and rudder and fin in opposition due to sideslip.

Figure 13-15. *Wings level engine-out flight.*

2. Engine inoperative flight using ailerons alone requires an 8–10° bank angle toward the operative engine. *[Figure 13-16]* This assumes no rudder input, the ball is displaced well toward the operative engine, and climb performance is greatly reduced by the large sideslip toward the operative engine. Due to the increased risk of loss of control, instructors should not normally demonstrate this.

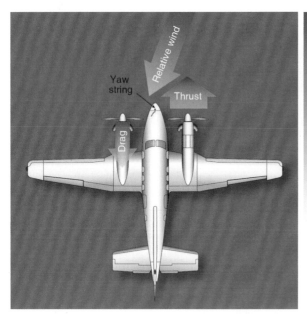
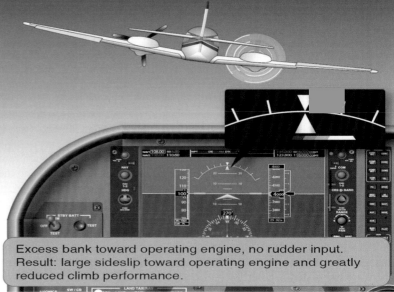

Excess bank toward operating engine, no rudder input. Result: large sideslip toward operating engine and greatly reduced climb performance.

Figure 13-16. *Excessive bank engine-out flight.*

3. Rudder and ailerons used together in the proper combination result in a bank of approximately 2° toward the operative engine. The ball is displaced approximately one-third to one-half toward the operative engine. The result is zero sideslip and maximum climb performance. *[Figure 13-17]* Any attitude other than zero sideslip increases drag, decreasing performance. V_{MC} under these circumstances is higher than published, as less than the 5° bank certification limit is employed.

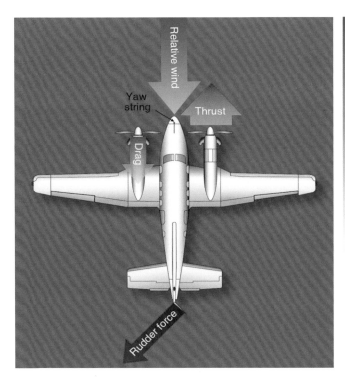
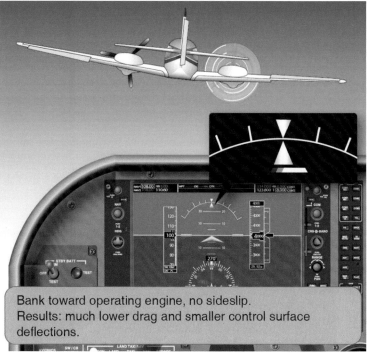

Figure 13-17. *Zero sideslip engine-out flight.*

When bank angle is plotted against climb performance for a hypothetical twin, zero sideslip results in the best (however marginal) climb performance or the least rate of descent. Whether the airplane can climb depends on the weight of the airplane, density altitude, and pilot technique. If the pilot uses zero bank (all rudder to counteract yaw), climb performance degrades as a result of moderate sideslip. Using bank angle alone (no rudder) severely degrades climb performance as a result of a large sideslip.

The precise condition of zero sideslip (bank angle and ball position) varies slightly from model to model and with available power and airspeed. If the airplane is not equipped with counter-rotating propellers, it also varies slightly with the engine failed due to P-factor. The foregoing zero sideslip recommendations apply to reciprocating engine multiengine airplanes flown at V_{YSE} with the inoperative engine feathered. The zero sideslip ball position for straight flight is also the zero sideslip position for turning flight.

The actual bank angle for zero sideslip varies among airplanes from one and one-half to two and one-half degrees. The position of the ball varies from one-third to one-half of a ball width from instrument center toward the operative engine.

During certain flight training scenarios, pilots and instructors simulate propeller feathering. *Zero thrust* means the pilot sets power on one engine such that drag from its rotating propeller equals that of a stopped feathered propeller. With an engine set to zero thrust (or feathered) and the airplane slowed to V_{YSE}, a climb with maximum power on the remaining engine reveals the precise bank angle and ball deflection required for zero sideslip and best climb performance. Again, if a yaw string were present, it aligns itself vertically on the windshield as an indication of zero sideslip. There are very minor changes from this attitude depending upon the engine failed (with non-counter-rotating propellers), power available, airspeed, and weight; but without more sensitive testing equipment, these changes are difficult to detect. The only significant difference would be the pitch attitude required to maintain V_{YSE} under different density altitude, power available, and weight conditions.

Low Altitude Engine Failure Scenarios

In OEI flight at low altitudes and airspeeds such as the initial climb after takeoff, pilots should operate the airplane so as to guard against the three major accident factors: (1) loss of directional control, (2) loss of performance, and (3) loss of flying speed. All have equal potential to be lethal. Loss of flying speed is not a factor, however, when the airplane is operated with due regard for directional control and performance.

A takeoff or go-around is the most critical time to suffer an engine failure. The airplane will be slow, close to the ground, and may even have landing gear and flaps extended. Altitude and time is minimal. Until feathered, the propeller of the failed engine is windmilling, producing a great deal of drag and yawing tendency. Airplane climb performance is marginal or even non-existent, and obstructions may lie ahead. An emergency contingency plan and safety brief should be clearly understood well before the takeoff roll commences. An engine failure before a predetermined airspeed or point results in an aborted takeoff. An engine failure after a certain airspeed and point, with the gear up, and climb performance assured result in a continued takeoff. With loss of an engine, it is paramount to maintain airplane control and comply with the manufacturer's recommended emergency procedures. Complete failure of one engine shortly after takeoff can be broadly categorized into one of three following scenarios.

Landing Gear Down

If the engine failure occurs prior to selecting the landing gear to the UP position [Figure 13-18]: Keep the nose as straight as possible, close both throttles, adjust pitch attitude to maintain adequate airspeed, and descend to the runway. Concentrate on a normal landing and do not force the aircraft on the ground. Land on the remaining runway or overrun. Depending upon how quickly the pilot reacts to the sudden yaw, the airplane may run off the side of the runway by the time action is taken. There are really no other practical options. As discussed earlier, the chances of maintaining directional control while retracting the flaps (if extended), landing gear, feathering the propeller, and accelerating are minimal. On some airplanes with a single-engine-driven hydraulic pump, failure of that engine means the only way to raise the landing gear is to allow the engine to windmill or to use a hand pump. This is not a viable alternative during takeoff.

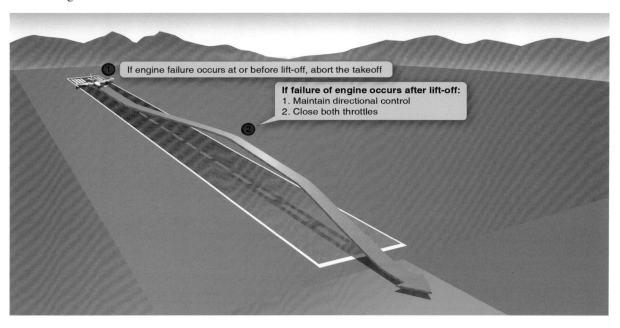

Figure 13-18. *Engine failure on takeoff, landing gear down.*

Landing Gear Control Selected Up, Single-Engine Climb Performance Inadequate

When operating near or above the single-engine ceiling and an engine failure is experienced shortly after lift-off, a landing needs to be accomplished on whatever essentially lies ahead. [Figure 13-19] There is also the option of continuing ahead, in a descent at V_{YSE} with the remaining engine producing power, as long as the pilot is not tempted to remain airborne beyond the airplane's performance capability. Remaining airborne and bleeding off airspeed in a futile attempt to maintain altitude is almost invariably fatal. Landing under control is paramount. The greatest hazard in a single-engine takeoff is attempting to fly when it is not within the performance capability of the airplane to do so. An accident is inevitable.

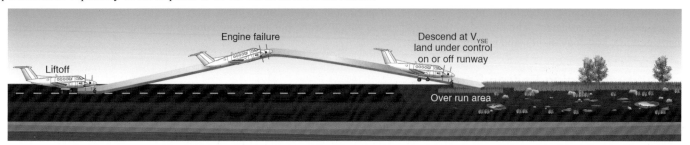

Figure 13-19. *Engine failure on takeoff, inadequate climb performance.*

Analysis of engine failures on takeoff reveals a very high success rate of off-airport engine inoperative landings when the airplane is landed under control. Analysis also reveals a very high fatality rate in stall spin accidents when the pilot attempts flight beyond the performance capability of the airplane.

As mentioned previously, if the airplane's landing gear retraction mechanism is dependent upon hydraulic pressure from a certain engine-driven pump, failure of that engine can mean a loss of hundreds of feet of altitude as the pilot either windmills the engine to provide hydraulic pressure to raise the gear or raises it manually with a backup pump.

Landing Gear Control Selected Up, Single-Engine Climb Performance Adequate

If the single-engine rate of climb is adequate, the procedures for continued flight should be followed. *[Figure 13-20]* There are four areas of concern: control, configuration, climb, and checklist.

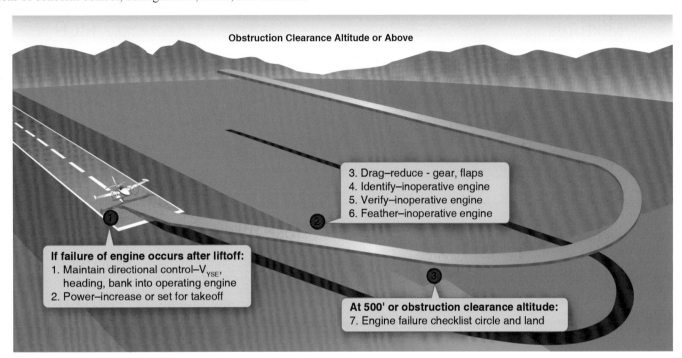

Figure 13-20. *Landing gear up—adequate climb performance.*

Control

The first consideration following engine failure during takeoff is to maintain control of the airplane. Maintaining directional control with prompt and often aggressive rudder application and STOPPING THE YAW is critical to the safety of flight. Ensure that airspeed stays above V_{MC}. If the yaw cannot be controlled with full rudder applied, reducing thrust on the operative engine is the only alternative. Attempting to correct the roll with aileron without first applying rudder increases drag and adverse yaw and further degrades directional control. After rudder is applied to stop the yaw, a slight amount of aileron should be used to bank the airplane toward the operative engine. This is the most efficient way to control the aircraft, minimize drag, and gain the most performance. Control forces, particularly on the rudder, may be high. The pitch attitude for V_{YSE} has to be lowered from that of V_Y. At least 5° and a maximum of 10° of bank toward the operative engine should be used initially to stop the yaw and maintain directional control. This initial bank input is held only momentarily, just long enough to establish or ensure directional control. Climb performance suffers when bank angles exceed approximately 2 or 3°, but obtaining and maintaining V_{YSE} and directional control are paramount. Trim should be adjusted to lower the control forces.

Configuration

The memory items from the engine failure after takeoff checklist should be promptly executed to configure the airplane for climb. *[Figure 13-21]* The specific procedures to follow are found in the AFM/POH and checklist for the particular airplane. Most direct the pilot to assume V_{YSE}, set takeoff power, retract the flaps and landing gear, identify, verify, and feather the failed engine. (On some airplanes, the landing gear is to be retracted before the flaps.)

The "identify" step is for the pilot to initially identify the failed engine. Confirmation on the engine gauges may or may not be possible, depending upon the failure mode. Identification should be primarily through the control inputs required to maintain straight flight, not the engine gauges. The "verify" step directs the pilot to retard the throttle of the engine thought to have failed. No change in performance when the suspected throttle is retarded is verification that the correct engine has been identified as failed. The corresponding propeller control should be brought fully aft to feather the engine.

Engine Failure After Takeoff

Airspeed	Maintain V_{YSE}
Mixtures	RICH
Propellers	HIGH RPM
Throttles	FULL POWER
Flaps	UP
Landing gear	UP
Identify	Determine failed engine
Verify	Close throttle of failed engine
Propeller	FEATHER
Trim tabs	ADJUST
Failed engine	SECURE
As soon as practical	LAND

Bold-faced items require immediate action and are to be accomplished from memory.

Figure 13-21. *Typical "engine failure after takeoff" emergency checklist.*

Climb

As soon as directional control is established and the airplane configured for climb, the bank angle should be reduced to that producing best climb performance. Without specific guidance for zero sideslip, a bank of 2° and one-third to one-half ball deflection on the slip/skid indicator toward the operative engine is suggested. V_{YSE} is maintained with pitch control. As turning flight reduces climb performance, climb should be made straight ahead or with shallow turns to avoid obstacles to an altitude of at least 400 feet AGL before attempting a return to the airport.

Checklist

Having accomplished the memory items from the engine failure after takeoff checklist, the printed copy should be reviewed as time permits. The securing failed engine checklist should then be accomplished. *[Figure 13-22]* Unless the pilot suspects an engine fire, the remaining items should be accomplished deliberately and without undue haste. Airplane control should never be sacrificed to execute the remaining checklists. The priority items have already been accomplished from memory.

Securing Failed Engine

Mixture .... IDLE CUT OFF

Magnetos .... OFF

Alternator .... OFF

Cowl flap .... CLOSE

Boost pump .... OFF

Fuel selector .... OFF

Prop sync .... OFF

Electrical load .... Reduce

Crossfeed .... Consider

Figure 13-22. *Typical "securing failed engine" emergency checklist.*

Other than closing the cowl flap of the failed engine, none of these items, if left undone, adversely affect airplane climb performance. There is a distinct possibility of actuating an incorrect switch or control if the procedure is rushed. The pilot should concentrate on flying the airplane and extracting maximum performance. If an ATC facility is available, an emergency should be declared.

The memory items in the engine failure after takeoff checklist may be redundant with the airplane's existing configuration. For example, in the third takeoff scenario, the gear and flaps were assumed to already be retracted, yet the memory items included gear and flaps. This is not an oversight. The purpose of the memory items is to either initiate the appropriate action or to confirm that a condition exists. Action on each item may not be required in all cases. The memory items also apply to more than one circumstance. In an engine failure from a go-around, for example, the landing gear and flaps would likely be extended when the failure occurred.

The three preceding takeoff scenarios all include the landing gear as a key element in the decision to land or continue. With the landing gear selector in the DOWN position, for example, continued takeoff and climb is not recommended. This situation, however, is not justification to retract the landing gear the moment the airplane lifts off the surface on takeoff as a normal procedure. The landing gear should remain selected down as long as there is usable runway or overrun available to land on. The use of wing flaps for takeoff virtually eliminates the likelihood of a single-engine climb until the flaps are retracted.

There are two time-tested memory aids the pilot may find useful in dealing with engine-out scenarios. The first, "dead foot—dead engine" is used to assist in identifying the failed engine. Depending on the failure mode, the pilot will not be able to consistently identify the failed engine in a timely manner from the engine gauges. In maintaining directional control, however, rudder pressure is exerted on the side (left or right) of the airplane with the operating engine. Thus, the "dead foot" is on the same side as the "dead engine." Variations on this saying include "idle foot—idle engine" and "working foot–working engine."

The second memory aid has to do with climb performance. The phrase "raise the dead" is a reminder that the best climb performance is obtained with a very shallow bank, about 2° toward the operating engine. Therefore, the inoperative, or "dead" engine should be "raised" with a very slight bank.

Not all engine failures result in complete power loss. If there is a performance loss when the throttle of the affected engine is retarded, some power is still available. In this case, the pilot may consider allowing the engine to run until the airplane reaches a safe altitude and airspeed for single-engine flight. While shutdown of a malfunctioning engine may prevent additional damage to the engine in certain circumstances, shutting down an engine that can still produce partial power may increase risk for an accident.

Engine Failure During Flight

Engine failures well above the ground are handled differently than those occurring at lower speeds and altitudes. Cruise airspeed allows better airplane control and altitude, which may permit time for a possible diagnosis and remedy of the failure. Maintaining airplane control, however, is still paramount. Airplanes have been lost at altitude due to apparent fixation on the engine problem to the detriment of flying the airplane.

Not all engine failures or malfunctions are catastrophic in nature (catastrophic meaning a major mechanical failure that damages the engine and precludes further engine operation). Many cases of power loss are related to fuel starvation, where restoration of power may be made with the selection of another tank. An orderly inventory of gauges and switches may reveal the problem. Carburetor heat or alternate air can be selected. The affected engine may run smoothly on just one magneto or at a lower power setting. Altering the mixture may help. If fuel vapor formation is suspected, fuel boost pump operation may be used to eliminate flow and pressure fluctuations.

Although it is a natural desire among pilots to save an ailing engine with a precautionary shutdown, the engine should be left running if there is any doubt as to needing it for further safe flight. Catastrophic failure accompanied by heavy vibration, smoke, blistering paint, or large trails of oil, on the other hand, indicate a critical situation. The affected engine should be feathered and the securing failed engine checklist completed. The pilot should divert to the nearest suitable airport and declare an emergency with ATC for priority handling.

Fuel crossfeed is a method of getting fuel from a tank on one side of the airplane to an operating engine on the other. Crossfeed is used for extended single-engine operation. If a suitable airport is close at hand, there is no need to consider crossfeed. If prolonged flight on a single-engine is inevitable due to airport non-availability, then crossfeed allows use of fuel that would otherwise be unavailable to the operating engine. It also permits the pilot to balance the fuel consumption to avoid an out-of-balance wing heaviness.

The AFM/POH procedures for crossfeed vary widely. Thorough fuel system knowledge is essential if crossfeed is to be conducted. Fuel selector positions and fuel boost pump usage for crossfeed differ greatly among multiengine airplanes. Prior to landing, crossfeed should be terminated and the operating engine returned to its main tank fuel supply.

If the airplane is above its single-engine absolute ceiling at the time of engine failure, it slowly loses altitude. The pilot should maintain V_{YSE} to minimize the rate of altitude loss. This "drift down" rate is greatest immediately following the failure and decreases as the single-engine ceiling is approached. Due to performance variations caused by engine and propeller wear, turbulence, and pilot technique, the airplane may not maintain altitude even at its published single-engine ceiling. Any further rate of sink, however, would likely be modest.

An engine failure in a descent or other low power setting can be deceiving. The dramatic yaw and performance loss is absent. At very low power settings, the pilot may not even be aware of a failure. If a failure is suspected, the pilot should advance both engine mixtures, propellers, and throttles significantly, to the takeoff settings if necessary, to correctly identify the failed engine. The power on the operative engine can always be reduced later.

Engine Inoperative Approach and Landing

The approach and landing with OEI is essentially the same as a two-engine approach and landing. The traffic pattern should be flown at similar altitudes, airspeeds, and key positions as a two-engine approach. The differences are the reduced power available and the fact that the remaining thrust is asymmetrical. A higher-than-normal power setting is necessary on the operative engine.

With adequate airspeed and performance, the landing gear can still be extended on the downwind leg. In which case it should be confirmed DOWN no later than abeam the intended point of landing. Performance permitting, initial extension of wing flaps (typically 10°) and a descent from pattern altitude can also be initiated on the downwind leg. The airspeed should be no slower than V_{YSE}. The direction of the traffic pattern, and therefore the turns, is of no consequence as far as airplane controllability and performance are concerned. It is perfectly acceptable to make turns toward the failed engine.

On the base leg, if performance is adequate, the flaps may be extended to an intermediate setting (typically 25°). If the performance is inadequate, as measured by decay in airspeed or high sink rate, delay further flap extension until closer to the runway. V_{YSE} is still the minimum airspeed to maintain.

On final approach, a normal 3° glidepath to a landing is desirable. Visual approach slope indicator (VASI) or other vertical path lighting aids should be utilized if available. Slightly steeper approaches may be acceptable. However, a long, flat, low approach should be avoided. Large, sudden power applications or reductions should also be avoided. Maintain V_{YSE} until the landing is assured, then slow to $1.3 V_{SO}$ or the AFM/POH recommended speed. The final flap setting may be delayed until the landing is assured or the airplane may be landed with partial flaps.

The airplane should remain in trim throughout. The pilot should be prepared, however, for a rudder trim change as the power of the operating engine is reduced to idle in the round out just prior to touchdown. With drag from only one windmilling propeller, the airplane tends to float more than on a two-engine approach. Precise airspeed control therefore is essential, especially when landing on a short, wet, and/or slippery surface.

Some pilots favor resetting the rudder trim to neutral on final and compensating for yaw by holding rudder pressure for the remainder of the approach. This eliminates the rudder trim change close to the ground as the throttle is closed during the round out for landing. This technique eliminates the need for groping for the rudder trim and manipulating it to neutral during final approach, which many pilots find to be highly distracting. AFM/POH recommendations or personal preference should be used.

A single-engine go-around on final approach may not be possible. As a practical matter in single-engine approaches, once the airplane is on final approach with landing gear and flaps extended, it is committed to land on the intended runway, on another runway, a taxiway, or grassy infield. Most light-twins do not have the performance to climb on one engine with landing gear and flaps extended. Considerable altitude is lost while maintaining V_{YSE} and retracting landing gear and flaps. Losses of 500 feet or more are not unusual. If the landing gear has been lowered with an alternate means of extension, retraction may not be possible, virtually negating any climb capability.

Multiengine Training Considerations

Flight training in a multiengine airplane can be safely accomplished if both the instructor and the learner consider the following factors.

- The participants should conduct a preflight briefing of the objectives, maneuvers, expected learner actions, and completion standards before the flight begins.

- A clear understanding exists as to how simulated emergencies will be introduced, and what action the learner is expected to take.

The introduction, practice, and testing of emergency procedures has always been a sensitive subject. Surprising a multiengine learner with an emergency without a thorough briefing beforehand creates a hazardous condition. Simulated engine failures, for example, can very quickly become actual emergencies or lead to loss of the airplane when approached carelessly. Stall-spin accidents in training for emergencies rival the number of stall-spin accidents from actual emergencies. The training risk normally gets mitigated by a briefing. Pulling circuit breakers is not recommended for training purposes and can lead to a subsequent gear up landing.

Many normal, abnormal, and emergency procedures can be introduced and practiced in the airplane as it sits on the ground without the engines running. In this respect, the airplane is used as a procedures trainer. The value of this training may be substantial. The engines do not have to be operating for real learning to occur. Upon completion of a training session, care should be taken to restore items to their proper positions.

Pilots who do not use a checklist effectively will be at a significant disadvantage in multiengine airplanes. Use of the checklist is essential to safe operation of airplanes, and it is risky to conduct a flight without one. The manufacturer's checklist or an aftermarket checklist that conforms to the manufacturer's procedures for the specific make, model, and model year may be used. If there is a procedural discrepancy between the checklist and the AFM/POH, then the AFM/POH always takes precedence.

Certain immediate action items (such as a response to an engine failure in a critical phase of flight) are best committed to memory. After they are accomplished, and as work load permits, the pilot can compare the action taken with a checklist.

Simulated engine failures during the takeoff ground roll may be accomplished with the mixture control. The simulated failure should be introduced at a speed no greater than 50 percent of V_{MC}. If a learner does not react promptly by retarding both throttles, the instructor can always pull the other mixture.

The FAA recommends that all in-flight simulated engine failures below 3,000 feet AGL, be introduced with a smooth reduction of the throttle. Thus, the engine is kept running and is available for instant use, if necessary. Smooth throttle reduction avoids abusing the engine and possibly causing damage. Simulation of inflight engine failures below V_{SSE} introduces a very high and unnecessary training risk.

If the engines are equipped with dynamic crankshaft counterweights, it is essential to make throttle reductions for simulated failures smoothly. Other areas leading to dynamic counterweight damage include high rpm and low manifold pressure combinations, over-boosting, and propeller feathering. Severe damage or repetitive abuse to counterweights will eventually lead to engine failure. Dynamic counterweights are found on larger, more complex engines—instructors may check with maintenance personnel or the engine manufacturer to determine if their airplane engines are so equipped.

When an instructor simulates an engine failure, the learner should respond with the appropriate memory items and retard the appropriate propeller control toward the FEATHER position. Assuming zero thrust will be set, the instructor promptly moves the propeller control forward and sets the appropriate manifold pressure and rpm. It is vital that the learner be kept informed of the instructor's intentions. At this point the instructor may say words to the effect, "I have the right engine; you have the left. I have set zero thrust and the right engine is simulated feathered." Any ambiguity as to who is operating what systems or controls increases the likelihood of an unintended outcome.

Following a simulated engine failure, the instructor cares for the "failed" engine just as the learner cares for the operative engine. If zero thrust is set to simulate a feathered propeller, the cowl flap is normally closed and the mixture leaned. An occasional clearing of the engine is also desirable. If possible, avoid high power applications immediately following a prolonged cool-down at a zero-thrust power setting. A competent flight instructor teaches the multiengine learner about the critical importance of feathering the propeller in a timely manner should an actual engine failure situation ever be encountered. A windmilling propeller, in many cases, has given the improperly trained multiengine pilot the mistaken perception that the engine is still developing useful thrust, resulting in a psychological reluctance to feather, as feathering results in cessation of propeller rotation. The flight instructor should spend ample time demonstrating the difference in the performance capabilities of the airplane with a simulated feathered propeller (zero thrust) as opposed to a windmilling propeller.

Actual and safe propeller feathering for training is performed at altitudes and positions where safe landings on established airports may be readily accomplished if the propeller will not unfeather. Plan unfeathering and restart to be completed no lower than 3,000 feet AGL. At certain elevations and with many popular multiengine training airplanes, this may be above the single-engine service ceiling, and level flight will not be possible.

Repeated feathering and unfeathering is hard on the engine and airframe, and is done as necessary to ensure adequate training. The FAA's Airman Certification Standards for a multiengine class rating contains a task for feathering and unfeathering of one propeller during flight in airplanes in which it is safe to do so.

While much of this chapter has been devoted to the unique flight characteristics of a multiengine airplane with one engine inoperative, the modern well-maintained reciprocating engine is remarkably reliable. When training in an airplane, initiation of a simulated engine inoperative emergency at low altitude normally occurs at a minimum of 400 feet AGL to mitigate the risk involved and only after the learner has successfully mastered engine inoperative procedures at higher altitudes. Initiating a simulated low altitude engine inoperative emergency in the airplane at extremely low altitude, immediately after liftoff, or below V_{SSE} creates a situation where there are non-existent safety margins.

For training in maneuvers that would be hazardous in flight, or for initial and recurrent qualification in an advanced multiengine airplane, consider a simulator training center or manufacturer's training course. Comprehensive training manuals and classroom instruction are available along with system training aids, audio/visuals, and flight training devices and simulators. Training under a wide variety of environmental and aircraft conditions is available through simulation. Emergency procedures that would be either dangerous or impossible to accomplish in an airplane can be done safely and effectively in a flight training device or simulator. The flight training device or simulator need not necessarily duplicate the specific make and model of airplane to be useful. Highly effective instruction can be obtained in training devices for other makes and models as well as generic training devices.

The majority of multiengine training is conducted in four-to-six place airplanes at weights significantly less than maximum. Single-engine performance, particularly, at low density altitudes, may be deceptively good. To experience the performance expected at higher weights, altitudes and temperatures, the instructor may occasionally artificially limit the amount of manifold pressure available on the operative engine. Airport operations above the single-engine ceiling can also be simulated in this matter. Avoid loading the airplane with passengers to practice emergencies at maximum takeoff weight since this practice creates an unnecessary training hazard.

The use of the touch-and-go landing and takeoff in multiengine flight training has always been somewhat controversial. The value of the learning experience may be offset by the hazards of reconfiguring the airplane for takeoff in extremely limited time as well as the loss of the follow-through ordinarily experienced in a full stop landing. Touch-and-goes are not recommended during initial aircraft familiarization in multiengine airplanes.

If touch-and-goes are to be performed at all, the learner and instructor responsibilities should be carefully briefed prior to each flight. Following touchdown, the learner will ordinarily maintain directional control while keeping the left hand on the yoke and the right hand on the throttles. The instructor resets the flaps and trim and announces when the airplane has been reconfigured. The multiengine airplane uses considerably more runway to perform a touch-and-go than a single-engine airplane. A full stop-taxi back landing is preferable during initial familiarization. Solo touch-and-goes in twins are strongly discouraged.

Chapter Summary

Small multiengine airplanes handle much like single-engine airplanes as long as both engines are functioning normally. A competent multiengine pilot, however, acquires the additional knowledge, risk mitigation strategies, and practical skills required to fly a multiengine airplane in case a loss of thrust from one engine actually occurs. In that case, the pilot will be able to take appropriate action leading to a safe outcome. Much of this chapter discussed loss of directional control. How to obtain the best performance with an inoperative engine was also described in detail. These two considerations correspond to the red radial line (V_{MC}) and the blue radial line (V_{YSE}) on the airspeed indicator. The actions a pilot takes when dealing with stalls, V_{MC}, or best performance vary greatly. Understanding these concepts, knowing how to mitigate the risks, and possessing the skills to handle an engine failure in a variety of situations, allows a pilot to enjoy the increased performance and safety provided when flying a multiengine airplane.

Airplane Flying Handbook (FAA-H-8083-3C)
Chapter 14: Transition to Tailwheel Airplanes

Introduction

Due to their design and structure, tailwheel airplanes (tailwheels) exhibit operational and handling characteristics different from those of tricycle-gear airplanes (nose-wheels). *[Figure 14-1]* A few aircraft, primarily antique and experimental, may have a tailskid instead of a tailwheel. The same principles discussed in this chapter usually apply to tailskid. In general, tailwheels are less forgiving of pilot error while in contact with the ground than are nose-wheels. This chapter focuses on the operational differences that occur during ground operations, takeoffs, and landings.

Figure 14-1. *The Piper Super Cub on the left is a popular tailwheel airplane. The airplane on the right is a Mooney M20, which is a nose-wheel (tricycle gear) airplane.*

Although still termed "conventional-gear airplanes," tailwheel designs are most likely to be encountered today by pilots who have first learned in nose-wheels. Therefore, tailwheel operations are approached as they appear to a pilot making a transition from nose-wheel designs.

Landing Gear

The main landing gear forms the principal support of the airplane on the ground. The tailwheel also supports the airplane, but steering and directional control are its primary functions. With the tailwheel-type airplane, the two main landing gear struts are attached to the airplane slightly ahead of the airplane's center of gravity (CG), so that the plane naturally rests in a nose-high attitude on the triangle created by the main gear and the tailwheel. This arrangement is responsible for the three major handling differences between nose-wheel and tailwheel airplanes. They center on directional instability, angle of attack (AOA), and crosswind weathervaning tendencies.

Proper usage of the rudder pedals is crucial for directional control while taxiing. Steering with the pedals may be accomplished through the forces of airflow or propeller slipstream acting on the rudder surface or through a direct mechanical linkage or a mechanical linkage acting through springs to turn the tailwheel. Initially, the pilot should taxi with the heels of the feet resting on the floor and the balls of the feet on the bottom of the rudder pedals. The feet should be slid up onto the brake pedals only when it is necessary to depress the brakes. This permits the simultaneous application of rudder and brake whenever needed. Some models of tailwheel airplanes are equipped with heel brakes rather than toe brakes. As in nose-wheel airplanes, brakes are used to slow and stop the aircraft and to increase turning authority when tailwheel steering inputs prove insufficient. Whenever used, brakes should be applied smoothly and evenly.

Instability

Because of the relative placement of the main gear and the CG, tailwheel aircraft are inherently unstable on the ground. As taxi turns are started, the aircraft begins to pivot on one or the other of the main wheels. From that point, with the CG aft of that pivot point, the forward momentum of the plane acts to continue and even tighten the turn without further steering inputs. Ordinarily, removal of rudder pressure does not stop a turn that has been started, and it is necessary to apply an opposite input (opposite rudder) to bring the aircraft back to straight-line travel. For this reason, many tailwheel airplanes are equipped with a centering spring(s) or similar device that returns the tailwheel to a center position upon relaxation of a rudder pedal input. However, this mechanism may not return the airplane to a straight line of travel from a tight turn.

If the initial rudder input is maintained after a turn has been started, the turn continues to tighten, an unexpected result for pilots accustomed to a nose-wheel. In consequence, it is common for pilots making the transition between the two types to experience difficulty in early taxi attempts. As long as taxi speeds are kept low, however, no serious problems result, and pilots typically adjust quickly to the technique of using rudder pressure to start a turn, then neutralizing the pedals as the turn continues, and finally using an opposite pedal input to stop the turn and regain straight-line travel.

Because of this inbuilt instability, the most important lesson that can be taught in tailwheel airplanes is to taxi and make turns at slow speeds.

Angle of Attack

A second strong contrast to nose-wheel airplanes, tailwheel aircraft make lift while on the ground anytime there is a relative headwind. The amount of lift obviously depends on the wind speed, but even at slow taxi speeds, the wings and ailerons are doing their best to aid in liftoff. This phenomenon requires care and management, especially during the takeoff and landing rolls, and is again unexpected by nose-wheel pilots making the transition.

Taxiing

On most tailwheel-type airplanes, directional control while taxiing is facilitated by the use of a steerable tailwheel, which operates along with the rudder. The tailwheel steering mechanism remains engaged when the tailwheel is operated through an arc of about 30° each side of center. Beyond that limit, the tailwheel breaks free and becomes full swiveling. In full swivel mode, the airplane can be pivoted within its own length, if desired. While taxiing, the steerable tailwheel should be used for making normal turns and the pilot's feet kept off the brake pedals to avoid unnecessary wear on the brakes.

When beginning to taxi, the brakes should be tested immediately for proper operation. This is done by first applying power to start the airplane moving slowly forward, then retarding the throttle and simultaneously applying pressure smoothly to both brakes. If braking action is unsatisfactory, the engine should be shut down immediately.

To turn the airplane on the ground, the pilot should apply rudder in the desired direction of turn and use whatever power or brake necessary to control the taxi speed. At very low taxi speeds, directional response is sluggish as surface friction acting on the tailwheel inhibits inputs through the steering springs. At normal taxi speeds, rudder inputs alone should be sufficient to start and stop most turns. During taxi, the AOA built in to the structure gives control placement added importance when compared to nose-wheel models.

When taxiing in a quartering headwind, the upwind wing can easily be lifted by gusting or strong winds unless ailerons are positioned to "kill" lift on that side (stick held into the wind). This is standard control positioning for both nose-wheel and tailwheel airplanes, so the difference lies only in the added tailwheel vulnerability created by the fuselage pitch attitude. At the same time, elevator should usually be held full back to add downward pressure to the tailwheel assembly and improve tailwheel steering response. However, in a strong quartering headwind a wing could lift, and the elevator may be held closer to neutral.

When taxiing with a quartering tailwind, this fuselage angle reduces the tendency of the wind to lift either wing. Nevertheless, the basic vulnerability to surface winds common to all tailwheel airplanes makes it essential to be aware of wind direction at all times, so holding the stick away from the crosswind is good practice (left aileron in a right quartering tailwind).

Elevator positioning in tailwinds is a bit more complex. Standard teaching tends to recommend full forward stick in any degree of tailwind, arguing that a tailwind striking the elevator when it is deflected full down increases downward pressure on the tailwheel assembly and increases directional control. Equally important, if the elevator were to remain deflected up, a strong tailwind can get under the control surface and lift the tail with unfortunate consequences for the propeller and engine.

While stick-forward positioning is essential in strong tailwinds, it is not likely to be an appropriate response when winds are light. The propeller wash in even lightly-powered airplanes is usually strong enough to overcome the effects of light tailwinds, producing a net headwind over the tail. This in turn suggests that back stick, not forward, does the most to help with directional control. If in doubt, it is best to sample the wind as you taxi and position the elevator where it will do the most good.

Weathervaning

Tailwheel airplanes have an exaggerated tendency to weathervane, or turn into the wind, when operated on the ground in crosswinds. This tendency is greatest when taxiing with a direct crosswind, a factor that makes maintaining directional control more difficult, sometimes requiring use of the brakes when tailwheel steering alone proves inadequate to counteract the weathervane effect.

Visibility

In the normal nose-high attitude, the engine cowling may be high enough to restrict the pilot's vision of the area directly ahead of the airplane while on the ground. Consequently, objects directly ahead are difficult, if not impossible to see. In aircraft that are completely blind ahead, all taxi movements should be started with a small turn to ensure no other plane or ground vehicle has positioned itself directly under the nose while the pilot's attention was distracted with getting ready to takeoff. In taxiing such an airplane, the pilot should alternately turn the nose from one side to the other (zigzag) or make a series of short S-turns. This should be done slowly, smoothly, positively, and cautiously.

Directional Control

After absorbing all the information presented to this point, the transitioning pilot may conclude that the best approach to maintaining directional control is to limit rudder inputs from fear of overcontrolling. Although intuitive, this is an incorrect assumption: the disadvantages built in to the tailwheel design sometimes require vigorous rudder inputs to maintain or retain directional control. The best approach is to understand the fact that tailwheel aircraft are not damaged from the use of too much rudder, but rather from rudder inputs held for too long.

Normal Takeoff Roll

Wing flaps should be lowered prior to takeoff if recommended by the manufacturer. After taxiing onto the runway, the airplane should be aligned with the intended takeoff direction, and the tailwheel positioned straight or centered. In airplanes equipped with a locking device, the tailwheel should be locked in the centered position. After releasing the brakes, the throttle should be smoothly and continuously advanced to takeoff power. The pilot should carefully avoid applying brake pressure during the takeoff roll.

After a brief period of acceleration, positive forward elevator should be applied to smoothly lift the tail. The goal is to achieve a pitch attitude that improves forward visibility and produces a smooth transition to climbing flight as the aircraft continues to accelerate.

It is important to note that nose-down pitch movement produces left yaw, the result of gyroscopic precession created by the propeller. The amount of force created by this precession is directly related to the rate the propeller axis is tilted when the tail is raised, so it is best to avoid an abrupt pitch change. Whether smooth or abrupt, the need to react to this yaw with rudder inputs emphasizes the increased directional demands common to tailwheel airplanes, a demand likely to be unanticipated by pilots transitioning from nose-wheel models.

As speed is gained on the runway, the added authority of the elevator naturally continues to pitch the nose forward. During this stage, the pilot should concentrate on maintaining a constant-pitch attitude by gradually reducing elevator deflection. At the same time, directional control should be maintained with smooth, prompt, positive rudder corrections. All this activity emphasizes the point that tailwheel planes start to "fly" long before leaving the runway surface.

Liftoff

When the appropriate pitch attitude is maintained throughout the takeoff roll, liftoff occurs when the AOA and airspeed combine to produce the necessary lift without any additional "rotation" input. The ideal takeoff attitude requires only minimum pitch adjustments shortly after the airplane lifts off to attain the desired climb speed.

All modern tailwheel aircraft can be lifted off in the three-point attitude. That is, the AOA with all three wheels on the ground does not exceed the critical AOA, and the wings will not be stalled. While instructive, this technique results in an unusually high pitch attitude and an AOA excessively close to stall, both inadvisable circumstances when flying only inches from the ground.

As the airplane leaves the ground, the pilot should continue to maintain straight flight and hold the proper pitch attitude. During takeoffs in strong, gusty winds, it is advisable to add an extra margin of speed before the airplane is allowed to leave the ground. A takeoff at the normal takeoff speed may result in a lack of positive control, or a stall, when the airplane encounters a sudden lull in strong, gusty wind or other turbulent air currents. In this case, the pilot should hold the airplane on the ground longer to attain more speed, then make a smooth, positive rotation to leave the ground.

Crosswind Takeoff

It is important to establish and maintain proper crosswind corrections prior to liftoff; that is, application of aileron deflection into the wind to keep the upwind wing from rising and rudder deflection as needed to prevent weathervaning.

Takeoffs made into strong crosswinds are the reason for maintaining a positive AOA (tail-low attitude) while accelerating on the runway. Because the wings are making lift during the takeoff roll, a strong upwind aileron deflection can bank the airplane into the wind and provide positive crosswind correction soon after the takeoff roll begins. The remainder of the takeoff roll is then made on the upwind main wheel while the pilot uses rudder to maintain the alignment of the longitudinal axis with the runway. As the airplane accelerates, the pilot smoothly decreases the pitch attitude and adjusts aileron and rudder control pressures to maintain the appropriate crosswind correction. If the pitch attitude remains excessively steep or if it is too flat, crosswind control during the ground roll becomes more difficult. As the aircraft leaves the runway, the wings can be leveled as appropriate drift correction (crab) is established.

Short-Field Takeoff

With the exception of flap settings and initial climb speed as recommended by the manufacturer, there is little difference between the techniques described above for normal takeoffs. After liftoff, the pitch attitude should be adjusted as required for obstacle clearance. However, note that manufacturers of some airplanes, especially of higher power, recommend a short-field technique with liftoff in a three-point attitude. Pilots should always review and follow the airplane manufacturer's recommended procedures.

Soft-Field Takeoff

Wing flaps may be lowered prior to starting the takeoff (if recommended by the manufacturer) to provide additional lift and transfer the airplane's weight from the wheels to the wings as early as possible. The airplane should be taxied onto the takeoff surface without stopping on a soft surface since mud or snow might bog the airplane down. The airplane should be kept in continuous motion with sufficient power while lining up for the takeoff roll. Due to the high power settings, it is usually best to have the elevator full up while taxiing onto the runway in soft conditions. There is not only the danger of the airplane bogging down, but also a danger of it tipping up onto its nose.

As the airplane is aligned with the proposed takeoff path, takeoff power is applied smoothly and as rapidly as the powerplant will accept without faltering. The tail should be kept very low to maintain the inherent positive AOA and to avoid any tendency of the airplane to nose over as a result of soft spots, tall grass, or deep snow.

When the airplane is held at a nose-high attitude throughout the takeoff run, the wings progressively relieve the wheels of more and more of the airplane's weight, thereby minimizing the drag caused by surface irregularities or adhesion. Once airborne, the airplane should be allowed to accelerate to climb speed in ground effect.

Landing

The difference between nose-wheel and tailwheel airplanes becomes apparent when discussing the touchdown and the period of deceleration to taxi speed. In the nose-wheel design, touchdown is followed quite naturally by a reduction in pitch attitude to bring the nose-wheel tire into contact with the runway. This pitch change reduces AOA, removes almost all wing lift, and rapidly transfers aircraft weight to the tires.

In tailwheel designs, this reduction of AOA and weight transfer are not practical and, as noted in the section on takeoffs, it is rare to encounter tailwheel planes designed so that the wings are beyond critical AOA in the three-point attitude. In consequence, the airplane continues to "fly" in the three-point attitude after touchdown, requiring careful attention to heading, roll, and pitch for an extended period.

Touchdown

Tailwheel airplanes are less forgiving of crosswind landing errors than nose-wheel models. It is important that touchdown occurs with the airplane's longitudinal axis parallel to the direction the airplane is moving along the runway. *[Figure 14-2]* Failure to accomplish this imposes side loads on the landing gear which leads to directional instability. To avoid side stresses and directional problems, the pilot should not allow the airplane to touch down while in a crab or while drifting.

There are two significantly different techniques used to manage tailwheel aircraft touchdowns: three-point and wheel landings. In the first, the airplane is held off the surface of the runway until the attitude needed to remain aloft matches the geometry of the landing gear. When touchdown occurs at this point, the main gear and the tailwheel make contact at the same time. In the second technique (wheel landings), the airplane is allowed to touch down earlier in the process in a lower pitch attitude, so that the main gear touch while the tail remains off the runway.

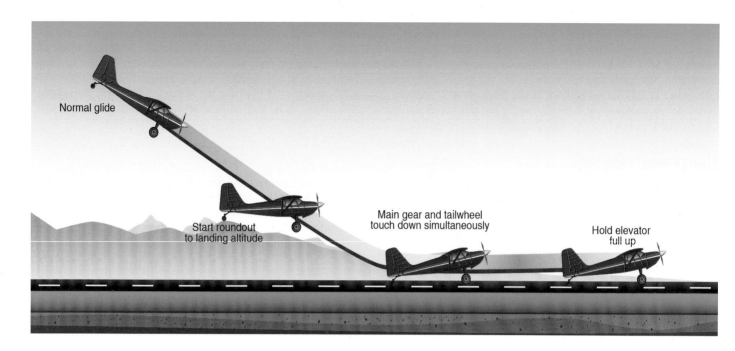

Figure 14-2. *Tailwheel touchdown.*

Three-Point Landing

As with all landings, success begins with an orderly arrival: airspeed, alignment, and configuration well in hand crossing the threshold. Round out (level-off) should be made with the main wheels about one foot off the surface. From that point forward, the technique is essentially the same that is used in nose-wheels: a gentle increase in AOA to maintain flight while slowing. In a tailwheel aircraft, however, the goal is to attain a much steeper fuselage angle than that commonly used in nose-wheel models; one that touches the tailwheels at the same time as the main wheels.

With the tailwheel on the surface, a further increase in pitch attitude is impossible, so the plane remains on the runway, albeit tenuously. With deceleration, weight shifts increasingly from wings to wheels, with the final result that the plane once again becomes a ground vehicle after shedding most of its speed.

There are two potential errors in attempting a three-point landing. In the first, the main wheels are allowed to make runway contact a little early with the tail still in the air. With the CG aft of the main wheels, the tail naturally drops when the main wheels touch, AOA increases, and the plane may become airborne again. This "skip" is easily managed by re-flaring and again trying to hold the plane off until reaching the three-point attitude. A large "skip" or bounce may result in being high above the runway with insufficient energy. In these circumstances, the pilot should execute a go-around.

In the second error, the plane is held off the ground a bit too long so that the in-flight pitch attitude is steeper than the three-point attitude. When touchdown is made in this attitude, the tail makes contact first. Provided this happens from no more than a foot off the surface, the result is undramatic: the tail touches, the plane pitches forward slightly onto the main wheels, and rollout proceeds normally.

In every case, once the tailwheel makes contact, the elevator control should be eased fully back to press the tailwheel on the runway. Without this elevator input, the AOA of the horizontal stabilizer develops enough lift to lighten pressure on the tailwheel and render it useless as a directional control with possibly unwelcomed consequences. This after-landing elevator input is quite foreign to nose-wheel pilots and needs to be stressed during transition training.

Note: Before the tailwheel is on the ground, application of full back elevator during the flare lowers the tail, increases the AOA, and quite naturally puts the plane in climbing flight.

Wheel Landing

In some wind conditions, the need to retain control authority may make it desirable to make contact with the runway at a higher airspeed than that associated with the three-point attitude. This necessitates landing in a flatter pitch attitude on the main wheels only, with the tailwheel still off the surface. *[Figure 14-3]* As noted, if the tail is off the ground, it tends to drop and put the plane airborne, so a soft touchdown and a slight relaxation of back elevator just after the wheels touch are key ingredients to a successful wheel landing.

Once the main wheels are on the surface, the tail should be permitted to drop on its own accord until it too makes ground contact. At this point, the elevator should be brought to the full aft position and deceleration should be allowed to proceed as in a three-point landing.

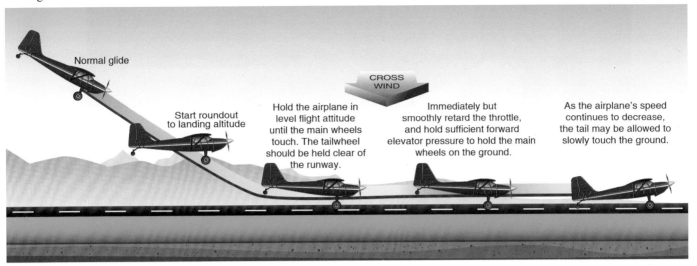

Figure 14-3. *Wheel landing.*

Within the figure:
- Normal glide
- Start roundout to landing altitude
- Hold the airplane in level flight attitude until the main wheels touch. The tailwheel should be held clear of the runway.
- CROSS WIND
- Immediately but smoothly retard the throttle, and hold sufficient forward elevator pressure to hold the main wheels on the ground.
- As the airplane's speed continues to decrease, the tail may be allowed to slowly touch the ground.

If the touchdown is made at too high a rate of descent, the tail is forced down by its own weight, resulting in a sudden increase in lift. If the pilot now pushes forward in an attempt to again make contact with the surface, a potentially dangerous pilot-induced oscillation may develop. It is far better to respond to a bounced wheel landing attempt by initiating a go-around or converting to a three-point landing if conditions permit.

Note: The only difference between three-point and wheel landings is the timing of the touchdown (early and later). There is no difference between the approach angles and airspeeds in the two techniques.

Crosswinds

As noted, it is highly desirable to eliminate crab and drift at touchdown. By far the best approach to crosswind management is a side-slip or wing-low touchdown. Landing in this attitude, only one main wheel makes initial contact, either in concert with the tailwheel in three-point landings or by itself in wheel landings. Many tailwheel pilots prefer completing a wheel landing in a crosswind, as the initial touchdown speed is higher than for a three-point landing, making the flight controls more effective. In addition, in some aircraft, the rudder effectiveness can be reduced by the blocking effect of the fuselage and flaps with the tail low and on the ground.

After-Landing Roll

The landing process should never be considered complete until the airplane decelerates to the normal taxi speed during the landing roll or has been brought to a complete stop when clear of the landing area. The pilot should be alert for directional control difficulties immediately upon and after touchdown, and the elevator control should be held back as far as possible and as firmly as possible until the airplane stops. This provides more positive control with tailwheel steering, tends to shorten the after-landing roll, and prevents bouncing and skipping.

Any difference between the direction the airplane is traveling and the direction it is headed (drift or crab) produces a moment about the pivot point of the wheels, and the airplane tends to swerve. Loss of directional control may lead to an aggravated, uncontrolled, tight turn on the ground, or a ground loop. The combination of inertia acting on the CG and ground friction of the main wheels during the ground loop may cause the airplane to tip enough for the outside wingtip to contact the ground and may even impose a sideward force that could collapse one landing gear leg. *[Figure 14-4]* In general, this combination of events is eliminated by landing straight and avoiding turns at higher than normal running speed.

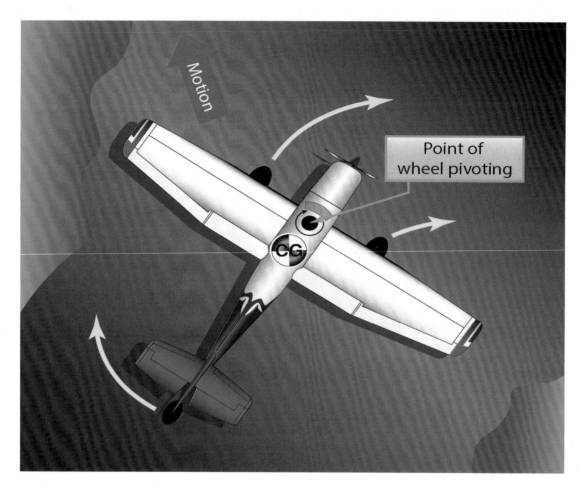

Figure 14-4. *Effect of CG on directional control.*

To use the brakes, the pilot should slide the toes or feet up from the rudder pedals to the brake pedals (or apply heel pressure in airplanes equipped with heel brakes). If rudder pressure is being held at the time braking action is needed, that pressure should not be released as the feet or toes are being slid up to the brake pedals because control may be lost before brakes can be applied. During the ground roll, the airplane's direction of movement may be changed by carefully applying pressure on one brake or uneven pressures on each brake in the desired direction. Caution should be exercised when applying brakes to avoid overcontrolling.

If a wing starts to rise, aileron control should be applied toward that wing to lower it. The amount required depends on speed because as the forward speed of the airplane decreases, the ailerons become less effective.

If available runway permits, the speed of the airplane should be allowed to dissipate in a normal manner by the friction and drag of the wheels on the ground. Brakes may be used if needed to help slow the airplane. After the airplane has been slowed sufficiently and has been turned onto a taxiway or clear of the landing area, it should be brought to a complete stop. Only after this is done should the pilot retract the flaps and perform other checklist items.

Crosswind After-Landing Roll

Particularly during the after-landing roll, special attention should be given to maintaining directional control by the use of rudder and tailwheel steering while keeping the upwind wing from rising by the use of aileron. Characteristically, an airplane has a greater profile or side area behind the main landing gear than forward of it. With the main wheels acting as a pivot point and the greater surface area exposed to the crosswind behind that pivot point, the airplane tends to turn or weathervane into the wind. *[Figure 14-5]* This weathervaning tendency is more prevalent in the tailwheel-type because the airplane's surface area behind the main landing gear is greater than in nose-wheel-type airplanes.

Pilots should be familiar with the crosswind component of each airplane they fly and avoid operations in wind conditions that exceed the capability of the airplane, as well as their own limitations. While the airplane is decelerating during the after-landing roll, more aileron should be applied to keep the upwind wing from rising. Since the airplane is slowing down, there is less airflow around the ailerons and they become less effective. At the same time, the relative wind is becoming more of a crosswind and exerting a greater lifting force on the upwind wing. Consequently, when the airplane is coming to a stop, the aileron control should be held fully toward the wind.

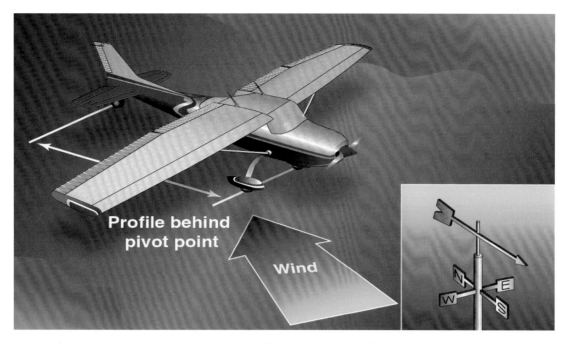

Figure 14-5. *Weathervaning tendency.*

Short-Field Landing

Upon touchdown, the airplane should be firmly held in a three-point attitude. This provides aerodynamic braking by the wings. Immediately upon touchdown and closing the throttle, the brakes should be applied evenly and firmly to minimize the after-landing roll. The airplane should be stopped within the shortest possible distance consistent with safety.

Soft-Field Landing

The tailwheel should touchdown simultaneously with or just before the main wheels and should then be held down by maintaining firm back-elevator pressure throughout the landing roll. This minimizes any tendency for the airplane to nose over and provides aerodynamic braking. The use of brakes on a soft field is not needed because the soft or rough surface itself provides sufficient reduction in the airplane's forward speed. Often, it is found that upon landing on a very soft field, the pilot needs to increase power to keep the airplane moving and from becoming stuck in the soft surface.

Ground Loop

A ground loop is an uncontrolled turn during ground operations that may occur during taxi, takeoff, or during the after-landing roll. Ground loops start with a swerve that is allowed to continue for too long. The swerve may be the result of side-load on landing, a taxi turn started with too much groundspeed, overcorrection, or even an uneven ground surface or a soft spot that retards one main wheel of the airplane.

Due to the inbuilt instability of the tailwheel design, the forces that lead to a ground loop accumulate as the angle between the fuselage and inertia, acting from the CG, increase. If allowed to develop, these forces may become great enough to tip the airplane to the outside of the turn until one wing strikes the ground.

To counteract the possibility of an uncontrolled turn, the pilot should counter any swerve with firm rudder input. In stronger swerves, differential braking is essential as tailwheel steering proves inadequate. It is important to note, however, that as corrections begin to become apparent, rudder and braking inputs need to be removed promptly to avoid starting yet another departure in the opposite direction.

Chapter Summary

This chapter focuses on the operational differences between tailwheel and nose-wheel airplanes that occur during ground operations, takeoffs, and landings. The chapter covers specific topics, such as landing gear, taxiing, visibility, liftoff, and landing. Comparisons are given as to how each react during the takeoff and landing, as well as situations that should be avoided. Pilots who use proper rudder control techniques should be able to transition to tailwheel airplanes without too much difficulty.

Airplane Flying Handbook (FAA-H-8083-3C)
Chapter 15: Transition to Turbopropeller-Powered Airplanes

Introduction

The turbopropeller-powered airplane flies and handles just like any other airplane of comparable size and weight, since the aerodynamics are the same. The major differences between flying a turboprop and other non-turbine-powered airplanes are found in the handling of the airplane's powerplant and its associated systems, which are unique to gas turbine engines. The turbopropeller-powered airplane also has the advantage of being equipped with a constant speed, full feathering and reversing propeller—something normally not found on piston-powered airplanes.

Gas Turbine Engine

Both piston (reciprocating) engines and gas turbine engines are internal combustion engines. They have a similar cycle of operation that consists of induction, compression, combustion, expansion, and exhaust. In a piston engine, each of these events is a separate distinct occurrence in each cylinder. Also in a piston engine, an ignition event occurs during each cycle in each cylinder. Unlike reciprocating engines, in gas turbine engines these phases of power occur simultaneously and continuously instead of successively one cycle at a time. Additionally, ignition occurs during the starting cycle and is continuous thereafter. The basic gas turbine engine contains four sections: intake, compression, combustion, and exhaust. *[Figure 15-1]*

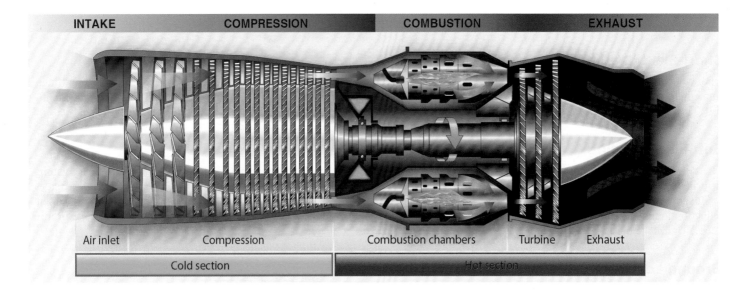

Figure 15-1. *Basic components of a gas turbine engine.*

To start a gas turbine engine, the compressor section is normally rotated by an electric starter. As compressor revolutions per minute (rpm) increase, air flowing through the inlet is compressed to a high pressure, delivered to the combustion section, and ignited. In gas turbine engines, not all of the compressed air is used to support combustion. Some of the compressed air bypasses the burner section within the engine to provide internal cooling. The fuel/air mixture in the combustion chamber burns in a continuous combustion process and produces a very high temperature, typically around 4,000° Fahrenheit (F). When this hot air mixes with bypass air, the temperature of the mixed air mass drops to 1,600 – 2,400 °F. The mixture of hot air and gases expands and passes through the turbine blades forcing the turbine section to rotate. The turbine drives the compressor section by means of a direct shaft, a concentric shaft, or a combination of both. After powering the turbine section, the combustion gases and bypass air flow out of the engine through the exhaust. Once the hot gases from the burner section provide sufficient power to maintain engine operation through the turbine, the starter is de-energized, and the starting sequence ends. Combustion continues until the engine is shut down by cutting off the fuel supply.

Note: Because compression produces heat and pressure, some pneumatic aircraft systems tap into the source of hot compressed air from the engine compressor (bleed air) and use it for engine anti-ice, airfoil anti-ice, aircraft pressurization, and other ancillary systems after further conditioning its internal pressure and temperature.

Turboprop Engines

The turbojet engine (discussed in more detail in the Transition to Jet-Powered Airplanes chapter) excels the reciprocating engine in top speed and altitude performance. On the other hand, the turbojet engine has limited takeoff and initial climb performance when compared to its overall performance. In the matter of takeoff and initial climb performance, the reciprocating engine with a constant speed propeller produces maximum thrust on takeoff. Turbojet engines are most efficient at high speeds and high altitudes, while propellers are most efficient at slow and medium speeds (less than 400 miles per hour (mph)). Propellers also improve takeoff and climb performance. The development of the turboprop engine was an attempt to combine the best characteristics of both the turbojet and propeller-driven reciprocating engine.

The turboprop engine offers several advantages over other types of engines, such as:

1. Light weight
2. Mechanical reliability due to relatively few moving parts
3. Simplicity of operation
4. Minimum vibration
5. High power per unit of weight
6. Use of propeller for takeoff and landing

Turboprop engines are most efficient at speeds between 250 and 400 mph and altitudes between 18,000 and 30,000 feet. They also perform well at the slow speeds required for takeoff and landing and are fuel efficient. The minimum specific fuel consumption of the turboprop engine is normally available in the altitude range of 25,000 feet up to the tropopause.

The power output of a piston engine is measured in horsepower and is determined primarily by rpm and manifold pressure. The power of a turboprop engine, however, is measured in shaft horsepower (shp). Shaft horsepower is determined by the rpm and the torque (twisting moment) applied to the propeller shaft. Since turboprop engines are gas turbine engines, some jet thrust is produced by exhaust leaving the engine. This thrust is added to the shaft horsepower to determine the total engine power or equivalent shaft horsepower (eshp). Jet thrust usually accounts for less than 10 percent of the total engine power.

Although the turboprop engine is more complicated and heavier than a turbojet engine of equivalent size and power, it delivers more thrust at low subsonic airspeeds. However, the advantages decrease as flight speed increases. In normal cruising speed ranges, the propulsive efficiency (output divided by input) of a turboprop decreases as speed increases.

The propeller of a typical turboprop engine is responsible for roughly 90 percent of the total thrust under sea level conditions on a standard day. The excellent performance of a turboprop during takeoff and climb is the result of the ability of the propeller to accelerate a large mass of air while the airplane is moving at a relatively low ground and flight speed. "Turboprop," however, should not be confused with "turbo supercharged" or similar terminology. All turbine engines have a similarity to normally aspirated (non-supercharged) reciprocating engines in that maximum available power decreases almost as a direct function of increased altitude.

Although power decreases as the airplane climbs to higher altitudes, engine efficiency in terms of specific fuel consumption (expressed as pounds of fuel consumed per horsepower per hour) is increased. Decreased specific fuel consumption plus the increased true airspeed at higher altitudes is a definite advantage of a turboprop engine.

All turbine engines should operate within their limiting temperatures, rotational speeds, and (in the case of turboprops) torque. Depending on the installation, the primary parameter for power setting might be temperature, torque, fuel flow, or rpm (either propeller rpm, gas generator (compressor) rpm, or both). In cold weather conditions, torque limits can be exceeded while temperature limits are still within acceptable range. In hot weather conditions, the maximum temperature limits may be exceeded without exceeding torque limits. In any weather, reaching one of these operating limits normally occurs before the pilot moves the throttles to the full forward position. The transitioning pilot should understand the importance of knowing and observing limits on turbine engines. An over temperature or over torque condition that lasts for more than a few seconds can destroy internal engine components.

Turboprop Engine Types
Fixed-Shaft

One type of turboprop engine is the fixed-shaft constant-speed type, such as the Garrett TPE331. [Figure 15-2] In this type engine, ambient air is directed to the compressor section through the engine inlet. An acceleration/diffusion process in the two-stage compressor increases air pressure and directs it rearward to a combustor. The combustor is made up of a combustion chamber, a transition liner, and a turbine plenum. Atomized fuel is added to the air in the combustion chamber. Air also surrounds the combustion chamber to provide for cooling and insulation of the combustor.

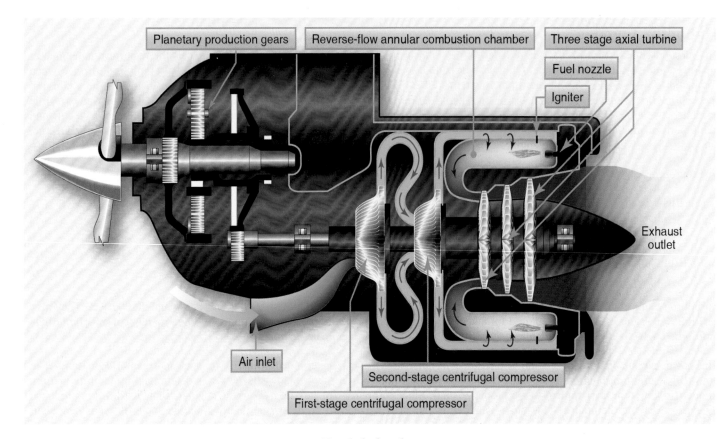

Figure 15-2. *Fixed-shaft turboprop engine.*

The gas mixture is initially ignited by high-energy igniter plugs, and the expanding combustion gases flow to the turbine. The energy of the hot, high-velocity gases is converted to torque on the main shaft by the turbine rotors. The reduction gear converts the high rpm—low torque of the main shaft to low rpm—high torque to drive the accessories and the propeller. The spent gases leaving the turbine are directed to the atmosphere by the exhaust pipe.

Most of the air passing through the engine provides internal cooling. Only about 10 percent of the air that passes through the engine is actually used in the combustion process. Up to approximately 20 percent of the compressed air may be bled off for the purpose of heating, cooling, cabin pressurization, and pneumatic systems. Over half the engine power is devoted to driving the compressor, and it is the compressor that can potentially produce very high drag in the case of a failed, windmilling engine.

In the fixed-shaft constant-speed engine, the engine rpm may be varied within a narrow range of 96 percent to 100 percent. During ground operation, the rpm may be reduced to 70 percent. In flight, the engine operates at a constant speed that is maintained by the governing section of the propeller. Power changes are made by increasing fuel flow and propeller blade angle rather than engine speed. An increase in fuel flow causes an increase in temperature and a corresponding increase in energy available to the turbine. The turbine absorbs more energy and transmits it to the propeller in the form of torque. The increased torque forces the propeller blade angle to be increased to maintain the constant speed. Turbine temperature is a very important factor to be considered in power production. It is directly related to fuel flow and thus to the power produced. It needs to be limited because of strength and durability of the material in the combustion and turbine section. The control system schedules fuel flow to produce specific temperatures and to limit those temperatures so that the temperature tolerances of the combustion and turbine sections are not exceeded. The engine is designed to operate for its entire life at 100 percent. All of its components, such as compressors and turbines, are most efficient when operated at or near the rpm design point.

Powerplant (engine and propeller) control is achieved by means of a power lever and a condition lever for each engine. *[Figure 15-3]* There is no mixture control and/or rpm lever as found on piston-engine airplanes.

On the fixed-shaft constant-speed turboprop engine, the power lever is advanced or retarded to increase or decrease forward thrust. The power lever is also used to provide reverse thrust. The condition lever sets the desired engine rpm within a narrow range between that appropriate for ground operations and flight.

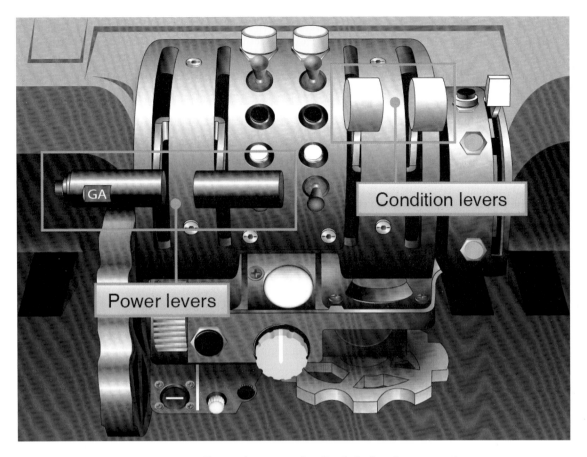

Figure 15-3. *Powerplant controls—fixed-shaft turboprop engine.*

Powerplant instrumentation in a fixed-shaft turboprop engine typically consists of the following basic indicators. *[Figure 15-4]*

1. Torque or horsepower
2. Interturbine temperature (ITT)
3. Fuel flow
4. RPM

Figure 15-4. *Powerplant instrumentation—fixed-shaft turboprop engine.*

Torque developed by the turbine section is measured by a torque sensor. The torque is then reflected on the instrument panel horsepower gauge calibrated in horsepower times 100. ITT is a measurement of the combustion gas temperature between the first and second stages of the turbine section. The gauge is calibrated in degrees Celsius (°C). Propeller rpm is reflected on a tachometer as a percentage of maximum rpm. Normally, a vernier indicator on the gauge dial indicates rpm in 1 percent graduations as well. The fuel flow indicator indicates fuel flow rate in pounds per hour.

Propeller feathering in a fixed-shaft constant-speed turboprop engine is normally accomplished with the condition lever. An engine failure in this type engine, however, results in a serious drag condition due to the large power requirements of the compressor being absorbed by the propeller. This could create a serious airplane control problem in twin-engine airplanes unless the failure is recognized immediately and the affected propeller feathered. For this reason, the fixed-shaft turboprop engine is equipped with negative torque sensing (NTS).

NTS is a condition wherein propeller torque drives the engine, and the propeller is automatically driven to high pitch to reduce drag. The function of the negative torque sensing system is to limit the torque the engine can extract from the propeller during windmilling and thereby prevent large drag forces on the airplane. The NTS system causes a movement of the propeller blades automatically toward their feathered position should the engine suddenly lose power while in flight. The NTS system is an emergency backup system in the event of sudden engine failure. It is not a substitution for the feathering device controlled by the condition lever.

Split-Shaft/Free Turbine Engine

In a free power-turbine engine, such as the Pratt & Whitney PT-6 engine, the propeller is driven by a separate turbine through reduction gearing. The propeller is not on the same shaft as the basic engine turbine and compressor. *[Figure 15-5]* Unlike the fixed-shaft engine, in the split-shaft engine the propeller can be feathered in flight or on the ground with the basic engine still running. The free power-turbine design allows the pilot to select a desired propeller governing rpm, regardless of basic engine rpm.

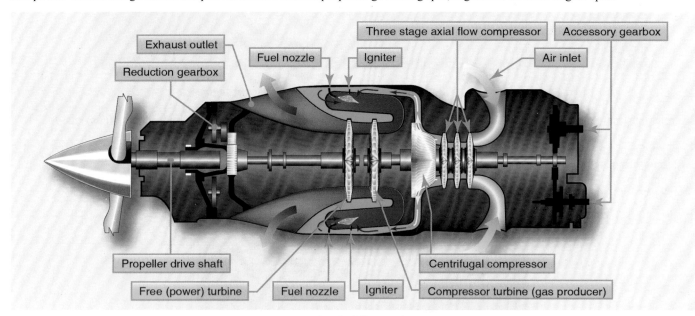

Figure 15-5. *Split shaft/free turbine engine.*

A typical free power-turbine engine has two independent counter-rotating turbines. One turbine drives the compressor, while the other drives the propeller through a reduction gearbox. The compressor in the basic engine consists of three axial flow compressor stages combined with a single centrifugal compressor stage. The axial and centrifugal stages are assembled on the same shaft and operate as a single unit.

Inlet air enters the engine via a circular plenum near the rear of the engine and flows forward through the successive compressor stages. The flow is directed outward by the centrifugal compressor stage through radial diffusers before entering the combustion chamber, where the flow direction is actually reversed. The gases produced by combustion are once again reversed to expand forward through each turbine stage. After leaving the turbines, the gases are collected in a peripheral exhaust scroll and are discharged to the atmosphere through two exhaust ports near the front of the engine.

A pneumatic fuel control system schedules fuel flow to maintain the power set by the gas generator power lever. Except in the beta range, propeller speed within the governing range remains constant at any selected propeller control lever position through the action of a propeller governor.

The accessory drive at the aft end of the engine provides power to drive fuel pumps, fuel control, oil pumps, a starter/generator, and a tachometer transmitter. At this point, the speed of the drive (N_1) is the true speed of the compressor side of the engine, approximately 37,500 rpm.

Powerplant (engine and propeller) operation is achieved by three sets of controls for each engine: the power lever, propeller lever, and condition lever. *[Figure 15-6]* The power lever serves to control engine power in the range from idle through takeoff power. Forward or aft motion of the power lever increases or decreases gas generator rpm (N_1) and thereby increases or decreases engine power. The propeller lever is operated conventionally and controls the constant-speed propellers through the primary governor. The propeller rpm range is normally from 1,500 to 1,900. The condition lever controls the flow of fuel to the engine. Like the mixture lever in a piston-powered airplane, the condition lever is located at the far right of the power quadrant. But the condition lever on a turboprop engine is really just an on/off valve for delivering fuel. There are HIGH IDLE and LOW IDLE positions for ground operations, but condition levers have no metering function. Leaning is not required in turbine engines; this function is performed automatically by a dedicated fuel control unit.

Figure 15-6. *Powerplant controls—split-shaft/free turbine engine.*

Engine instruments in a split-shaft/free turbine engine typically consist of the following basic indicators.
[Figure 15-7]

1. ITT indicator
2. Torquemeter
3. Propeller tachometer
4. N₁ (gas generator) tachometer
5. Fuel flow indicator
6. Oil temperature/pressure indicator

Figure 15-7. *Engine instruments—split shaft/free turbine engine.*

The ITT indicator gives an instantaneous reading of engine gas temperature between the compressor turbine and the power turbines. The torquemeter responds to power lever movement and gives an indication in foot-pounds (ft/lb) of the torque being applied to the propeller. Because in the free turbine engine the propeller is not attached physically to the shaft of the gas turbine engine, two tachometers are justified—one for the propeller and one for the gas generator. The propeller tachometer is read directly in revolutions per minute. The N_1 or gas generator is read in percent of rpm. In the Pratt & Whitney PT-6 engine, it is based on a figure of 37,000 rpm at 100 percent. Maximum continuous gas generator is limited to 38,100 rpm or 101.5 percent N_1.

The ITT indicator and torquemeter are used to set takeoff power. Climb and cruise power are established with the torquemeter and propeller tachometer while observing ITT limits. Gas generator (N_1) operation is monitored by the gas generator tachometer. Proper observation and interpretation of these instruments provide an indication of engine performance and condition.

Reverse Thrust and Beta Range Operations

The thrust that a propeller provides is a function of the angle of attack (AOA) at which the air strikes the blades, and the speed at which this occurs. The AOA varies with the pitch angle of the propeller.

Forward pitch produces forward thrust—higher pitch angles being required at higher airplane speeds. *[Figure 15-8A]* So called "flat pitch," shown in *Figure 15-8B,* is the blade position offering minimum resistance to rotation and no net thrust for moving the airplane.

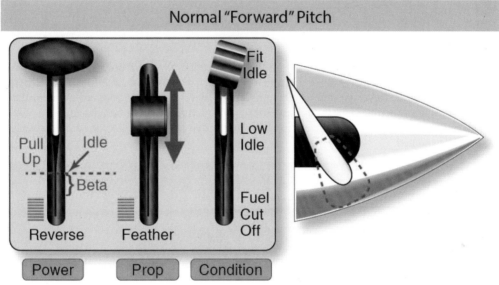

Figure 15-8A. *Propeller forward pitch angle characteristics.*

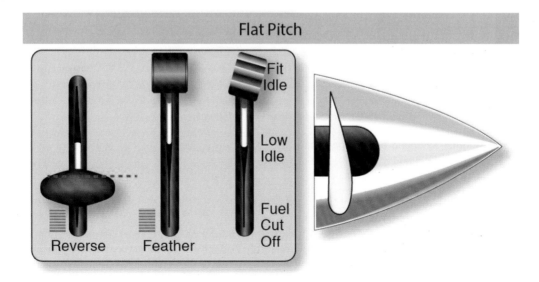

Figure 15-8B. *Propeller flat pitch characteristics.*

The "feathered" position is the highest pitch angle obtainable. *[Figure 15-8C]* The feathered position produces no forward thrust. The propeller is generally placed in feather only in case of in-flight engine failure to minimize drag and prevent the air from using the propeller as a turbine.

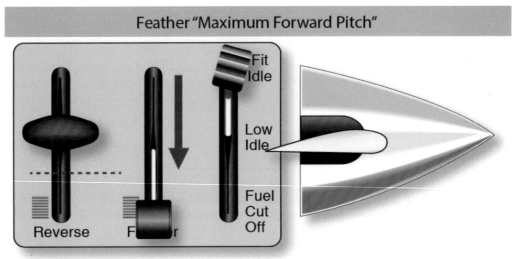

Figure 15-8C. *Propeller feather (maximum forward pitch angle) characteristics.*

In the "reverse" pitch position, the engine/propeller turns in the same direction as in the normal (forward) pitch position, but the propeller blade angle is positioned to the other side of flat pitch. *[Figure 15-8D]* In reverse pitch, air is pushed away from the airplane rather than being drawn over it. Reverse pitch results in braking action, rather than forward thrust of the airplane. It is used for backing away from obstacles when taxiing, controlling taxi speed, or to aid in bringing the airplane to a stop during the landing roll. Reverse pitch does not mean reverse rotation of the engine. The engine delivers power just the same, no matter which side of flat pitch the propeller blades are positioned.

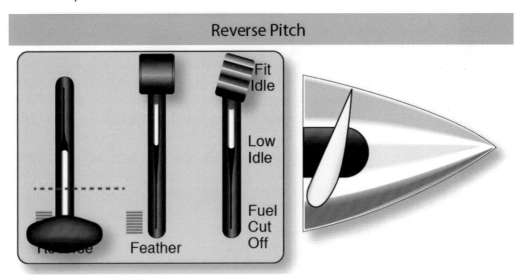

Figure 15-8D. *Propeller reverse pitch characteristics.*

With a turboprop engine, in order to obtain enough power for flight, the power lever is placed somewhere between flight idle (in some engines referred to as "high idle") and maximum. The power lever directs signals to a fuel control unit to manually select fuel. The propeller governor selects the propeller pitch needed to keep the propeller/engine on speed. This is referred to as the propeller governing or "alpha" mode of operation. When positioned aft of flight idle, however, the power lever directly controls propeller blade angle. This is known as the "beta" range of operation.

The beta range of operation consists of power lever positions from flight idle to maximum reverse. Beginning at power lever positions just aft of flight idle, propeller blade pitch angles become progressively flatter with aft movement of the power lever until they go beyond maximum flat pitch and into negative pitch, resulting in reverse thrust. While in a fixed-shaft/constant-speed engine, the engine speed remains largely unchanged as the propeller blade angles achieve their negative values. On the split-shaft PT-6 engine, as the negative 5° position is reached, further aft movement of the power lever also results in a progressive increase in engine (N_1) rpm until a maximum value of about negative 11° of blade angle and 85 percent N_1 are achieved.

Operating in the beta range and/or with reverse thrust requires specific techniques and procedures depending on the particular airplane make and model. Specific engine parameters and limitations for operations within this area should be adhered to. It is essential that a pilot transitioning to turboprop airplanes becomes knowledgeable and proficient in these areas, which are unique to turbine-engine powered airplanes.

Turboprop Airplane Electrical Systems

The typical turboprop airplane electrical system is a 28-volt direct current (DC) system, which receives power from one or more batteries and a starter/generator for each engine. The batteries are either lead-acid, nickel-cadmium (NiCad), or Lithium-ion. When battery voltage is low, its ability to turn the compressor for engine start is greatly diminished, and the possibility of engine damage due to a hot start increases. Therefore, it is essential to check the battery's condition before every engine start. The different battery types have different operating characteristics depending on the specific aircraft installation and operational environment.

The DC generators used in turboprop airplanes double as starter motors and are called "starter/generators." The starter/generator uses electrical power to produce mechanical torque to start the engine and then uses the engine's mechanical torque to produce electrical power after the engine is running. Some of the DC power produced is changed to 28 volt 400 cycle alternating current (AC) power for certain avionic, lighting, and indicator synchronization functions. This is accomplished by an electrical component called an inverter.

The distribution of DC and AC power throughout the system is accomplished through the use of power distribution buses. These "buses" as they are called are actually common terminals from which individual electrical circuits get their power. *[Figure 15-9]*

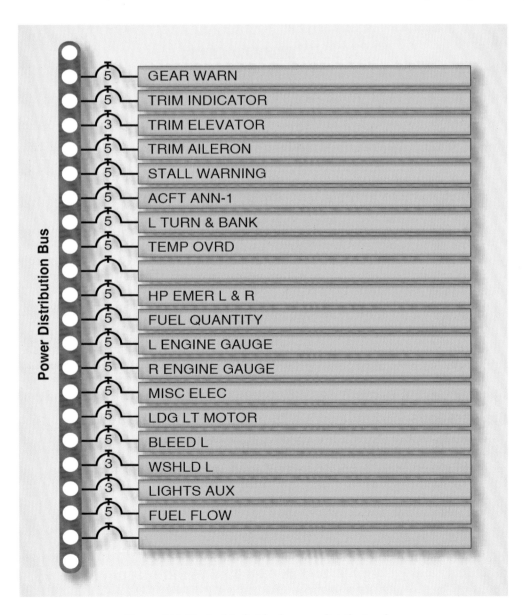

Figure 15-9. *Typical individual power distribution bus.*

Buses are usually named for what they power (avionics bus, for example) or for where they get their power (right generator bus, battery bus). The distribution of DC and AC power is often divided into functional groups (buses) that give priority to certain equipment during normal and emergency operations. Main buses serve most of the airplane's electrical equipment. Essential buses feed power to equipment having top priority. *[Figure 15-10]*

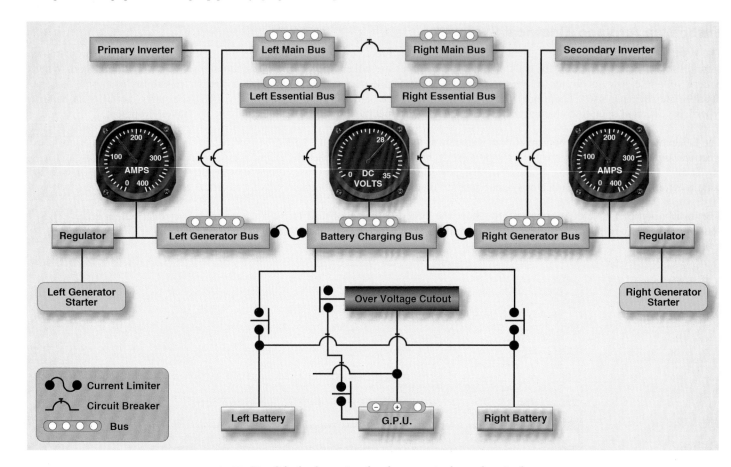

Figure 15-10. *Simplified schematic of turboprop airplane electrical system.*

Multiengine turboprop airplanes normally have several power sources—at least one generator per engine and at least one battery for the airplane. The electrical systems are usually designed so that any bus can be energized by any of the power sources. For example, a typical system has a left and right engine generator-powered bus. While these buses are normally isolated, they may be fed from other power sources. However, in the event of a short-circuit, the bus remains isolated. Pilots should refer to the appropriate checklist when an electrical fault occurs.

Power distribution buses are protected from short circuits and other malfunctions by a type of fuse called a current limiter. In the case of excessive current supplied by any power source, the current limiter opens the circuit and thereby isolates that power source and separates the affected bus from the system. If this occurs, pilots should refer to the appropriate checklist.

Operational Considerations

As previously stated, a turboprop airplane flies just like any other piston engine airplane of comparable size and weight. It is the operation of the engines and airplane systems that makes the turboprop airplane different from its piston engine counterpart. Pilot errors in engine and/or systems operation are common causes of aircraft damage or loss of aircraft control. There are two engine-related issues that should be considered when a pilot transitions to turboprop operations.

The first issue concerns the split-shaft/free turbine engine, where power output lags for several seconds when the pilot moves the power lever from flight idle to a high power setting. This delay may surprise a pilot who has only flown airplanes with a piston engine (or a fixed-shaft turboprop). Certain operations such as firefighting and agricultural application require maneuvering close to the ground while operating at or near flight idle. Although smooth power applications are still the rule, the pilot should be aware that a greater physical movement of the power levers is required as compared to throttle movement in a piston engine. The pilot should understand the lag and anticipate and lead the power changes more than in the past and should keep in mind that the last 30 percent of engine rpm represents the majority of the engine thrust. Below that setting, the application of power has very little effect.

A second consideration for transitioning pilots concerns turbine engine heat sensitivity. A turbine engine cannot tolerate an over temperature condition for more than a very seconds without experiencing serious damage. Engine temperatures get hotter during starting than at any other time. Thus, turbine engines have minimum rotational speeds for introducing fuel into the combustion chambers during startup. Vigilant monitoring of temperature and acceleration on the part of the pilot remain crucial until the engine is running at a stable speed. Successful engine starting depends on assuring the correct minimum battery voltage before initiating start or employing a ground power unit (GPU) of adequate output.

After fuel is introduced to the combustion chamber during the start sequence, "light-off" and its associated heat rise occur very quickly. Engine temperatures may approach the maximum in a matter of 2 or 3 seconds before the engine stabilizes and temperatures fall into the normal operating range. During this time, the pilot should watch for any tendency of the temperatures to exceed limitations and be prepared to cut off fuel to the engine.

An engine tendency to exceed maximum starting temperature limits is termed a hot start. The temperature rise may be preceded by unusually high initial fuel flow, which may be the first indication the pilot has that the engine start is not proceeding normally. Serious engine damage occurs if the hot start is allowed to continue.

A condition where the engine is accelerating more slowly than normal is termed a hung start or false start. During a hung start/false start, the engine may stabilize at an engine rpm that is not high enough for the engine to continue to run without help from the starter. This is usually the result of low battery power or the starter not turning the engine fast enough for it to start properly.

Takeoffs in turboprop airplanes are not made by automatically pushing the power lever full forward to the stops. As stated earlier, depending on conditions, takeoff power may be limited by either torque or by engine temperature. Normally, the power lever position on takeoff is somewhat aft of full forward.

Takeoff and departure in a turboprop airplane (especially a twin-engine cabin-class airplane) should be accomplished in accordance with a standard takeoff and departure "profile" developed for the particular make and model. *[Figure 15-11]* The takeoff and departure profile should be in accordance with the airplane manufacturer's recommended procedures as outlined in the Federal Aviation Administration (FAA)-approved Airplane Flight Manual and/or the Pilot's Operating Handbook (AFM/POH). The increased complexity of turboprop airplanes makes the standardization of procedures a necessity for safe and efficient operation. The transitioning pilot should review the profile procedures before each takeoff to form a mental picture of the takeoff and departure process.

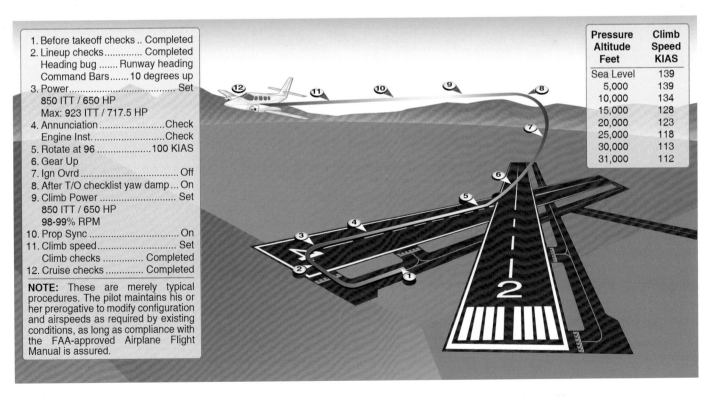

| 1. Before takeoff checks .. Completed |
| 2. Lineup checks .... Completed |
| Heading bug .... Runway heading |
| Command Bars .... 10 degrees up |
| 3. Power .... Set |
| 850 ITT / 650 HP |
| Max: 923 ITT / 717.5 HP |
| 4. Annunciation .... Check |
| Engine Inst. .... Check |
| 5. Rotate at 96 .... 100 KIAS |
| 6. Gear Up |
| 7. Ign Ovrd .... Off |
| 8. After T/O checklist yaw damp ... On |
| 9. Climb Power .... Set |
| 850 ITT / 650 HP |
| 98-99% RPM |
| 10. Prop Sync .... On |
| 11. Climb speed .... Set |
| Climb checks .... Completed |
| 12. Cruise checks .... Completed |

NOTE: These are merely typical procedures. The pilot maintains his or her prerogative to modify configuration and airspeeds as required by existing conditions, as long as compliance with the FAA-approved Airplane Flight Manual is assured.

Pressure Altitude Feet	Climb Speed KIAS
Sea Level	139
5,000	139
10,000	134
15,000	128
20,000	123
25,000	118
30,000	113
31,000	112

Figure 15-11. *Example of a typical turboprop airplane takeoff and departure profile.*

For any given high-horsepower operation, the pilot can expect that the engine temperature will climb as altitude increases at a constant power. On a warm or hot day, maximum temperature limits may be reached at a rather low altitude, making it impossible to maintain high horsepower to higher altitudes. Also, the engine's compressor section has to work harder with decreased air density. Power capability is reduced by high-density altitude and power use may have to be modulated to keep engine temperature within limits.

In a turboprop airplane, the pilot can close the throttles(s) at any time without concern for cooling the engine too rapidly. Consequently, rapid descents with the propellers in low pitch can be dramatically steep. Like takeoffs and departures, approach and landing should be accomplished in accordance with a standard approach and landing profile. *[Figure 15-12]* However, when flying an airplane equipped with a split shaft/free turbine engine, the pilot should anticipate the demand for power and account for any lag in "spool-up" time.

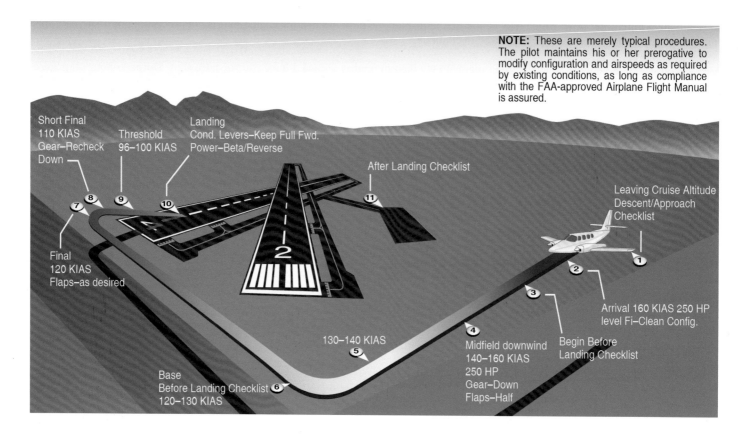

Figure 15-12. *Example of a typical turboprop airplane arrival and landing profile.*

A stabilized approach is an essential part of the approach and landing process. In a stabilized approach, the airplane, depending on design and type, is placed in a stabilized descent on a glidepath ranging from 2.5 to 3.5°. The speed is stabilized at some reference from the AFM/POH—usually 1.25 to 1.30 times the stall speed in approach configuration. The descent rate is stabilized from 500 fpm to 700 fpm until the landing flare.

Landing some turboprop airplanes (as well as some piston twins) can result in a hard, premature touchdown if the engines are idled too soon. This is because large propellers spinning rapidly in low pitch create considerable drag. In such airplanes, it may be preferable to maintain power throughout the landing flare and touchdown. Once firmly on the ground, propeller beta range operation dramatically reduces the need for braking in comparison to piston airplanes of similar weight.

Training Considerations

The medium and high altitudes at which turboprop airplanes are flown provide an entirely different environment in terms of regulatory requirements, airspace structure, physiological requirements, and even meteorology. The pilot transitioning to turboprop airplanes, particularly those who are not familiar with operations in the high/medium altitude environment, should approach turboprop transition training with this in mind. Thorough ground training should cover all aspects of high/medium altitude flight, including the flight environment, weather, flight planning and navigation, physiological aspects of high-altitude flight, oxygen and pressurization system operation, and high-altitude emergencies.

Flight training should prepare the pilot to demonstrate a comprehensive knowledge of airplane performance, systems, emergency procedures, and operating limitations, along with a high degree of proficiency in performing all flight maneuvers and in-flight emergency procedures. The training outline below covers information used by pilots to operate safely at high altitudes.

Ground Training

1. High-Altitude Flight Environment
    a. Airspace and Reduced Vertical Separation Minimum (RVSM) Operations
    b. Title 14 Code of Federal Regulations (14 CFR) part 91, section 91.211, Requirements for Use of Supplemental Oxygen

2. Weather
    a. Atmosphere
    b. Winds and clear air turbulence
    c. Icing

3. Flight Planning and Navigation
    a. Flight planning
    b. Weather charts
    c. Navigation
    d. Navigation aids (NAVAIDs)
    e. High Altitude Redesign (HAR)
    f. RNAV/Required Navigation Performance (RNP) and Receiver Autonomous Integrity Monitoring (RAIM) prediction

4. Physiological Training
    a. Respiration
    b. Hypoxia
    c. Effects of prolonged oxygen use
    d. Decompression sickness
    e. Vision
    f. Altitude chamber (optional)

5. High-Altitude Systems and Components
    a. Oxygen and oxygen equipment
    b. Pressurization systems
    c. High-altitude components

6. Aerodynamics and Performance Factors
    a. Acceleration and deceleration
    b. Gravity (G)-forces
    c. Mach Tuck and Mach Critical (turbojet airplanes)
    d. Swept-wing concept

7. Emergencies
    a. Decompression
    b. Donning of oxygen masks
    c. Failure of oxygen mask or complete loss of oxygen supply/system
    d. In-flight fire
    e. Flight into severe turbulence or thunderstorms
    f. Compressor stalls

Flight Training

1. Preflight Briefing

2. Preflight Planning
    a. Weather briefing and considerations
    b. Course plotting
    c. Airplane Flight Manual (AFM)
    d. Flight plan

3. Preflight Inspection
    a. Functional test of oxygen system, including the verification of supply and pressure, regulator operation,
       oxygen flow, mask fit, and pilot and air traffic control (ATC) communication using mask microphones

4. Engine Start Procedures, Run-up, Takeoff, and Initial Climb

5. Climb to High Altitude and Normal Cruise Operations While Operating Above 25,000 Feet Mean Sea Level (MSL)

6. Emergencies
    a. Simulated rapid decompression, including the immediate donning of oxygen masks
    b. Emergency descent

7. Planned Descents

8. Shutdown Procedures

9. Postflight Discussion

Chapter Summary

Transitioning from a non-turbopropeller airplane to a turbopropeller-powered airplane is discussed in this chapter. The major differences are introduced specifically handling, powerplant, and the associated systems. Turbopropeller electrical systems and operational considerations are explained to include starting procedures and high temperature considerations. Training considerations are also discussed and a sample training syllabus is given to show the topics that a pilot should become proficient in when transitioning to a turbopropeller-powered airplane.

Airplane Flying Handbook (FAA-H-8083-3C)
Chapter 16: Transition to Jet-Powered Airplanes

Introduction

This chapter contains an overview of jet-powered airplane operations. The information contained in this chapter provides a useful preparation for, and a supplement to, structured jet airplane qualification training. This chapter provides information on major differences a pilot may encounter when transitioning to jet-powered airplanes. The major differences between jet-powered airplanes and piston-powered airplanes have been addressed in several distinct areas: differences in aerodynamics, systems, and pilot operating procedures. For airplane-specific information, a pilot should refer to the FAA-approved Airplane Flight Manual for that airplane.

Ground Safety

Stepping out on the ramp in the vicinity of jet airplanes requires special caution. There is no propeller to indicate visually whether a jet engine is running. It is easy to inadvertently stray into danger since, even at idle, jet engines are a threat. Enough air is being sucked into the intake to pull a nearby person into the fan. The air coming from the exhaust is hot and moving fast enough to blow a person down.

Pilots operating jet-powered airplanes should exercise caution during taxi and when adding power to start moving. Adding too much power can pull damaging debris up off the ground or cause damage well behind the aircraft. Jet blast when taxiing into parking areas may affect any loose ground equipment.

Jet Engine Basics

A jet engine is a gas turbine with basic cycle of operation; that is, induction, compression, combustion, expansion, and exhaust. Air passes through the intake and enters the compressor section, which is made up of a series of fan blades or "stages." The first stage, visible from the front of the engine, is the largest diameter and has the biggest blades. Each subsequent stage contains smaller diameter and thinner blades of increasing pitch. The compression in each stage raises the air temperature and pressure. The high-pressure hot air enters the combustion chamber where fuel is added. During engine start, igniters set the fuel air mixture on fire, after which the fire is self-sustaining. The rapidly expanding air flows to the turbine section, which like the compressor section, consists of a series of fan blade stages. The turbine section extracts a portion of the available energy from the airflow to turn a shaft, which drives the compressor. The remaining energy causes rapid air expansion in the nozzle of the tail pipe, accelerates the gas to a high velocity, and produces thrust. *[Figure 16-1]*

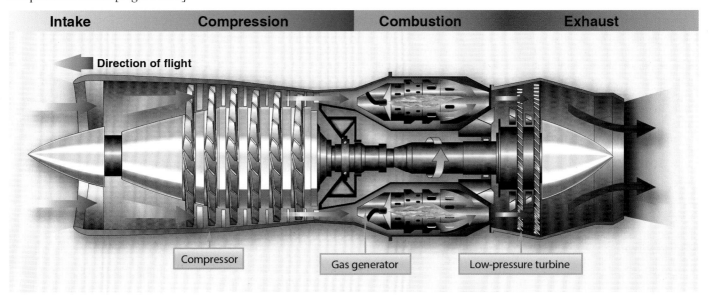

Figure 16-1. *Basic turbojet engine.*

The large first stage design of a turbofan engine, a ducted fan, diverts some of the air around the engine core. This cooler bypass air produces some of the thrust. The amount of air that bypasses the core compared to the amount compressed for combustion determines a turbofan's bypass ratio. In a turbofan engine, the compressor and turbine sections divide into sub-sections. Each sub-section in the turbine section connects to a specific sub-section of the compressor section via a split-spool shaft. *[Figure 16-2]*

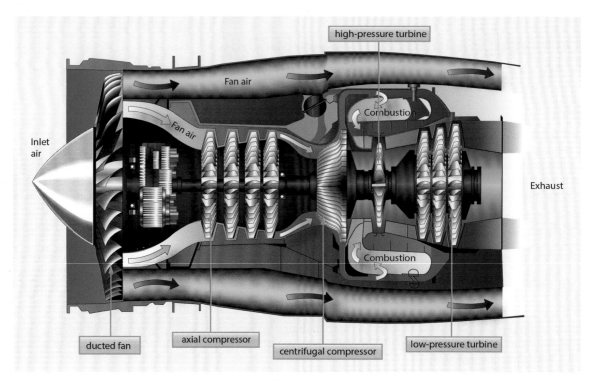

Figure 16-2. *Turbofan engine.*

Air drawn into the engine for the gas generator is further compressed and constitutes the core airflow. While a turbojet engine uses the entire gas generator's output to produce thrust in the form of a high-velocity exhaust gas jet, the lower velocity and cooler bypass air produces some of the thrust produced by a turbofan engine.

The turbofan engine design increases the thrust of the jet engine, particularly at lower speeds and altitudes. Although less efficient at higher altitudes, the turbofan engine increases acceleration, decreases the takeoff roll, improves initial climb performance, and often has the effect of decreasing fuel consumption.

Operating the Jet Engine

In a jet engine, the amount of fuel injected into the combustion chamber controls thrust. Because most engine control functions are automatic, the power controls on most turbojet-powered and turbofan-powered airplanes consist of just one thrust lever for each engine. The thrust lever links to a fuel control and/or electronic engine computer that meters fuel flow based on revolutions per minute (rpm), internal temperatures, ambient conditions, and other factors.

Typically in jet airplanes, there are flight deck indications for the rotation speed of each major engine section. Each engine section rotates at many thousands of rpm. For ease of interpretation, the indications read as percent of rpm rather than actual rpm. Depending on the make and model, there are usually indications for fuel flow, as well as for gas temperatures and pressures. The associated engine indications have different names according to their location.

As in any gas turbine engine, exceeding temperature or rpm limits, even for a few seconds, may result in serious damage to turbine blades and other components. The pilot should monitor the temperature of turbine gases and rotation speeds as needed. Modern aircraft are designed to prevent exceedances and alert the pilot of an impending or actual exceedance. Older designs rely more on the pilot to prevent any exceedances.

Setting Power

When setting power, the pilot normally uses pressure or rpm indications to set maximum allowable thrust. However, the forward movement of the thrust levers should be stopped for any limitation (e.g., pressure, rpm, or temperature).

Thrust to Thrust Lever Relationship

In a jet engine, thrust output changes much more per increment of throttle movement at high engine speeds. If the power setting is already high, it normally takes a small amount of movement to change the power output. This is a significant difference for the pilot transitioning to jet-powered airplanes. In a situation where significantly more thrust is needed and the jet engine is at low rpm, inching the thrust lever forward will have little effect. It this situation, the pilot needs to make a smooth and significant thrust lever position change to increase the power.

Variation of Thrust with RPM

Jets operate most efficiently in the 85 percent to 100 percent range. At idle rpm of approximately 55 percent to 60 percent, they produce a relatively small amount of thrust. An increase in rpm from 90 to 100 percent may increase thrust by as much as the total available at 70 percent. *[Figure 16-3]*

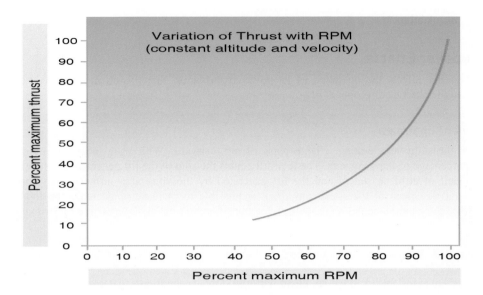

Figure 16-3. *Variation of thrust with rpm.*

Slow Acceleration of the Jet Engine

Acceleration of a piston engine from idle to full power is relatively rapid. The acceleration on different jet engines can vary considerably, but it is usually much slower. In some cases, the transition to full power could take up to 10 seconds. *[Figure 16-4]* Pilots should anticipate the need for adding power from low power settings.

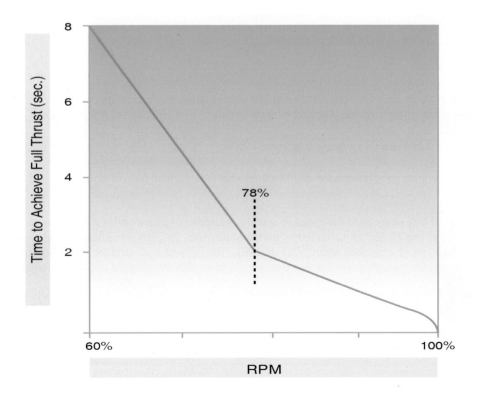

Figure 16-4. *Typical jet engine acceleration times.*

Jet Engine Efficiency

The efficiency of the jet engine increases in the cold temperatures found at high altitudes. The fuel consumption of jet engines decreases as the outside air temperature decreases for constant engine rpm and true airspeed (TAS). Thus, by flying at a high altitude, the airplane operates with improved fuel economy and speed. At high altitudes, engines may be operating close to rpm or temperature limits, and excess thrust may not be available. Therefore, pilots should accomplish all maneuvering within the limits of available thrust, stability, and controllability.

Absence of Propeller Effects

The absence of a propeller affects the operation of jet-powered airplanes. Specific effects include the absence of lift from the propeller slipstream and the absence of propeller drag.

Absence of Propeller Slipstream

A propeller produces thrust by accelerating a large mass of air rearward. With wing-mounted engines, this air passes over a comparatively large percentage of the wing area. The total lift equals the sum of the lift generated by the wing area not in the wake of the propeller (as a result of airplane speed) and the lift generated by the wing area influenced by the propeller slipstream. By increasing or decreasing the speed of the slipstream air, it is possible to increase or decrease the total lift on the wing without changing airspeed. Since the jet airplane has no propellers, the transitioning pilot should note the following:

1. Lift is not increased instantly by adding power.
2. The stall speed is not decreased by adding power.

The lack of ability to produce instant lift in the jet, along with the slow acceleration of jet engines, necessitates a stabilized approach where landing configuration, constant airspeed, controlled rate of descent, and stable power settings are maintained until over the threshold of the runway. This allows for better engine response when making minor changes in the approach speed or rate of descent and improves go-around performance.

Absence of Propeller Drag

When the throttles are closed on a piston-powered airplane, the propellers create significant drag. Airspeed or altitude is immediately decreased. The effect of reducing power to idle on the jet engine, however, produces no such drag effect. In fact, at an idle power setting, the jet engine still produces forward thrust. While this can be an advantage in certain descent profiles, it is a handicap when it is necessary to lose speed quickly. The lack of propeller drag, along with the aerodynamically clean airframe of the jet, are new to most pilots, and slowing the airplane down is one of the initial problems encountered by pilots transitioning into jets. In level flight at idle power, it takes about 1 mile to lose 10 knots of airspeed.

Speed Margins

Maximum speeds in jet airplanes are expressed differently and always define the maximum operating speed of the airplane, which is comparable to the VNE of the piston airplane. These maximum speeds in a jet airplane are referred to as:

- V_{MO}—maximum operating speed expressed in terms of knots.

- M_{MO}—maximum operating speed expressed as a Mach number (the decimal ratio of true airspeed to the speed of sound).

Mach number is the ratio of true airspeed to the speed of sound. The speed of sound varies with temperature. At low/warm altitudes, the speed of sound is so high that an aircraft is limited by indicated airspeed. At high/cold altitudes, the speed of sound is lower so the aircraft is limited by Mach. To observe both limits V_{MO} and M_{MO}, the pilot of a jet airplane needs both an airspeed indicator and a Mach indicator. In most jet airplanes, these are combined into a single display for airspeed and Mach number, as appropriate.

It looks much like a conventional airspeed display with the addition of a "barber pole" that automatically moves so as to indicate the applicable speed limit at all times. *[Figure 16-5]*

A jet airplane can easily exceed its speed limitations. The handling qualities of a jet may change significantly at speeds higher than the maximum allowed.

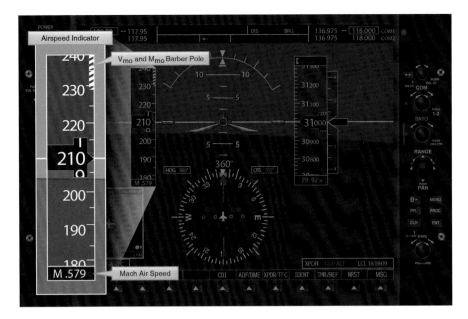

Figure 16-5. *Jet airspeed indicator.*

High-speed airplanes designed for subsonic flight are limited to some Mach number below the speed of sound. Shock waves (and the adverse effects associated with them) can occur when the airplane speed is substantially below Mach 1.0. The Mach number at which some portion of the airflow over the wing first equals Mach 1.0 is termed the critical Mach number (M_{CR}).

There is no particular problem associated with the acceleration of the airflow up to the critical Mach number, the point where Mach 1.0 airflow begins. However, a shock wave is formed at the point where the airflow suddenly returns to subsonic flow. This shock wave becomes more severe and moves aft on the wing as airflow velocity increases. Eventually, flow separation occurs behind the well-developed shock wave. *[Figure 16-6]*

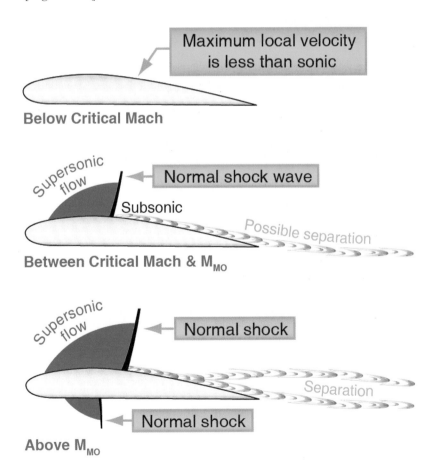

Figure 16-6. *Transonic flow patterns.*

If airplane speed progresses sufficiently beyond M_{MO}, the separation of air behind the shock wave may result in severe buffeting and possible loss of control or "upset." Because of the accompanying changes to the center of lift, the airplane may exhibit pitch change tendencies.

With increased speed and the aft movement of the shock wave, the wing's center of pressure moves aft causing the start of a nose-down tendency or "tuck." Mach tuck develops gradually, and the condition should not be allowed to progress to where there is no longer enough elevator authority to prevent entry into a steep, sometimes unrecoverable, dive. An alert pilot should respond to excessive airspeed, buffeting, or warning devices before the onset of extreme nose-down forces.

Due to the critical aspects of high-altitude/high-Mach flight, most jet airplanes capable of operating in the Mach ranges use some form of automated Mach tuck compensation. If the system becomes inoperative, the airplane is typically limited to a reduced maximum Mach number.

Mach Buffet

Mach buffet arises when airflow separates on the upper surface of a wing behind a shock wave. All other things being equal, shock wave strength increases as the local airflow speed ahead of the shock wave increases. Mach buffet is a function of the speed of the airflow over the wing—not necessarily the forward speed of the airplane, and the shock wave strength, rather than a stall, creates the airflow separation.

Mach buffet may result from two different conditions in cruise. At high-speed cruise, a shock wave that becomes too strong as the airflow speeds up over the upper surface causes a buffet. At low-speed cruise, the flow has a greater turn to make to follow the wing's upper surface. The air speeds up to do that and may exceed Mach 1 over the upper surface.

The shock wave position is different between the two situations. At high speed and a lower AOA, the shock wave tends to move aft. So when the flow separates behind the shock, that separated flow acts over a small range of the chord. In some cases, the separated flow acting on a small surface area may produce a little buzz. At low-speed cruise, the true airspeed is still high, but the shock wave does not move as far aft as it does in high-speed cruise. The separated flow behind the shock wave acts over a larger portion of the chord, which leads to a more significant effect on aircraft control.

The altitude at which an airplane flying at M_{MO} would experience buffeting with any increase in AOA determines the absolute or aerodynamic ceiling. This is the altitude where:

- If an airplane flew any faster, it would exceed M_{MO} leading to high-speed Mach buffet.
- If an airplane flew any slower, it would require an angle of attack leading to low-speed Mach buffet.

This region of the airplane's flight envelope is known as "coffin corner." Conceivably, a buffet could be the first indication of an issue at altitude, and pilots should understand the cause of any buffet in order to respond appropriately.

An increase in load factor (G factor) will raise the low-end buffet speed. For example, a jet airplane flying at 51,000 feet altitude at 1.0 G and a speed of 0.73 Mach that experiences a 1.4 G load, may encounter low-speed buffet. Consequently, a maximum cruising flight altitude and speed should be selected, which will allow sufficient margin for maneuvering and turbulence. The pilot should know the manufacturer's recommended turbulence penetration speed for the particular make and model airplane. This speed normally gives the greatest margin between the high-speed and low-speed buffets.

Low-Speed Flight

The jet airplane wing, designed primarily for high-speed flight, has relatively poor low-speed characteristics. As opposed to the normal piston-powered airplane, the jet wing has less area relative to the airplane's weight, a lower aspect ratio (long chord/short span), and thin airfoil shape—all of which amount to the need for speed to generate enough lift. The swept wing is additionally penalized at low speeds because its effective lift is proportional to airflow speed that is perpendicular to the leading edge.

In a typical piston-engine airplane, V_{MD} (minimum drag) in the clean configuration is normally at a speed of about 1.3 V_S. *[Figure 16-7]* Flight below V_{MD} in a piston-engine airplane is well identified and predictable. In contrast, in a jet airplane, flight in the area of V_{MD} (typically 1.5 – 1.6 V_S) does not normally produce any noticeable changes in flying qualities other than a lack of speed stability—a condition where a decrease in speed leads to an increase in drag, which leads to a further decrease in speed, which creates the potential for a speed divergence. A pilot who is not aware of a developing speed divergence may find a serious sink rate developing at a constant power setting, while pitch attitude appears to be normal. The fact that lack of speed stability may lead to a sinking flightpath, is one of the most important aspects of jet-airplane flying.

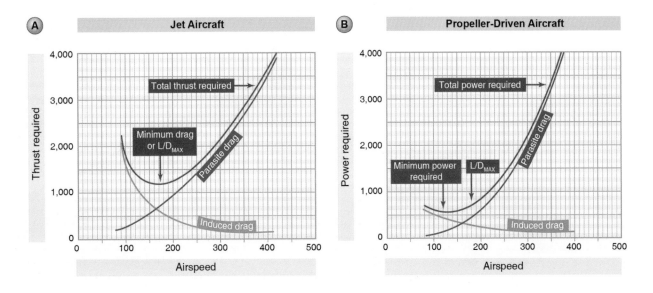

Figure 16-7. *Thrust and power required curves (jet aircraft vs. propeller-driven aircraft).*

Stalls

The stalling characteristics of the swept wing jet airplane can vary considerably from those of the normal straight wing airplane. The greatest difference noticeable to the pilot is the lift developed vs. angle of attack. An increase in angle of attack of the straight wing produces a substantial and constantly increasing lift vector up to its maximum coefficient of lift, and soon thereafter flow separation (stall) occurs with a rapid deterioration of lift.

By contrast, the swept wing produces a much more gradual buildup of lift with a less well-defined maximum coefficient. This less-defined peak also means that a swept wing may not have as dramatic a loss of lift at angles of attack beyond its maximum lift coefficient. However, these high-lift conditions are accompanied by high drag, which may result in a high rate of descent. *[Figure 16-8]*

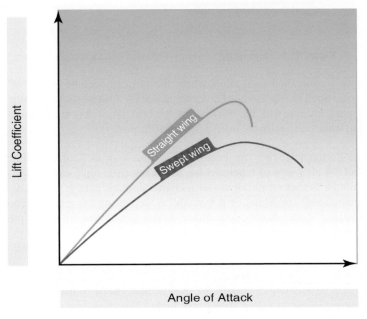

Figure 16-8. *Stall versus angle of attack—swept wing versus straight wing.*

If a simple, straight-wing airplane's airfoil is swept, a natural tendency arises that it will stall at the wing tips first. This is because the boundary layer tends to flow spanwise toward the tips. *[Figure 16-9]* The tendency for tip stall allowing the center of lift to move forward is greatest when wing sweep and taper are combined. To discourage a swept wing from stalling at the wingtips, manufacturers modify the wing spanwise with twist, changes in airfoil section, inclusion of vortex generators, or a combination of those modifications. This helps a pilot retain roll control initially if a stall is entered inadvertently.

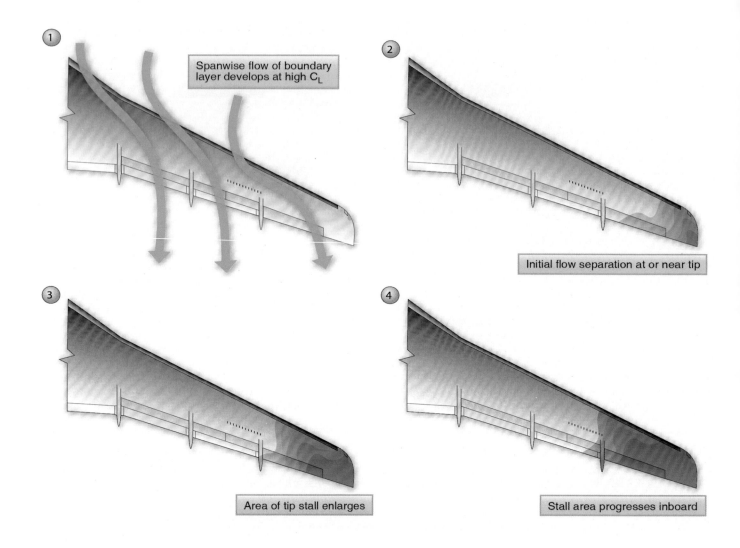

Figure 16-9. *Unmodified swept wing stall characteristics.*

Some T-tail configurations are prone to so-called deep stalls where the tail can become immersed in the wing wake at very high angles of attack and lose effectiveness. Such a situation can be accompanied by a high-rate of descent. Since high angles of attack can occur at any pitch attitude, even a pitch attitude with the nose below the horizon, it may seem counterintuitive in such a situation to take the appropriate recovery action, which is to push the nose down even further.

Deep stalls may be unrecoverable. Fortunately, they are easily avoided as long as published limitations are observed. On those airplanes susceptible to deep stalls (not all swept or tapered wing airplanes are), sophisticated stall warning systems such as stick shakers are standard equipment. A stick pusher (if installed), as its name implies, acts to automatically reduce the airplane's AOA before the airplane reaches a dangerous stall condition, or it may aid in recovering the airplane from a stall if an airplane's natural aerodynamic characteristics do so weakly. Pilots should avoid situations that would activate a stick pusher when close to the ground.

Pilots undergoing training in jet airplanes are taught to recover at the first indication of an impending stall instead of going beyond those initial cues and into a full stall. Normally, this is indicated by aural stall warning devices, annunciators, or activation of the airplane's stick shaker. Stick shakers normally activate around 107 percent of the actual stall speed. In response to a stall warning, the proper action is for the pilot to apply a nose-down input until the stall warning stops (pitch trim may be necessary). Then, the wings are rolled level, followed by adjusting thrust to return to normal flight. The elapsed time will be small between these actions, particularly at low altitude where significant available thrust exists. It is important to understand that reducing AOA eliminates the stall, but added thrust will allow the descent to be stopped once the wing is flying again. Note that airplanes without vortex generators may stall with little to no buffet.

At high altitudes the stall recovery technique is the same. A pilot will need to reduce the AOA by lowering the nose until the stall warning stops. However, after the AOA has been reduced to where the wing is again developing efficient lift, the airplane will still likely need to accelerate to a desired airspeed. At high altitudes where the available thrust is significantly less than at lower altitudes, recovery may require significant pitch down to regain airspeed. As such, several thousand feet or more of altitude loss may occur during the recovery. The above discussion covers most airplanes; however, the stall recovery procedures for a particular make and model airplane may differ, as recommended by the manufacturer, and are contained in the FAA-approved Airplane Flight Manual for that airplane.

Drag Devices

Jet airplanes have higher glide ratios than piston-powered airplanes. Due to their low drag design, jets take more time and distance to descend or reduce speed. Therefore, jet airplanes are often equipped with drag devices, such as spoilers and speed brakes.

The primary purpose of spoilers is to spoil lift. The most common type of spoiler consists of one or more rectangular plates that lie flush with the upper surface of each wing. They are installed approximately parallel to the lateral axis of the airplane and are hinged along the leading edges. When deployed, spoilers deflect up against the relative wind, which interferes with the flow of air about the wing. *[Figure 16-10]* This both spoils lift and increases drag. Spoilers are usually installed forward of the flaps but not in front of the ailerons so as not to interfere with roll control. Some aircraft use spoilers to augment roll control.

Figure 16-10. *Spoilers.*

When flight and ground spoilers are deployed after landing, most of the wing's lift is destroyed. This action transfers the airplane's weight to the landing gear so that the wheel brakes are more effective. A secondary beneficial effect of deploying spoilers on landing is that they create considerable drag, adding to the overall aerodynamic braking.

The primary purpose of speed brakes is to produce drag. Spoilers may also serve as speed brakes, or they may be panels attached to the fuselage. Deploying speed brakes results in a rapid decrease in airspeed and/or an increased rate of descent. Typically, speed brakes can be deployed at any time during flight. There is usually a certain amount of noise and buffeting associated with the use of speed brakes, along with an obvious penalty in fuel consumption. Pilots can minimize the use of speed brakes with proper descent and approach planning. Procedures for the use of spoilers and/or speed brakes in various situations are contained in the FAA-approved AFM for the particular airplane.

Thrust Reversers

Jet airplanes have high kinetic energy during the landing roll because of weight and speed. This energy is difficult to dissipate because a jet airplane has low drag with the nose-wheel on the ground, and the engines continue to produce forward thrust with the power levers at idle. While wheel brakes serve as the primary means to stop the airplane, reverse thrust, when available, assists in deceleration.

Certain thrust reverser designs effectively reverse the flow of the exhaust gases. The flow does not completely reverse. Typically, the final path of the exhaust gases is about 45° from straight ahead. This, together with the losses from the flow paths, reduces reverse thrust efficiency. If the pilot uses less than maximum rpm in reverse, the reverse thrust is further reduced.

Normally, a jet engine has one of two types of thrust reversers: a target reverser or a cascade reverser. *[Figure 16-11]* Target reversers are simple clamshell doors that swivel from the stowed position at the engine tailpipe to redirect thrust to a more forward direction.

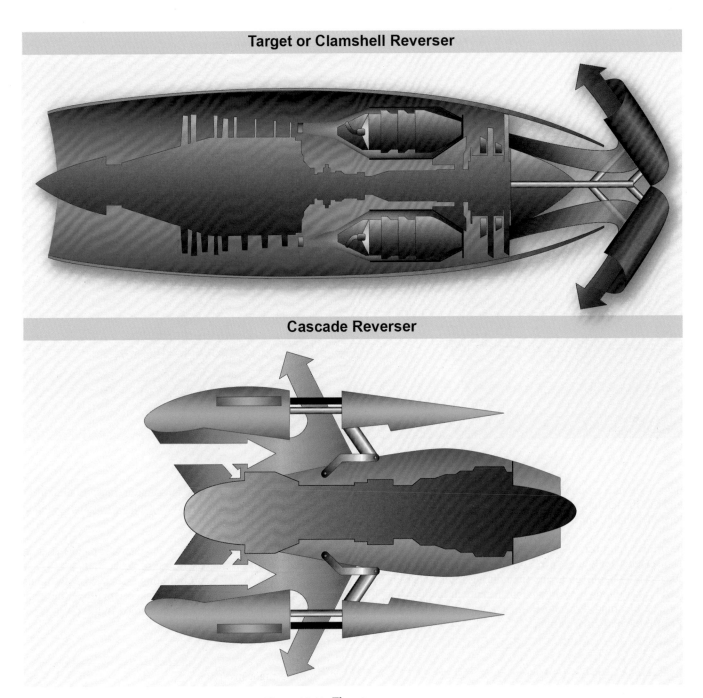

Figure 16-11. *Thrust reversers.*

Cascade reversers are normally found on turbofan engines and are often designed to reverse only the fan air portion. Blocking doors in the shroud obstruct forward fan thrust and redirect it through cascade vanes to generate reverse thrust.

On most installations, the pilot selects reverse thrust with the thrust levers at idle by pulling up the reverse levers to a detent. Doing so positions the reversing mechanisms for operation but leaves the engines at idle rpm. Further upward and backward movement of the reverse levers increases engine power. Reverse is canceled by closing the reverse levers to the idle reverse position, then dropping them fully back to the forward idle position. This last movement selects the stowed position, and the reversers return to the forward thrust position.

Reverse thrust is more effective at high speed than at low speed. For maximum reverse thrust efficiency, the pilot should use it as soon as is prudent after touchdown. The pilot should remember that some airplanes tend to pitch nose-up when reverse is selected on landing and this effect, particularly when combined with the nose-up pitch effect from the spoilers, can cause the airplane to leave the ground again momentarily. On these types, the airplane should be firmly on the ground with the nose-wheel down before reverse is selected. Other types of airplanes have no change in pitch, and reverse idle may be selected after the main gear is down and before the nose-wheel is down. Since reverse thrust may affect directional control, runway surface conditions (e.g., contamination), factor into the use of reverse thrust. Specific procedures for reverse thrust operation for a particular airplane/engine combination are contained in the FAA-approved AFM for that airplane.

There is a significant difference between reverse pitch on a propeller and reverse thrust from a jet engine. Idle reverse on a propeller produces a large amount of drag. On a jet engine, however, selecting idle reverse produces very little reverse thrust. In a jet airplane, the pilot should select reverse, apply reverse thrust as appropriate, and remain within any AFM limitations.

It is essential that pilots understand not only the normal procedures and limitations of thrust reverser use, but also the procedures for coping with uncommanded reverse. While thrust reverser systems are designed to prevent unintentional deployment, an uncommanded or inadvertent deployment of thrust reversers, while airborne, is an emergency. The systems normally contain several lock systems: one to keep reversers from operating in the air, another to prevent operation with the thrust levers out of the idle detent, and/or an "auto-stow" circuit to command reverser stowage any time thrust reverser deployment would be inappropriate, such as during takeoff and while airborne.

Pilot Sensations in Jet Flying

Pilots transitioning into jets may notice these general sensations:

1. response differences
2. increased control sensitivity
3. increased tempo of flight

In some flight conditions, airspeed changes may occur more slowly than in a propeller airplane. At high altitudes, the reduction in available thrust reduces the ability to accelerate. The long spool-up time required from low throttle settings also may affect acceleration. Finally, the clean aerodynamic design of a jet can result in more gradual deceleration when thrust is reduced.

The lack of propeller effects results in less drag at low power settings. Other changes the transitioning pilot should notice include the lack of effective slipstream over the lifting and control surfaces, and the lack of propeller torque effect.

Even though moving the power levers has less effect at low power settings, the pilot should change power settings smoothly. To slow the airplane, the transitioning pilot may also need to learn when to use available drag devices appropriately.

Transitioning pilots should learn power setting management for different situations. Power settings for desired performance vary because of significant changes in airplane weight as fuel is consumed. Therefore, the pilot needs to use a variety of cues to achieve desired performance. For example, airspeed trend information provides feedback for power required.

Power changes may result in a pitching tendency. These characteristics should be noticed and compensated for.

The jet airplane will differ regarding pitch tendencies with the lowering of flaps, landing gear, and drag devices. With experience, the jet airplane pilot will learn to anticipate the pitch change required for a particular operation. Most jet airplanes are equipped with a thumb operated pitch trim button on the control wheel. The usual method of operating the trim button is to apply several small, intermittent applications of trim in the direction desired rather than holding the trim button for longer periods of time, which can lead to overcontrolling.

The variation of pitch attitudes flown in a jet airplane also results from high thrust, flight characteristics of the low aspect ratio, and the swept wing. Flight at higher pitch attitudes requires greater reliance on the flight instruments for airplane control since outside references may be absent. Proficiency in attitude instrument flying, therefore, is essential to successful transition to jet airplane flying.

Control sensitivity will differ amongst various airplanes. Because of the higher speeds flown, the control surfaces are more effective and a variation of just a few degrees in pitch attitude in a jet can result in over twice the rate of altitude change that would be experienced in a slower airplane. The sensitive pitch control in jet airplanes is one of the first flight differences that the pilot may notice, and the transitioning pilot may have a tendency to overcontrol pitch during initial training flights. Accurate and smooth control is one of the first techniques the transitioning pilot should master. Rather than gripping the yoke with the hand at high speeds, just using fingertips will result in smoother control inputs.

The pilot flying a swept wing jet airplane should understand that it is normal to fly at higher angles of attack. Depending on weight, density altitude, and available thrust, the pitch angle on takeoff may seem high. It is also not unusual to have a noticeable nose-up pitch on an approach to a landing.

Jet Airplane Takeoff and Climb

The following information is generic in nature and, since most civilian jet airplanes require a minimum flight crew of two pilots, assumes a two-pilot crew. If any of the following information conflicts with FAA-approved AFM procedures for a particular airplane, the AFM procedures take precedence. Also, if any of the following procedures differ from the FAA-approved procedures developed for use by a specific air operator and/or for use in an FAA-approved training center or pilot school curriculum, the FAA-approved procedures for that operator and/or training center/pilot school take precedence.

V-Speeds

The following are speeds that affect the jet airplane's takeoff performance. The jet airplane pilot should understand how to use these speeds when planning for takeoff.

- V_S—stalling speed or minimum steady flight speed at which the airplane is controllable.

- V_1—critical engine failure speed or takeoff decision speed. It is the speed at which the pilot is to continue the takeoff in the event of an engine failure or other serious emergency. At speeds less than V_1, it is considered safer to stop the aircraft within the accelerate-stop distance. It is also the minimum speed in the takeoff, following a failure of the critical engine at VEF, at which the pilot can continue the takeoff and achieve the required height above the takeoff surface within the takeoff distance.

- V_{EF}—speed used during certification at which the critical engine is assumed to fail.

- V_R—rotation speed, or speed at which the rotation of the airplane is initiated to takeoff attitude. This speed cannot be less than V_1 or less than $1.05 \times V_{MCA}$ (minimum control speed in the air). On a single-engine takeoff, it also allows for the acceleration to V_2 at the 35-foot height at the end of the runway.

- V_{LOF}—lift-off speed, or speed at which the airplane first becomes airborne. This is an engineering term used when the airplane is certificated to meet certain requirements. The pilot takes this speed into consideration if the AFM lists it.

- V_2—takeoff safety speed, or a referenced airspeed obtained after lift-off at which the required one-engine-inoperative climb performance can be achieved.

Takeoff Roll

After confirming the runway and position match expectations, the airplane should be aligned in the center of the runway. When runway length is limited, the brakes should be held while the thrust levers are brought to a power setting specified in the AFM and the engines allowed to stabilize. The engine instruments should be checked for proper operation before the brakes are released or the power increased further. This procedure assures symmetrical thrust during the takeoff roll and aids in prevention of overshooting the desired takeoff thrust setting. After brake release, the power levers should be set to the pre-computed takeoff power setting and takeoff thrust adjustments made prior to reaching 60 knots. The final engine power adjustments are normally made by the pilot not flying. Retarding a thrust lever would only be necessary in case an engine exceeds any limitation.

Takeoff data, including V_1/V_R and V_2 speeds, takeoff power settings, and required field length should be computed prior to each takeoff. For any make and model without an FMS, the data should be recorded on a takeoff data card. This data is based on airplane weight, runway length available, runway gradient, field temperature, field barometric pressure, wind, icing conditions, and runway condition. Both pilots should review the takeoff data entered in an FMS or separately compute the takeoff data and cross-check with the takeoff data card. If takeoff plans change while taxiing, the pilot or crew should recalculate the takeoff data.

Captain's Briefing

I will advance the thrust levers.

Follow me through on the thrust levers.

Monitor all instruments and warning lights on the takeoff roll and call out any discrepancies or malfunctions observed prior to V_1, and I will abort the takeoff. Stand by to arm thrust reversers on my command.

Give me a visual and oral signal for the following:
- 80 knots, and I will disengage nosewheel steering.
- V_1, and I will move my hand from thrust to yoke.
- V_R, and I will rotate.

In the event of engine failure at or after V_1, I will continue the takeoff roll to V_R, rotate and establish V_2 climb speed. I will identify the inoperative engine, and we will both verify. I will accomplish the shutdown, or have you do it on my command.

I will expect you to stand by on the appropriate emergency checklist.

I will give you a visual and oral signal for gear retraction and for power settings after the takeoff.

Our VFR emergency procedure is to.............................
Our IFR emergency procedure is to.............................

Figure 16-12. *Sample captain's briefing.*

A captain's briefing is an essential part of crew resource management (CRM) procedures and should be accomplished prior to takeoff. *[Figure 16-12]*

If sufficient runway length is available, a "rolling" takeoff may be made without stopping at the end of the runway. Using this procedure, as the airplane rolls onto the runway, the thrust levers should be smoothly advanced to the recommended intermediate power setting and the engines allowed to stabilize, and then proceed as in the static takeoff outlined above. Rolling takeoffs can also be made from the end of the runway by advancing the thrust levers from idle as the brakes are released.

During the takeoff roll, the pilot flying should concentrate on directional control of the airplane. This is made somewhat easier because there is no torque-produced yawing in a jet as there is in a propeller-driven airplane. The airplane should be maintained exactly on centerline with the wings level. This automatically aids the pilot when contending with an engine failure. If a crosswind exists, the wings should be kept level by displacing the control wheel into the crosswind. During the takeoff roll, the primary responsibility of the pilot not flying is to closely monitor the aircraft systems and to call out the proper V speeds as directed in the captain's briefing.

Slight forward pressure should be held on the control column to keep the nose-wheel rolling firmly on the runway. If nose-wheel steering is being utilized, the pilot flying should monitor the nose-wheel steering to about 80 knots (or V_{MCG} for the particular airplane) while the pilot not flying applies the forward pressure. After reaching V_{MCG}, the pilot flying should bring his or her left hand up to the control wheel. The pilot's other hand should be on the thrust levers until at least V_1 speed is attained. Although the pilot not flying maintains a check on the engine instruments throughout the takeoff roll, the pilot flying (pilot-in-command) makes the decision to continue or reject a takeoff for any reason. A decision to reject a takeoff requires immediate retarding of thrust levers.

The takeoff and climb-out should be accomplished in accordance with a standard takeoff and departure profile developed for the particular make and model airplane. *[Figure 16-13]*

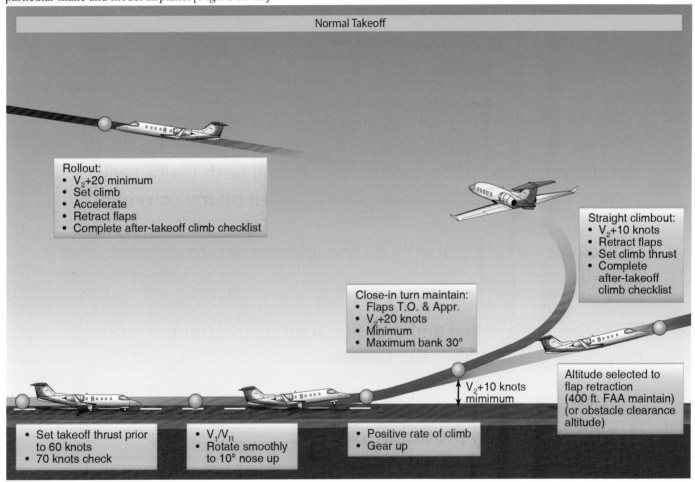

Figure 16-13. *Sample takeoff and departure profile.*

The pilot not flying should call out V_1. After passing V_1 speed on the takeoff roll, it is no longer mandatory for the pilot flying to keep a hand on the thrust levers. The point for abort has passed, and both hands may be placed on the control wheel. As the airspeed approaches V_R, the control column should be moved to a neutral position. As the pre-computed V_R speed is attained, the pilot not flying should make the appropriate call-out, and the pilot flying should smoothly rotate the airplane to the appropriate takeoff pitch attitude.

Rejected Takeoff

Every takeoff could potentially result in a rejected takeoff (RTO) for a variety of reasons: engine failure, fire or smoke, unsuspected equipment on the runway, bird strike, blown tires, direct instructions from the governing ATC authority, or recognition of a significant abnormality (split-airspeed indications, activation of a warning horn, etc.).

Ill-advised rejected takeoff decisions by flight crews and improper pilot technique during the execution of a rejected takeoff contribute to a majority of takeoff-related commercial aviation accidents worldwide. Statistically, although only 2 percent of rejected takeoffs are in this category, high-speed aborts above 120 knots account for the vast majority of RTO overrun accidents. A brief moment of indecision may mean the difference between running out of runway and coming to a safe halt after an aborted takeoff.

It is paramount to remember that FAA-approved takeoff data for any aircraft is based on aircraft performance demonstrated in ideal conditions, using a clean, dry runway, and maximum braking (reverse thrust is not used to compute stopping distance). In reality, stopping performance can be degraded by an array of factors as diversified as:

- Reduced runway friction (grooved/non-grooved)
- Mechanical runway contaminants (rubber, oily residue, debris)
- Natural contaminants (standing water, snow, slush, ice, dust)
- Wind direction and velocity
- Low air density
- Flap configuration
- Bleed air configuration
- Underinflated or failing tires
- Penalizing MEL or CDL items
- Deficient wheel brakes or RTO auto-brakes
- Inoperative anti-skid
- Pilot technique and individual proficiency

Taking pilot response times into account, the go/no-go decision should be made before V_1 so that deceleration can begin no later than V_1. If braking has not begun by V_1, the decision to continue the takeoff is made by default. Delaying the RTO maneuver by just one second beyond V_1 increases the speed 4 to 6 knots on average. Knowing that crews require 3 to 7 seconds to identify an impending RTO and execute the maneuver, it stands to reason that a decision should be made prior to V_1 in order to ensure a successful outcome of the rejected takeoff. This prompted the FAA to expand on the regulatory definition of V_1 and to introduce a couple of new terms through the publication of Advisory Circular (AC) 120-62, "Takeoff Safety Training Aid."

The expanded definition of V_1 is as follows:

a.) V_1—the speed selected for each takeoff, based upon approved performance data and specified conditions, which represents:

1.) The maximum speed by which a rejected takeoff assures that a safe stop can be completed within the remaining runway or runway and stopway;

2.) The minimum speed which assures that a takeoff can be safely completed within the remaining runway, or runway and clearway, after failure of the most critical engine at the designated speed; and

3.) The single speed which permits a successful stop or continued takeoff when operating at the minimum allowable field length for a particular weight.

b.) Minimum V_1—the minimum permissible V_1 speed for the reference conditions from which the takeoff can be safely completed from a given runway, or runway and clearway, after the critical engine had failed at the designated speed.

c.) Maximum V_1—the maximum possible V_1 speed for the reference conditions at which a rejected takeoff can be initiated and the airplane stopped within the remaining runway, or runway and stopway.

d.) Reduced V_1—a V_1 less than maximum V_1 or the normal V_1, but more than the minimum V_1, selected to reduce the RTO stopping distance required.

The main purpose for using a reduced V_1 is to properly adjust the RTO stopping distance in light of the degraded stopping capability associated with wet or contaminated runways, while adding approximately 2 seconds of recognition time for the crew.

Most aircraft manufacturers recommend that operators identify a "low-speed" regime (i.e., 80 knots and below) and a "high-speed" regime (i.e., 100 knots and above) of the takeoff run. In the "low-speed" regime, pilots should abort takeoff for any malfunction or abnormality (actual or suspected). In the "high-speed" regime, takeoff should only be rejected because of catastrophic malfunctions or life-threatening situations. Pilots should weigh the threat against the risk of overshooting the runway during an RTO maneuver. Standard operating procedures (SOPs) should be tailored to include a speed call-out during the transition from low-speed to high-speed regime, the timing of which serves to remind pilots of the impending critical window of decision-making, to provide them with a last opportunity to crosscheck their instruments, to verify their airspeed, and to confirm that adequate takeoff thrust is set, while at the same time performing a pilot incapacitation check through the "challenge and response" ritual.

Brakes provide the most effective stopping force, but experience has shown that the initial tendency of a flight crew is to use normal after-landing braking during a rejected takeoff. Delaying the intervention of the primary deceleration force during an RTO maneuver, when every second counts, increases stopping distance. Instead of braking after the throttles are retarded and the spoilers are deployed (normal landing), pilots should apply maximum braking immediately while simultaneously retarding the throttles, with spoiler extension and thrust reverser deployment following in short sequence. Differential braking applied to maintain directional control also diminishes the effectiveness of the brakes. A blown tire will eliminate any kind of braking action on that particular tire, and could also lead to the failure of adjacent tires.

In order to better assist flight crews in making a split-second go/no-go decision during a high-speed takeoff run, and avoid an unnecessary high-speed RTO, some commercial aircraft manufacturers have gone as far as inhibiting aural or visual malfunction warnings of non-critical equipment beyond a preset speed. The purpose is to prevent an overreaction by the crew and a tendency to select a risky high-speed RTO maneuver over a safer takeoff with a non-critical malfunction. Indeed, the successful outcome of a rejected takeoff, one that concludes without damage or injury, may be influenced by equipment characteristics.

In summary, a rejected takeoff should be perceived as an emergency. RTO safety could be vastly improved by:

- Developing SOPs aiming to advance the expanded FAA definitions of takeoff decision speed and their practical application, including the use of progressive callouts to identify transition from low-speed to high-speed regime.

- Promoting recognition of emergency versus abnormal situations through enhanced CRM training.

- Encouraging crews to carefully consider factors that may affect or even compromise available performance data.

- Expanding practical training in the proper use of brakes, throttles, spoilers, and reverse thrust during RTO demonstrations.

- Encouraging aircraft manufacturers to eliminate non-critical malfunction warnings during the takeoff roll at preset speeds.

Rotation and Lift-Off

Rotation and lift-off in a jet airplane requires planning, precision, and a fine control touch. The objective is to initiate the rotation to takeoff pitch attitude exactly at V_R so that the airplane accelerates through V_{LOF} and attains V_2 speed at 35 feet AGL. Rotation to the proper takeoff attitude too soon may extend the takeoff roll or cause an early lift-off, which results in a lower rate of climb and a divergence from the predicted flightpath. A late rotation, on the other hand, results in a longer takeoff roll, exceeding V_2 speed, and a takeoff and climb path below the predicted path.

Each airplane has its own specific takeoff pitch attitude that remains constant regardless of weight. The takeoff pitch attitude in a jet airplane is normally between 10° and 15° nose up. The rotation to takeoff pitch attitude should be made smoothly but deliberately and at a constant rate. Depending on the particular airplane, the pilot should plan on a rate of pitch attitude increase of approximately 2.5° to 3° per second.

In training, it is common for the pilot to overshoot V_R and then overshoot V_2 because the pilot not flying calls for rotation at or just past V_R. The pilot flying may visually verify V_R and then rotate late. If the airplane leaves the ground at or above V_2, the excess airspeed may be of little concern on a normal takeoff. However, a delayed rotation can be critical when runway length or obstacle clearance is limited. On some airplanes, the rapidly increasing airspeed may cause the achieved flightpath to fall below the engine-out scheduled flightpath unless flying correct speeds. Rotation at the right speed and rate to the right attitude gets the airplane off the ground at the right speed and within the right distance.

Initial Climb

Once the proper pitch attitude is attained, the pilot should maintain it. Takeoff power is also maintained and the airspeed allowed to accelerate. Landing gear retraction should be accomplished after a positive rate of climb has been established and confirmed. In some airplanes gear retraction may temporarily increase the airplane drag while landing gear doors open. Premature gear retraction may cause the airplane to settle back toward the runway surface. In addition, the vertical speed indicator and the altimeter may not show a positive climb until the airplane is 35 to 50 feet above the runway due to ground effect.

The pilot should hold the climb pitch attitude as the airplane accelerates to flap retraction speed. However, the flaps should not be retracted until obstruction clearance altitude or 400 feet AGL has been passed. Ground effect and landing gear drag reduction result in rapid acceleration during this phase of the takeoff and climb. Airspeed, altitude, climb rate, attitude, and heading should be monitored carefully. As the airplane develops a steady climb, longitudinal stick forces can be trimmed out. If making a power reduction, the pilot should reduce the pitch attitude simultaneously if needed and monitor the airplane airspeed and rate of climb so as to preclude an inadvertent reduction in desired performance or a descent.

Speed is limited to 250 KIAS below 10,000 feet MSL in the United States unless otherwise authorized by the Administrator (14 CFR part 91, section 91.117(a)). At or above that altitude, the best rate of climb speed is published in the AFM. If asked to increase rate of climb, increasing pitch slightly will have the desired effect as airspeed bleeds off. If the airplane slows to L/D_{MAX}, the airplane is at its best angle of climb speed, but the rate of climb is less than it was at best rate of climb speed. Trading airspeed for altitude and a temporary increased rate of climb is referred to as a "zoom climb." This type of climb provides an increased rate of climb for a few thousand feet, but it ultimately reduces overall climb performance.

Jet Airplane Descent and Approach

The smoothest and most fuel-efficient descent would be to reduce power to flight idle and slow to L/D_{MAX}. In this scenario, the pilot would descend, level off to decelerate, configure for landing, intercept the final approach, and continue a gradual deceleration until setting power for a stabilized descent on final. Traffic and time considerations almost always require deviation from this example, and the typical descent profile has three descent segments with two speed reductions in between.

Descent Planning

For a typical idle power descent, the top of descent (TOD), point A in *figure 16-14*, is determined by altitude, adjusted for wind. Jet descent profiles normally approximate a 3 degree path, with some time/distance required for deceleration in level flight. While exact distances will vary, having a descent plan will put the pilot well ahead of the jet and in a better position to monitor the automation.

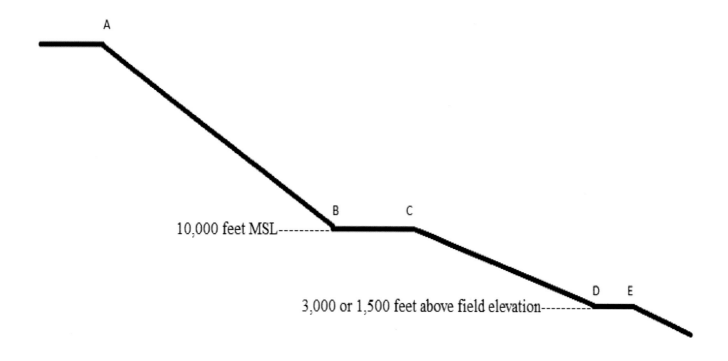

Figure 16-14. *Typical descent profile.*

For a straight-in VFR approach to an airport without factoring wind, an estimate for TOD may be calculated by multiplying the planned descent (in thousands of feet) by 3 and adding any distance needed for speed reductions in level flight (losing about 10 KIAS per mile when level). If flying at 35,000 feet above airport elevation, a cruise descent would start approximately 120 miles from the airport (35 times 3, plus about 15 miles for speed reduction, in stages, from cruise speed in this example). *[Figure 16-14]* Normally, cruise Mach is maintained until increasing air density causes indicated airspeed to increase to the desired descent speed, which usually occurs just below 30,000 feet. If arriving at point B at 10,000 feet MSL about 40 miles from the airport for deceleration to 250 knots, the pilot would resume a descent about 35 miles from the airport, continuing to 1,500 feet about 15 miles from the runway. The approach would continue with deceleration and flap extension so as to start the final descent 5 miles from the runway. There, the pilot extends the landing gear and selects landing flaps by 1,000 feet AGL, and brings the power up by 500 feet AGL to maintain the appropriate speed for a stabilized approach.

Variables that affect the TOD calculation include:

- Head/tail wind component (adjust distance 1 mile for each 10 knots of wind at cruise altitude),

- Field elevation,

- Terrain considerations,

- Runway alignment on arrival,

- ATC vectors and speed restrictions,

- Type of approach.

Descent Energy Management

While descending, the pilot can check the progress periodically. Estimating using round numbers keeps the calculation simple. Passing 25,000 feet should occur at 75 miles out plus or minus corrections; 20,000 feet should be at 60 miles, etc. If there is a deviation from the desired altitude/distance target, the energy state needs to be adjusted.

As discussed in Chapter 4, *Using Energy Management to Master Altitude and Airspeed Control*, there are two forms of energy in an airplane: potential energy in the form of altitude, and kinetic energy in the form of speed. In the normal operating regime at speeds above L/D_{MAX}, increasing speed increases total drag, while a decreasing speed will decrease total drag.

At idle power and at speeds above L/D_{MAX}, increasing speed increases the rate of descent. Sample data for a particular make and model might look like the following:

- 210 KIAS = 1,000 feet per minute

- 250 KIAS = 1,500 feet per minute

- 300 KIAS = 3,000 feet per minute

The exponential increase in parasite drag at higher speeds has a significant impact on both the rate of descent and the descent angle. Using the sample numbers, a 20% increase in airspeed from 210 to 250 knots, results in a 50% increase in the descent rate. However, a 20% increase in airspeed from 250 to 300 knots results in a 100% increase in the descent rate. Therefore, when at a higher altitude than desired in a descent, lowering the nose to increase speed will increase the descent angle and get the aircraft back to the desired path. Conversely, if lower than planned in descent, raising the nose to decrease speed will reduce descent angle until back on the desired path. Often, just a 10-knot change in speed allows for a smooth and gradual correction.

If speed adjustment is not an option, power can be added to correct a low-energy state, or the speed brakes used to correct a high-energy state. Numerous power fluctuations or repeated deployment and stowing of speed brakes is an indication of either pilot failure to adequately plan and/or manage the descent, or a poorly designed arrival procedure.

If a different descent speed from that planned is used during a descent, an adjustment should be made to the top of descent point. If ahead of schedule, leaving cruise altitude sooner, setting flight idle, and descending at a slower speed will burn less fuel. Conversely, if running late and willing to burn some extra fuel, the pilot can leave cruise later and descend at a higher speed. In all cases, the pilot should check progress during the descent and continue to adjust as necessary.

Planned descent speed will affect the position of the planned top of descent point. *[Figure 16-15]* In this example, both jets fly past point X at the same cruise speed and altitude with plans to arrive at point Y at 10,000 feet and 250 knots. In both cases, the aircraft would then be in a position to set up a continued descent. The 250-knot descent requires a few miles for deceleration and gives a shallower descent path. The 300-knot descent allows staying at altitude longer, descending at a steeper angle, and then leveling off to slow to 250 knots. The jet that descended at 300 knots arrives first at point Y but burns more fuel. While not depicted, an inefficient descent plan would start the descent at point X, maintain 300 knots, and require power to maintain that airspeed on a shallow descent path.

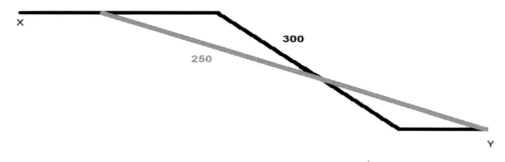

Figure 16-15. *Effect of speed on descent path.*

Descending prior to the planned TOD point will increase time to destination and fuel consumption. When given a descent clearance prior to the planned TOD, it is acceptable to ask ATC if the descent can be done at the pilot's discretion. If authorized to do so, this option allows for maintaining speed and altitude until reaching the calculated top of descent point. If an immediate descent is required, a descent at 1,000 feet per minute is usually acceptable until reaching the desired path. If a descent clearance has not been received by the planned TOD point, a speed reduction will reduce the airplane's kinetic and total energy while potential energy remains constant. When the clearance is received, a slightly steeper descent at the onset allows for a desired increase in kinetic energy at the expense of altitude and an appropriate descent rate such that the airplane follows the steeper desired path with acceptable energy distribution.

Jet Engine Landing

14 CFR part 25, section 25.125 defines the horizontal distance needed in order to land a jet airplane. The regulation describes the landing profile as the horizontal distance required to land and come to a complete stop from a point 50 feet above the landing surface. Manufacturers determine the landing distance on a dry, level runway at standard temperatures without using thrust reversers, auto brakes, or auto-land systems as a baseline. The pilot uses the landing weight and environmental conditions to determine the actual expected landing requirement based on the FAA-approved data in the AFM. As an accepted safety practice, pilots normally add a 40% cushion for landing on a dry runway. Dividing the usable runway length by 1.67 should give a number equal to or greater than the landing distance calculated from AFM data. For a wet runway, the distance should be increased by an additional 15%. *[Figure 16-16]*

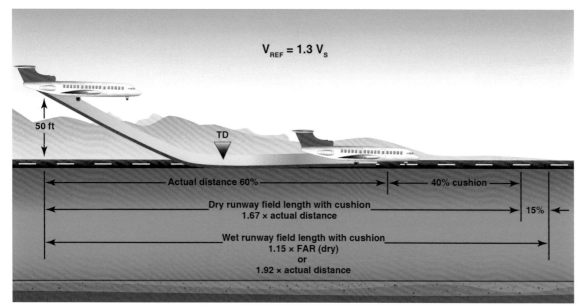

Figure 16-16. *FAR landing field length required.*

Simply put, the pilot divides the length of an intended runway by 1.67 or 1.92, as appropriate, to determine the minimum distance that should be available for landing. With this safety margin, it works out that the minimum dry runway field length should be at least 1.4 times the calculated air and ground distance needed, and the wet runway landing field length should be at least 1.61 times the calculated air and ground distance needed. Careful flight planning allows a pilot to determine how much load in terms of fuel, passengers, or cargo can be carried to a particular runway while still maintaining the desired safety margin. Depending on the destination, the load might need to be limited in order to protect the safety margin when landing. This is often complex, since fuel load has its own safety implications.

Certified landing field length requirements are computed for the stop made with speed brakes deployed and maximum wheel braking. Reverse thrust is not used in establishing the certified landing distances. However, reversers should definitely be used, if available.

Landing Speeds

As in the takeoff planning, there are certain speeds that should be taken into consideration when landing a jet airplane. The speeds are as follows:

- V_{SO}—stall speed in the landing configuration.

- V_{REF}—1.3 times the stall speed in the landing configuration.

- Approach climb—the speed that guarantees adequate performance in a go-around situation with an inoperative engine.

- Landing climb—the speed that guarantees adequate performance in arresting the descent and making a go-around from the final stages of landing with the airplane in the full landing configuration and maximum takeoff power available on all engines.

Pilots may need to perform traffic pattern takeoffs and landings. Pilots should use speeds recommended by the manufacturer while maneuvering in the traffic pattern prior to slowing to the final approach target speed in relation to V_{REF}. The speeds should be calculated for every landing and posted where they are visible to both pilots.

The approach and landing sequence in a jet airplane should be accomplished in accordance with an approach and landing profile developed for the particular airplane. *[Figure 16-17]*

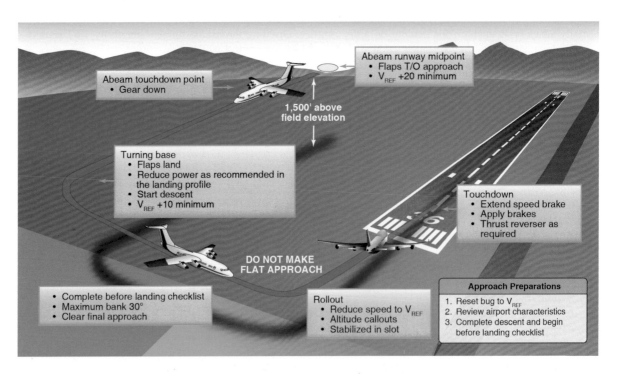

Figure 16-17. *Typical approach and landing profile.*

16-20

Significant Differences

A safe approach in any type of airplane culminates in a particular position, speed, and height over the runway threshold. That final flight condition is the target window at which the entire approach aims. Propeller-powered airplanes are able to approach that target from wider angles, greater speed differentials, and a larger variety of glidepath angles. Jet airplanes are not as responsive to power and course corrections, so the final approach should be more stable, more deliberate, and more constant in order to reach the window accurately.

The transitioning pilot should understand that in spite of their impressive performance capabilities, there are many reasons why jet airplanes are less forgiving than piston-engine airplanes during approaches and when correcting approach errors.

- There is no propeller slipstream to produce immediate extra lift at constant airspeed. There is no such thing as salvaging a misjudged glidepath with a sudden burst of power. Added lift can only be achieved by accelerating the airframe.

- Propeller slipstream is not available to lower the power-on stall speed. There is virtually no difference between power-on and power-off stall speed. It is not possible in a jet airplane to jam the thrust levers forward to avoid a stall.

- Jet engine response at low rpm is slower. This characteristic requires that the approach be flown at a stable speed and power setting on final so that sufficient power is available quickly if needed.

- Jet airplanes are consistently heavier and have faster approach speeds than a comparably sized propeller airplane. Since greater force is required to overcome momentum for speed changes or course corrections, the typical jet responds less quickly than the propeller airplane and requires careful planning and stable conditions throughout the approach.

- When the speed does increase or decrease, there is little tendency for the jet airplane to re-acquire the original speed. The pilot needs to make speed adjustments promptly in order to remain on speed.

- Drag increases faster than lift and produces a high sink rate at low speeds. Jet airplane wings typically have a large increase in drag in the approach configuration. When a sink rate does develop, the only immediate remedy is to increase pitch attitude (AOA). Because drag increases faster than lift, that pitch change rapidly contributes to an even greater sink rate unless a significant amount of power is promptly applied.

These flying characteristics of jet airplanes make a stabilized approach an absolute necessity.

Stabilized Approach

The performance charts and the limitations contained in the FAA-approved AFM are predicated on momentum values that result from programmed speeds and weights. Runway length limitations assume an exact 50-foot threshold height at an exact speed of 1.3 times V_{SO}. That "window" is critical and is a prime reason for the stabilized approach. Performance figures also assume that once through the target threshold window, the airplane touches down in a target touchdown zone approximately 1,000 feet down the runway, after which maximum stopping capability is used.

The basic elements to the stabilized approach are listed below as follows:

- The airplane should be in the landing configuration by 1,000 feet AGL in the approach. The landing gear should be down, landing flaps selected, trim set, and fuel balanced. Ensuring that these tasks are completed helps keep the number of variables to a minimum during the final approach.

- The airplane should be on profile before descending below 1,000 feet. Configuration, trim, speed, and glidepath should be at or near the optimum parameters early in the approach to avoid distractions and conflicts as the airplane nears the threshold window. An optimum glidepath angle of about 3° should be established and maintained.

- Indicated airspeed should be between zero and 10 knots above the target airspeed by 500 feet AGL. There are strong relationships between trim, speed, and power in most jet airplanes, and it is important to stabilize the speed in order to minimize those other variables.

- The optimum descent rate is dependent upon ground speed. A rule of thumb is to multiply half of ground speed by 10. For example, a 130-knot ground speed should result in a (65 times 10) 650 feet per minute descent rate. Typical descent rates fall between 500 and 700 feet per minute. An excessive vertical speed may indicate a problem with the approach.

Every approach should be evaluated at 500 feet. In a typical jet airplane, this is approximately 1 minute from touchdown. If the approach is not stabilized at that height, a go-around should be initiated. *[Figure 16-18]*

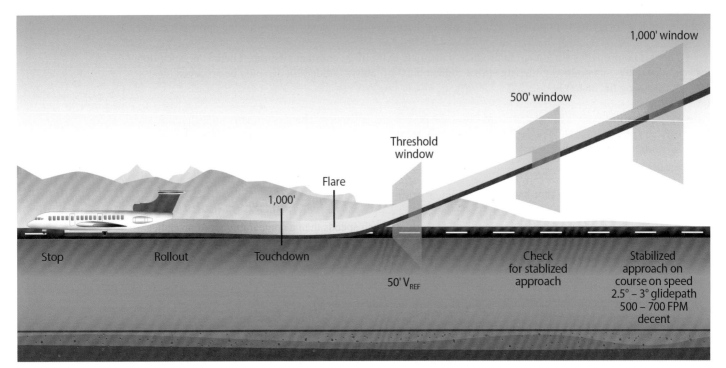

Figure 16-18. *Stabilized approach.*

Approach Speed

Any speed deviation on final approach should be detected immediately and corrected. With experience, the pilot is able to detect the onset of an increasing or decreasing airspeed trend, which normally can be corrected with a small adjustment. It is imperative the pilot does not allow the airspeed to decrease below V_{REF} or a high sink rate can develop. If an increasing sink rate is detected, it should be countered by increasing the AOA and simultaneously increasing thrust to counter the extra drag. The degree of correction depends on how much the sink rate needs to be reduced. For small amounts, smooth and gentle, almost anticipatory corrections are sufficient. For large sink rates, drastic corrective measures would be required that, even if successful, would destabilize the approach.

A common error in the performance of approaches in jet airplanes is excess approach speed. Excess approach speed carried through the threshold window and onto the runway increases the minimum stopping distance required by 20–30 feet per knot for a dry runway and 40–50 feet for a wet runway. Worse yet, the excess speed increases the chances of an extended flare, which increases the distance to touchdown by approximately 250 feet for each excess knot in speed.

Proper speed control on final approach is of primary importance. The pilot should anticipate the need for speed adjustment so that only small adjustments are required, and the airplane arrives at the approach threshold window exactly on speed.

Glidepath Control

The optimum glidepath angle is about 3°. On visual approaches, pilots may have a tendency to make flat approaches. A flat approach, however, increases landing distance and should be avoided. For example, an approach angle of 2° instead of a recommended 3° adds 500 feet to landing distance.

A more common error is excessive height over the threshold. This could be the result of an unstable approach or a stable but high approach. It also may occur during a nonprecision instrument approach where the missed approach point is close to or at the runway threshold. Regardless of the cause, excessive height over the threshold most likely results in a touchdown beyond the normal aiming point. An extra 50 feet of height over the threshold adds approximately 1,000 feet to the landing distance. The airplane should arrive at the approach threshold window exactly on altitude (50 feet above the runway).

The Flare

The flare reduces the approach rate of descent to a more acceptable rate for touchdown. Unlike light airplanes, a jet airplane should be flown onto the runway rather than "held off" the surface as speed dissipates. A jet airplane is aerodynamically clean even in the landing configuration, and its engines still produce residual thrust at idle rpm. Holding it off during the flare in an attempt to make a smooth landing greatly increases landing distance. A firm landing is normal and desirable. A firm landing does not mean a hard landing, but rather a deliberate or positive landing.

For most airports, the airplane passes over the end of the runway with the landing gear 30–45 feet above the surface, depending on the landing flap setting and the location of the touchdown zone. It takes 5–7 seconds from the time the airplane passes the end of the runway until touchdown. The flare is initiated by increasing the pitch attitude just enough to reduce the sink rate to 100–200 fpm when the landing gear is approximately 15 feet above the runway surface. In most jet airplanes, this requires a pitch attitude increase of only 1° to 3°. The thrust is smoothly reduced to idle as the flare progresses.

The normal speed bleed off during the time between passing the end of the runway and touchdown is just a few knots. Most of the decrease occurs during the flare when thrust is reduced. If the flare is extended (held off) while an additional speed is bled off, hundreds or even thousands of feet of runway may be used up. *[Figure 16-19]* The extended flare also results in additional pitch attitude, which may lead to a tail strike. It is, therefore, essential to fly the airplane onto the runway at the target touchdown point, even if the speed is excessive. A deliberate touchdown should be planned and practiced on every flight. A positive touchdown helps prevent an extended flare.

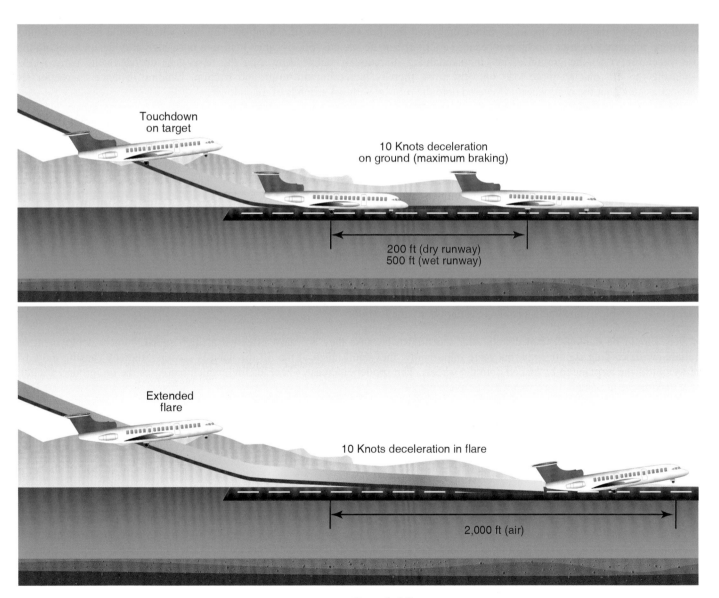

Figure 16-19. *Extended flare.*

Pilots should learn the flare characteristics of each model of airplane they fly. The visual reference cues observed from each airplane are different because window geometry and visibility are different. The geometric relationship between the pilot's eye and the landing gear is different for each make and model. It is essential that the flare maneuver be initiated at the proper height—not too high and not too low.

Beginning the flare too high or reducing the thrust too early may result in the airplane floating beyond the target touchdown point or may include a rapid pitch up as the pilot attempts to prevent a high sink rate touchdown. This can lead to a tail strike. The flare that is initiated too late may result in a hard touchdown.

Proper thrust management through the flare is also important. In many jet airplanes, the engines produce a noticeable effect on pitch trim when the thrust setting is changed. A rapid change in the thrust setting requires a quick elevator response. If the thrust levers are moved to idle too quickly during the flare, the pilot may need to make rapid changes in pitch control. If the thrust levers are moved more slowly, the elevator input can be more easily coordinated.

Touchdown and Rollout

A proper approach and flare positions the airplane to touch down in the touchdown target zone, which is usually about 1,000 feet beyond the runway threshold. Once the main wheels have contacted the runway, the pilot should maintain directional control and initiate the stopping process on the runway that remains in front of the airplane. The runway distance available to stop is longest if the touchdown was on target. The energy to be dissipated is least if there is no excess speed.

At the point of touchdown, the airplane represents a very large mass that is moving at a relatively high speed. The large total energy gets dissipated by the brakes, the aerodynamic drag, and the thrust reversers (if available). The nose-wheel should be lowered onto the ground immediately after touchdown because a jet airplane decelerates poorly when held in a nose-high attitude, and placing the nose-wheel tire(s) on the ground assists in maintaining directional control. Lowering the nose gear decreases the wing AOA, decreasing the lift, placing more load onto the tires, thereby increasing tire-to-ground friction. Landing distance charts for jet airplanes assume that the nose-wheel is lowered onto the runway within 4 seconds of touchdown.

There are only three forces available for stopping the airplane: wheel braking, reverse thrust, and aerodynamic braking. Of the three, the brakes are most effective and therefore the most important stopping force for most landings. When the runway is very slippery, reverse thrust and drag may be the dominant forces. Both reverse thrust and aerodynamic drag are most effective at high speeds. Neither is affected by runway surface conditions. Brakes, on the other hand, are most effective at low speed. The landing rollout distance depends on the touchdown speed, what forces are applied, and when they are applied. The pilot controls the what and when factors, but the maximum braking force may be limited by tire-to-ground friction.

The pilot should begin braking as soon after touchdown and wheel spin-up as possible, and smoothly continue the braking until stopped or a safe taxi speed is reached. However, caution should be used if the airplane is not equipped with a functioning anti-skid system. In such a case, heavy braking can cause the wheels to lock and the tires to skid.

Both directional control and braking utilize tire ground friction. They share the maximum friction force the tires can provide. Increasing either subtracts from the other. Understanding tire ground friction, how runway contamination affects it, and how to use the friction available to maximum advantage is important to a jet pilot.

Spoilers should be deployed immediately after touchdown because they are most effective at high speed. Timely deployment of spoilers increases drag significantly, but more importantly, they spoil much of the lift the wing is creating, thereby causing more of the weight of the airplane to be loaded onto the wheels. The spoilers increase wheel loading, which increases the tire ground friction force making the maximum tire braking forces available.

Like spoilers, thrust reversers are most effective at high speeds and should be deployed quickly after touchdown. However, the pilot should not command significant reverse thrust until the nose-wheel is on the ground. If the reversers deploy asymmetrically resulting in an uncontrollable yaw toward the side with more reverse thrust, the pilot needs whatever nose-wheel steering is available to maintain directional control. When runway length is not a factor, using idle reverse thrust may be adequate.

Jet Airplane Systems and Maintenance

All FAA-certificated jet airplanes are certificated under Title 14 of the Code of Federal Regulations (14 CFR) part 25, which contains the airworthiness standards for transport category airplanes. The FAA-certificated jet airplane is a highly sophisticated machine with proven levels of performance and guaranteed safety margins. The jet airplane's performance and safety margins can only be realized, however, if the airplane is operated in strict compliance with the procedures and limitations contained in the FAA-approved AFM for the particular airplane. Furthermore, in accordance with 14 CFR part 91, section 91.213(a), a turbine-powered airplane does not qualify to takeoff with inoperable instruments or equipment installed unless, among other requirements, an approved Minimum Equipment List (MEL) exists for that aircraft, and the aircraft is operated under all applicable conditions and limitations contained in the MEL (section 91.213(a)(5)).

Minimum Equipment List

The MEL serves as a reference guide for dispatchers and pilots to determine whether takeoff of an aircraft with inoperative instruments or equipment is authorized under the provisions of applicable regulatory requirements.

The operator models the MEL after the FAA's Master MEL (MMEL) for each type of aircraft and the Administrator approves the MEL before its implementation. The MEL includes a "General" section, comprised of definitions, general policies, as well as operational procedures for flight crews and maintenance personnel. Each aircraft component addressed in the MEL is listed in an alphabetical index for quick reference. A table of contents further divides the manual in different chapters, each numbered for its corresponding aircraft system designation (i.e., the electrical system, also designated as system number 24, would be found in chapter 24 of the MEL).

Pilots may defer repair of items on those aircraft systems and components allowed by the approved MEL. Per 14 CFR part 91, section 91.213(a)(3)(ii), an MEL must provide for the operation of the aircraft with the instruments and equipment in an inoperable condition. If particular items do not allow for safe operation, they do not appear on the MEL and takeoff is not authorized until the item is adequately repaired or replaced (section 91.213(a)). In cases where repairs may temporarily be deferred, operation or dispatch of an aircraft whose systems have been impaired is often subject to limitations or other conditional requirements explicitly stated in the MEL. Such conditional requirements may be of an operational nature, a mechanical nature, or both.

Mechanical conditions outlined in the MEL may require precautionary pre-flight checks, partial repairs prior to departure, or the isolation of selected elements of the deficient aircraft system (or related interacting systems), as well as the securing of other system components to avoid further degradation in flight. The MEL may contain either a step-by-step description of required partial maintenance actions or a list of numerical references to the Maintenance Procedures Manual (MPM) where each corrective procedure is explained in detail. Procedures performed to ensure the aircraft can be safely operated are categorized as either pperations procedures or maintenance procedures. The MEL will denote which by indicating an "O" or an "M" as appropriate.

If operational and mechanical conditions can be met, an authorized person makes an entry in the aircraft MEL Deferral Record and issues a temporary placard. This authorizes the operation for a limited time before permanent repairs take place. The placard is affixed by maintenance personnel or the flight crew onto or next to the instrument or control mechanism to remind the flight crew of any limitations.

The MEL only applies while the aircraft sits on the ground awaiting departure or takeoff. It is essentially a dispatching reference tool used in support of all applicable Federal Aviation Regulations. If dispatchers are not required by the operator's certificate, flight crews still need to refer to the MEL before dispatching themselves to ensure that the flight is planned and conducted within the operating limits set forth in the MEL. Once the aircraft leaves the ground, any mechanical failures should be addressed using the appropriate checklists and approved AFM, not the MEL. Although a pilot may refer to the MEL for background information and documentation, actions in flight should be based strictly on instructions provided by the AFM (i.e., Abnormal or Emergency sections).

Configuration Deviation List

A Configuration Deviation List (CDL) is used in the same manner as an MEL but it differs in that it addresses missing external parts of the aircraft rather than failing internal systems and their constituent parts. They typically include elements, such as service doors, power receptacle doors, slat track doors, landing gear doors, APU ram air doors, flaps fairings, nose-wheel spray deflectors, position light lens covers, slat segment seals, static dischargers, etc.

Chapter Summary

Some of the differences when transitioning from props to jets include:
- Engine intake suction and exhaust create a ground hazard.
- There is no propeller-induced lift when power increases.
- Engine spool up time from low power settings is longer.
- Swept wing stalls begin at the tips.
- Higher speeds require smaller and smoother flight control inputs.
- Descents require more planning and optimally occur at idle power.
- When descending at speeds above L/D_{MAX}, increasing speed increases rate of descent and descent angle.

There are many considerations for a pilot when transitioning to turbojet-powered airplanes. In addition to the information found in this chapter and type specific information that will be found in an FAA-approved Airplane Flight Manual, a pilot can find basic aerodynamic information for swept wing jets, considerations for operating at high altitudes, and airplane upset causes and general recovery procedures in the Airplane Upset Recovery Training Aid, Supplement, pages 1-14, and all of Section 2 found at www.faa.gov/other_visit/aviation_industry/airline_operators/training/media/ap_upsetrecovery_book.pdf.

Airplane Flying Handbook (FAA-H-8083-3C)
Chapter 17: Transition to Light Sport Airplanes (LSA)

Introduction

The light-sport aircraft (LSA) concept broadens the access of flight to more people. LSA have been defined as a simple-to-operate, easy-to-fly aircraft; however, "simple-to-operate" and "easy-to-fly" do not negate the need for proper and effective training. This chapter introduces the light-sport category of airplanes and places emphasis on transition to a light-sport airplane.

Even though light-sport airplane flight may appear simple to an experienced pilot, a transition to a light-sport airplane should include the same methodical training approach as transitioning into any other airplane. A pilot seeking a transition into light-sport airplane flying should follow a systematic, structured light-sport airplane training course under the guidance of a competent instructor with recent experience in the specific training airplane.

Light-Sport Aircraft Background

Several groups were instrumental in the development and success of the LSA concept. These included the Federal Aviation Administration (FAA), Light Aircraft Manufacturers Association, American Society for Testing and Materials (ASTM) International, and countless individuals who promoted the concept since the early 1990s. In 2004, the FAA released a rule that created a light-sport classification for airplane, gyroplane, lighter-than-air, weight-shift-control, glider, and powered parachute. *[Figure 17-1]*

Figure 17-1. *The LSA category covers a wide variety of aircraft including: A) airplane, B) gyroplane, C) lighter-than-air, D) weight-shift-control, E) glider, and F) powered parachute.*

The primary concept of the LSA is built around a defined set of standards found in 14 CFR part 1, section 1.1:

- Powered (if powered) by single reciprocating engine.
- Fixed landing gear (except seaplanes and gliders).
- Fixed pitch or ground adjustable propeller.
- Maximum takeoff weight of 1,320 pounds for landplane, 1,430 for seaplane.
- Maximum of two occupants.
- Non-pressurized cabin.
- Maximum speed in level flight at maximum continuous power of 120 knots calibrated airspeed (CAS).
- Maximum stall speed of 45 knots. *[Figure 17-2]*

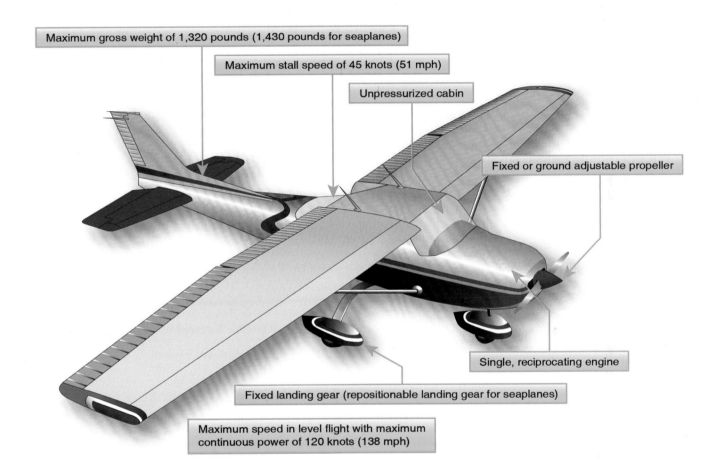

Figure 17-2. *Light-sport airplane.*

The LSA category includes standard, special, and experimental designations. Some standard airworthiness certificated aircraft (i.e., a Piper J-2 or J-3) may meet the Title 14 of the Code of Federal Regulation (14 CFR) part 1, section 1 definition of LSA. Type certificated aircraft that continue to meet the section 1.1 definition of LSA may be flown by a pilot who holds a sport pilot certificate with the appropriate endorsement for the aircraft (14 CFR part 61, section 61.315(a)). The sport pilot certificate is discussed later in this chapter. Aircraft that are specifically manufactured for the LSA market are included in either the Special (S-LSA) or Experimental (E-LSA) designations. An approved S-LSA is manufactured in a ready-to-fly condition and an E-LSA is either a kit or plans-built aircraft based on an approved S-LSA model.

It is important to note that S-LSAs or E-LSAs are not type certificated by the FAA and are not required to meet any airworthiness requirements of 14 CFR part 23. Instead, S-LSAs and E-LSAs are designed and manufactured in accordance with ASTM Committee F-37 Industry Consensus Standards. Therefore, LSA designs are not subjected to the scrutiny, demands, and testing of FAA standard airworthiness certification. Industry Consensus Standards are intended to be less costly and less restrictive than 14 CFR part 23 certification requirements and, as a result, LSA manufacturers have greater latitude with their designs. ASTM Industry Consensus Standards were accepted by the FAA in 2005, which established FAA-accepted industry-developed standards for the design and manufacture of aircraft for the first time.

ASTM Industry Consensus Standards for LSA cover the following areas:

- Design and performance
- Required equipment
- Quality assurance
- Production acceptance tests
- Aircraft operating instructions
- Maintenance and inspection procedures
- Identification and recording of major repairs and major alterations
- Continued airworthiness
- Manufacturer's assembly instructions (E-LSA aircraft)

Using the ASTM Industry Consensus Standards, an LSA manufacturer can design and manufacture their aircraft and assess its compliance to the consensus standards. The manufacturer then, through evaluation services offered by a designated airworthiness representative, completes the process by submitting the required paperwork to the FAA. Upon approval, an LSA manufacturer is permitted to sell ready-to-fly S-LSA aircraft.

Light-Sport Airplane Synopsis

- The airplane must meet the weight, speed, and other criteria listed in 14 CFR part 1, section 1 that define an LSA.

- Airplanes under the S-LSA certification commonly find use in sport and recreation, flight training, and aircraft rental.

- E-LSA-certified airplanes may be used for sport and recreation and flight instruction for the owner of the airplane. However, E-LSA certification is not the same as Experimental Amateur-Built aircraft certification. E-LSA certification is based on an approved S-LSA airplane.

- FAA policy allows sport pilots with an airplane rating to fly certain airplanes (i.e., a Piper J-2 or J-3) that continue to meet the 14 CFR part 1, section 1 LSA definition even though the airplane was originally issued a standard airworthiness type certificate.

- No person may operate the aircraft unless it has been registered by its owner, if eligible for registration, per 14 CFR part 47, section 47.3(b).

- United States or foreign manufacturers can be authorized.

- FAA policy allows holders of a sport pilot certificate or higher level pilot certificate (recreational, private, commercial, or ATP) to pilot sport aircraft.

- LSAs may be operated by VFR at night if the aircraft is equipped with the instruments and equipment specified in 14 CFR part 91, section 91.205(c), and if night operations are allowed by the airplane's operating limitations. However, sport pilots may not fly at night (14 CFR 61.315(c)(5)).

- LSAs may be operated between sunset and sunrise by a recreational pilot as the sole occupant of the aircraft, in accordance with 14 CFR part 61, section 61.101(i)(3) for the purpose of obtaining additional certificates or ratings and while under the supervision of an authorized instructor and provided the flight or surface visibility is at least 5 statute miles.

Sport Pilot Certificate

In addition to the LSA rules, the FAA created a new sport pilot certificate in 2004 that lowered the minimum training time requirements, in comparison to other pilot certificates, for newly certificated pilots wishing to exercise privileges only in LSA aircraft. Pilots who hold recreational, private, commercial, or airline transport pilot certificates may pilot light-sport airplanes provided they possess the appropriate category/class rating and a U.S. driver's license or medical certificate that meets the requirements for the aircraft displayed in the 14 CFR part 61, section 61.303 table. For example, a commercial pilot rated in airplane multiengine land and rotorcraft gyroplane is qualified to fly a light-sport gyroplane as pilot in command (PIC) if also holding a medical certificate or a U.S. driver's license. However, that pilot is not qualified to act as PIC of a light-sport airplane (sport airplanes are single-engine) without supervision from an authorized instructor.

Pilots holding higher level certificates with the appropriate category and class ratings may fly LSAs as long as the pilot holds a valid U.S. driver's license as evidence of medical eligibility. However, if the pilot's most recent medical certificate was denied, revoked, suspended, or withdrawn, a U.S. driver's license is not sufficient. The pilot would then need to hold a valid FAA medical certificate to fly an LSA.

Transition Training Considerations
Flight Schools

The LSA category has created new business opportunities due to low fuel usage, reliability, and low maintenance costs. Many owners and operators of flight schools use S-LSAs for flight instruction and rental.

When considering a transition to LSA, a pilot should look for a flight school that has experience in LSA instruction and can provide quality instruction. Personally touring a school and soliciting feedback from other pilots that have transitioned into LSAs may help find an appropriate school. Some questions to be asked include the following:

- How many pilots has the flight school transitioned into LSAs and how many LSAs are available for instruction?
- What are the flight school's rental, insurance, and safety policies?
- How is maintenance accomplished and by whom?
- How are records maintained and how is scheduling accomplished?

Flight Instructors

The flight instructor is an important link in a successful LSA transition. A transitioning pilot should choose a flight instructor that has verifiable experience in LSA instruction. The Sport Pilot rule allows for a Sport Pilot flight instructor certificate, the flight instructor-S. 14 CFR part 61, section 61.413 limits a flight instructor-S to instruction in LSAs—a flight instructor-S cannot give instruction in a non-LSA airplane (e.g., a Cessna 150). While FAA policy allows a flight instructor certificated as a flight instructor-A to give instruction in both a light-sport airplane and a non-light-sport airplane, a flight instructor-S with teaching experience in LSA might provide better instruction than a flight instructor-A who has minimal teaching experience in light-sport airplanes.

A transitioning LSA pilot should have an opportunity to review the curriculum, syllabus, lesson plans, as well as the process for tracking progress through the training program. Depending on the transitioning pilot's experience, currency, and type of airplane typically flown, the flight instructor should make appropriate adjustments to any LSA training curriculum. A suggested LSA transition training outline is presented:

- CFR review as pertaining to LSAs and sport pilots
- Pilot's Operating Handbook (POH) review
- LSA maintenance
- LSA weather considerations
- Wake turbulence avoidance
- Performance and limitations
- Operation of systems
- Ground operations
- Preflight inspection
- Before takeoff check
- Normal and crosswind takeoff and climb
- Normal and crosswind approach and landing
- Soft-field takeoff, climb, approach, and landing
- Short-field takeoff, climb, approach, and landing
- Go-around/rejected landing
- Steep turns
- Stalls and spin awareness
- Emergency approach and landing
- Systems and equipment malfunctions
- After landing, parking, and securing

LSA Maintenance

LSAs should be treated with the same level of care as any standard airworthiness certificated airplane. However, S-LSAs have greater latitude pertaining to who may conduct maintenance as compared to standard airworthiness certificated airplanes. S-LSAs may be maintained and inspected by:

1. An LSA Repairman with a Maintenance rating; or,
2. An FAA-certificated Airframe and Powerplant Mechanic (A&P); or,
3. As specified by the aircraft manufacturer; or
4. As permitted, owners performing limited maintenance on their S-LSA.

The airplane maintenance manual includes the specific information for repair and maintenance on inspections, repair, and authorization for repairs and maintenance. Most often, S-LSA inspections can be signed off by an FAA-certificated A&P or LSA repairman with a Maintenance rating rather than an A&P with Inspection Authorization (IA); however, the aircraft maintenance manual provides the procedures to follow. The FAA does not issue Airworthiness Directives (ADs) for S-LSAs or E-LSAs. If an FAA-certified component is installed on an LSA, the FAA issues any pertaining ADs for that specific component. Manufacturer safety directives are not distributed by the FAA. S-LSA owners should comply with:

- Safety directives (alerts, bulletins, and notifications) issued by the LSA manufacturer
- ADs if any FAA-certificated components are installed
- Safety alerts (immediate action)
- Service bulletins (recommending future action)
- Safety notifications (informational)

S-LSA compliance with maintenance requirements provides greater latitude for owners and operators of these airplanes. Because of the options in complying with the maintenance requirements, pilots who are transitioning to LSAs should understand how maintenance is accomplished; who is providing the maintenance services; and verify that all compliance requirements have been met.

Airframe and Systems
Construction

LSAs may be constructed using wood, tube and fabric, metal, composite, or any combination of materials. In general, the manufacturer selects materials and design to keep the airplane lightweight while maintaining the structural requirements. Composite LSAs tend to be sleek and modern looking with clean lines as molding of the various components allows designers great flexibility in shaping the airframe. Other LSAs are authentic-looking renditions of early aviation airplanes with fabric covering a framework of steel tubes. Of course, LSAs may be anything in between using both metal and composite construction. *[Figure 17-3]* A pilot transitioning into LSA should understand the types of construction and the typical concerns for each type of construction:

- Steel tube and fabric—while the techniques of steel tube and fabric construction hails back to the early days of aviation, this construction method has proven to be lightweight, strong, and inexpensive to build and maintain. Advances in fabric technology continue to make this method of covering airframes an excellent choice. Fabric can be limited in its life span if not properly maintained. Fabric should be free from tears, well-painted with little to no fading, and should easily spring back when lightly pressed.

- Aluminum—an aluminum-fabricated airplane has been a favorite choice for decades. Pilots should be quite familiar with this type of construction. Generally, airframes tend to be lightly rounded structures dotted with rivets and fasteners. This construction is easily inspected due to the wide-spread experience with aluminum structures. Any corrosion, working rivets, dents, and cracks should be identified during a pilot's preflight inspection.

- Composite—a composite airplane is principally made from structural epoxies and cloth-like fabrics, such as bi-directional and uni-directional fiberglass cloths, and specialty cloths like carbon fiber. Airframe components, such as wing and fuselage halves, are made in molds that result in a sculpted, mirror-like finish. Generally, composite construction has few fasteners, such as protruding rivets and bolts. Pilots should become acquainted with inspection concerns such as looking for hair-line cracks and delamination.

Figure 17-3. *LSA can be constructed using both metal and composites.*

Engines

LSAs use a variety of engines that range from FAA-certificated to non-FAA-certificated. Engine technology varies significantly from conventional air-cooled to high revolutions per minute (rpm)/water-cooled designs. *[Figure 17-4]* These different technologies present a transitioning pilot new training opportunities and challenges. Since most light-sport airplanes use non-FAA-certificated engines, a transitioning pilot should fully understand the engine controls, procedures, and limitations. In most light-sport airplanes, engines are water-cooled, 4-cycle, and carbureted. These engines have much higher operating rpm and require a gear-box to reduce propeller rpm to the proper range. Because of the higher engine operating rpm, vibration and noise signatures are quite different in most light-sport airplanes when compared to most standard type certificated designs.

Figure 17-4. *A water-cooled 4-cycle engine.*

Instrumentation

In addition to advanced airframe and engine technology, LSAs often have advanced flight and engine instrumentation. Installation of electronic flight instrumentation systems (EFIS) provides attitude, airspeed, altimeter, vertical speed, direction, moving map, navigation, terrain awareness, traffic, weather, engine data, etc., all on one or two liquid crystal displays. *[Figure 17-5]* EFIS has become a cost-effective replacement for traditional mechanical gyros and instruments. Compared to mechanical instrumentation systems, EFIS requires almost no maintenance. There are tremendous advantages to EFIS systems as long as the pilot is correctly trained in their use. EFIS systems can cause a "heads down" syndrome and loss of situation awareness if the pilot is not trained to quickly and properly configure, access, program, and interpret the information provided. If EFIS is installed, transition training should include instruction in the use of the specific EFIS in the training airplane. In some cases, EFIS manufacturers or third party products are available for the pilot to practice EFIS operations on a personal computer as opposed to learning their functions in flight.

Figure 17-5. *An electronic flight instrumentation system provides attitude, airspeed, altimeter, vertical speed, direction, moving map, navigation, terrain awareness, traffic, weather, and engine data all on one or two liquid crystal displays.*

Weather Considerations

Managing weather factors is important for all aircraft but becomes more significant as the weight of the airplane decreases. Smaller, lighter weight airplanes are more easily affected by strong winds (especially crosswinds), turbulence, terrain influences, and other hazardous conditions. *[Figures 17-6 and 17-7]* LSA Pilots should carefully consider any hazardous weather and effectively use an appropriate set of personal minimums to mitigate flight risk. Some LSAs have a maximum recommend wind velocity regardless of wind direction. *[Figure 17-8]* While this is not a limitation, it would be prudent to heed any factory recommendations.

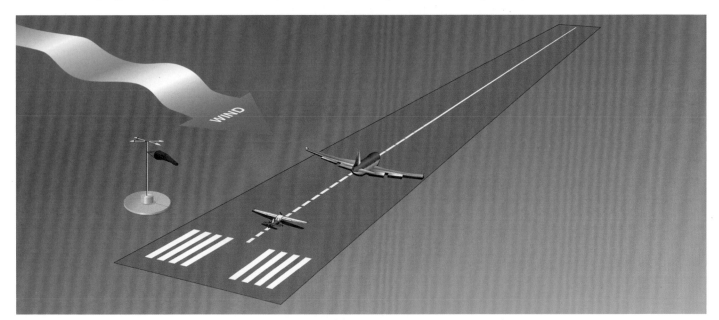

Figure 17-6. *Crosswind landing.*

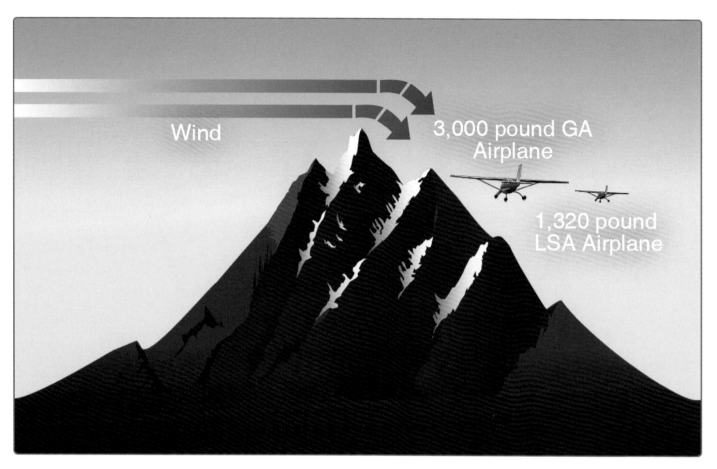

Figure 17-7. *Moderate mountain winds can create severe turbulence for LSA.*

Maximum Demonstrated Crosswind Velocity
Takeoff or landing..12 knots

Maximum Recommended Wind Velocity
All operations...22 knots

Figure 17-8. *Example of wind limitations that an LSA may have.*

Due to an LSA's lighter weight, even greater distances from convective weather should be considered the norm. While low-level winds that enter and exit a thunderstorm should be avoided by all airplanes, operations in the vicinity of convection should not be attempted in lightweight airplanes. Since it is not always possible to fly in clear, calm air, pilots of lighter weight LSAs should carefully manage all weather-related risks. For example, some consideration should be given to flight activity that crosses varying terrain boundaries, such as grass or water to hard surfaces. Differential heating can cause lighter weight airplanes to experience sinking and lifting to a greater degree than heavier airplanes. Careful planning, knowledge, experience, and an understanding of the flying environment assists in mitigating weather-related risks.

Flight Environment

The skills used to fly LSAs resemble those pilots use when flying any airplane, but the techniques may vary. This section outlines areas that are unique to light-sport airplanes. Most skills learned in a standard airworthiness type certificated airplane are transferable to LSAs; however, since LSAs can vary significantly in performance, equipment, systems, and construction, pilots should seek competent flight instruction and refer to the airplane's POH for detailed and specific information prior to flight.

Preflight

The preflight inspection of any airplane is critical to mitigating flight risks. A pilot transitioning into an LSA should allow adequate time to become familiar with the airplane prior to a first flight. First, the pilot and flight instructor should review the POH and cover the airplane's limitations, systems, performance, weight and balance, normal procedures, emergency procedures, and handling requirements. *[Figure 17-9]*

Figure 17-9. *Pilot's Operating Handbook for a LSA.*

Inside the Airplane

Transitioning pilots find an LSA very familiar when conducting a preflight inspection; however, some preflight differences are worth pointing out. For example, many LSAs do not have adjustable seats but rather adjustable rudder pedals. *[Figure 17-10]* Often, LSA seats are in a fixed position. LSA manufacturers have implemented various systems for rudder pedal position adjustment. Some manufacturers use a simple removable pin while others use a knob near the rudder pedals to adjust position. Shorter pilots may find that the adjustment range may not be sufficient, and an appropriate seat cushion may be needed to experience the proper range of rudder pedal movement. In addition, seats in some LSAs are in a semi-reclined position. The first time a pilot sits in a semi-reclining seat, it may seem somewhat unusual. A pilot should take time to get comfortable.

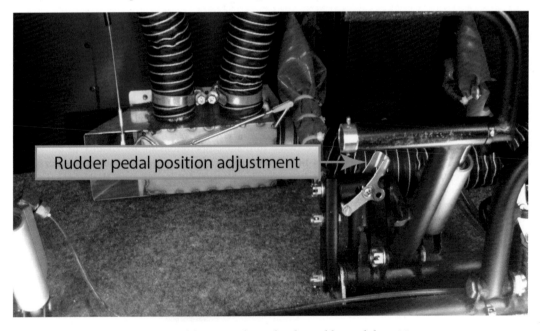

Figure 17-10. *Adjustment lever for the rudder pedal position.*

Transitioning pilots should become familiar with the flight and engine controls. These may vary significantly from airplane model to airplane model. Some light-sport airplanes use a conventional control stick while others use a yoke. One manufacturer has combined the two types of controls in what has been termed a "stoke." While this control may seem unique, it provides a completely natural feel for flight control. *[Figure 17-11]* Regardless of the flight controls, the pilot should perform a full range of motion check of the flight controls. This means full forward to full forward left to full aft left to full aft right and then full forward right. Verify that each control surface moves freely and smoothly. On some LSAs, aileron control geometry, in an attempt to minimize adverse yaw, moves ailerons in a highly differential manner; a pilot may see very little "down" aileron when compared to the "up" aileron. Pilots should always verify the direction of control surface movement.

Figure 17-11. *Stoke flight control with conventional engine controls.*

Elevator trim on many LSAs is electrically actuated with no mechanical trim adjustment available. *[Figure 17-12]* Depending on the airplane, trim position indication may be displayed on the EFIS, an LED display, or with a mechanical indicator. On electric trim systems, as it is with any airplane, it is important to ensure that the trim position is correctly set prior to takeoff. Because trim positioning/indicting systems vary widely in light-sport airplanes, pilots should fully understand not only how to position the trim, but also how to respond to a trim-run-away condition. Part of the preflight inspection should include actuating the trim switch in both nose-up and nose-down directions, verifying that the trim disconnect (if equipped) is properly functioning, and then properly setting the takeoff trim position.

Figure 17-12. *Trim control.*

Depending on the engine manufacturer, the engine controls may be completely familiar to a transitioning pilot (throttle, mixture, and carburetor heat); however, some engines have no mixture control or carburetor heat. Instead, there could be a throttle, a choke control, and carburetor preheater.

Regardless, a pilot should become familiar with the specific engine installed and its operation. A transitioning pilot also needs to become comfortable with the difference between conventional engine control knobs and those found in LSAs. In standard airworthiness airplanes, control knobs are reasonably standardized; however, LSAs may use controls that are much larger or smaller in size.

If the LSA is equipped with an EFIS, the manufacturer's EFIS Pilot Guide should be available for reference. In addition, the airplane POH likely has specific EFIS preflight procedures that should be completed. These checks are to verify that all internal tests are passed, that no red "Xs" are displayed, and that appropriate annunciators are illuminated. Some systems have a "reversionary" mode where the information from one display can be sent to another display. For example, should the Primary Flight Display (PFD) fail, information can be routed to the Multi-Function Display (MFD). Not all LSA EFIS systems are equipped with a MFD or reversionary capability, so it is important for a transitioning pilot to understand the system and its limitations.

Fuel level in any airplane should be checked both visually and via the fuel level instrument or sight gauges. In LSAs, fuel level quantities can be shown using a variety of systems. Some models may have conventional float activated indictors while others may have the fuel level display on the EFIS with low-fuel alarm capability. It is not uncommon for a light-sport airplane to have advanced EFIS technology for attitude and navigation information, but have a simple sight gauge for fuel level indication. Fuel tank selection can also vary from simple on/off valves to a left/right selector. Fuel starvation remains a leading factor in aircraft accidents, which should be a reminder that when transitioning into a new airplane, time spent understanding the fuel system is time well spent.

A ballistic parachute is a popular safety feature on certain LSAs. *[Figure 17-13]* These devices have been shown to be well worth their cost in the unlikely event of a catastrophic failure or some other unsurvivable emergency. This system rockets a parachute into a deployed state such that the parachute slowly lowers the aircraft. The preflight inspections of these systems require a check of the mounts, safety pin and flag, and the activation handle and cable. Because most standard airworthiness type certificated airplanes do not have these systems installed, LSA training should cover the operation and limitations of the system.

Figure 17-13. *A ballistic recovery parachute is a popular safety feature available on some LSA.*

Outside the Airplane

Transitioning pilots should feel comfortable and in a familiar setting when preflighting the outside of an LSA. Some unique areas worthy of notation are presented below.

Propellers of LSAs may range from a conventional metal propeller to composite or wood. If a transitioning pilot is principally familiar with metal propellers, time should be spent with the LSA flight instructor covering the type of propeller installed. Many LSA propellers are composite and have a ground adjustable pitch adjustment. There may be more areas to check with these types of propellers. For example, on ground adjustable propellers, ensure that the blades are tight against the hub by snugly twisting the blade at the root to verify that there is no rotation of the blade at the hub.

Many LSAs are equipped with water-cooled engines. LSAs may be tightly cowled, which reduces drag. A liquid-cooled engine minimizes the need for cylinder cooling inlets, which further reduces drag and improves performance. This does present a new system for a transitioning pilot to check. Preflighting this system requires that the radiator, coolant hoses, and expansion tank are checked for condition, freedom from leaks, and coolant level requirements.

Split flaps may be used on some LSA designs. *[Figure 17-14]* These flaps hinge down from underneath the wing and inspecting these flaps requires the pilot to crouch and twist low for inspection. A suitable handheld mirror can facilitate inspection without undue twisting and bending. In an attempt to keep complexity to a minimum, flap control is typically a handle that actuates the flaps. A pilot should verify that the flaps extend and retract smoothly.

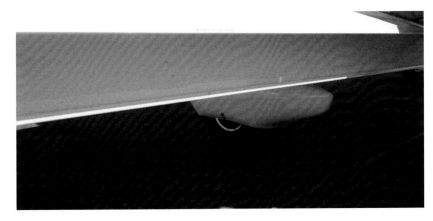

Figure 17-14. *Split flap.*

Before Start and Starting Engine

Once a pilot has completed the preflight inspection of the LSA, the pilot should properly seat themselves in the airplane ensuring that the rudder pedals can be exercised with full-range movement without over-reaching. Seat belts should be checked for proper position and security. The pilot should continue to use the POH checklists. Starting newer generation LSA engines can be quite simple and only require the pull of the choke and a twist of the ignition switch. If the LSA is equipped with a standard certificated engine, starting procedures are normal and routine. The canopy or doors of an LSA may have quite different latching mechanisms than standard airworthiness airplanes. Practice latching and unlatching the doors or canopy to ensure that understanding is complete. Having a gull-wing door or sliding canopy "pop" open in flight can become an emergency in seconds.

Taxi

LSAs may have a full-castoring or steerable nose-wheel, or a tailwheel if equipped with conventional gear. In order to taxi a full-castoring nose-wheel equipped airplane, the use of differential brakes is required. This type of nose-wheel can require practice to keep the airplane on the centerline while minimizing brake application or damage to the tires. If the taxi speed is too slow, application of a brake can cause the aircraft to pivot to a stop, rather than adjust in direction. This results in excessive brake and tire wear. If the speed is too fast, excessive brake wear is likely.

Transitioning into an LSA with conventional gear (tailwheel) should occur initially during no-wind conditions. Due to its light weight the airplane, requires the development of the proper flight control responses before operations in any substantial wind.

Takeoff and Climb

Takeoff and climb performance of LSAs can be spirited as they typically have a high horsepower-to-weight ratio and accelerate quickly. Due to design requirements for low stall speeds, LSAs typically have low rotation and climb speeds with impressive climb rates. Like other airplanes, the pilot should be flying the published speeds as given in the airplane's POH. Stick (yoke or stoke) forces tend to be light, which may lead a transitioning pilot to initially over-control. The key is to relax, have reasonable patience, and input only appropriate flight control pressures needed to get the required response. If a transitioning pilot is inducing excessive control inputs, they should minimize flight control pressures, set attitudes based on outside references, and allow the airplane to settle.

During climbs, visibility over the nose may be difficult in some LSAs. As always, it is important to properly clear the airspace for traffic and other hazards. Occasionally lowering the airplane's nose to get a good look out toward the horizon is important for managing flight safety. Shallow banked turns in both directions of 10° to 20° also allow for clearing. Because flight control pressures tend to be light, it is easy to get in the habit of flying with a light-sport airplane out of trim. This is to be avoided. Trim off any flight control pressures. This allows the pilot to focus as much time as possible looking outside.

Cruise

After leveling off at cruise altitude, the airplane should be allowed to accelerate to cruise speed, reduce power to cruise rpm, adjust pitch, and then trim off any flight control pressures. *[Figure 17-15]* The first time a transitioning pilot sees cruise rpm setting of 4,800 rpm (or as recommended), they may have a sense that the engine is turning too fast; however, remember that the engine has gear-reduction drive and the propeller is turning much slower. If the LSA is equipped with a standard aircraft engine, rpm should be in a range comparable to airplanes the transitioning pilot is used to. The pilot should refer to the Cruise Checklist to ensure that the airplane is properly configured.

Figure 17-15. *EFIS indication of level cruise flight.*

In slower cruise flight, stick forces are likely to be light; therefore, correction to pitch and roll attitudes should be made with light pressures. Excess pressure used to correct a deviation may cause a series of pilot-induced oscillations. The pilot should use fingertip pressures only and not use a wrapped palm of the hand. Stick forces can change dramatically as airspeed changes. For example, what could be considered light control pressures at 80 knots may become quite stiff at 100 knots. A flight instructor-S or flight instructor-A experienced in the light-sport airplane is able to demonstrate this effect, which is dependent on the specific model of LSA.

LSA maneuvers such as steep turns, slow flight, and stalls are typical. These maneuvers should be practiced as part of a good transition training program. Steep turns in LSA airplanes tend to be quite easy to perform with precision. Light flight control pressures, stick mounted trim (if installed), and highly differential ailerons (if part of the airplane's design), make the maneuvers seem simpler than in heavier airplanes. Basic aerodynamics applies to any airplane. Factors, such as over-banking tendency, are still prevalent and should be expected.

Slow flight in LSAs is accomplished at slower airspeeds than standard airworthiness airplanes since stall speeds tend to be well below the 45-knot limit. Practicing slow flight demonstrates the unique capability of LSAs. Recovery from power-off stalls involves lowering the nose. Application of power puts the airplane back flying. However, a pilot should understand that control pressures tend to be light, and an aggressive forward movement of the elevator is generally not required. In addition, proper application of rudder to compensate for propeller forces is required, and retraction of any flap should be completed prior to reaching VFE, which occurs quickly if full power and nose down pitch attitude are maintained. Power-on stalls can result in a very high nose-up attitude unless the airplane is adequately slowed down prior to the maneuver. In addition, some manufacturers limit pitch attitudes to 30° during power-on stalls. If aggressive pitch attitudes are coupled with uncoordinated rudder inputs, spin entry is likely to be quick and aggressive.

Depending on the LSA design, especially those airplanes which use control tubes rather than wires and pulleys, flight in turbulence may couple motion to the stick rather distinctively. If a transitioning pilot's flight experience is only with airplanes that have control cables and pulleys, the first flight in turbulence may be disconcerting; however, once the pilot becomes familiar with the control sensations induced by the turbulence, it only becomes another means for the pilot to feel the airplane.

Approach and Landing

Approach and landing in an LSA is routine and comfortable. Speeds in the pattern tend to be in the 60-knot range. Flap limit airspeeds tend to be lower in LSAs than standard airworthiness airplanes, so managing airspeed is important. Light control forces require smooth application of control pressures to avoid over-controlling. Pitch and power are the same in an LSA as in a standard airworthiness airplane.

The weight limit of light-sport airplanes makes crosswind landings an important subject to focus on. The pilot should realize that strong gusty crosswind conditions may exceed the airplane's control capability resulting in loss of control during the landing. Manufacturers place a maximum demonstrated crosswind speed in the POH, and until sufficient practice and experience is gained in the airplane, a transitioning pilot should have personal minimums that do not approach the manufacturer's demonstrated crosswind speed. Control application does not change for crosswind technique in an LSA. However, the LSA's weight, slow landing speeds, and light control forces can result in a pilot making control deflections that exceed those necessary to compensate for the crosswind.

Emergencies

While an LSA is designed to be simple, a complete knowledge of its systems is needed such that a transitioning pilot is able to respond properly to any emergency.

The airplane's POH describes the appropriate responses to the various emergency situations that may be encountered. *[Figure 17-16]* Consider a few examples: the EFIS is displaying a red "X" across the airspeed tape, electric trim runaway, or control system failure. The pilot should be able to respond to immediate action items from memory and locate emergency procedures quickly. In the example of trim runaway, the pilot needs to quickly assess the trim runaway condition, locate and depress the trim disconnect (if installed), or pull the trim power circuit breaker. Then depending on control forces required to maintain pitch attitude, the pilot may need to make a no-flap landing due to the flap pitching moments. If the EFIS "blanks" out and POH recovery procedures do not reset the EFIS, an LSA pilot may have to be prepared to land without airspeed, altitude, or vertical speed information. An effective training program covers these emergency procedures.

CESSNA
Model 162
Garmin G300

SECTION 3
EMERGENCY PROCEDURES

INTRODUCTION

▌Section 3 provides checklist and amplified procedures for coping with emergencies that may occur. Emergencies caused by airplane or engine malfunctions are extremely rare if proper preflight inspections and maintenance are practiced. Enroute weather emergencies can be minimized or eliminated by careful flight planning and good judgment when unexpected weather is encountered. However, should an emergency arise, the basic guidelines described in this section should be considered and applied as necessary to correct the problem. In any emergency situation, the most important task is continued control of the airplane and maneuver to execute a successful landing.

Emergency procedures associated with optional or supplemental
▌equipment are found in Section 9, Supplements.

AIRSPEEDS FOR EMERGENCY OPERATIONS

ENGINE FAILURE AFTER TAKEOFF
    Wing flaps UP...70 KIAS
    Wing flaps 10" - FULL ....65 KIAS

▌MAXIMUM OPERATING MANEUVERING SPEED
    1320 pounds...89 KIAS
    1200 pounds...85 KIAS
    1100 pounds...80 KIAS

▌DESIGN MANEUVERING SPEED ....102 KIAS

▌MAXIMUM GLIDE..70 KIAS

PRECAUTIONARY LANDING WITH ENGINE POWER........................... 60 kias

LANDING WITHOUT ENGINE POWER
    Wing flaps UP...70 KIAS
    Wing flaps 10" - FULL ....65 KIAS

Figure 17-16. *Example of a POH Emergency Procedures section.*

Post-Flight

After the airplane has been shutdown, tied-down, and secured, the pilot should conduct a complete post-flight inspection. Any squawks or discrepancies should be noted and reported to maintenance. Transitioning pilots should insist on a training debriefing where critique and planning for the next lesson takes place. Documentation of the pilot's progress should be noted in the student's records.

Key Points

LSAs with an open flight deck, easy build characteristics, low cost, and simplicity of operation and maintenance tend to be less aerodynamic and incur more drag. When combined with their low mass and inertia, these LSAs tend to decelerate rapidly when power is reduced. When attempting a crosswind landing in a high-drag LSA, a rapid reduction in airspeed prior to touchdown may result in a loss of rudder and/or aileron control, which may push the aircraft off of the runway heading. To avoid loss of control, maintain airspeed during the approach. When power is reduced, it may be necessary to lower the nose of the aircraft to a fairly low pitch attitude in order to maintain airspeed.

If the pilot makes a power-off approach to landing, the approach angle will be high and the landing flare will need to be close to the ground with minimum float. This is because the aircraft will lose airspeed quickly in the flare and will not float like a more efficiently designed aircraft. Too low of an airspeed during the landing flare may lead to insufficient energy to arrest the descent and may result in a hard landing. Maintaining power during the approach will result in a reduced angle of attack and will extend the landing flare allowing more time to make adjustments to the aircraft during the landing. Always remember that rapid power reductions require an equally rapid reduction in pitch attitude to maintain airspeed.

In the event of an engine failure in an LSA, quickly transition to the required nose-down flight attitude in order to maintain airspeed. For example, if the aircraft has a power-off glide angle of 30 degrees below the horizon, position the aircraft to a nose-down 30 degree attitude as quickly as possible. The higher the pitch attitude is when the engine failure occurs, the quicker the aircraft will lose airspeed and the more likely the aircraft is to stall. Should a stall occur, decrease the aircraft's pitch attitude rapidly in order to increase airspeed to allow for a recovery. Stalls that occur at low altitudes are especially dangerous because the closer to the ground the stall occurs, the less time there is to recover. For this reason, when climbing at a low altitude, excessive pitch attitude is discouraged.

Chapter Summary

LSAs are a category of small, lightweight aircraft that may include advanced systems, such a parachutes, EFIS, and composite construction. While the transition is not difficult, it does require a properly designed transition training program led by a competent flight instructor-S or flight instructor-A. Safety is of utmost importance when it comes to any flight activity. In order to properly assess the hazards of flight and mitigate flight risk, a pilot needs to develop the appropriate knowledge, risk management, and skill, to effectively and safely pilot an LSA.

Airplane Flying Handbook (FAA-H-8083-3C)
Chapter 18: Emergency Procedures

Introduction

This chapter describes certain abnormal and emergency situations that may occur in flight. The key to successful management of an emergency situation, and/or preventing an abnormal situation from progressing into a true emergency, is a thorough familiarity with, and adherence to, the procedures developed by the airplane manufacturer. The following guidelines are generic and are not meant to replace the airplane manufacturer's recommended procedures contained in the Federal Aviation Administration (FAA) approved Airplane Flight Manual and/or Pilot's Operating Handbook (AFM/POH). Rather, they are meant to enhance the pilot's general knowledge in the area of abnormal and emergency operations. If any of the guidance in this chapter conflicts in any way with the manufacturer's recommended procedures for a particular make and model airplane, the manufacturer's recommended procedures take precedence.

Emergency Landings

This section contains information on emergency landing techniques in small fixed-wing airplanes. The guidelines that are presented apply to the more adverse terrain conditions for which no practical training is possible. The objective is to instill in the pilot the knowledge that almost any terrain can be considered "suitable" for a survivable crash landing if the pilot knows how to use the airplane structure for self-protection and the protection of passengers.

Types of Emergency Landings

The different types of emergency landings are defined as follows:

- Forced landing—an immediate landing, on or off an airport, necessitated by the inability to continue further flight. A typical example of which is an airplane forced down by engine failure.

- Precautionary landing—a premeditated landing, on or off an airport, when further flight is possible but inadvisable. Examples of conditions that may call for a precautionary landing include deteriorating weather, being lost, fuel shortage, and gradually developing engine trouble.

- Ditching—a forced or precautionary landing on water.

A precautionary landing, generally, is less hazardous than a forced landing because the pilot has more time for terrain selection and the planning of the approach. In addition, the pilot can use power to compensate for errors in judgment or technique. The pilot should be aware that too many situations calling for a precautionary landing are allowed to develop into immediate forced landings, when the pilot uses wishful thinking instead of reason, especially when dealing with a self-inflicted predicament. The non-instrument-rated pilot trapped by weather, or the pilot facing imminent fuel exhaustion who does not give any thought to the feasibility of a precautionary landing, accepts an extremely hazardous alternative.

Psychological Hazards

There are several factors that may interfere with a pilot's ability to act promptly and properly when faced with an emergency. Some of these factors are listed below.

- Reluctance to accept the emergency situation—a pilot who allows the mind to become paralyzed at the thought that the airplane will be on the ground in a very short time, regardless of the pilot's actions or hopes, is severely handicapped in the handling of the emergency. An unconscious desire to delay the dreaded moment may lead to such errors as: failure to lower the nose to maintain flying speed, delay in the selection of the most suitable landing area within reach, and indecision in general. Desperate attempts to correct whatever went wrong at the expense of airplane control fall into the same category.

- Undue concern about getting hurt—fear is a vital part of the self-preservation mechanism. However, when fear leads to panic, we invite that which we want most to avoid. The survival records favor pilots who maintain their composure and know how to apply the general concepts and procedures that have been developed through the years. The success of an emergency landing is as much a matter of the mind as of skills.

- Desire to save the airplane—the pilot who has been conditioned during training to expect to find a relatively safe landing area, whenever the flight instructor closed the throttle for a simulated forced landing, may ignore all basic rules of airmanship to avoid a touchdown in terrain where airplane damage is unavoidable. Typical consequences are: making a 180° turn back to the runway when available altitude is insufficient; stretching the glide without regard for minimum control speed in order to reach a more appealing field; and accepting an approach and touchdown situation that leaves no margin for error. The desire to save the airplane, regardless of the risks involved, may be influenced by two other factors: the pilot's financial stake in the airplane and the certainty that an undamaged airplane implies no bodily harm. There are times, however, when a pilot should be more interested in sacrificing the airplane so that the occupants can safely walk away from it.

Basic Safety Concepts
General

A pilot who is faced with an emergency landing in terrain that makes extensive airplane damage inevitable should keep in mind that the avoidance of crash injuries is largely a matter of: (1) keeping the vital structure (cabin area) relatively intact by using dispensable structure (i.e., wings, landing gear, fuselage bottom) to absorb the violence of the stopping process before it affects the occupants and (2) avoiding forceful bodily contact with interior structure. Avoiding forcible contact with interior structure is a matter of seat and body security. Unless the occupant decelerates at the same rate as the surrounding structure, no benefit is realized from its relative intactness. The occupant is brought to a stop violently in the form of a secondary collision.

The advantage of sacrificing dispensable structure is demonstrated daily on the highways. A head-on car impact against a tree at 20 miles per hour (mph) is less hazardous for a properly restrained driver than a similar impact against the driver's door. Accident experience shows that the extent of crushable structure between the occupants and the principal point of impact on the airplane has a direct bearing on the severity of the transmitted crash forces and, therefore, on survivability.

Dispensable airplane structure is not the only available energy absorbing medium in an emergency situation. Vegetation, trees, and even manmade structures may be used for this purpose. Cultivated fields with dense crops, such as mature corn and grain, are almost as effective in bringing an airplane to a stop with repairable damage as an emergency arresting device on a runway. *[Figure 18-1]* Brush and small trees provide considerable cushioning and braking effect without destroying the airplane. When dealing with natural and manmade obstacles with greater strength than the dispensable airplane structure, the pilot should plan the touchdown in such a manner that only nonessential structure is "used up" in the principal slowing-down process.

Figure 18-1. *Using vegetation to absorb energy.*

The overall severity of a deceleration process is governed by speed (groundspeed) and stopping distance. The most critical of these is speed; doubling the groundspeed means quadrupling the total destructive energy and vice versa. Even a small change in groundspeed at touchdown—be it as a result of wind or pilot technique—affects the outcome of a controlled crash. It is important that the actual touchdown during an emergency landing be made at the lowest possible controllable airspeed, using all available aerodynamic devices.

Most pilots instinctively—and correctly—look for the largest available flat and open field for an emergency landing. Actually, very little stopping distance is required if the speed can be dissipated uniformly; that is, if the deceleration forces can be spread evenly over the available distance. This concept is designed into the arresting gear of aircraft carriers that provides a nearly constant stopping force from the moment of hookup.

The typical light airplane is designed to provide protection in crash landings that expose the occupants to nine times the acceleration of gravity (9G) in a forward direction. Assuming a uniform 9G deceleration, at 50 mph the required stopping distance is about 9.4 feet. While at 100 mph, the stopping distance is about 37.6 feet—about four times as great. *[Figure 18-2]* Although these figures are based on an ideal deceleration process, it is interesting to note what can be accomplished in an effectively used short stopping distance. Understanding the need for a firm but uniform deceleration process in very poor terrain enables the pilot to select touchdown conditions that spread the breakup of dispensable structure over a short distance, thereby reducing the peak deceleration of the cabin area.

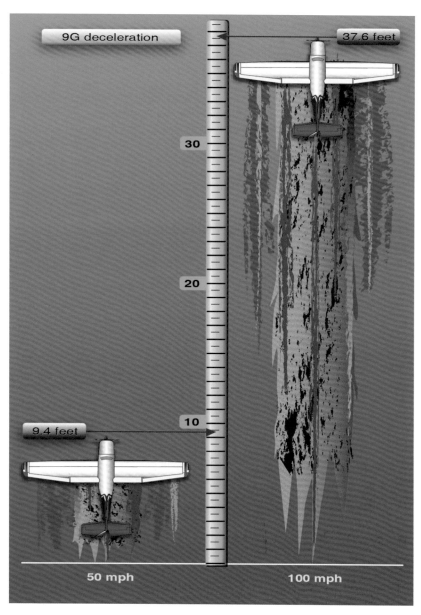

Figure 18-2. *Stopping distance vs. groundspeed.*

Attitude and Sink Rate Control

The most critical and often the most inexcusable error that can be made in the planning and execution of an emergency landing, even in ideal terrain, is the loss of initiative over the airplane's attitude and sink rate at touchdown. When the touchdown is made on flat, open terrain, an excessive nose-low pitch attitude brings the risk of "sticking" the nose in the ground. Steep bank angles just before touchdown should also be avoided, as they increase the stalling speed and the likelihood of a wingtip strike.

Since the airplane's vertical component of velocity is immediately reduced to zero upon ground contact, it should be kept well under control. A flat touchdown at a high sink rate (well in excess of 500 feet per minute (fpm)) on a hard surface can be injurious without destroying the cabin structure, especially during gear-up landings in low-wing airplanes. A rigid bottom construction of these airplanes may preclude adequate cushioning by structural deformation. Similar impact conditions may cause structural collapse of the overhead structure in high-wing airplanes. On soft terrain, an excessive sink rate may cause digging in of the lower nose structure and severe forward deceleration.

Terrain Selection

A pilot's choice of emergency landing sites is governed by:

- The route selected during preflight planning
- The height above the ground when the emergency occurs
- Excess airspeed (excess airspeed can be converted into distance and/or altitude)

The only time the pilot has a very limited choice is during the low and slow portion of the takeoff. However, even under these conditions, the ability to change the impact heading only a few degrees may ensure a survivable crash.

If beyond gliding distance of a suitable open area, the pilot should judge the available terrain for its energy absorbing capability. If the emergency starts at a considerable height above the ground, the pilot should be more concerned about first selecting the desired general area than a specific spot. Terrain appearances from altitude can be very misleading and considerable altitude may be lost before the best spot can be pinpointed. For this reason, the pilot should not hesitate to discard the original plan for one that is obviously better. However, as a general rule, the pilot should not change his or her mind more than once; a well-executed crash landing in poor terrain can be less hazardous than an uncontrolled touchdown on an established field.

Airplane Configuration

Since flaps improve maneuverability at slow speed, and lower the stalling speed, their use during final approach is recommended when time and circumstances permit. However, the associated increase in drag and decrease in gliding distance call for caution in the timing and the extent of their application; premature use of flap and dissipation of altitude may jeopardize an otherwise sound plan.

A hard and fast rule concerning the position of a retractable landing gear at touchdown cannot be given. In rugged terrain and trees, or during impacts at high sink rate, an extended gear would definitely have a protective effect on the cabin area. However, this advantage has to be weighed against the possible side effects of a collapsing gear, such as a ruptured fuel tank. As always, the manufacturer's recommendations as outlined in the AFM/POH should be followed.

When a normal touchdown is assured, and ample stopping distance is available, a gear-up landing on level, but soft terrain or across a plowed field may result in less airplane damage than a gear-down landing. *[Figure 18-3]* Deactivation of the airplane's electrical system before touchdown reduces the likelihood of a post-crash fire.

However, the battery master switch should not be turned off until the pilot no longer has any need for electrical power to operate vital airplane systems. Positive airplane control during the final part of the approach has priority over all other considerations, including airplane configuration and checklist tasks. The pilot should attempt to exploit the power available from an irregularly running engine; however, it is generally better to switch the engine and fuel off just before touchdown. This not only ensures the pilot's initiative over the situation, but a cooled-down engine reduces the fire hazard considerably.

Approach

When the pilot has time to maneuver, the planning of the approach should be governed by the following three factors:

- Wind direction and velocity
- Dimensions and slope of the chosen field
- Obstacles in the final approach path

Figure 18-3. *Intentional gear-up landing.*

These three factors are seldom compatible. When compromises have to be made, the pilot should aim for a wind/obstacle/terrain combination that permits a final approach with some margin for error in judgment or technique. A pilot who overestimates the gliding range may be tempted to stretch the glide across obstacles in the approach path. For this reason, it is sometimes better to plan the approach over an unobstructed area, regardless of wind direction. Experience shows that a collision with obstacles at the end of a ground roll or slide is much less hazardous than striking an obstacle at flying speed before the touchdown point is reached.

Terrain Types

Since an emergency landing on suitable terrain resembles a situation in which the pilot should be familiar through training, only the more unusual situations are discussed.

Confined Areas

The natural preference to set the airplane down on the ground should not lead to the selection of an open spot between trees or obstacles where the ground cannot be reached without making a steep descent.

Once the intended touchdown point is reached, and the remaining open and unobstructed space is very limited, it may be better to force the airplane down on the ground than to delay touchdown until it stalls (settles). An airplane decelerates faster after it is on the ground than while airborne. Thought may also be given to the desirability of ground-looping or retracting the landing gear in certain conditions.

A river or creek can be an inviting alternative in otherwise rugged terrain. The pilot should ensure that the water or creek bed can be reached without snagging the wings. The same concept applies to road landings with one additional reason for caution: manmade obstacles on either side of a road may not be visible until the final portion of the approach.

When planning the approach across a road, it should be remembered that most highways and even rural dirt roads are paralleled by power or telephone lines. Only a sharp lookout for the supporting structures or poles may provide timely warning.

Trees (Forest)

Although a tree landing is not an attractive prospect, the following general guidelines help to make the experience survivable.

- Use the normal landing configuration (full flaps, gear down).

- Keep the groundspeed low by heading into the wind.

- Make contact at minimum indicated airspeed, but not below stall speed, and "hang" the airplane in the tree branches in a nose-high landing attitude. Involving the underside of the fuselage and both wings in the initial tree contact provides a more even and positive cushioning effect, while preventing penetration of the windshield. *[Figure 18-4]*

- Avoid direct contact of the fuselage with heavy tree trunks.

- Low, closely spaced trees with wide, dense crowns (branches) close to the ground are much better than tall trees with thin tops; the latter allow too much free fall height (a free fall from 75 feet results in an impact speed of about 40 knots, or about 4,000 fpm).

- Ideally, initial tree contact should be symmetrical; that is, both wings should meet equal resistance in the tree branches. This distribution of the load helps to maintain proper airplane attitude. It may also preclude the loss of one wing, which invariably leads to a more rapid and less predictable descent to the ground.

- If heavy tree trunk contact is unavoidable once the airplane is on the ground, it is best to involve both wings simultaneously by directing the airplane between two properly spaced trees. Do not attempt this maneuver, however, while still airborne.

Figure 18-4. *Tree landing.*

Water (Ditching) and Snow

A well-executed water landing normally involves less deceleration violence than a poor tree landing or a touchdown on extremely rough terrain. Also, an airplane that is ditched at minimum speed and in a normal landing attitude does not immediately sink upon touchdown. Intact wings and fuel tanks (especially when empty) provide floatation for at least several minutes, even if the cabin may be just below the water line in a high-wing airplane.

Loss of depth perception may occur when landing on a wide expanse of smooth water with the risk of flying into the water or stalling in from excessive altitude. To avoid this hazard, the airplane should be "dragged in" when possible. Use no more than intermediate flaps on low-wing airplanes. The water resistance of fully extended flaps may result in asymmetrical flap failure and slowing of the airplane. Keep a retractable gear up unless the AFM/POH advises otherwise.

A landing in snow should be executed like a ditching, in the same configuration and with the same regard for loss of depth perception (white out) in reduced visibility and on wide-open terrain.

Engine Failure After Takeoff (Single-Engine)

A number of variables and pilot actions factor into a successful emergency landing shortly after takeoff. When an engine failure occurs during the initial climb, the pilot should lower the nose of the airplane and establish the proper glide attitude. What happens next if the engine does not restart? Does the pilot select a field directly ahead (or slightly to the side of the takeoff path) or should the pilot turn back toward the point of departure? There's not much time to decide and a lot to consider.

Continuing straight ahead or making a slight turn gives the pilot time to establish a safe landing attitude, and the landing occurs under control and as slowly as possible (assuming a takeoff made into a headwind). This minimizes the risk of injury and usually represents the option with the lowest risk—i.e. the safest option. Turning back requires a more complex analysis and consideration of risk. At some urban airports, there may be numerous hazards in the departure path. In that case, the pilot might turn back, but only if certain the airplane can reach the field from its current position and the pilot has trained and practiced the turn back maneuver.

Turning back to an airport after a low-altitude engine failure, also known as "the impossible turn," presents many challenges, and a pilot who attempts to turn back without due consideration and training will need considerable luck to prevent disaster. If the airplane strikes the ground during the turn, cartwheeling could occur. If the pilot does not lower the nose sufficiently during the turn, an accelerated stall and fatal crash may occur. Even after executing a successful turn, a return to the airport often results in a downwind approach. The increased groundspeed could rush a pilot not properly trained for landing downwind. The increased groundspeed and associated increase in kinetic energy also raise the likelihood of serious injury if unable to make the field.

If considering a turn back to the runway following an engine failure on takeoff, the pilot should know the expected altitude loss during the turn for the specific make and model airplane as well as whether the airplane can physically glide back to the field after executing the turn. Traditionally, the FAA has given the following example. An airplane has taken off and climbed to an altitude of 300 feet above ground level (AGL) when the engine fails. *[Figure 18-5]* After a typical 4-second reaction time, the pilot elects to turn back to the runway. Using a standard rate (3° change in direction per second) turn, it takes 1 minute to turn 180°. At a glide speed of 65 knots, the radius of the turn is 2,100 feet, so at the completion of the turn, the airplane is 4,200 feet to one side of the runway. The pilot needs to turn another 45° to head the airplane toward the runway. By this time, the total change in direction is 225° equating to 75 seconds plus the 4-second reaction time. If the airplane in a power-off glide descends at approximately 1,000 fpm, it has descended 1,316, feet placing it 1,016 feet below the runway.

The preceding example illustrates why a turn back, if attempted, requires a turn with a higher bank angle. A standard rate or shallow turn consumes too much time, requires too much distance, and generates an unacceptable solution.

Training for a turn back includes practicing turns in both directions at a safe altitude in the make and model flown after simulating an engine failure from a climb. Practice should result in consistent altitude loss and the ability to avoid an accelerated stall when executing a gliding steep turn. Pilots should be alert for and respond appropriately to any stall warning and reduce wing loading during the turn as necessary. There will be some observed variation in altitude loss during training. The pilot should anticipate that during an actual emergency, the expected altitude loss could end up at the high end of the range observed while practicing. Success in training involves the demonstrated ability to evaluate the effect of climb performance of the airplane, determine the better direction to turn back (usually into a crosswind), predict the altitude above ground after the turn, know the distance to the landing zone, and know if the glide performance of the airplane will allow the pilot to make the field. Some airplanes cannot usually make the return successfully, some can make the return under certain conditions, and some can usually return. The pilot should not attempt a turn back unless a successful turn back will result.

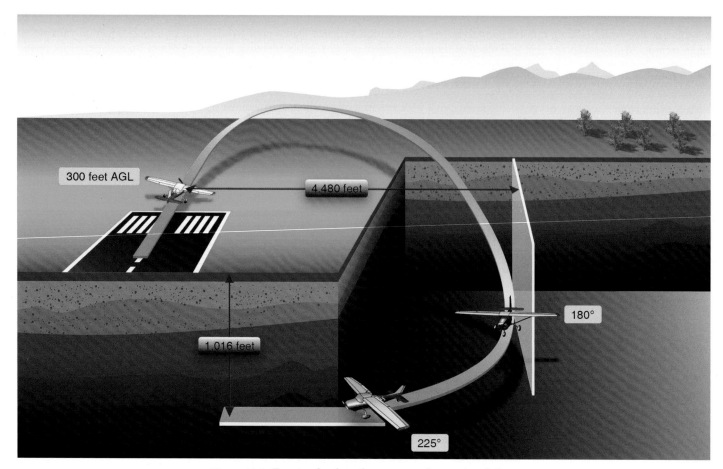

Figure 18-5. *Turning back to the runway after engine failure.*

A turn back to the departure runway may require more than a 180° change in direction. There could also be cases where turning back results in overshooting the runway, and the pilot needs to sense the aiming point within seconds after completing a turn back and make any necessary adjustments to achieve the best possible outcome. A turn back at low altitudes presents an unacceptable risk for student pilots, low-time pilots, untrained pilots, pilots without adequate proficiency, and pilots flying airplanes with insufficient glide performance to return to the field. However, experienced pilots interested in knowing when and how to make an emergency turn back after takeoff should use the services of an authorized flight instructor who can explain and demonstrate the practicality (or impracticality) of "the impossible turn" in the specific make and model used during training.

Emergency Descents

An emergency descent is a maneuver for descending as rapidly as possible to a lower altitude or to the ground for an emergency landing. *[Figure 18-6]* The need for this maneuver may result from an uncontrollable fire, a sudden loss of cabin pressurization, or any other situation demanding an immediate and rapid descent. The objective is to descend the airplane as soon and as rapidly as possible while not exceeding any structural limitations of the airplane. Simulated emergency descents should be made in a turn to check for other air traffic below and to look around for a possible emergency landing area. A radio call announcing descent intentions may be appropriate to alert other aircraft in the area. When initiating the descent, a bank of approximately 30 to 45° should be established to maintain positive load factors (G forces) on the airplane.

Emergency descent training should be performed as recommended by the manufacturer, including the configuration and airspeeds. Except when prohibited by the manufacturer, the power should be reduced to idle, and the propeller control (if equipped) should be placed in the low pitch (or high revolutions per minute (rpm)) position. This allows the propeller to act as an aerodynamic brake to help prevent an excessive airspeed buildup during the descent. The landing gear and flaps should be extended as recommended by the manufacturer. This provides maximum drag so that the descent can be made as rapidly as possible, without excessive airspeed. The pilot should not allow the airplane's airspeed to pass the never-exceed speed (V_{NE}), the maximum landing gear extended speed (V_{LE}), or the maximum flap extended speed (V_{FE}), as applicable. In the case of an engine fire, a high airspeed descent could blow out the fire. However, the weakening of the airplane structure is a major concern and descent at low airspeed would place less stress on the airplane. If the descent is conducted in turbulent conditions, the pilot also needs to comply with the design maneuvering speed (V_A) limitations. The descent should be made at the maximum allowable airspeed consistent with the procedure used. This provides increased drag and a high rate of descent. The recovery from an emergency descent should be initiated at a high enough altitude to ensure a safe recovery back to level flight or a precautionary landing.

Figure 18-6. *Emergency descent.*

When the descent is established and stabilized during training and practice, the descent should be terminated. In airplanes with piston engines, prolonged practice of emergency descents should be avoided to prevent excessive cooling of the engine cylinders.

In-Flight Fire

A fire in-flight demands immediate and decisive action. The pilot should be familiar with the procedures outlined to meet this emergency contained in the AFM/POH for the particular airplane. For the purposes of this handbook, in-flight fires are classified as in-flight engine fires, electrical fires, and cabin fires.

Engine Fire

An in-flight engine compartment fire is usually caused by a failure that allows a flammable substance, such as fuel, oil, or hydraulic fluid, to come in contact with a hot surface. This may be caused by a mechanical failure of the engine itself, an engine-driven accessory, a defective induction or exhaust system, or a broken line. Engine compartment fires may also result from maintenance errors, such as improperly installed/fastened lines and/or fittings resulting in leaks.

Engine compartment fires can be indicated by smoke and/or flames coming from the engine cowling area. They can also be indicated by discoloration, bubbling, and/or melting of the engine cowling skin in cases where flames and/or smoke are not visible to the pilot. By the time a pilot becomes aware of an in-flight engine compartment fire, it usually is well developed. Unless the airplane manufacturer directs otherwise in the AFM/POH, the first step on discovering a fire should be to shut off the fuel supply to the engine by placing the mixture control in the idle cut off position and the fuel selector shutoff valve to the OFF position. The ignition switch should be left ON in order to use up the fuel that remains in the fuel lines and components between the fuel selector/shutoff valve and the engine. This procedure may starve the engine compartment of fuel and cause the fire to die naturally. If the flames are snuffed out, no attempt should be made to restart the engine.

If the engine compartment fire is oil-fed, as evidenced by thick black smoke, as opposed to a fuel-fed fire, which produces bright orange flames, the pilot should consider stopping the propeller rotation by feathering or other means, such as (with constant-speed propellers) placing the pitch control lever to the minimum rpm position and raising the nose to reduce airspeed until the propeller stops rotating. This procedure stops an engine-driven oil (or hydraulic) pump from continuing to pump the flammable fluid that is feeding the fire.

Some light airplane emergency checklists direct the pilot to shut off the electrical master switch. However, the pilot should consider that unless the fire is electrical in nature, or a crash landing is imminent, deactivating the electrical system prevents the use of panel radios for transmitting distress messages and also causes air traffic control (ATC) to lose transponder returns.

Pilots of powerless single-engine airplanes are left with no choice but to make a forced landing. Pilots of twin-engine airplanes may elect to continue the flight to the nearest airport. However, consideration should be given to the possibility that a wing could be seriously impaired and lead to structural failure. Even a brief but intense fire could cause dangerous structural damage. In some cases, the fire could continue to burn under the wing (or engine cowling in the case of a single- engine airplane) out of view of the pilot. Engine compartment fires that appear to have been extinguished have been known to rekindle with changes in airflow pattern and airspeed.

The pilot should be familiar with the airplane's emergency descent procedures. The pilot should also bear in mind the following:

- The airplane may be severely structurally damaged to the point that its ability to remain under control could be lost at any moment.

- The airplane may still be on fire and susceptible to explosion.

- The airplane is expendable and the only thing that matters is the safety of those on board.

Electrical Fires

The initial indication of an electrical fire is usually the distinct odor of burning insulation. Once an electrical fire is detected, the pilot should attempt to identify the faulty circuit by checking circuit breakers, instruments, avionics, and lights. If the faulty circuit cannot be readily detected and isolated, and flight conditions permit, the battery master switch and alternator/generator switches should be turned off to remove the possible source of the fire. However, any materials that have been ignited may continue to burn.

If electrical power is absolutely essential for the flight, an attempt may be made to identify and isolate the faulty circuit by:

1. Turning the electrical master switch OFF.
2. Turning all individual electrical switches OFF.
3. Turning the master switch back ON.
4. Selecting electrical switches that were ON before the fire indication one at a time, permitting a short time lapse after each switch is turned on to check for signs of odor, smoke, or sparks.

This procedure, however, has the effect of recreating the original problem. The most prudent course of action is to land as soon as possible.

Cabin Fire

Cabin fires generally result from one of three sources: (1) careless smoking on the part of the pilot and/or passengers; (2) electrical system malfunctions; or (3) heating system malfunctions. A fire in the cabin presents the pilot with two immediate demands: attacking the fire and getting the airplane safely on the ground as quickly as possible. A fire or smoke in the cabin should be controlled by identifying and shutting down the faulty system. In many cases, smoke may be removed from the cabin by opening the cabin air vents. This should be done only after the fire extinguisher (if available) is used. Then the cabin air control can be opened to purge the cabin of both smoke and fumes. If smoke increases in intensity when the cabin air vents are opened, they should be immediately closed. This indicates a possible fire in the heating system, nose compartment baggage area (if so equipped), or that the increase in airflow is feeding the fire.

On pressurized airplanes, the pressurization air system removes smoke from the cabin; however, if the smoke is intense, it may be necessary to either depressurize at altitude, if oxygen is available for all occupants, or execute an emergency descent.

In unpressurized single-engine and light twin-engine airplanes, the pilot can attempt to expel the smoke from the cabin by opening the foul weather windows. These windows should be closed immediately if the fire becomes more intense. If the smoke is severe, the passengers and crew should use oxygen masks if available, and the pilot should initiate an immediate descent. The pilot should also be aware that on some airplanes, lowering the landing gear and/or wing flaps can aggravate a cabin smoke problem.

Flight Control Malfunction/Failure
Total Flap Failure

The inability to extend the wing flaps necessitates a no-flap approach and landing. In light airplanes, a no-flap approach and landing is not particularly difficult or dangerous. However, there are certain factors that should be considered in the execution of this maneuver. A no-flap landing requires substantially more runway than normal. The increase in required landing distance could be as much as 50 percent.

When flying in the traffic pattern with the wing flaps retracted, the airplane should be flown in a relatively nose-high attitude to maintain altitude, as compared to flight with flaps extended. Losing altitude can be more of a problem without the benefit of the drag normally provided by flaps. A wider, longer traffic pattern may be required in order to avoid the necessity of diving to lose altitude and consequently building up excessive airspeed.

On final approach, a nose-high attitude can make it difficult to see the runway. This situation, if not anticipated, can result in serious errors in judgment of height and distance. Approaching the runway in a relatively nose-high attitude can also cause the perception that the airplane is close to a stall. This may cause the pilot to lower the nose abruptly and risk touching down on the nose-wheel.

With the flaps retracted and the power reduced for landing, the airplane is slightly less stable in the pitch and roll axes. Without flaps, the airplane tends to float considerably during roundout. The pilot should avoid the temptation to force the airplane onto the runway at an excessively high speed. Neither should the pilot flare excessively because without flaps, this might cause the tail to strike the runway.

Asymmetric (Split) Flap

An asymmetric "split" flap situation is one in which one flap deploys or retracts while the other remains in position. The problem is indicated by a pronounced roll toward the wing with the least flap deflection when wing flaps are extended/retracted.

The roll encountered in a split flap situation is countered with opposite aileron. The yaw caused by the additional drag created by the extended flap requires substantial opposite rudder resulting in a cross-control condition. Almost full aileron may be required to maintain a wings-level attitude, especially at the reduced airspeed necessary for approach and landing. The pilot should not attempt to land with a crosswind from the side of the deployed flap because the additional roll control required to counteract the crosswind may not be available.

The approach to landing with a split flap condition should be flown at a higher than normal airspeed. The pilot should not risk an asymmetric stall and subsequent loss of control by flaring excessively. Rather, the airplane should be flown onto the runway so that the touchdown occurs at an airspeed consistent with a safe margin above flaps-up stall speed.

Loss of Elevator Control

In many airplanes, the elevator is controlled by two cables: a "down" cable and an "up" cable. Normally, a break or disconnect in only one of these cables does not result in a total loss of elevator control. In most airplanes, a failed cable results in a partial loss of pitch control. In the failure of the "up" elevator cable (the "down" elevator being intact and functional), the control yoke moves aft easily but produces no response. Forward yoke movement, however, beyond the neutral position produces a nose-down attitude. Conversely, a failure of the "down" elevator cable, forward movement of the control yoke produces no effect. The pilot, however, has partial control of pitch attitude with aft movement.

When experiencing a loss of up-elevator control, the pilot can retain pitch control by:

- Applying considerable nose-up trim
- Pushing the control yoke forward to attain and maintain desired attitude
- Increasing forward pressure to lower the nose and relaxing forward pressure to raise the nose
- Releasing forward pressure to flare for landing

When experiencing a loss of down-elevator control, the pilot can retain pitch control by:

- Applying considerable nose-down trim
- Pulling the control yoke aft to attain and maintain attitude
- Releasing back pressure to lower the nose and increasing back pressure to raise the nose
- Increasing back pressure to flare for landing

Trim mechanisms can be useful in the event of an in-flight primary control failure. For example, if the linkage between the cabin and the elevator fails in flight, leaving the elevator free to weathervane in the wind, the trim tab can be used to raise or lower the elevator within limits. The trim tabs are not as effective as normal linkage control in conditions such as low airspeed, but they do have some positive effect—usually enough to bring about a safe landing.

If an elevator becomes jammed, resulting in a total loss of elevator control movement, various combinations of power and flap extension offer a limited amount of pitch control. A successful landing under these conditions, however, can be problematic.

Landing Gear Malfunction

Once the pilot has confirmed that the landing gear has in fact malfunctioned and that one or more gear legs refuses to respond to the conventional or alternate methods of gear extension contained in the AFM/POH, a gear-up landing is considered inevitable. The pilot should select an airport with crash and rescue facilities, if possible. The pilot should not hesitate to request that emergency equipment is standing by.

When selecting a landing surface, the pilot should consider that a smooth, hard-surface runway usually causes less damage than a rough, unimproved grass strip. A hard surface does, however, create sparks that can ignite fuel. If the airport is so equipped, the pilot can request that the runway surface be foamed. The pilot should consider burning off excess fuel. This reduces landing speed and fire potential.

If the landing gear malfunction is limited to one main landing gear leg, the pilot should consume as much fuel from that side of the airplane as practicable, thereby reducing the weight of the wing on that side. The reduced weight makes it possible to delay the unsupported wing from contacting the surface during the landing roll until the last possible moment. Reduced impact speeds result in less damage.

If only one landing gear leg fails to extend, the pilot has the option of landing on the available gear legs or landing with all the gear legs retracted. Landing on only one main gear usually causes the airplane to veer strongly in the direction of the faulty gear leg after touchdown. If the landing runway is narrow and/or ditches and obstacles line the runway edge, maximum directional control after touchdown is a necessity. In this situation, a landing with all three gear retracted may be the safest course of action.

If the pilot elects to land with one main gear retracted (and the other main gear and nose gear down and locked), the landing should be made in a nose-high attitude with the wings level. As airspeed decays, the pilot should apply whatever aileron control is necessary to keep the unsupported wing airborne as long as possible. [Figure 18-7] Once the wing contacts the surface, the pilot can anticipate a strong yaw in that direction. The pilot should be prepared to use full opposite rudder and aggressive braking to maintain some degree of directional control.

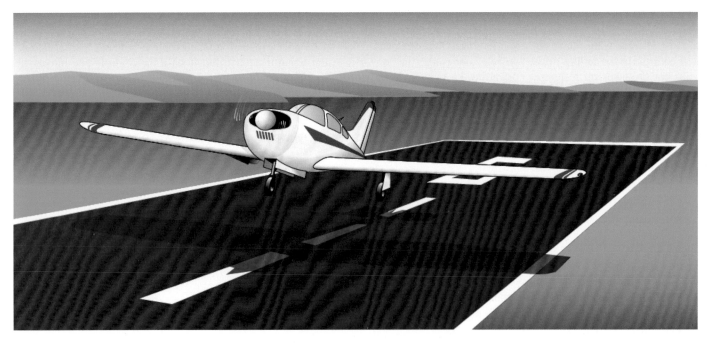

Figure 18-7. *Landing with one main gear retracted.*

When landing with a retracted nose-wheel (and the main gear extended and locked), the pilot should hold the nose off the ground until almost full up-elevator has been applied. [Figure 18-8] The pilot should then release back pressure in such a manner that the nose settles slowly to the surface. Applying and holding full up-elevator results in the nose abruptly dropping to the surface as airspeed decays, possibly resulting in burrowing and/or additional damage. Brake pressure should not be applied during the landing roll unless absolutely necessary to avoid a collision with obstacles.

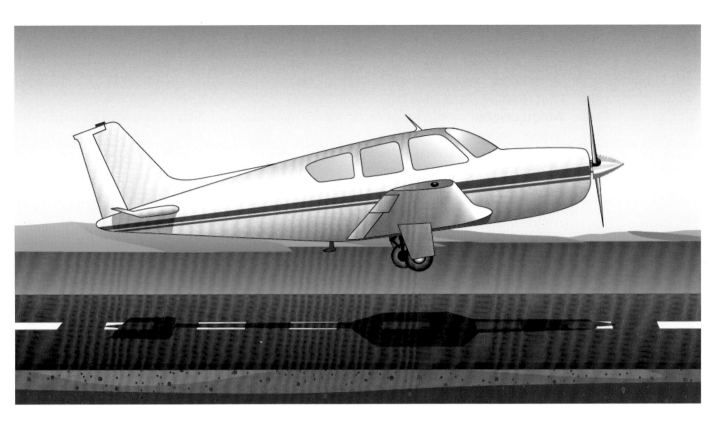

Figure 18-8. *Landing with nose-wheel retracted.*

If the landing occurs with only the nose gear extended, the initial contact should be made on the aft fuselage structure with a nose-high attitude. This procedure helps prevent porpoising and/or wheelbarrowing. The pilot should then allow the nose-wheel to gradually touchdown, using nose-wheel steering as necessary for directional control.

System Malfunctions
Electrical System

The loss of electrical power can deprive the pilot of numerous critical systems, and therefore should not be taken lightly even in day/visual flight rules (VFR) conditions. Most in-flight failures of the electrical system are located in the generator or alternator. Once the generator or alternator system goes off line, the electrical source in a typical light airplane is a battery. If a warning light or ammeter indicates the probability of an alternator or generator failure in an airplane with only one generating system, however, the pilot may have very little time available from the battery.

The rating of the airplane battery provides a clue as to how long it may last. With batteries, the higher the amperage load, the faster any available stored energy gets consumed. Thus, a 25-amp hour battery could produce 5 amps per hour for 5 hours, but if the load were increased to 10 amps, it might last only 2 hours. A 40-amp load might discharge the battery fully in about 10 or 15 minutes. Much depends on the battery condition at the time of the system failure. If the battery has been in service for a few years, its power may be reduced substantially because of internal resistance. Or if the system failure was not detected immediately, much of the stored energy may have already been used. It is essential, therefore, that the pilot immediately shed non-essential loads when the generating source fails. *[Figure 18-9]* The pilot should then plan to land at the nearest suitable airport.

What constitutes an "emergency" load following a generating system failure cannot be predetermined because the actual circumstances are always somewhat different—for example, whether the flight is VFR or instrument flight rules (IFR), conducted in day or at night, in clouds or in the clear. Distance to nearest suitable airport can also be a factor.

The pilot should remember that the electrically-powered (or electrically-selected) landing gear and flaps do not function properly on the power left in a partially-depleted battery. Landing gear and flap motors use power at rates much greater than most other types of electrical equipment. The result of selecting these motors on a partially-depleted battery may well result in an immediate total loss of electrical power.

Electrical Loads for Light Single	Number of units	Total Amperes
A. Continuous Load		
Pitot Heating (Operating)	1	3.30
Wingtip Lights	4	3.00
Heater Igniter	1	1-20
**Navigation Receivers	1-4	1-2 each
**Communications Receivers	1-2	1-2 each
Fuel Indicator	1	0.40
Instrument Lights (overhead)	2	0.60
Engine Indicator	1	0.30
Compass Light	1	0.20
Landing Gear Indicator	1	0.17
Flap Indicator	1	0.17
B. Intermittent Load		
Starter	1	100.00
Landing Lights	2	17.80
Heater Blower Motor	1	14.00
Flap Motor	1	13.00
Landing Gear Motor	1	10.00
Cigarette Lighter	1	7.50
Transceiver (keyed)	1	5-7
Fuel Boost Pump	1	2.00
Cowl Flap Motor	1	1.00
Stall Warning Horn	1	1.50

** Amperage for radios varies with equipment.
In general, the more recent the model, the less amperage required.
NOTE: Panel and indicator lights usually draw less than one amp.

Figure 18-9. *Electrical load for light single.*

If the pilot expects an imminent and complete in-flight loss of electrical power, the following steps should be taken:

• Shed all but the most necessary electrically-driven equipment.

• Understand that any loss of electrical power is critical in a small airplane—notify ATC of the situation immediately. Request radar vectors for a landing at the nearest suitable airport.

• If landing gear or flaps are electrically controlled or operated, plan the arrival well ahead of time. Expect to make a no-flap landing and anticipate a manual landing gear extension.

Pitot-Static System

The source of the pressure for operating the airspeed indicator, the vertical speed indicator (VSI), and the altimeter is the pitot-static system. The major components of the pitot-static system are the impact pressure chamber and lines and the static pressure chamber and lines, each of which are subject to total or partial blockage by ice, dirt, and/or other foreign matter. Blockage of the pitot-static system adversely affects instrument operation. *[Figure 18-10]*

Partial static system blockage is insidious in that it may go unrecognized until a critical phase of flight. During takeoff, climb, and level-off at cruise altitude the altimeter, airspeed indicator, and VSI may operate normally. No indication of malfunction may be present until the airplane begins a descent.

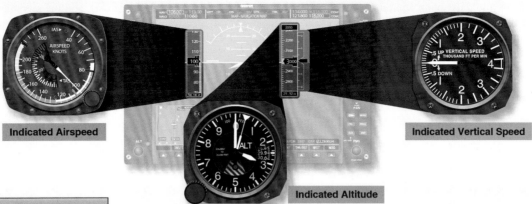

Indicated Airspeed

Indicated Vertical Speed

Indicated Altitude

Effect of Blocked Pitot/Static Sources on Airspeed, Altimeter, and Vertical Speed Indications	Indicated Airspeed	Indicated Altitude	Indicated Vertical Speed
Pitot source blocked	Increases with altitude gain, decreases with altitude loss.	Unaffected	Unaffected
One static source blocked	Inaccurate while sideslipping; very sensitive in turbulence.		
Both static sources blocked	Decreases with altitude gain, increases with altitude loss.	Does not change with actual gain or loss of altitude.	Does not change with actual variations in vertical speed.
Both static and pitot sources blocked	All indications remain constant, regardless of actual changes in airspeed, altitude, and vertical speed.		

Figure 18-10. *Effects of blocked pitot-static sources.*

If the static reference system is severely restricted, but not entirely blocked, as the airplane descends, the static reference pressure at the instruments begins to lag behind the actual outside air pressure. While descending, the altimeter may indicate that the airplane is higher than actual because the obstruction slows the airflow from the static port to the altimeter. The VSI confirms the altimeter's information regarding rate of change because the reference pressure is not changing at the same rate as the outside air pressure. The airspeed indicator, unable to tell whether it is experiencing more airspeed pitot pressure or less static reference pressure, indicates a higher airspeed than actual. To the pilot, the instruments indicate that the airplane is too high, too fast, and descending at a rate much lower than desired.

If the pilot levels off and then begins a climb, the altitude indication may still lag. The VSI indicates that the airplane is not climbing as fast as actual. The indicated airspeed, however, may begin to decrease at an alarming rate. The least amount of pitch-up attitude may cause the airspeed needle to indicate dangerously near stall speed.

Managing a static system malfunction requires that the pilot know and understand the airplane's pitot-static system. If a system malfunction is suspected, the pilot should confirm it by opening the alternate static source. This should be done while the airplane is climbing or descending. If the instrument needles move significantly when this is done, a static pressure problem exists and the alternate source should be used during the remainder of the flight.

Failure of the pitot-static system may also have serious consequences for Electronic Flight Instrument Systems (EFIS). To satisfy the requirements of Title 14 of the Code of Federal Regulations (14 CFR) part 23, section 23.2615(b)(2), information essential for continued safe flight and landing will be available to the flightcrew in a timely manner after any single failure or probable combination of failures. However, many of the light aircraft equipped with glass displays typically share the same pitot-static inputs for the backup instrumentation. Since both systems are receiving the same input signals, both could fail if affected by obstructed or blocked pitot tubes and static ports and create a difficult situation for a pilot flying in IMC. Some manufacturers combine both the air data computer (ADC) and the attitude and heading reference system (AHRS) functions so that a blockage of the input system may also affect the attitude display.

With conventional instrumentation, the design and operation are similar regardless of aircraft or manufacturer. By comparing information between the six conventional instruments, pilots are able to diagnose common failure modes. Instrument failure indications of conventional instruments and electronic flight displays may be entirely different, and electronic systems failure indications are not standardized. With the wide diversity in system design of glass displays, the primary display and the backup display may respond differently to any interruption of data input, and both displays may function differently than conventional instruments under the same conditions.

It is imperative for pilots to obtain equipment-specific information in reference to both the aircraft and the avionics that fully prepare them to interpret and properly respond to equipment malfunctions of electronic flight instrument displays. Rapidly changing equipment, complex systems, and the difficulty or inability to simulate failure modes and functions can impose training limitations. Pilots still should be able to respond to equipment malfunctions in a timely manner without impairing other critical flight tasks should the need arise.

Abnormal Engine Instrument Indication

The AFM/POH for the specific airplane contains information that should be followed in the event of any abnormal engine instrument indications. The table shown in *Figure 18-11* offers generic information on some of the more commonly experienced in-flight abnormal engine instrument indications, their possible causes, and corrective actions.

Malfunction	Probable Cause	Corrective Action
Loss of rpm during cruise flight (non-altitude engines)	Carburetor or induction icing or air filter clogging	Apply carburetor heat. If dirty filter is suspected and non-filtered air is available, switch selector to unfiltered position.
Loss of manifold pressure during cruise flight	Same as above	Same as above.
	Turbocharger failure	Possible exhaust leak. Shut down engine or use lowest practicable power setting. Land as soon as possible.
Gain of manifold pressure during cruise flight	Throttle has opened, propeller control has decreased rpm, or improper method of power reduction	Readjust throttle and tighten friction lock. Reduce manifold pressure prior to reducing rpm.
High oil temperature	Oil congealed in cooler	Reduce power. Land. Preheat engine.
	Inadequate engine cooling	Reduce power. Increase airspeed.
	Detonation or preignition	Observe cylinder head temperatures for high reading. Reduce manifold pressure. Enrich mixture.
	Forthcoming internal engine failure	Land as soon as possible or feather propeller and stop engine.
	Defective thermostatic oil cooler control	Land as soon as possible. Consult maintenance personnel.
Low oil temperature	Engine not warmed up to operating temperature	Warm engine in prescribed manner.
High oil pressure	Cold oil	Same as above.
	Possible internal plugging	Reduce power. Land as soon as possible.
Low oil pressure	Broken pressure relief valve	Land as soon as possible or feather propeller and stop engine.
	Insufficient oil	Same as above.
	Burned out bearings	Same as above.
Fluctuating oil pressure	Low oil supply, loose oil lines, defective pressure relief valve	Same as above.
High cylinder head temperature	Improper cowl flap adjustment	Adjust cowl flaps.
	Insufficient airspeed for cooling	Increase airspeed.
	Improper mixture adjustment	Adjust mixture.
	Detonation or preignition	Reduce power, enrich mixture, increase cooling airflow.
Low cylinder head temperature	Excessive cowl flap opening	Adjust cowl flaps.
	Excessively rich mixture	Adjust mixture control.
	Extended glides without clearing engine	Clear engine long enough to keep temperatures at minimum range.
Ammeter indicating discharge	Alternator or generator failure	Shed unnecessary electrical load. Land as soon as practicable.
Load meter indicating zero	Same as above	Same as above
Surging rpm and overspeeding	Defective propeller	Adjust propeller rpm.
	Defective engine	Consult maintenance.
	Defective propeller governor	Adjust propeller control. Attempt to restore normal operation.
	Defective tachometer	Consult maintenance.
	Improper mixture setting	Readjust mixture for smooth operation.
Loss of airspeed in cruise flight with manifold pressure and rpm constant	Possible loss of one or more cylinders	Land as soon as possible.
Rough running engine	Improper mixture control setting	Adjust mixture for smooth operation
	Defective ignition or valves	Consult maintenance personnel.
	Detonation or preignition	Reduce power, enrich mixture, open cowl flaps to reduce cylinder head temp. Land as soon as practicable.
	Induction air leak	Reduce power. Consult maintenance.
	Plugged fuel nozzle (fuel injection)	Same as above.
	Excessive fuel pressure or fuel flow	Lean mixture control.
Loss of fuel pressure	Engine-driven pump failure No fuel	Turn on boost pumps. Switch tanks, turn on fuel.

Figure 18-11. *Commonly experienced in-flight abnormal engine instrument indications, their possible causes, and corrective actions.*

Door Opening In-Flight

In most instances, the occurrence of an inadvertent door opening is not of great concern to the safety of a flight, but rather, the pilot's reaction at the moment the incident happens. A door opening in flight may be accompanied by a sudden loud noise, sustained noise level, and possible vibration or buffeting. If a pilot allows himself or herself to become distracted to the point where attention is focused on the open door rather than maintaining control of the airplane, loss of control may result even though disruption of airflow by the door is minimal.

In the event of an inadvertent door opening in flight or on takeoff, the pilot should adhere to the following:

- Concentrate on flying the airplane. Particularly in light single and twin-engine airplanes; a cabin door that opens in flight seldom if ever compromises the airplane's ability to fly. There may be some handling effects, such as roll and/or yaw, but in most instances these can be easily overcome.

- If the door opens after lift-off, do not rush to land. Climb to normal traffic pattern altitude, fly a normal traffic pattern, and make a normal landing.

- Do not release the seat belt and shoulder harness in an attempt to reach the door. Leave the door alone. Land as soon as practicable, and close the door once safely on the ground.

- Remember that most doors do not stay wide open. They usually bang open and then settle partly closed. A slip towards the door may cause it to open wider; a slip away from the door may push it closed.

- Do not panic. Try to ignore the unfamiliar noise and vibration. Also, do not rush. Attempting to get the airplane on the ground as quickly as possible may result in steep turns at low altitude.

- Complete all items on the landing checklist.

- Remember that accidents are almost never caused by an open door. Rather, an open door accident is caused by the pilot's distraction or failure to maintain control of the airplane.

Inadvertent VFR Flight Into IMC

It is beyond the scope of this handbook to incorporate a course of training in basic attitude instrument flying. This information is contained in the Instrument Flying Handbook (FAA-H-8083-15). Certain pilot certificates and/or associated ratings require training in instrument flying and a demonstration of specific instrument flying tasks on the practical test.

Pilots and flight instructors should refer to the Instrument Flying Handbook (FAA-H-8083-15) for guidance in the performance of these tasks and to the appropriate airman certification standards (ACS) for information on the evaluation of tasks performed for the particular certificate level and/or rating. The pilot should remember, however, that unless these tasks are practiced on a continuing and regular basis, skill erosion begins almost immediately. In a very short time, the pilot's assumed level of confidence is much higher than the performance he or she is actually able to demonstrate should the need arise.

Accident statistics show that the pilot who has not been trained in attitude instrument flying, or one whose instrument skills have eroded, lose control of the airplane in about 10 minutes once forced to rely solely on instrument references. The purpose of this section is to provide guidance on practical emergency measures to maintain airplane control for a limited period of time in the event a VFR pilot encounters instrument meteorological conditions (IMC). The main goal is not precision instrument flying; rather, it is to help the VFR pilot keep the airplane under adequate control until suitable visual references are regained.

The first steps necessary for surviving an encounter with IMC by a VFR pilot are as follows:

- Recognition and acceptance of the seriousness of the situation and the need for immediate remedial action
- Maintaining control of the airplane
- Obtaining the appropriate assistance to get the airplane safely on the ground

Recognition

Anytime a VFR pilot is unable to maintain airplane attitude control by reference to the natural horizon, the condition is considered to be IMC regardless of the circumstances or the prevailing weather conditions. Whether the cause is inadvertent or intentional, the VFR pilot is, in effect, in IMC if unable to navigate or establish geographical position by visual reference to landmarks on the surface. These situations should be accepted by the pilot involved as a genuine emergency requiring appropriate action.

Pilots should understand that unless they are trained, qualified, and current in the control of an airplane solely by reference to flight instruments, they will be unable to do so for any length of time. Many hours of VFR flying using the attitude indicator as a reference for airplane control may lull pilots into a false sense of security based on an overestimation of their personal ability to control the airplane solely by instrument references. In VFR conditions, even though the pilot believes the instrument references will be easy to use, the pilot also receives an overview of the natural horizon and may subconsciously rely on it more than the attitude indicator. If the natural horizon were to suddenly disappear, the untrained instrument pilot would be subject to vertigo, spatial disorientation, and inevitable control loss.

Maintaining Airplane Control

Once the pilot recognizes and accepts the situation, he or she should understand that the only way to control the airplane safely is by using and trusting the flight instruments. Attempts to control the airplane partially by reference to flight instruments while searching outside of the airplane for visual confirmation of the information provided by those instruments results in inadequate airplane control. This may be followed by spatial disorientation and complete control loss.

The most important point to be stressed is that the pilot should not panic. The task at hand may seem overwhelming, and the situation may be compounded by extreme apprehension. However, the pilot should make a conscious effort to relax. The pilot needs to understand the most important concern—in fact the only concern at this point—is to keep the wings level. An uncontrolled turn or bank usually leads to difficulty in achieving the objectives of any desired flight condition, but good bank control has the effect of making pitch control much easier.

The pilot should remember that a person cannot feel control pressures with a tight grip on the controls. Relaxing and learning to "control with the eyes and the brain," instead of only the muscles usually takes considerable conscious effort.

The pilot needs to believe what the flight instruments show about the airplane's attitude regardless of what the natural senses tell. The vestibular sense (motion sensing by the inner ear) can and will confuse the pilot. Because of inertia, the sensory areas of the inner ear cannot detect slight changes in airplane attitude, nor can they accurately sense attitude changes that occur at a uniform rate over a period of time. On the other hand, false sensations are often generated, leading the pilot to believe the attitude of the airplane has changed when, in fact, it has not. These false sensations result in the pilot experiencing spatial disorientation.

Attitude Control

An airplane is, by design, an inherently stable platform and, except in turbulent air, maintains approximately straight-and-level flight if properly trimmed and left alone. It is designed to maintain a state of equilibrium in pitch, roll, and yaw. The pilot should be aware, however, that a change about one axis affects the stability of the others. The typical light airplane exhibits a good deal of stability in the yaw axis, slightly less in the pitch axis, and even lesser still in the roll axis. The key to emergency airplane attitude control, therefore, is to:

- Trim the airplane with the elevator trim so that it maintains hands-off level flight at cruise airspeed.

- Resist the tendency to over-control the airplane. Fly the attitude indicator with fingertip control. No attitude changes should be made unless the flight instruments indicate a definite need for a change.

- Make all attitude changes smooth and small, yet with positive pressure. Remember that a small change as indicated on the horizon bar corresponds to a proportionately much larger change in actual airplane attitude.

- Make use of any available aid in attitude control, such as autopilot or wing leveler.

The primary instrument for attitude control is the attitude indicator. *[Figure 18-12]* Once the airplane is trimmed so that it maintains hands-off level flight at cruise airspeed, that airspeed need not vary until the airplane is slowed for landing. All turns, climbs, and descents can and should be made at this airspeed. Straight flight is maintained by keeping the wings level using "fingertip pressure" on the control wheel. Any pitch attitude change should be made by using no more than one bar width up or down.

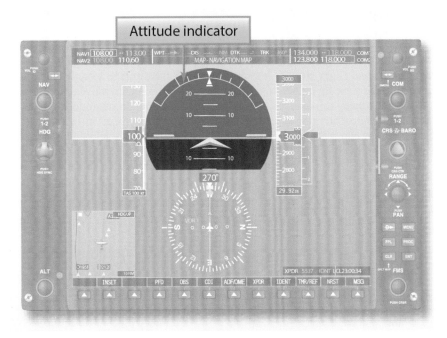

Figure 18-12. *Attitude indicator.*

Turns

Turns are perhaps the most potentially dangerous maneuver for the untrained instrument pilot for two reasons:

- The normal tendency of the pilot to over-control, leading to steep banks and the possibility of a "graveyard spiral."

- The inability of the pilot to cope with the instability resulting from the turn.

When a turn is to be made, the pilot should anticipate and cope with the relative instability of the roll axis. The smallest practical bank angle should be used—in any case no more than 10° bank angle. *[Figure 18-13]* A shallow bank takes very little vertical lift from the wings resulting in little if any deviation in altitude. It may be helpful to turn a few degrees and then return to level flight if a large change in heading is necessary. Repeat the process until the desired heading is reached. This process may relieve the progressive overbanking that often results from prolonged turns.

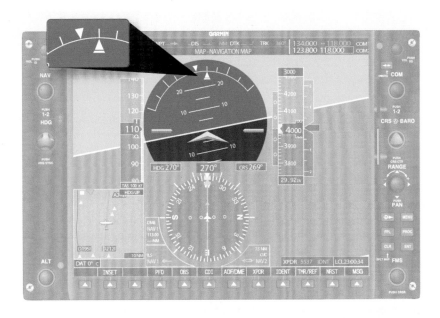

Figure 18-13. *Level turn.*

Climbs

If a climb is necessary, the pilot should raise the miniature airplane on the attitude indicator no more than one bar width and apply power. *[Figure 18-14]* The pilot should not attempt to attain a specific climb speed but accept whatever speed results. The objective is to deviate as little as possible from level flight attitude in order to disturb the airplane's equilibrium as little as possible. If the airplane is properly trimmed, it assumes a nose-up attitude on its own commensurate with the amount of power applied. Torque and P-factor cause the airplane to have a tendency to bank and turn to the left. This should be anticipated and compensated for. If the initial power application results in an inadequate rate of climb, power should be increased in increments of 100 rpm or 1 inch of manifold pressure until the desired rate of climb is attained. Maximum available power is seldom necessary. The more power that is used, the more the airplane wants to bank and turn to the left. Resuming level flight is accomplished by first decreasing pitch attitude to level on the attitude indicator using slow but deliberate pressure, allowing airspeed to increase to near cruise value and then decreasing power.

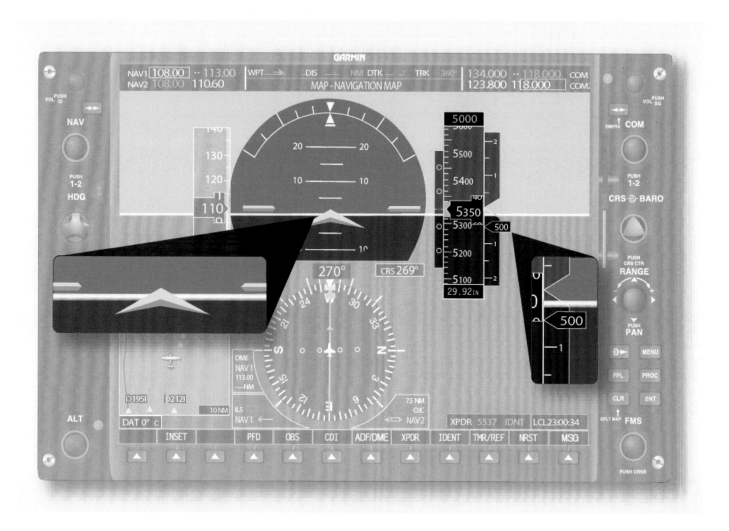

Figure 18-14. *Level climb.*

Descents

Descents are very much the opposite of the climb procedure if the airplane is properly trimmed for hands-off straight-and-level flight. In this configuration, the airplane requires a certain amount of thrust to maintain altitude. The pitch attitude is controlling the airspeed. The engine power, therefore, (translated into thrust by the propeller) is maintaining the selected altitude. Following a power reduction, however slight, there is an almost imperceptible decrease in airspeed. However, even a slight change in speed results in less down load on the tail, whereupon the designed nose heaviness of the airplane causes it to pitch down just enough to maintain the airspeed for which it was trimmed. The airplane then descends at a rate directly proportionate to the amount of thrust that has been removed. Power reductions should be made in increments of 100 rpm or 1 inch of manifold pressure and the resulting rate of descent should never exceed 500 fpm. The wings should be held level on the attitude indicator, and the pitch attitude should not exceed one bar width below level. *[Figure 18-15]*

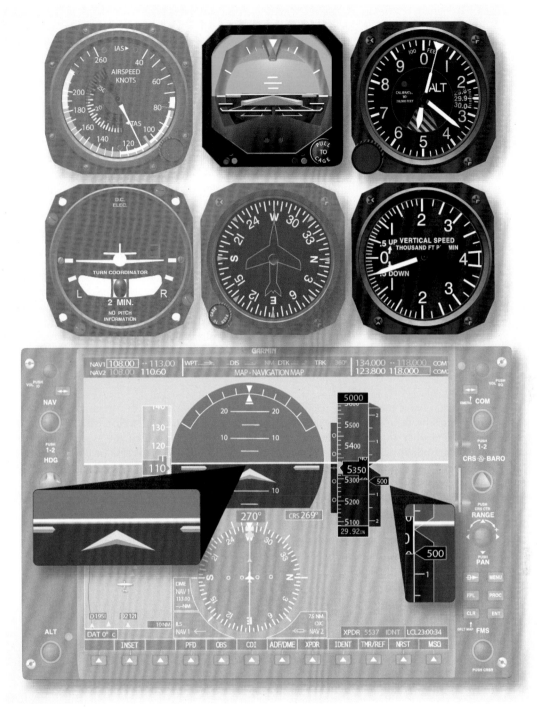

Figure 18-15. *Level descent.*

Combined Maneuvers

Combined maneuvers, such as climbing or descending turns, should be avoided if at all possible by an untrained instrument pilot. Combining maneuvers only compounds the problems encountered in individual maneuvers and increases the risk of control loss. The objective is to keep the airplane under control by maintaining as much of the airplane's natural equilibrium as possible. Deviating as little as possible from straight-and-level flight attitude makes this much easier.

When being assisted by ATC, the pilot may detect a sense of urgency while being directed to change heading and/or altitude. This sense of urgency reflects a normal concern for safety on the part of the controller. Nevertheless, the pilot should not let this prompting lead to rushing into a maneuver that could result in loss of control. It's reasonable to ask the controller to slow down, if this becomes an issue.

Transition to Visual Flight

One of the most difficult tasks a trained and qualified instrument pilot contends with is the transition from instrument to visual flight prior to landing. For the untrained instrument pilot, these difficulties are magnified.

The difficulties center around acclimatization and orientation. On an instrument approach, the trained instrument pilot prepares in advance for the transition to visual flight. The pilot has a mental picture of what to expect when the transition to visual flight is made and will quickly acclimate to the new environment. Geographical orientation also begins before the transition, as the pilot visualizes where the airplane is in relation to the airport/runway.

In an ideal situation, the transition to visual flight is made with ample time, at a sufficient altitude above terrain, and to visibility conditions sufficient to accommodate acclimatization and geographical orientation. This, however, is not always the case. The untrained instrument pilot may find the visibility still limited, the terrain completely unfamiliar, and altitude above terrain such that a "normal" airport traffic pattern and landing approach is not possible. Additionally, the pilot is most likely under considerable self-induced psychological pressure to get the airplane on the ground. The pilot should take this into account and, if possible, allow time to become acclimatized and geographically oriented before attempting an approach and landing, even if it means flying straight and level for a time or circling the airport. This is especially true at night.

Emergency Response Systems

Airplanes may have installed systems that provide alternatives in certain emergency situations. For example, ballistic parachute systems, if installed, may be deployed in an emergency allowing an airplane to descend slowly enough toward the ground such that occupants usually survive the resulting impact with minor or no injuries. Airplanes may also have an Emergency Autoland (EAL) system, which can take over control of the aircraft when necessary for a safe outcome.

Ballistic Parachutes

Deployment of an airplane ballistic parachute system results in the loss of the airframe, but deploying such systems within an acceptable flight regime prevents injuries and saves lives. Pilots need to understand and follow the procedures for arming and disarming these systems before and after flight, and understand the conditions under which the system would be deployed. For example, a catastrophic loss of controllability due to a collision or mechanical failure, actual loss of control, or pilot incapacitation would qualify. Pilots should brief passengers with access to any deployment mechanism regarding the conditions for a safe deployment. Generally, the passenger would deploy the system only if the pilot were incapacitated. At a minimum, the pilot should also brief the passengers regarding the basic sequence of steps for deployment. Pilots should study the information provided by manufacturers and suppliers of these systems and follow the guidance provided.

The system design may include airplane components designed to absorb the forces of vertical impact. The design of landing gear and seats maximize the protection afforded to the occupants and extend the time over which impact forces are absorbed. Once on the ground, there are hazards associated with a deployed parachute and the effect of surface winds, and the occupants should know the procedures for evacuation.

Autoland

If the EAL senses erratic flying, it stabilizes the aircraft, and checks for pilot responsiveness. Without further input, the EAL initiates an emergency descent. Without pilot responsiveness after an emergency descent, EAL initiates the process for an automated landing. The system also allows for manual activation by a pilot or a passenger.

Once activated, the EAL system transmits automated radio broadcasts on the aircraft's last selected frequency and on Guard (121.5 MHz) to alert controllers or pilots in the area of the EAL aircraft's imminent arrival to the selected runway. The system repeatedly transmits the call sign and intention to divert to a particular airport and runway using a recognizable non-human synthesized voice. Additionally, EAL sets the transponder to squawk 7700 to indicate an emergency. After the initial broadcast, the system pauses for 25 seconds to allow air traffic control (ATC) to communicate with potential conflicting traffic. Once the EAL aircraft is within 12 miles of the selected runway and at or below 12,000 feet MSL, it broadcasts on the tower frequency or Common Traffic Advisory Frequency (CTAF), and continues to broadcast its position via ADS-B. It announces its call sign, "pilot incapacitation," its position relative to the destination airport, gives the airport and airport identifier, and the time to landing on a specific runway at that airport. The system makes a similar "one-minute out" broadcast prior to landing.

The EAL system selects a suitable landing airfield based on several factors. These factors include weather, wind, runway length, and towered/non-towered airport status. EAL only considers airports with an area navigation (RNAV) or Global Positioning System (GPS) approach, selects towered airports over non-towered airports where possible, and uses runway requirements that depend on the aircraft type. EAL systems also utilize obstacle and a terrain databases. If the system loses GPS coverage, the airplane continues straight flight without attempting to land until GPS coverage resumes.

Currently EAL system capabilities do not include detecting and avoiding other aircraft; receiving or reacting to ATC instructions or Notices to Airmen (NOTAMs); avoiding military operations areas (MOAs), special use airspace (SUA), Restricted Areas, or Temporary Flight Restrictions (TFRs); or turning on aircraft lights.

Chapter Summary

This chapter provided general guidance and recommended procedures that may apply to light single-engine airplanes involved in certain emergency situations. The information presented is intended to enhance the general knowledge of emergency operations with the clear understanding that the manufacturer's recommended emergency procedures take precedence.

Information was provided concerning failure of the pitot-static system in aircraft with EFIS. The redundancy of backup systems for IFR flight may be less than desired if both the primary and backup instrumentation may receive signal data input from the same pitot-static source. The failure indications of EFIS may be entirely different from conventional instruments making recognition of system malfunction much more difficult for the pilot. Lack of system standardization compounds the problem making equipment specific information and knowledge an important asset when analyzing electronic display malfunctions. The inability to simulate certain failure modes during training and evaluation could make the pilot less prepared for an actual emergency. As electronic avionics become more advanced, the training and proficiency needed to safely operate these systems should receive careful analysis.

Airplane Flying Handbook (FAA-H-8083-3A)

Glossary

Numbers and Symbols

14 CFR. (Title 14 of the Code of Federal Regulations. Federal regulations pertaining to aviation activity. Previously known as Federal Aviation Regulations.

100-Hour Inspection. An inspection, identical in scope to an annual inspection. Must be conducted every 100 hours of flight on aircraft of under 12,500 pounds that are used for hire.

A

Absolute altitude. The vertical distance of an airplane above the terrain or above ground level (AGL).

Absolute ceiling. The altitude at which a climb is no longer possible.

Accelerate-go distance. The distance required to accelerate to V_1 with all engines at takeoff power, experience an engine failure at V_1 and continue the takeoff on the remaining engine(s). The runway required includes the distance required to climb to 35 feet by which time V_2 speed must be attained.

Accelerate-stop distance. The distance required to accelerate to V_1 with all engines at takeoff power, experience an engine failure at V_1, and abort the takeoff and bring the airplane to a stop using braking action only (use of thrust reversing is not considered).

Acceleration. Force involved in overcoming inertia, and which may be defined as a change in velocity per unit of time.

Accessories. Components that are used with an engine, but are not a part of the engine itself. Units such as magnetos, carburetors, generators, and fuel pumps are commonly installed engine accessories.

Adjustable stabilizer. A stabilizer that can be adjusted in flight to trim the airplane, thereby allowing the airplane to fly hands-off at any given airspeed.

Adverse yaw. A condition of flight in which the nose of an airplane tends to yaw toward the outside of the turn. This is caused by the higher induced drag on the outside wing, which is also producing more lift. Induced drag is a by-product of the lift associated with the outside wing.

Aerodynamic ceiling. The point (altitude) at which, as the indicated airspeed decreases with altitude, it progressively merges with the low speed buffet boundary where pre-stall buffet occurs for the airplane at a load factor of 1.0 G.

Aerodynamics. The science of the action of air on an object, and with the motion of air on other gases. Aerodynamics deals with the production of lift by the aircraft, the relative wind, and the atmosphere.

Ailerons. Primary flight control surfaces mounted on the trailing edge of an airplane wing, near the tip. Ailerons control roll about the longitudinal axis.

Air start. The act or instance of starting an aircraft's engine while in flight, especially a jet engine after flameout.

Aircraft energy management. The process of planning, monitoring and controlling altitude and airspeed in relation to the airplane's energy state. Note that this definition is concerned with managing mechanical energy (altitude and airspeed) and addresses the safety (flight control) side of energy management. It does not address the efficiency (aircraft performance) side of energy management, which is concerned with how efficiently the engine generates mechanical energy from fuel and how efficiently the airframe spends that energy in flight.

Aircraft logbooks. Journals containing a record of total operating time, repairs, alterations or inspections performed, and all Airworthiness Directive (AD) notes complied with. A maintenance logbook should be kept for the airframe, each engine, and each propeller.

Airfoil. An airfoil is any surface, such as a wing, propeller, rudder, or even a trim tab, which provides aerodynamic force when it interacts with a moving stream of air.

Airmanship. A sound acquaintance with the principles of flight, the ability to operate an airplane with competence and precision both on the ground and in the air, and the exercise of sound judgment that results in optimal operational safety and efficiency.

Airplane Flight Manual (AFM). A document developed by the airplane manufacturer and approved by the Federal Aviation Administration (FAA). It is specific to a particular make and model airplane by serial number and it contains operating procedures and limitations.

Airplane Owner/Information Manual. A document developed by the airplane manufacturer containing general information about the make and model of an airplane. The airplane owner's manual is not FAA-approved and is not specific to a particular serial numbered airplane. This manual is not kept current, and therefore cannot be substituted for the AFM/POH.

Airworthiness Certificate. A certificate issued by the FAA to all aircraft that have been proven to meet the minimum standards set down by the Code of Federal Regulations.

Airworthiness Directive. A regulatory notice sent out by the FAA to the registered owner of an aircraft informing the owner of a condition that prevents the aircraft from continuing to meet its conditions for airworthiness. Airworthiness Directives (AD) must be complied with within the required time limit, and the fact of compliance, the date of compliance, and the method of compliance must be recorded in the aircraft's maintenance records.

Airworthiness. A condition in which the aircraft conforms to its type certificated design including supplemental type certificates and field-approved alterations. The aircraft must also be in a condition for safe flight as determined by annual, 100-hour, preflight and any other required inspections.

Alpha mode of operation. The operation of a turboprop engine that includes all of the flight operations, from takeoff to landing. Alpha operation is typically between 95 percent to 100 percent of the engine operating speed.

Alternate air. A device which opens, either automatically or manually, to allow induction airflow to continue should the primary induction air opening become blocked.

Alternate static source. A manual port that when opened allows the pitot static instruments to sense static pressure from an alternate location should the primary static port become blocked.

Alternator/generator. A device that uses engine power to generate electrical power.

Altimeter. A flight instrument that indicates altitude by sensing pressure changes.

Altitude (AGL). The actual height above ground level (AGL) at which the aircraft is flying.

Altitude (MSL). The actual height above mean sea level (MSL) at which the aircraft is flying.

Altitude chamber. A device that simulates high altitude conditions by reducing the interior pressure. The occupants will suffer from the same physiological conditions as flight at high altitude in an unpressurized aircraft.

Altitude engine. A reciprocating aircraft engine having a rated takeoff power that is producible from sea level to an established higher altitude.

Angle of attack. The acute angle between the chord line of the airfoil and the direction of the relative wind.

Angle of incidence. The angle formed by the chord line of the wing and a line parallel to the longitudinal axis of the airplane.

Annual inspection. Except as provided in regulation, no person may operate an aircraft unless, within the preceding 12 calendar months, it has had an annual inspection, per 14 CFR part 91, section 91.409(a). This inspection is normally performed every 12 calendar months by an A&P technician holding an Inspection Authorization (14 CFR part 65, section 65.95(a)(2)).

Anti-icing. The prevention of the formation of ice on a surface. Ice may be prevented by using heat or by covering the surface with a chemical that prevents water from reaching the surface. Anti-icing should not be confused with deicing, which is the removal of ice after it has formed on the surface.

Attitude indicator. An instrument which uses an artificial horizon and miniature airplane to depict the position of the airplane in relation to the true horizon. The attitude indicator senses roll as well as pitch, which is the up and down movement of the airplane's nose.

Attitude. The position of an aircraft as determined by the relationship of its axes and a reference, usually the earth's horizon.

Autokinesis. This is caused by staring at a single point of light against a dark background for more than a few seconds. After a few moments, the light appears to move on its own.

Autopilot. An automatic flight control system which keeps an aircraft in level flight or on a set course. Automatic pilots can be directed by the pilot, or they may be coupled to a radio navigation signal.

Axes of an aircraft. Three imaginary lines that pass through an aircraft's center of gravity. The axes can be considered as imaginary axles around which the aircraft turns. The three axes pass through the center of gravity at 90° angles to each other. The axis from nose to tail is the longitudinal axis, the axis that passes from wingtip to wingtip is the lateral axis, and the axis that passes vertically through the center of gravity is the vertical axis.

Axial flow compressor. A type of compressor used in a turbine engine in which the airflow through the compressor is essentially linear. An axial-flow compressor is made up of several stages of alternate rotors and stators. The compressor ratio is determined by the decrease in area of the succeeding stages.

B

Back side of the power curve. Flight regime in which flight at a higher airspeed requires a lower power setting and a lower airspeed requires a higher power setting in order to maintain altitude.

Balked landing. A go-around.

Ballast. Removable or permanently installed weight in an aircraft used to bring the center of gravity into the allowable range.

Balloon. The result of a too aggressive flare during landing causing the aircraft to climb.

Basic empty weight (GAMA). Basic empty weight includes the standard empty weight plus optional and special equipment that has been installed.

Best angle of climb (V_X). The speed at which the aircraft will produce the most gain in altitude in a given distance.

Best glide. The airspeed in which the aircraft glides the furthest for the least altitude lost when in non-powered flight.

Best rate of climb (V_Y). The speed at which the aircraft will produce the most gain in altitude in the least amount of time.

Blade face. The flat portion of a propeller blade, resembling the bottom portion of an airfoil.

Bleed air. Compressed air tapped from the compressor stages of a turbine engine by use of ducts and tubing. Bleed air can be used for deice, anti-ice, cabin pressurization, heating, and cooling systems.

Bleed valve. In a turbine engine, a flapper valve, a pop off valve, or a bleed band designed to bleed off a portion of the compressor air to the atmosphere. Used to maintain blade angle of attack and provide stall-free engine acceleration and deceleration.

Boost pump. An electrically driven fuel pump, usually of the centrifugal type, located in one of the fuel tanks. It is used to provide fuel to the engine for starting and providing fuel pressure in the event of failure of the engine driven pump. It also pressurizes the fuel lines to prevent vapor lock.

Buffeting. The beating of an aerodynamic structure or surface by unsteady flow, gusts, etc.; the irregular shaking or oscillation of a vehicle component owing to turbulent air or separated flow.

Bus bar. An electrical power distribution point to which several circuits may be connected. It is often a solid metal strip having a number of terminals installed on it.

Bus tie. A switch that connects two or more bus bars. It is usually used when one generator fails and power is lost to its bus. By closing the switch, the operating generator powers both buses.

Bypass air. The part of a turbofan's induction air that bypasses the engine core.

Bypass ratio. The ratio of the mass airflow in pounds per second through the fan section of a turbofan engine to the mass airflow that passes through the gas generator portion of the engine. Or, the ratio between fan mass airflow (lb/sec.) and core engine mass airflow (lb/sec.).

C

Cabin pressurization. A condition where pressurized air is forced into the cabin simulating pressure conditions at a much lower altitude and increasing the aircraft occupants comfort.

Calibrated airspeed (CAS). Indicated airspeed corrected for installation error and instrument error. Although manufacturers attempt to keep airspeed errors to a minimum, it is not possible to eliminate all errors throughout the airspeed operating range. At certain airspeeds and with certain flap settings, the installation and instrument errors may total several knots. This error is generally greatest at low airspeeds. In the cruising and higher airspeed ranges, indicated airspeed and calibrated airspeed are approximately the same. Refer to the airspeed calibration chart to correct for possible airspeed errors.

Cambered. The camber of an airfoil is the characteristic curve of its upper and lower surfaces. The upper camber is more pronounced, while the lower camber is comparatively flat. This causes the velocity of the airflow immediately above the wing to be much higher than that below the wing.

Carburetor ice. Ice that forms inside the carburetor due to the temperature drop caused by the vaporization of the fuel. Induction system icing is an operational hazard because it can cut off the flow of the fuel/air charge or vary the fuel/air ratio.

Carburetor. 1. Pressure: A hydromechanical device employing a closed feed system from the fuel pump to the discharge nozzle. It meters fuel through fixed jets according to the mass airflow through the throttle body and discharges it under a positive pressure. Pressure carburetors are distinctly different from float-type carburetors, as they do not incorporate a vented float chamber or suction pickup from a discharge nozzle located in the venturi tube. 2. Float-type: Consists essentially of a main air passage through which the engine draws its supply of air, a mechanism to control the quantity of fuel discharged in relation to the flow of air, and a means of regulating the quantity of fuel/air mixture delivered to the engine cylinders.

Cascade reverser. A thrust reverser normally found on turbofan engines in which a blocker door and a series of cascade vanes are used to redirect exhaust gases in a forward direction.

Center of gravity (CG). The point at which an airplane would balance if it were possible to suspend it at that point. It is the mass center of the airplane, or the theoretical point at which the entire weight of the airplane is assumed to be concentrated. It may be expressed in inches from the reference datum, or in percent of mean aerodynamic chord (MAC). The location depends on the distribution of weight in the airplane.

Center-of-gravity limits. The specified forward and aft points within which the CG must be located during flight. These limits are indicated on pertinent airplane specifications.

Center-of-gravity range. The distance between the forward and aft CG limits indicated on pertinent airplane specifications.

Centrifugal flow compressor. An impeller-shaped device that receives air at its center and slings air outward at high velocity into a diffuser for increased pressure. Also referred to as a radial outflow compressor.

Chart Supplements. A listing of data on record with the FAA on all open-to-the-public airports, seaplane bases, heliports, military facilities and selected private use airports specifically requested by the Department of Defense (DOD) for which a DOD instrument approach procedure has been published in the U.S. Terminal Procedures Publication, airport sketches, NAVAIDs, communications data, weather data sources, airspace, special notices, VFR waypoints, Airport Diagrams and operational procedures.

Chord line. An imaginary straight line drawn through an airfoil from the leading edge to the trailing edge.

Circuit breaker. A circuit-protecting device that opens the circuit in case of excess current flow. A circuit breaker differs from a fuse in that it can be reset without having to be replaced.

Clear air turbulence. Turbulence not associated with any visible moisture.

Climb gradient. The ratio between distance traveled and altitude gained.

Cockpit resource management. Techniques designed to reduce pilot errors and manage errors that do occur utilizing cockpit human resources. The assumption is that errors are going to happen in a complex system with error-prone humans.

Coefficient of lift. See lift coefficient.

Coffin corner. The flight regime where any increase in airspeed will induce high speed Mach buffet and any decrease in airspeed will induce low speed Mach buffet.

Combustion chamber. The section of the engine into which fuel is injected and burned.

Common traffic advisory frequency (CTAF). The common frequency used by airport traffic to announce position reports in the vicinity of the airport.

Complex aircraft. An aircraft with retractable landing gear, flaps, and a controllable-pitch propeller, or one that is turbine-powered.

Compression ratio. 1. In a reciprocating engine, the ratio of the volume of an engine cylinder with the piston at the bottom center to the volume with the piston at top center. 2. In a turbine engine, the ratio of the pressure of the air at the discharge to the pressure of air at the inlet.

Compressor bleed air. See bleed air.

Compressor bleed valves. See bleed valve.

Compressor section. The section of a turbine engine that increases the pressure and density of the air flowing through the engine.

Compressor stall. In gas turbine engines, a condition in an axial-flow compressor in which one or more stages of rotor blades fail to pass air smoothly to the succeeding stages. A stall condition is caused by a pressure ratio that is incompatible with the engine rpm Compressor stall will be indicated by a rise in exhaust temperature or rpm fluctuation, and if allowed to continue, may result in flameout and physical damage to the engine.

Compressor surge. A severe compressor stall across the entire compressor that can result in severe damage if not quickly corrected. This condition occurs with a complete stoppage of airflow or a reversal of airflow.

Condition lever. In a turbine engine, a powerplant control that controls the flow of fuel to the engine. The condition lever sets the desired engine rpm within a narrow range between that appropriate for ground and flight operations.

Configuration. This is a general term, which normally refers to the position of the landing gear and flaps.

Constant speed propeller. A controllable-pitch propeller whose pitch is automatically varied in flight by a governor to maintain a constant rpm in spite of varying air loads.

Control touch. The ability to sense the action of the airplane and its probable actions in the immediate future, with regard to attitude and speed variations, by sensing and evaluation of varying pressures and resistance of the control surfaces transmitted through the cockpit flight controls.

Controllability. A measure of the response of an aircraft relative to the pilot's flight control inputs.

Controllable-pitch propeller. A propeller in which the blade angle can be changed during flight by a control in the cockpit.

Conventional landing gear. Landing gear employing a third rear-mounted wheel. These airplanes are also sometimes referred to as tailwheel airplanes.

Coordinated flight. Application of all appropriate flight and power controls to prevent slipping or skidding in any flight condition.

Coordination. The ability to use the hands and feet together subconsciously and in the proper relationship to produce desired results in the airplane.

Core airflow. Air drawn into the engine for the gas generator.

Cowl flaps. Devices arranged around certain air-cooled engine cowlings which may be opened or closed to regulate the flow of air around the engine.

Crab. A flight condition in which the nose of the airplane is pointed into the wind a sufficient amount to counteract a crosswind and maintain a desired track over the ground.

Crazing. Small fractures in aircraft windshields and windows caused from being exposed to the ultraviolet rays of the sun and temperature extremes.

Critical altitude. The maximum altitude under standard atmospheric conditions at which a turbocharged engine can produce its rated horsepower.

Critical angle of attack. The angle of attack at which a wing stalls regardless of airspeed, flight attitude, or weight.

Critical engine. The engine whose failure has the most adverse effect on directional control.

Cross controlled. A condition where aileron deflection is in the opposite direction of rudder deflection.

Crossfeed. A system that allows either engine on a twin- engine airplane to draw fuel from any fuel tank.

Crosswind component. The wind component, measured in knots, at 90° to the longitudinal axis of the runway.

Current limiter. A device that limits the generator output to a level within that rated by the generator manufacturer.

D

Datum (reference datum). An imaginary vertical plane or line from which all measurements of moment arm are taken. The datum is established by the manufacturer. Once the datum has been selected, all moment arms and the location of CG range are measured from this point.

Decompression sickness. A condition where the low pressure at high altitudes allows bubbles of nitrogen to form in the blood and joints causing severe pain. Also known as the bends.

Deicer boots. Inflatable rubber boots attached to the leading edge of an airfoil. They can be sequentially inflated and deflated to break away ice that has formed over their surface.

Deicing. Removing ice after it has formed.

Delamination. The separation of layers.

Density altitude. This altitude is pressure altitude corrected for variations from standard temperature. When conditions are standard, pressure altitude and density altitude are the same. If the temperature is above standard, the density altitude is higher than pressure altitude. If the temperature is below standard, the density altitude is lower than pressure altitude. This is an important altitude because it is directly related to the airplane's performance.

Designated pilot examiner (DPE). An individual designated by the FAA to administer practical tests to pilot applicants.

Detonation. The sudden release of heat energy from fuel in an aircraft engine caused by the fuel-air mixture reaching its critical pressure and temperature. Detonation occurs as a violent explosion rather than a smooth burning process.

Dewpoint. The temperature at which air can hold no more water.

Differential ailerons. Control surface rigged such that the aileron moving up moves a greater distance than the aileron moving down. The up aileron produces extra parasite drag to compensate for the additional induced drag caused by the down aileron. This balancing of the drag forces helps minimize adverse yaw.

Diffusion. Reducing the velocity of air causing the pressure to increase.

Directional stability. Stability about the vertical axis of an aircraft, whereby an aircraft tends to return, on its own, to flight aligned with the relative wind when disturbed from that equilibrium state. The vertical tail is the primary contributor to directional stability, causing an airplane in flight to align with the relative wind.

Ditching. Emergency landing in water.

Downwash. Air deflected perpendicular to the motion of the airfoil.

Drag curve. A visual representation of the amount of drag of an aircraft at various airspeeds.

Drag. An aerodynamic force on a body acting parallel and opposite to the relative wind. The resistance of the atmosphere to the relative motion of an aircraft. Drag opposes thrust and limits the speed of the airplane.

Drift angle. Angle between heading and track.

Ducted-fan engine. An engine-propeller combination that has the propeller enclosed in a radial shroud. Enclosing the propeller improves the efficiency of the propeller.

Dutch roll. A combination of rolling and yawing oscillations that normally occurs when the dihedral effects of an aircraft are more powerful than the directional stability. Usually dynamically stable but objectionable in an airplane because of the oscillatory nature.

Dynamic hydroplaning. A condition that exists when landing on a surface with standing water deeper than the tread depth of the tires. When the brakes are applied, there is a possibility that the brake will lock up and the tire will ride on the surface of the water, much like a water ski. When the tires are hydroplaning, directional control and braking action are virtually impossible. An effective anti-skid system can minimize the effects of hydroplaning.

Dynamic stability. The property of an aircraft that causes it, when disturbed from straight-and-level flight, to develop forces or moments that restore the original condition of straight and level.

E

Electrical bus. See bus bar.

Electrohydraulic. Hydraulic control which is electrically actuated.

Elevator. The horizontal, movable primary control surface in the tail section, or empennage, of an airplane. The elevator is hinged to the trailing edge of the fixed horizontal stabilizer.

Emergency locator transmitter. A small, self-contained radio transmitter that will automatically, upon the impact of a crash, transmit an emergency signal on 121.5, 243.0, or 406.0 MHz.

Empennage. The section of the airplane that consists of the vertical stabilizer, the horizontal stabilizer, and the associated control surfaces.

Energy balance equation: According to this equation, the net transfer of mechanical energy into and out of the airplane (a function of thrust minus drag) is always equal to the change in its total mechanical energy (a function of altitude and airspeed).

Energy distribution error. An energy error where the total mechanical energy is correct, but the distribution between potential (altitude) and kinetic energy (airspeed) is not correct relative to the intended altitude-speed profile. When this error occurs, the pilot will observe that altitude and airspeed deviate in opposite directions (e.g., higher and slower than desired; or lower and faster than desired). An example would be an airplane on final approach that is above the desired glide slope and at a slower airspeed than desired.

Energy error. An altitude and/or airspeed deviation from an intended target expressed in terms of energy. Depending on the airplane's total amount of energy and its distribution between altitude and airspeed, energy errors are classified as total energy errors, energy distribution errors, or a combination of both errors.

Energy exchange. Trading one form of energy (e.g., altitude) for another form (e.g., airspeed).

Energy height or total specific energy (E_S). Measured in units of height (e.g., feet), it represents the airplane's total energy per unit weight. It is found by dividing the sum of potential energy and kinetic energy by the airplane's weight. It also represents the maximum height that an airplane would reach from its current altitude, if it were to trade all its speed for altitude.

Energy state. The airplane's total mechanical energy *and* its distribution between altitude and airspeed.

Energy system. A flying airplane is an *open* energy system. That means that the airplane can gain energy from some source (e.g., fuel) and lose energy to the environment (e.g., surrounding air). In addition, energy can be added to or removed from the airplane's total mechanical energy stored as altitude and airspeed.

Engine pressure ratio (EPR). The ratio of turbine discharge pressure divided by compressor inlet pressure that is used as an indication of the amount of thrust being developed by a turbine engine.

Environmental systems. In an aircraft, the systems, including the supplemental oxygen systems, air conditioning systems, heaters, and pressurization systems, which make it possible for an occupant to function at high altitude.

Equilibrium. A condition that exists within a body when the sum of the moments of all of the forces acting on the body is equal to zero. In aerodynamics, equilibrium is when all opposing forces acting on an aircraft are balanced (steady, unaccelerated flight conditions).

Equivalent shaft horsepower (ESHP). A measurement of the total horsepower of a turboprop engine, including that provided by jet thrust.

Exhaust gas temperature (EGT). The temperature of the exhaust gases as they leave the cylinders of a reciprocating engine or the turbine section of a turbine engine.

Exhaust manifold. The part of the engine that collects exhaust gases leaving the cylinders.

Exhaust. The rear opening of a turbine engine exhaust duct. The nozzle acts as an orifice, the size of which determines the density and velocity of the gases as they emerge from the engine.

F

False horizon. An optical illusion where the pilot confuses a row of lights along a road or other straight line as the horizon.

False start. See hung start.

Feathering propeller (feathered). A controllable pitch propeller with a pitch range sufficient to allow the blades to be turned parallel to the line of flight to reduce drag and prevent further damage to an engine that has been shut down after a malfunction.

Fixation. A psychological condition where the pilot fixes attention on a single source of information and ignores all other sources.

Fixed-shaft turboprop engine. A turboprop engine where the gas producer spool is directly connected to the output shaft.

Fixed-pitch propellers. Propellers with fixed blade angles. Fixed-pitch propellers are designed as climb propellers, cruise propellers, or standard propellers.

Flaps. Hinged portion of the trailing edge between the ailerons and fuselage. In some aircraft, ailerons and flaps are interconnected to produce full-span "flaperons." In either case, flaps change the lift and drag on the wing.

Flat pitch. A propeller configuration when the blade chord is aligned with the direction of rotation.

Flicker vertigo. A disorienting condition caused from flickering light off the blades of the propeller.

Flight director. An automatic flight control system in which the commands needed to fly the airplane are electronically computed and displayed on a flight instrument. The commands are followed by the human pilot with manual control inputs or, in the case of an autopilot system, sent to servos that move the flight controls.

Flight idle. Engine speed, usually in the 70-80 percent range, for minimum flight thrust.

Floating. A condition when landing where the airplane does not settle to the runway due to excessive airspeed.

Force (F). The energy applied to an object that attempts to cause the object to change its direction, speed, or motion. In aerodynamics, it is expressed as F, T (thrust), L (lift), W (weight), or D (drag), usually in pounds.

Form drag. The part of parasite drag on a body resulting from the integrated effect of the static pressure acting normal to its surface resolved in the drag direction.

Forward slip. A slip in which the airplane's direction of motion continues the same as before the slip was begun. In a forward slip, the airplane's longitudinal axis is at an angle to its flightpath.

Free power turbine engine. A turboprop engine where the gas producer spool is on a separate shaft from the output shaft. The free power turbine spins independently of the gas producer and drives the output shaft.

Friction drag. The part of parasitic drag on a body resulting from viscous shearing stresses over its wetted surface.

Frise-type aileron. Aileron having the nose portion projecting ahead of the hinge line. When the trailing edge of the aileron moves up, the nose projects below the wing's lower surface and produces some parasite drag, decreasing the amount of adverse yaw.

Fuel control unit. The fuel-metering device used on a turbine engine that meters the proper quantity of fuel to be fed into the burners of the engine. It integrates the parameters of inlet air temperature, compressor speed, compressor discharge pressure, and exhaust gas temperature with the position of the cockpit power control lever.

Fuel efficiency. Defined as the amount of fuel used to produce a specific thrust or horsepower divided by the total potential power contained in the same amount of fuel.

Fuel heater. A radiator-like device which has fuel passing through the core. A heat exchange occurs to keep the fuel temperature above the freezing point of water so that entrained water does not form ice crystals, which could block fuel flow.

Fuel injection. A fuel metering system used on some aircraft reciprocating engines in which a constant flow of fuel is fed to injection nozzles in the heads of all cylinders just outside of the intake valve. It differs from sequential fuel injection in which a timed charge of high-pressure fuel is sprayed directly into the combustion chamber of the cylinder.

Fuel load. The expendable part of the load of the airplane. It includes only usable fuel, not fuel required to fill the lines or that which remains trapped in the tank sumps.

Fuel tank sump. A sampling port in the lowest part of the fuel tank that the pilot can utilize to check for contaminants in the fuel.

Fuselage. The section of the airplane that consists of the cabin and/or cockpit, containing seats for the occupants and the controls for the airplane.

G

Gas generator. The basic power producing portion of a gas turbine engine and excluding such sections as the inlet duct, the fan section, free power turbines, and tailpipe. Each manufacturer designates what is included as the gas generator, but generally consists of the compressor, diffuser, combustor, and turbine.

Gas turbine engine. A form of heat engine in which burning fuel adds energy to compressed air and accelerates the air through the remainder of the engine. Some of the energy is extracted to turn the air compressor, and the remainder accelerates the air to produce thrust. Some of this energy can be converted into torque to drive a propeller or a system of rotors for a helicopter.

Glide ratio. The ratio between distance traveled and altitude lost during non-powered flight.

Glidepath. The path of an aircraft relative to the ground while approaching a landing.

Global position system (GPS). A satellite-based radio positioning, navigation, and time-transfer system.

Go-around. Terminating a landing approach.

Governing range. The range of pitch a propeller governor can control during flight.

Governor. A control which limits the maximum rotational speed of a device.

Gross weight. The total weight of a fully loaded aircraft including the fuel, oil, crew, passengers, and cargo.

Ground adjustable trim tab. A metal trim tab on a control surface that is not adjustable in flight. Bent in one direction or another while on the ground to apply trim forces to the control surface.

Ground effect. A condition of improved performance encountered when an airplane is operating very close to the ground. When an airplane's wing is under the influence of ground effect, there is a reduction in upwash, downwash, and wingtip vortices. As a result of the reduced wingtip vortices, induced drag is reduced.

Ground idle. Gas turbine engine speed usually 60-70 percent of the maximum rpm range, used as a minimum thrust setting for ground operations.

Ground loop. A sharp, uncontrolled change of direction of an airplane on the ground.

Ground power unit (GPU). A type of small gas turbine whose purpose is to provide electrical power, and/or air pressure for starting aircraft engines. A ground unit is connected to the aircraft when needed. Similar to an aircraft-installed auxiliary power unit.

Ground track. The aircraft's path over the ground when in flight.

Groundspeed (GS). The actual speed of the airplane over the ground. It is true airspeed adjusted for wind. Groundspeed decreases with a headwind, and increases with a tailwind.

Gust penetration speed. The speed that gives the greatest margin between the high and low Mach speed buffets.

Gyroscopic precession. An inherent quality of rotating bodies, which causes an applied force to be manifested 90° in the direction of rotation from the point where the force is applied.

H

Hand propping. Starting an engine by rotating the propeller by hand.

Heading bug. A marker on the heading indicator that can be rotated to a specific heading for reference purposes, or to command an autopilot to fly that heading.

Heading indicator. An instrument which senses airplane movement and displays heading based on a 360° azimuth, with the final zero omitted. The heading indicator, also called a directional gyro, is fundamentally a mechanical instrument designed to facilitate the use of the magnetic compass. The heading indicator is not affected by the forces that make the magnetic compass difficult to interpret.

Heading. The direction in which the nose of the aircraft is pointing during flight.

Headwind component. The component of atmospheric winds that acts opposite to the aircraft's flightpath.

High performance aircraft. An aircraft with an engine of more than 200 horsepower.

Horizon. The line of sight boundary between the earth and the sky.

Horsepower. The term, originated by inventor James Watt, means the amount of work a horse could do in one second. One horsepower equals 550 foot-pounds per second, or 33,000 foot-pounds per minute.

Hot start. In gas turbine engines, a start which occurs with normal engine rotation, but exhaust temperature exceeds prescribed limits. This is usually caused by an excessively rich mixture in the combustor. The fuel to the engine must be terminated immediately to prevent engine damage.

Hung start. In gas turbine engines, a condition of normal light off but with rpm remaining at some low value rather than increasing to the normal idle rpm This is often the result of insufficient power to the engine from the starter. In the event of a hung start, the engine should be shut down.

Hydraulics. The branch of science that deals with the transmission of power by incompressible fluids under pressure.

Hydroplaning. A condition that exists when landing on a surface with standing water deeper than the tread depth of the tires. When the brakes are applied, there is a possibility that the brake will lock up and the tire will ride on the surface of the water, much like a water ski. When the tires are hydroplaning, directional control and braking action are virtually impossible. An effective anti-skid system can minimize the effects of hydroplaning.

Hypoxia. A lack of sufficient oxygen reaching the body tissues.

I

Igniter plugs. The electrical device used to provide the spark for starting combustion in a turbine engine. Some igniters resemble spark plugs, while others, called glow plugs, have a coil of resistance wire that glows red hot when electrical current flows through the coil.

Impact ice. Ice that forms on the wings and control surfaces or on the carburetor heat valve, the walls of the air scoop, or the carburetor units during flight. Impact ice collecting on the metering elements of the carburetor may upset fuel metering or stop carburetor fuel flow.

Inclinometer. An instrument consisting of a curved glass tube, housing a glass ball, and damped with a fluid similar to kerosene. It may be used to indicate inclination, as a level, or, as used in the turn indicators, to show the relationship between gravity and centrifugal force in a turn.

Indicated airspeed (IAS). The direct instrument reading obtained from the airspeed indicator, uncorrected for variations in atmospheric density, installation error, or instrument error. Manufacturers use this airspeed as the basis for determining airplane performance. Takeoff, landing, and stall speeds listed in the AFM or POH are indicated airspeeds and do not normally vary with altitude or temperature.

Indicated altitude. The altitude read directly from the altimeter (uncorrected) when it is set to the current altimeter setting.

Induced drag. That part of total drag which is created by the production of lift. Induced drag increases with a decrease in airspeed.

Induction manifold. The part of the engine that distributes intake air to the cylinders.

Inertia. The opposition which a body offers to a change of motion.

Initial climb. This stage of the climb begins when the airplane leaves the ground and a pitch attitude has been established to climb away from the takeoff area.

Instrument Flight Rules (IFR). Rules that govern the procedure for conducting flight in weather conditions below VFR weather minimums. The term "IFR" also is used to define weather conditions and the type of flight plan under which an aircraft is operating.

Integral fuel tank. A portion of the aircraft structure, usually a wing, which is sealed off and used as a fuel tank. When a wing is used as an integral fuel tank, it is called a "wet wing."

Intercooler. A device used to reduce the temperature of the compressed air before it enters the fuel metering device. The resulting cooler air has a higher density, which permits the engine to be operated with a higher power setting.

Internal combustion engine. An engine that produces power as a result of expanding hot gases from the combustion of fuel and air within the engine itself. A steam engine where coal is burned to heat up water inside the engine is an example of an external combustion engine.

International Standard Atmosphere (ISA). Standard atmospheric conditions consisting of a temperature of 59 °F (15 °C), and a barometric pressure of 29.92 "Hg. (1013.2 mb) at sea level. ISA values can be calculated for various altitudes using a standard lapse rate of approximately 2 °C per 1,000 feet.

Interstage turbine temperature (ITT). The temperature of the gases between the high pressure and low pressure turbines.

Inverter. An electrical device that changes DC to AC power.

Irreversible Deceleration and/or Sink Rate. Unrecoverable depletion of mechanical energy as a result of continuous loss of airspeed and/or altitude coupled with insufficient excess power available under a given flight condition. Failure to recover above a certain critical AGL altitude results in the airplane hitting the ground regardless of what the pilot does.

J

Jet-powered airplane. An aircraft powered by a turbojet or turbofan engine.

K

Kinesthesia. The sensing of movements by feel.

Kinetic energy. Amount of energy due to the airspeed, expressed as $\frac{1}{2}mV^2$ where m = airplane's mass and V = airspeed.

L

Lateral axis. An imaginary line passing through the center of gravity of an airplane and extending across the airplane from wingtip to wingtip.

Lateral stability (rolling). The stability about the longitudinal axis of an aircraft. Rolling stability or the ability of an airplane to return to level flight due to a disturbance that causes one of the wings to drop.

Lead-acid battery. A commonly used secondary cell having lead as its negative plate and lead peroxide as its positive plate. Sulfuric acid and water serve as the electrolyte.

Leading edge devices. High lift devices which are found on the leading edge of the airfoil. The most common types are fixed slots, movable slats, and leading edge flaps.

Leading edge flap. A portion of the leading edge of an airplane wing that folds downward to increase the camber, lift, and drag of the wing. The leading-edge flaps are extended for takeoffs and landings to increase the amount of aerodynamic lift that is produced at any given airspeed.

Leading edge. The part of an airfoil that meets the airflow first.

Licensed empty weight. The empty weight that consists of the airframe, engine(s), unusable fuel, and undrainable oil plus standard and optional equipment as specified in the equipment list. Some manufacturers used this term prior to GAMA standardization.

Lift coefficient. A coefficient representing the lift of a given airfoil. Lift coefficient is obtained by dividing the lift by the free-stream dynamic pressure and the representative area under consideration.

Lift. One of the four main forces acting on an aircraft. On a fixed-wing aircraft, an upward force created by the effect of airflow as it passes over and under the wing.

Lift/drag ratio (L/D). The efficiency of an airfoil section. It is the ratio of the coefficient of lift to the coefficient of drag for any given angle of attack.

Lift-off. The act of becoming airborne as a result of the wings lifting the airplane off the ground, or the pilot rotating the nose up, increasing the angle of attack to start a climb.

Limit load factor. Amount of stress, or load factor, that an aircraft can withstand before structural damage or failure occurs.

Load factor. The ratio of the load supported by the airplane's wings to the actual weight of the aircraft and its contents. Also referred to as G-loading.

Longitudinal axis. An imaginary line through an aircraft from nose to tail, passing through its center of gravity. The longitudinal axis is also called the roll axis of the aircraft. Movement of the ailerons rotates an airplane about its longitudinal axis.

Longitudinal stability (pitching). Stability about the lateral axis. A desirable characteristic of an airplane whereby it tends to return to its trimmed angle of attack after displacement.

M

Mach buffet. Airflow separation behind a shock-wave pressure barrier caused by airflow over flight surfaces exceeding the speed of sound.

Mach compensating device. A device to alert the pilot of inadvertent excursions beyond its certified maximum operating speed.

Mach critical. The Mach speed at which some portion of the airflow over the wing first equals Mach 1.0. This is also the speed at which a shock wave first appears on the airplane.

Mach tuck. A condition that can occur when operating a swept-wing airplane in the transonic speed range. A shock wave could form in the root portion of the wing and cause the air behind it to separate. This shock-induced separation causes the center of pressure to move aft. This, combined with the increasing amount of nose down force at higher speeds to maintain left flight, causes the nose to "tuck." If not corrected, the airplane could enter a steep, sometimes unrecoverable dive.

Mach. Speed relative to the speed of sound. Mach 1 is the speed of sound.

Magnetic compass. A device for determining direction measured from magnetic north.

Main gear. The wheels of an aircraft's landing gear that supports the major part of the aircraft's weight.

Maneuverability. Ability of an aircraft to change directions along a flightpath and withstand the stresses imposed upon it.

Maneuvering speed (V_A). The maximum speed where full, abrupt control movement can be used without overstressing the airframe.

Manifold pressure (MP). The absolute pressure of the fuel/air mixture within the intake manifold, usually indicated in inches of mercury.

Maximum allowable takeoff power. The maximum power an engine is allowed to develop for a limited period of time; usually about one minute.

Maximum landing weight. The greatest weight that an airplane normally is allowed to have at landing.

Maximum ramp weight. The total weight of a loaded aircraft, including all fuel. It is greater than the takeoff weight due to the fuel that will be burned during the taxi and run-up operations. Ramp weight may also be referred to as taxi weight.

Maximum takeoff weight. The maximum allowable weight for takeoff.

Maximum weight. The maximum authorized weight of the aircraft and all of its equipment as specified in the Type Certificate Data Sheets (TCDS) for the aircraft.

Maximum zero fuel weight (GAMA). The maximum weight, exclusive of usable fuel.

Minimum controllable airspeed. An airspeed at which any further increase in angle of attack, increase in load factor, or reduction in power, would result in an immediate stall.

Minimum drag speed (L/D_{MAX}). The point on the total drag curve where the lift-to-drag ratio is the greatest. At this speed, total drag is minimized.

Mixture. The ratio of fuel to air entering the engine's cylinders.

M_{MO}. Maximum operating speed expressed in terms of a decimal of Mach speed.

Moment arm. The distance from a datum to the applied force.

Moment index (or index). A moment divided by a constant such as 100, 1,000, or 10,000. The purpose of using a moment index is to simplify weight and balance computations of airplanes where heavy items and long arms result in large, unmanageable numbers.

Moment. The product of the weight of an item multiplied by its arm. Moments are expressed in pound-inches (lb-in). Total moment is the weight of the airplane multiplied by the distance between the datum and the CG.

Movable slat. A movable auxiliary airfoil on the leading edge of a wing. It is closed in normal flight but extends at high angles of attack. This allows air to continue flowing over the top of the wing and delays airflow separation.

Mushing. A flight condition caused by slow speed where the control surfaces are marginally effective.

N

N_1, N_2, N_3. Spool speed expressed in percent rpm. N_1 on a turboprop is the gas producer speed. N_1 on a turbofan or turbojet engine is the fan speed or low pressure spool speed. N_2 is the high pressure spool speed on engine with 2 spools and medium pressure spool on engines with 3 spools with N_3 being the high pressure spool.

Nacelle. A streamlined enclosure on an aircraft in which an engine is mounted. On multiengine propeller-driven airplanes, the nacelle is normally mounted on the leading edge of the wing.

Negative static stability. The initial tendency of an aircraft to continue away from the original state of equilibrium after being disturbed.

Negative torque sensing (NTS). A system in a turboprop engine that prevents the engine from being driven by the propeller. The NTS increases the blade angle when the propellers try to drive the engine.

Neutral static stability. The initial tendency of an aircraft to remain in a new condition after its equilibrium has been disturbed.

Nickel-cadmium battery (NiCad). A battery made up of alkaline secondary cells. The positive plates are nickel hydroxide, the negative plates are cadmium hydroxide, and potassium hydroxide is used as the electrolyte.

Normal category. An airplane that has a seating configuration, excluding pilot seats, of nine or less, a maximum certificated takeoff weight of 12,500 pounds or less, and intended for nonacrobatic operation.

Normalizing (turbonormalizing). A turbocharger that maintains sea level pressure in the induction manifold at altitude.

O

Octane. The rating system of aviation gasoline with regard to its antidetonating qualities.

Overboost. A condition in which a reciprocating engine has exceeded the maximum manifold pressure allowed by the manufacturer. Can cause damage to engine components.

Overspeed. A condition in which an engine has produced more rpm than the manufacturer recommends, or a condition in which the actual engine speed is higher than the desired engine speed as set on the propeller control.

Overtemp. A condition in which a device has reached a temperature above that approved by the manufacturer or any exhaust temperature that exceeds the maximum allowable for a given operating condition or time limit. Can cause internal damage to an engine.

Overtorque. A condition in which an engine has produced more torque (power) than the manufacturer recommends, or a condition in a turboprop or turboshaft engine where the engine power has exceeded the maximum allowable for a given operating condition or time limit. Can cause internal damage to an engine.

P

Parasite drag. That part of total drag created by the design or shape of airplane parts. Parasite drag increases with an increase in airspeed.

Payload (GAMA). The weight of occupants, cargo, and baggage.

P-factor. A tendency for an aircraft to yaw to the left due to the descending propeller blade on the right producing more thrust than the ascending blade on the left. This occurs when the aircraft's longitudinal axis is in a climbing attitude in relation to the relative wind. The P-factor would be to the right if the aircraft had a counterclockwise rotating propeller.

Pilot's Operating Handbook (POH). A document developed by the airplane manufacturer and contains the FAA-approved Airplane Flight Manual (AFM) information.

Piston engine. A reciprocating engine.

Pitch. The rotation of an airplane about its lateral axis, or on a propeller, the blade angle as measured from plane of rotation.

Pivotal altitude. A specific altitude at which, when an airplane turns at a given groundspeed, a projecting of the sighting reference line to a selected point on the ground will appear to pivot on that point.

Pneumatic system. The power system in an aircraft used for operating such items as landing gear, brakes, and wing flaps with compressed air as the operating fluid.

Porpoising. Oscillating around the lateral axis of the aircraft during landing.

Position lights. Lights on an aircraft consisting of a red light on the left wing, a green light on the right wing, and a white light on the tail. CFRs require that these lights be displayed in flight from sunset to sunrise.

Positive static stability. The initial tendency to return to a state of equilibrium when disturbed from that state.

Potential energy. Amount of energy due to the altitude, expressed as *mgh*, where m = airplane's mass, and g = gravitational constant, and h = altitude.

Power available. The airplane's rate of energy gain due to maximum available engine thrust at a given airspeed. Expressed as *TV*, where *T* = engine thrust, and *V* = airspeed. Usually measured in horsepower, foot-pound per minute, or foot-pound per second.

Power distribution bus. See bus bar.

Power lever. The cockpit lever connected to the fuel control unit for scheduling fuel flow to the combustion chambers of a turbine engine.

Power required. The airplane's rate of energy loss due to total drag at a given airspeed. Expressed as *DV*, where *D* = total drag, and *V* = airspeed. Usually measured in horsepower, foot-pound per minute, or foot-pound per second.

Power. Implies work rate or units of work per unit of time, and as such, it is a function of the speed at which the force is developed. The term "power required" is generally associated with reciprocating engines.

Powerplant. A complete engine and propeller combination with accessories.

Practical slip limit. The maximum slip an aircraft is capable of performing due to rudder travel limits.

Precession. The tilting or turning of a gyro in response to deflective forces causing slow drifting and erroneous indications in gyroscopic instruments.

Preignition. Ignition occurring in the cylinder before the time of normal ignition. Preignition is often caused by a local hot spot in the combustion chamber igniting the fuel/air mixture.

Pressure altitude. The altitude indicated when the altimeter setting window (barometric scale) is adjusted to 29.92. This is the altitude above the standard datum plane, which is a theoretical plane where air pressure (corrected to 15 °C) equals 29.92 "Hg. Pressure altitude is used to compute density altitude, true altitude, true airspeed, and other performance data.

Profile drag. The total of the skin friction drag and form drag for a two-dimensional airfoil section.

Propeller blade angle. The angle between the propeller chord and the propeller plane of rotation.

Propeller lever. The control on a free power turbine turboprop that controls propeller speed and the selection for propeller feathering.

Propeller slipstream. The volume of air accelerated behind a propeller producing thrust.

Propeller synchronization. A condition in which all of the propellers have their pitch automatically adjusted to maintain a constant rpm among all of the engines of a multiengine aircraft.

Propeller. A device for propelling an aircraft that, when rotated, produces by its action on the air, a thrust approximately perpendicular to its plane of rotation. It includes the control components normally supplied by its manufacturer.

R

Ramp weight. The total weight of the aircraft while on the ramp. It differs from takeoff weight by the weight of the fuel that will be consumed in taxiing to the point of takeoff.

Rate of turn. The rate in degrees/second of a turn.

Reciprocating engine. An engine that converts the heat energy from burning fuel into the reciprocating movement of the pistons. This movement is converted into a rotary motion by the connecting rods and crankshaft.

Reduction gear. The gear arrangement in an aircraft engine that allows the engine to turn at a faster speed than the propeller.

Region of reverse command. Flight regime in which flight at a higher airspeed requires a lower power setting and a lower airspeed requires a higher power setting in order to maintain altitude.

Registration certificate. A State and Federal certificate that documents aircraft ownership.

Relative wind. The direction of the airflow with respect to the wing. If a wing moves forward horizontally, the relative wind moves backward horizontally. Relative wind is parallel to and opposite the flightpath of the airplane.

Reverse thrust. A condition where jet thrust is directed forward during landing to increase the rate of deceleration.

Reversing propeller. A propeller system with a pitch change mechanism that includes full reversing capability. When the pilot moves the throttle controls to reverse, the blade angle changes to a pitch angle and produces a reverse thrust, which slows the airplane down during a landing.

Roll. The motion of the aircraft about the longitudinal axis. It is controlled by the ailerons.

Roundout (flare). A pitch-up during landing approach to reduce rate of descent and forward speed prior to touchdown.

Rudder. The movable primary control surface mounted on the trailing edge of the vertical fin of an airplane. Movement of the rudder rotates the airplane about its vertical axis.

Ruddervator. A pair of control surfaces on the tail of an aircraft arranged in the form of a V. These surfaces, when moved together by the control wheel, serve as elevators, and when moved differentially by the rudder pedals, serve as a rudder.

Runway centerline lights. Runway centerline lights are installed on some precision approach runways to facilitate landing under adverse visibility conditions. They are located along the runway centerline and are spaced at 50-foot intervals. When viewed from the landing threshold, the runway centerline lights are white until the last 3,000 feet of the runway. The white lights begin to alternate with red for the next 2,000 feet, and for the last 1,000 feet of the runway, all centerline lights are red.

Runway centerline markings. The runway centerline identifies the center of the runway and provides alignment guidance during takeoff and landings. The centerline consists of a line of uniformly spaced stripes and gaps.

Runway edge lights. Runway edge lights are used to outline the edges of runways during periods of darkness or restricted visibility conditions. These light systems are classified according to the intensity or brightness they are capable of producing: they are the High Intensity Runway Lights (HIRL), Medium Intensity Runway Lights (MIRL), and the Low Intensity Runway Lights (LIRL). The HIRL and MIRL systems have variable intensity controls, whereas the LIRLs normally have one intensity setting.

Runway end identifier lights (REIL). One component of the runway lighting system. These lights are installed at many airfields to provide rapid and positive identification of the approach end of a particular runway.

Runway incursion. Any occurrence at an airport involving an aircraft, vehicle, person, or object on the ground that creates a collision hazard or results in loss of separation with an aircraft taking off, intending to takeoff, landing, or intending to land.

Runway threshold markings. Runway threshold markings come in two configurations. They either consist of eight longitudinal stripes of uniform dimensions disposed symmetrically about the runway centerline, or the number of stripes is related to the runway width. A threshold marking helps identify the beginning of the runway that is available for landing. In some instances, the landing threshold may be displaced.

S

Safety (SQUAT) switch. An electrical switch mounted on one of the landing gear struts. It is used to sense when the weight of the aircraft is on the wheels.

Scan. A procedure used by the pilot to visually identify all resources of information in flight.

Sea level. A reference height used to determine standard atmospheric conditions and altitude measurements.

Segmented circle. A visual ground based structure to provide traffic pattern information.

Service ceiling. The maximum density altitude where the best rate-of-climb airspeed will produce a 100 feet-per-minute climb at maximum weight while in a clean configuration with maximum continuous power.

Servo tab. An auxiliary control mounted on a primary control surface, which automatically moves in the direction opposite the primary control to provide an aerodynamic assist in the movement of the control.

Shaft horse power (SHP). Turboshaft engines are rated in shaft horsepower and calculated by use of a dynamometer device. Shaft horsepower is exhaust thrust converted to a rotating shaft.

Shock waves. A compression wave formed when a body moves through the air at a speed greater than the speed of sound.

Sideslip. A slip in which the airplane's longitudinal axis remains parallel to the original flightpath, but the airplane no longer flies straight ahead. Instead, the horizontal component of wing lift forces the airplane to move sideways toward the low wing.

Single engine absolute ceiling. The altitude that a twin engine airplane can no longer climb with one engine inoperative.

Single engine service ceiling. The altitude that a twin engine airplane can no longer climb at a rate greater than 50 fpm with one engine inoperative.

Skid. A condition where the tail of the airplane follows a path outside the path of the nose during a turn.

Slip. An intentional maneuver to decrease airspeed or increase rate of descent, and to compensate for a crosswind on landing. A slip can also be unintentional when the pilot fails to maintain the aircraft in coordinated flight.

Specific excess power (P_S). Measured in feet per minute or feet per second, it represents rate of energy change—the ability of an airplane to climb or accelerate from a given flight condition. Available specific excess power is found by dividing the difference between power available and power required by the airplane's weight.

Specific fuel consumption. Number of pounds of fuel consumed in 1 hour to produce 1 HP.

Speed brakes. A control system that extends from the airplane structure into the airstream to produce drag and slow the airplane.

Speed instability. A condition in the region of reverse command where a disturbance that causes the airspeed to decrease causes total drag to increase, which in turn, causes the airspeed to decrease further.

Speed sense. The ability to sense instantly and react to any reasonable variation of airspeed.

Speed. The distance traveled in a given time.

Spin. An aggravated stall that results in what is termed an "autorotation" wherein the airplane follows a downward corkscrew path. As the airplane rotates around the vertical axis, the rising wing is less stalled than the descending wing creating a rolling, yawing, and pitching motion.

Spiral instability. A condition that exists when the static directional stability of the airplane is very strong as compared to the effect of its dihedral in maintaining lateral equilibrium.

Spiraling slipstream. The slipstream of a propeller-driven airplane rotates around the airplane. This slipstream strikes the left side of the vertical fin, causing the airplane to yaw slightly. Vertical stabilizer offset is sometimes used by aircraft designers to counteract this tendency.

Split shaft turbine engine. See free power turbine engine.

Spoilers. High-drag devices that can be raised into the air flowing over an airfoil, reducing lift and increasing drag. Spoilers are used for roll control on some aircraft. Deploying spoilers on both wings at the same time allows the aircraft to descend without gaining speed. Spoilers are also used to shorten the ground roll after landing.

Spool. A shaft in a turbine engine which drives one or more compressors with the power derived from one or more turbines.

Stabilator. A single-piece horizontal tail surface on an airplane that pivots around a central hinge point. A stabilator serves the purposes of both the horizontal stabilizer and the elevator.

Stability. The inherent quality of an airplane to correct for conditions that may disturb its equilibrium, and to return or to continue on the original flightpath. It is primarily an airplane design characteristic.

Stabilized approach. A landing approach in which the pilot establishes and maintains a constant angle glidepath towards a predetermined point on the landing runway. It is based on the pilot's judgment of certain visual cues, and depends on the maintenance of a constant final descent airspeed and configuration.

Stall strips. A spoiler attached to the inboard leading edge of some wings to cause the center section of the wing to stall before the tips. This assures lateral control throughout the stall.

Stall. A rapid decrease in lift caused by the separation of airflow from the wing's surface brought on by exceeding the critical angle of attack. A stall can occur at any pitch attitude or airspeed.

Standard atmosphere. At sea level, the standard atmosphere consists of a barometric pressure of 29.92 inches of mercury ("Hg) or 1013.2 millibars, and a temperature of 15 °C (59 °F). Pressure and temperature normally decrease as altitude increases. The standard lapse rate in the lower atmosphere for each 1,000 feet of altitude is approximately 1 "Hg and 2 °C (3.5 °F). For example, the standard pressure and temperature at 3,000 feet mean sea level (MSL) is 26.92 "Hg (29.92 – 3) and 9 °C (15 °C – 6 °C).

Standard day. See standard atmosphere.

Standard empty weight (GAMA). This weight consists of the airframe, engines, and all items of operating equipment that have fixed locations and are permanently installed in the airplane; including fixed ballast, hydraulic fluid, unusable fuel, and full engine oil.

Standard weights. These have been established for numerous items involved in weight and balance computations. These weights should not be used if actual weights are available.

Standard-rate turn. A turn at the rate of 3° per second which enables the airplane to complete a 360° turn in 2 minutes.

Starter/generator. A combined unit used on turbine engines. The device acts as a starter for rotating the engine, and after running, internal circuits are shifted to convert the device into a generator.

Static stability. The initial tendency an aircraft displays when disturbed from a state of equilibrium.

Station. A location in the airplane that is identified by a number designating its distance in inches from the datum. The datum is, therefore, identified as station zero. An item located at station +50 would have an arm of 50 inches.

Stick puller. A device that applies aft pressure on the control column when the airplane is approaching the maximum operating speed.

Stick pusher. A device that applies an abrupt and large forward force on the control column when the airplane is nearing an angle of attack where a stall could occur.

Stick shaker. An artificial stall warning device that vibrates the control column.

Stress risers. A scratch, groove, rivet hole, forging defect, or other structural discontinuity that causes a concentration of stress.

Subsonic. Speed below the speed of sound.

Supercharger. An engine- or exhaust-driven air compressor used to provide additional pressure to the induction air so the engine can produce additional power.

Supersonic. Speed above the speed of sound.

Supplemental Type Certificate (STC). A certificate authorizing an alteration to an airframe, engine, or component that has been granted an approved type certificate.

Swept-wing. A wing planform in which the tips of the wing are farther back than the wing root.

T

Tailwheel aircraft. See conventional landing gear.

Takeoff roll (ground roll). The total distance required for an aircraft to become airborne.

Target reverser. A thrust reverser in a jet engine in which clamshell doors swivel from the stowed position at the engine tailpipe to block all of the outflow and redirect some component of the thrust forward.

Taxiway lights. Omnidirectional lights that outline the edges of the taxiway and are blue in color.

Taxiway turnoff lights. Flush lights which emit a steady green color.

Tetrahedron. A large, triangular-shaped, kite-like object installed near the runway. Tetrahedrons are mounted on a pivot and are free to swing with the wind to show the pilot the direction of the wind as an aid in takeoffs and landings.

Throttle. The valve in a carburetor or fuel control unit that determines the amount of fuel-air mixture that is fed to the engine.

Thrust line. An imaginary line passing through the center of the propeller hub, perpendicular to the plane of the propeller rotation.

Thrust reversers. Devices which redirect the flow of jet exhaust to reverse the direction of thrust.

Thrust. The force which imparts a change in the velocity of a mass. This force is measured in pounds but has no element of time or rate. The term, thrust required, is generally associated with jet engines. A forward force which propels the airplane through the air.

Timing. The application of muscular coordination at the proper instant to make flight, and all maneuvers incident thereto, a constant smooth process.

Tire cord. Woven metal wire laminated into the tire to provide extra strength. A tire showing any cord must be replaced prior to any further flight.

Torque meter. An indicator used on some large reciprocating engines or on turboprop engines to indicate the amount of torque the engine is producing.

Torque sensor. See torque meter.

Torque. 1. A resistance to turning or twisting. 2. Forces that produce a twisting or rotating motion. 3. In an airplane, the tendency of the aircraft to turn (roll) in the opposite direction of rotation of the engine and propeller.

Total drag. The sum of the parasite and induced drag.

Total energy error. An energy error where the total amount of mechanical energy is not correct. The airplane has too much or too little total energy relative to the intended altitude-speed profile. When this error occurs, the pilot will observe that altitude and airspeed deviate in the *same* direction (e.g., higher and faster than desired; or lower and slower than desired). An example would be an airplane on final approach that is above the desired glide slope and at a faster airspeed than desired.

Total mechanical energy. Sum of the energy in altitude (potential energy) and the energy in airspeed (kinetic energy).

Touchdown zone lights. Two rows of transverse light bars disposed symmetrically about the runway centerline in the runway touchdown zone.

Track. The actual path made over the ground in flight.

Trailing edge. The portion of the airfoil where the airflow over the upper surface rejoins the lower surface airflow.

Transition liner. The portion of the combustor that directs the gases into the turbine plenum.

Transonic. At the speed of sound.

Transponder. The airborne portion of the secondary surveillance radar system. The transponder emits a reply when queried by a radar facility.

Tricycle gear. Landing gear employing a third wheel located on the nose of the aircraft.

Trim tab. A small auxiliary hinged portion of a movable control surface that can be adjusted during flight to a position resulting in a balance of control forces.

Triple spool engine. Usually a turbofan engine design where the fan is the N_1 compressor, followed by the N_2 intermediate compressor, and the N_3 high pressure compressor, all of which rotate on separate shafts at different speeds.

Tropopause. The boundary layer between the troposphere and the mesosphere which acts as a lid to confine most of the water vapor, and the associated weather, to the troposphere.

Troposphere. The layer of the atmosphere extending from the surface to a height of 20,000 to 60,000 feet depending on latitude.

True airspeed (TAS). Calibrated airspeed corrected for altitude and nonstandard temperature. Because air density decreases with an increase in altitude, an airplane has to be flown faster at higher altitudes to cause the same pressure difference between pitot impact pressure and static pressure. Therefore, for a given calibrated airspeed, true airspeed increases as altitude increases; or for a given true airspeed, calibrated airspeed decreases as altitude increases.

True altitude. The vertical distance of the airplane above sea level—the actual altitude. It is often expressed as feet above mean sea level (MSL). Airport, terrain, and obstacle elevations on aeronautical charts are true altitudes.

T-tail. An aircraft with the horizontal stabilizer mounted on the top of the vertical stabilizer, forming a T.

Turbine blades. The portion of the turbine assembly that absorbs the energy of the expanding gases and converts it into rotational energy.

Turbine outlet temperature (TOT). The temperature of the gases as they exit the turbine section.

Turbine plenum. The portion of the combustor where the gases are collected to be evenly distributed to the turbine blades.

Turbine rotors. The portion of the turbine assembly that mounts to the shaft and holds the turbine blades in place.

Turbine section. The section of the engine that converts high pressure high temperature gas into rotational energy.

Turbocharger. An air compressor driven by exhaust gases, which increases the pressure of the air going into the engine through the carburetor or fuel injection system.

Turbofan engine. A turbojet engine in which additional propulsive thrust is gained by extending a portion of the compressor or turbine blades outside the inner engine case. The extended blades propel bypass air along the engine axis but between the inner and outer casing. The air is not combusted but does provide additional thrust.

Turbojet engine. A jet engine incorporating a turbine-driven air compressor to take in and compress air for the combustion of fuel, the gases of combustion being used both to rotate the turbine and create a thrust producing jet.

Turboprop engine. A turbine engine that drives a propeller through a reduction gearing arrangement. Most of the energy in the exhaust gases is converted into torque, rather than its acceleration being used to propel the aircraft.

Turbulence. An occurrence in which a flow of fluid is unsteady.

Turn coordinator. A rate gyro that senses both roll and yaw due to the gimbal being canted. Has largely replaced the turn-and-slip indicator in modern aircraft.

Turn-and-slip indicator. A flight instrument consisting of a rate gyro to indicate the rate of yaw and a curved glass inclinometer to indicate the relationship between gravity and centrifugal force. The turn-and-slip indicator indicates the relationship between angle of bank and rate of yaw. Also called a turn-and-bank indicator.

Turning error. One of the errors inherent in a magnetic compass caused by the dip compensating weight. It shows up only on turns to or from northerly headings in the Northern Hemisphere and southerly headings in the Southern Hemisphere. Turning error causes the compass to lead turns to the north or south and lag turns away from the north or south.

U

Ultimate load factor. In stress analysis, the load that causes physical breakdown in an aircraft or aircraft component during a strength test, or the load that according to computations, should cause such a breakdown.

Unfeathering accumulator. Tanks that hold oil under pressure which can be used to unfeather a propeller.

UNICOM. A nongovernment air/ground radio communication station which may provide airport information at public use airports where there is no tower or FSS.

Unusable fuel. Fuel that cannot be consumed by the engine. This fuel is considered part of the empty weight of the aircraft.

Useful load. The weight of the pilot, copilot, passengers, baggage, usable fuel, and drainable oil. It is the basic empty weight subtracted from the maximum allowable gross weight. This term applies to general aviation aircraft only.

Utility category. An airplane that has a seating configuration, excluding pilot seats, of nine or less, a maximum certificated takeoff weight of 12,500 pounds or less, and intended for limited acrobatic operation.

V

V_1. Critical engine failure speed or takeoff decision speed. It is the speed at which the pilot is to continue the takeoff in the event of an engine failure or other serious emergency. At speeds less than V_1, it is considered safer to stop the aircraft within the accelerate-stop distance. It is also the minimum speed in the takeoff, following a failure of the critical engine at V_{EF}, at which the pilot can continue the takeoff and achieve the required height above the takeoff surface within the takeoff distance.

V_2. takeoff safety speed, or a referenced airspeed obtained after lift-off at which the required one engine-inoperative climb performance can be achieved.

V_A. The design maneuvering speed. This is the "rough air" speed and the maximum speed for abrupt maneuvers. If during flight, rough air or severe turbulence is encountered, reduce the airspeed to maneuvering speed or less to minimize stress on the airplane structure. It is important to consider weight when referencing this speed. For example, V_A may be 100 knots when an airplane is heavily loaded, but only 90 knots when the load is light.

Vapor lock. A condition in which air enters the fuel system and it may be difficult, or impossible, to restart the engine. Vapor lock may occur as a result of running a fuel tank completely dry, allowing air to enter the fuel system. On fuel-injected engines, the fuel may become so hot it vaporizes in the fuel line, not allowing fuel to reach the cylinders.

V-bars. The flight director displays on the attitude indicator that provide control guidance to the pilot.

Vector. A force vector is a graphic representation of a force and shows both the magnitude and direction of the force.

Velocity. The speed or rate of movement in a certain direction.

Vertical axis. An imaginary line passing vertically through the center of gravity of an aircraft. The vertical axis is called the z-axis or the yaw axis.

Vertical card compass. A magnetic compass that consists of an azimuth on a vertical card, resembling a heading indicator with a fixed miniature airplane to accurately present the heading of the aircraft. The design uses eddy current damping to minimize lead and lag during turns.

Vertical speed indicator (VSI). An instrument that uses static pressure to display a rate of climb or descent in feet per minute. The VSI can also sometimes be called a vertical velocity indicator (VVI).

Vertical stability. Stability about an aircraft's vertical axis. Also called yawing or directional stability.

V_{FE}. The maximum speed with the flaps extended. The upper limit of the white arc.

V_{FO}. The maximum speed that the flaps can be extended or retracted.

VFR Terminal Area Charts (1:250,000). Depict Class B airspace which provides for the control or segregation of all the aircraft within the Class B airspace. The chart depicts topographic information and aeronautical information which includes visual and radio aids to navigation, airports, controlled airspace, restricted areas, obstructions, and related data.

V-G diagram. A chart that relates velocity to load factor. It is valid only for a specific weight, configuration, and altitude and shows the maximum amount of positive or negative lift the airplane is capable of generating at a given speed. Also shows the safe load factor limits and the load factor that the aircraft can sustain at various speeds.

Visual approach slope indicator (VASI). The most common visual glidepath system in use. The VASI provides obstruction clearance within 10° of the extended runway centerline, and to 4 nautical miles (NM) from the runway threshold.

Visual Flight Rules (VFR). Code of Federal Regulations that govern the procedures for conducting flight under visual conditions.

V_{LE}. Landing gear extended speed. The maximum speed at which an airplane can be safely flown with the landing gear extended.

V_{LO}. Landing gear operating speed. The maximum speed for extending or retracting the landing gear if using an airplane equipped with retractable landing gear.

V_{LOF}. Lift-off speed. The speed at which the aircraft departs the runway during takeoff.

V_{MC}. Minimum control airspeed. This is the minimum flight speed at which a twin-engine airplane can be satisfactorily controlled when an engine suddenly becomes inoperative and the remaining engine is at takeoff power.

V_{MD}. Minimum drag speed.

V_{MO}. Maximum operating speed expressed in knots.

V_{NE}. Never-exceed speed. Operating above this speed is prohibited since it may result in damage or structural failure. The red line on the airspeed indicator.

V_{NO}. Maximum structural cruising speed. Do not exceed this speed except in smooth air. The upper limit of the green arc.

V_P. Minimum dynamic hydroplaning speed. The minimum speed required to start dynamic hydroplaning.

V_R. Rotation speed. The speed that the pilot begins rotating the aircraft prior to lift-off.

V$_{S0}$. Stalling speed or the minimum steady flight speed in the landing configuration. In small airplanes, this is the power-off stall speed at the maximum landing weight in the landing configuration (gear and flaps down). The lower limit of the white arc.

V$_{S1}$. Stalling speed or the minimum steady flight speed obtained in a specified configuration. For most airplanes, this is the power-off stall speed at the maximum takeoff weight in the clean configuration (gear up, if retractable, and flaps up). The lower limit of the green arc.

V-speeds. Designated speeds for a specific flight condition.

V$_{SSE}$. Safe, intentional one-engine inoperative speed. The minimum speed to intentionally render the critical engine inoperative.

V-tail. A design which utilizes two slanted tail surfaces to perform the same functions as the surfaces of a conventional elevator and rudder configuration. The fixed surfaces act as both horizontal and vertical stabilizers.

V$_X$. Best angle-of-climb speed. The airspeed at which an airplane gains the greatest amount of altitude in a given distance. It is used during a short-field takeoff to clear an obstacle.

V$_{XSE}$. Best angle of climb speed with one engine inoperative. The airspeed at which an airplane gains the greatest amount of altitude in a given distance in a light, twin-engine airplane following an engine failure.

V$_Y$. Best rate-of-climb speed. This airspeed provides the most altitude gain in a given period of time.

V$_{YSE}$. Best rate-of-climb speed with one engine inoperative. This airspeed provides the most altitude gain in a given period of time in a light, twin-engine airplane following an engine failure.

W

Wake turbulence. Wingtip vortices that are created when an airplane generates lift. When an airplane generates lift, air spills over the wingtips from the high pressure areas below the wings to the low pressure areas above them. This flow causes rapidly rotating whirlpools of air called wingtip vortices or wake turbulence.

Waste gate. A controllable valve in the tailpipe of an aircraft reciprocating engine equipped with a turbocharger. The valve is controlled to vary the amount of exhaust gases forced through the turbocharger turbine.

Weathervane. The tendency of the aircraft to turn into the relative wind.

Weight and balance. The aircraft is said to be in weight and balance when the gross weight of the aircraft is under the max gross weight, and the center of gravity is within limits and will remain in limits for the duration of the flight.

Weight. A measure of the heaviness of an object. The force by which a body is attracted toward the center of the earth (or another celestial body) by gravity. Weight is equal to the mass of the body times the local value of gravitational acceleration. One of the four main forces acting on an aircraft. Equivalent to the actual weight of the aircraft. It acts downward through the aircraft's center of gravity toward the center of the earth. Weight opposes lift.

Wheelbarrowing. A condition caused when forward yoke or stick pressure during takeoff or landing causes the aircraft to ride on the nose-wheel alone.

Wind correction angle. Correction applied to the course to establish a heading so that track will coincide with course.

Wind direction indicators. Indicators that include a wind sock, wind tee, or tetrahedron. Visual reference will determine wind direction and runway in use.

Wind shear. A sudden, drastic shift in wind speed, direction, or both that may occur in the horizontal or vertical plane.

Windmilling. When the air moving through a propeller creates the rotational energy.

Windsock. A truncated cloth cone open at both ends and mounted on a freewheeling pivot that indicates the direction from which the wind is blowing.

Wing area. The total surface of the wing (square feet), which includes control surfaces and may include wing area covered by the fuselage (main body of the airplane), and engine nacelles.

Wing span. The maximum distance from wingtip to wingtip.

Wing twist. A design feature incorporated into some wings to improve aileron control effectiveness at high angles of attack during an approach to a stall.

Wing. Airfoil attached to each side of the fuselage and are the main lifting surfaces that support the airplane in flight.

Wingtip vortices. The rapidly rotating air that spills over an airplane's wings during flight. The intensity of the turbulence depends on the airplane's weight, speed, and configuration. It is also referred to as wake turbulence. Vortices from heavy aircraft may be extremely hazardous to small aircraft.

Y

Yaw string. A string on the nose or windshield of an aircraft in view of the pilot that indicates any slipping or skidding of the aircraft.

Yaw. Rotation about the vertical axis of an aircraft.

Z

Zero fuel weight. The weight of the aircraft to include all useful load except fuel.

Zero sideslip. A maneuver in a twin-engine airplane with one engine inoperative that involves a small amount of bank and slightly uncoordinated flight to align the fuselage with the direction of travel and minimize drag.

Zero thrust (simulated feather). An engine configuration with a low power setting that simulates a propeller feathered condition.

Airplane Flying Handbook (FAA-H-8083-3C)

Index

A

G

H

I

W

Y